Complex Analysis and Algebraic Geometry

Complex Analysis and Algebraic Geometry

A Collection of Papers Dedicated to K. Kodaira

edited by

W. L. Baily, Jr. and T. Shioda
University of Chicago *University of Tokyo*

Cambridge University Press
Cambridge
London New York Melbourne

Copublished by the Syndics of the Cambridge University Press
The Pitt Building, Trumpington Street, Cambridge CB2 1RP
Bentley House, 200 Euston Road, London NW1 2DB
32 East 57th Street, New York, NY 10022, USA
296 Beaconsfield Parade, Middle Park, Melbourne 3206, Australia
and Iwanami Shoten, Publishers, 2–5–5 Hitotsubashi, Chiyoda-ku, Tokyo, Japan

First published 1977

Printed in Great Britain at the University Press, Cambridge

Library of Congress cataloguing in publication data

Main entry under title:

Complex analysis and algebraic geometry.

Bibliography: p.

Includes index.

CONTENTS: Ueno, K. and Shioda, T. Introduction – Artin, M. Coverings of the rational double points in characteristic p – Bombieri, E. and Mumford, D. Enriques' classification of surfaces in char. p, II – Hirzebruch, F. and Zagier, D. Classification of Hilbert modular surfaces [etc.]

1. Geometry, Algebraic – Addresses, essays, lectures.
2. Surfaces – Addresses, essays, lectures.
3. Complex manifolds – Addresses, essays, lectures.
4. Kodaira, Kunihiko, 1915–   I. Kodaira, Kunihiko, 1915–
II. Baily, Walter L.   III. Shioda, T., 1940–
QA564.C565   516′.35   77–78886

ISBN 0 521 21777 6

To Professor Kunihiko Kodaira,

whose profound influence and inspiration have greatly enriched the subjects of complex analysis and algebraic geometry, this volume is warmly dedicated by his friends and students.

Foreword

This volume has been written for Professor Kunihiko Kodaira by his friends and students to mark the occasion of his sixtieth birthday.

As is well known, Kodaira has made decisive contributions to algebraic geometry and complex analysis, i. e., to the theory of algebraic varieties and of complex manifolds. Recently the Collected Works of Kodaira (three volumes) have been published by Iwanami Shoten, Publishers (Tokyo) and Princeton University Press (Princeton). In these three volumes, one may see how this remarkable mathematician has developed his ideas, establishing many results of fundamental importance in various branches of mathematics. Among the most distinguished works of Kodaira in algebraic geometry and complex manifolds are those in the theory of harmonic integrals and its application to algebraic and Kähler varieties, Kodaira's vanishing theorem and the projective imbedding of Hodge varieties, deformation theory (with D. Spencer), and the theory of compact, complex analytic surfaces.

It might perhaps be more appropriate to the occasion to offer a discussion of the mathematical works of Professor Kodaira ; however, there is a survey by one of the present editors in the preface of the said Collected Works. Therefore, we have included instead an introduction which gives a brief account of the development of complex analysis and algebraic geometry in these thirty years. Hopefully it will serve as background material for the subjects treated in the present volume, especially to those readers not previously familiar with this area.

The articles in this volume present some recent developments in complex analysis and algebraic geometry. This book is divided, roughly, into three parts. Part I includes topics in the theory of algebraic surfaces and analytic surfaces. Part II includes topics in moduli and classification problems, as well as in the structure theory of certain complex manifolds. Part III includes various topics in algebraic geometry, analysis, and arithmetic. In most of these articles the influence of Kodaira will be clearly visible. In fact, he has provided a strong influence in attracting many mathematicians, especially some in the younger generation, to this lively branch of mathematics, and the present volume may be regarded in part as a natural outgrowth of this influence.

In conclusion, we feel an expression of gratitude is in order for others who have worked on this volume. We wish to thank warmly those of our colleagues who have helped in reading manuscripts and proofs and who include K. Akao, E. Horikawa, S. Iitaka, and M. Inoue. We are also very grateful to K. Ueno who wrote the introduction in cooperation with one of the editors. Finally, we wish to express our appreciation to Mr. H. Arai of Iwanami Shoten who has helped us in many ways.

W. L. Baily, Jr.
T. Shioda

List of Contributors

Kazuo Akao	Department of Mathematics, Gakushuin University, Tokyo, Japan
Aldo Andreotti	Istituto Matematico, Università di Pisa, Pisa, Italy
Michael Artin	Department of Mathematics, Massachusetts Institute of Technology, Cambridge, Massachusetts, U. S. A.
Walter L. Baily, Jr.	Department of Mathematics, University of Chicago, Illinois, U. S. A.
Enrico Bombieri	Scuola Normale Superiore, Pisa, Italy
Takao Fujita	Department of Mathematics, College of General Education, University of Tokyo, Komaba, Tokyo, Japan
Hubert Goldschmidt	Institute for Advanced Study, Princeton, New Jersey, U. S. A.
Friedrich Hirzebruch	Mathematisches Institut der Universität Bonn, Bonn, Federal Republic of Germany
Eiji Horikawa	Department of Mathematics, University of Tokyo, Hongo, Tokyo, Japan
Jun-ichi Igusa	Department of Mathematics, The Johns Hopkins University, Baltimore, Maryland, U. S. A.
Shigeru Iitaka	Department of Mathematics, University of Tokyo, Hongo, Tokyo, Japan
Hiroshi Inose	Department of Mathematics, University of Tokyo, Hongo, Tokyo, Japan
Masahisa Inoue	Department of Mathematics, Aoyamagakuin University, Tokyo, Japan
Arnold Kas	Department of Mathematics, Oregon State University, Corvallis, Oregon, U. S. A.
Masaki Kashiwara	Department of Mathematics, Nagoya University, Nagoya, Japan
Masahide Kato	Department of Mathematics, Rikkyo University, Tokyo, Japan

Sōichi Kawai	Department of Mathematics, Rikkyo University, Tokyo, Japan
Yoichi Miyaoka	Department of Mathematics, Tokyo Metropolitan University, Tokyo, Japan
David Mumford	Department of Mathematics, Harvard University, Cambridge, Massachusetts, U. S. A.
Mauro Nacinovich	Istituto Matematico, Università di Pisa, Pisa, Italy
Iku Nakamura	Department of Mathematics, Nagoya University, Nagoya, Japan
Yukihiko Namikawa	Department of Mathematics, Nagoya University, Nagoya, Japan
Fumio Sakai	Department of Mathematics, Kochi University, Kochi, Japan
Nobuo Sasakura	Department of Mathematics, Tokyo Metropolitan University, Tokyo, Japan
Tetsuji Shioda	Department of Mathematics, University of Tokyo, Hongo, Tokyo, Japan
Donald Spencer	Department of Mathematics, Princeton University, Princeton, New Jersey, U. S. A.
Tatsuo Suwa	Department of Mathematics, Hokkaido University, Sapporo, Japan
Kenji Ueno	Department of Mathematics, Kyoto University, Kyoto, Japan
A. Van de Ven	Mathematisch Instituut der Rijksuniversiteit te Leiden, Leiden, The Netherlands
G. van der Geer	Mathematisch Instituut der Rijksuniversiteit te Leiden, Leiden, The Netherlands
Don Zagier	Mathematisches Institut der Universität Bonn, Bonn, Federal Republic of Germany

Contents

Part III

Introduction

K. Ueno and T. Shioda

Algebraic geometry and complex analysis (i. e. the theory of complex manifolds) have made remarkable progress in the last several decades. Let us briefly review a part of their development as an introduction to various problems treated in the articles in this volume.[1)]

1. First, in the 1940's, a big reform began in algebraic geometry. One of the main aims was to establish solid foundations for this fascinating branch of mathematics, whose underpinnings had become the subject of criticism because of the highly intuitive, rather than rigorous, treatment of its early pioneers, the Italian algebraic geometers. From the algebraic (or rather "abstract") viewpoint, this was carried out by Zariski, Weil and others (cf. Zariski (3, vol. I), Weil (1, 2)). Their methods are based on abstract algebra, making no use of topological or transcendental methods, and thus are applicable not only to algebraic varieties over the field of complex numbers, but also to those over an arbitrary ground field. In particular, Weil's method, including the abstract theory of abelian varieties, opened a new way for the application of algebraic geometry to number theory; we mention here only Weil's proof (2) of the Riemann hypothesis for curves over a finite field. The reader is referred to the 1950 Congress talks of Weil (3) and Zariski (2) for their basic ideas.

As for the transcendental method of studying algebraic varieties of arbitrary dimension, Hodge (1, 2) initiated the theory of harmonic integrals, by which he obtained Lefschetz' results on the primitive decomposition of the cohomology groups, as well as the so-called Hodge decomposition of the cohomologies on a Kähler variety into components of type (p, q). This was further developed by Kodaira, de Rham, Weil and others, and applied to various classical problems in algebraic geometry (cf. Kodaira (0, vol. I), de Rham (1), Weil (5)). In particular, Kodaira proved the Riemann-Roch theorem for adjoint systems of ample linear systems, Severi's conjecture on the arithmetic genus, and the completeness of characteristic systems, etc., while Weil gave a reconstruction of the theory of theta-functions. In the meantime methods employing coherent analytic sheaves and their cohomology groups were introduced in the theory of several complex variables by H. Cartan (1) (inspired by the works of J. Leray and K. Oka), and soon

1) The references quoted in this introduction will be limited only to the basic ones, and the interested reader will find further references through those given here or through the related articles in this volume.

these became fundamental tools in the various related fields.

On the one hand, the theory of coherent analytic sheaves, combined with the potential theoretic method, produced many important results on compact complex manifolds. Kodaira and Spencer (0, vol. II) investigated the sheaf $\Omega^p(F)$ of germs of holomorphic p-forms with coefficients in a line bundle F over a Kähler manifold X and established the so-called Dolbeault isomorphism : $H^q(X, \Omega^p(F)) \simeq H^{p,q}(F)$, where $H^{p,q}(F)$ denotes the space of harmonic forms of type (p, q) with coefficients in F. In particular, this shows the finite dimensionality of the cohomology groups $H^q(X, \Omega^p(F))$ (cf. Cartan-Serre (2)). Putting $h^{p,q} = \dim H^q(X, \Omega^p)$, one obtains Hodge's symmetry : $h^{p,q} = h^{q,p}$ for X Kähler (hence, for X projective), which does not necessarily hold for non-Kähler X. Serre (1) proved the duality theorem which states that $H^q(X, \mathcal{O}(V))$ and $H^{n-q}(X, \Omega^n \otimes \mathcal{O}(\hat{V}))$ are dual vector spaces, where $\mathcal{O}(V)$ denotes the sheaf of germs of holomorphic sections of a vector bundle V on X and $\hat{V}$ denotes the dual vector bundle of V ($n = \dim X$). Kodaira (1) obtained the famous vanishing theorem (now named after him) by a differential geometric method : given a line bundle F on a Kähler manifold X, the cohomology group $H^q(X, \mathcal{O}(F))$ vanishes for $q \geq 1$, if $F-K$ is positive, K being the canonical bundle. Using it, he proved in (2) that a compact complex manifold is biholomorphic to a non-singular projective variety if and only if it carries a Hodge metric (i. e. a Kähler metric whose fundamental class belongs to an integral cohomology class), solving in definitive a problem raised by Hodge (2).

On the other hand, Serre applied the method of sheaves to algebraic geometry, independently of potential theory, and obtained important results such as the comparison theorems of algebraic and analytic geometry (2). Moreover Serre (3) established a new foundation for abstract algebraic geometry using the theory of coherent algebraic sheaves ; in particular, he verified the cohomological characterization of affine varieties and the abstract version of the Serre duality. Also, some new phenomena in characteristic p were discovered : Igusa (1, 2) showed that the completeness of the characteristic linear system does not hold in general (cf. Mumford (2)), and Serre (4) found an example where the symmetry $h^{p,q} = h^{q,p}$ fails to hold.

Another important achievement in the middle 1950's is the Riemann-Roch theorem, which is formulated in terms of sheaves as follows. The alternating sum $\Sigma(-1)^q \dim H^q(X, \mathcal{O}(V))$ of the dimensions of cohomology groups of the sheaf $\mathcal{O}(V)$ of holomorphic sections of a vector bundle V on X is expressed by means of a certain polynomial of Chern classes of V and the Todd genus of X. When X is a non-singular projective variety, this was first proven by Hirzebruch (1) by using Thom's cobordism theory. It was then generalized by Grothendieck to the case of arbitrary characteristic by defining Chern classes in terms of the Chow ring of algebraic cycles on X (cf. Borel-Serre (1)). Later Atiyah and Singer (1) obtained the index theorem for elliptic operators on a compact manifold, and deduced from it the Riemann-Roch theorem on an arbitrary compact complex manifold. We refer the reader to the excellent treatise of Hirzebruch (1) on these subjects.

2. In the next ten years, starting from the late 1950's, abstract algebraic geometry underwent another reformulation. Namely, Grothendieck introduced the concept of scheme, generalizing that of algebraic variety as a ringed space, and clarified the relationship of algebra and geometry in full generality (cf. Grothendieck (1), (2), (3)). The new methods had a strong influence on various parts of algebraic geometry, and offered new techniques for classical problems. We shall mention the following important results as examples. a) Hironaka (1) achieved the resolution of singularities of an algebraic variety of an arbitrary dimension in characteristic 0, generalizing earlier results of Zariski for dimensions ≤ 3. (Also he later succeeded (2) in the resolution of singularities of complex spaces.) b) Mumford (1) proved the existence of moduli schemes for curves and abelian varieties by reconsidering classical invariant theory. c) Artin and Grothendieck (3) constructed the theory of étale cohomology, which allows one to attach to an algebraic variety in arbitrary characteristic certain cohomology groups with coefficients in characteristic 0. The original program of Grothendieck (4) to apply this theory to the proof of Weil's conjecture (4) on the zeta function of a variety over a finite field has recently been carried out by Deligne (2).

During the same period of time, complex analysis was further developed on the foundations laid in the preceding era, and some new features of complex manifolds were revealed. Generalizing the idea of Riemann on the moduli of compact Riemann surfaces, Kodaira and Spencer developed the general theory of deformation of complex structures (Kodaira (3)). Given a compact complex manifold X, a complex analytic family of deformations of X is, by definition, a proper smooth morphism $\pi : \mathcal{X} \to S$ of connected complex spaces such that $\pi^{-1}(o) = X$ for some point $o \in S$; each fibre $\pi^{-1}(s)$ $(s \in S)$ is called a deformation of X. There is a natural linear map $\rho : T_o(S) \to H^1(X, \Theta)$, called the Kodaira-Spencer map, where $T_o(S)$ is the Zariski tangent space of S at o and Θ is the sheaf of germs of holomorphic vector fields on X. The family π is called effective at o if ρ is injective, and complete (at o) if ρ is surjective (or equivalently, if any other small deformation of X is induced by π). A fundamental theorem of Kodaira, Nirenberg and Spencer asserts that, if $H^2(X, \Theta) = 0$, then there exists a family $\pi : \mathcal{X} \to S$ with ρ bijective and S nonsingular. The existence of a locally complete and effective family for an arbitrary X was proved by Kuranishi (1). These results are based on the theory of elliptic partial differential equations. Kodaira and Spencer (4) also obtained the principle of upper semi-continuity asserting that, given a differentiable family of compact complex manifolds X_t equipped with holomorphic vector bundles V_t, the dimension of the cohomology of X_t with coefficients in the sheaf $\mathcal{O}(V_t)$ is an upper semi-continuous function of the parameter t. This result was later generalized by Grauert (1) as the so-called proper mapping theorem. There are many extensions of deformation theory to other types of structures. For the algebraic aspects of deformations and moduli problems, we refer the reader to the recent report of Seshadri (1).

Next we shall review the theory of surfaces in some detail. The theory of algebraic surfaces was begun already in the late 19th century and studied in detail about the beginning of this century, notably by M. Noether, Poincaré, Picard, Castelnuovo and Enriques, and has been one of the richest sources of modern algebraic geometry. After Lefschetz (1) introduced topological methods in the 1920's and Zariski (1) wrote an extensive monograph on algebraic surfaces in 1935, the theory was taken up by Zariski (3, vol. II), Kodaira (0, vol. III) and Šafarevič and others (1). Zariski studied among others, the problem of minimal models of algebraic surfaces by algebraic methods. Kodaira (4, 5, 6) made deep investigations on compact complex analytic surfaces and established the classification and structure theory of surfaces, incorporating the classification of algebraic surfaces due to Castelnuovo and Enriques (1), and providing comprehensive proofs for the results of the latter. His method makes use of all the techniques on complex manifolds developed earlier by him and also of the generalized Riemann-Roch theorem due to Atiyah-Singer. Let S denote a surface (i. e. a compact complex manifold of dimension 2) and let K be the canonical bundle of S. The plurigenus $P_m(S)=\dim H^0(S, \mathcal{O}(mK))$ $(m\geq 1)$ and the irregularity $q(S)=\dim H^1(S, \mathcal{O})$ are birational (or rather bimeromorphic) invariants of S, and Castelnuovo and Enriques classified algebraic surfaces by means of these numerical invariants. The m-th pluricanonical mapping $\Phi_{mK}: S\rightarrow \boldsymbol{P}^N (N=P_m-1)$ is defined by $\Phi_{mK}(z)=(\varphi_0(z):\cdots:\varphi_N(z))$, where $z\in S$ and $\{\varphi_i\}$ is a basis of $H^0(S, \mathcal{O}(mK))$. We set

$$\kappa(S)=\begin{cases}\max_m \dim \Phi_{mK}(S) & (\text{if } P_m\geq 1 \text{ for some } m\geq 1)\\ -\infty & (\text{otherwise}).\end{cases}$$

$\kappa(S)$ is called the Kodaira dimension of S. If $\kappa(S)=2$, then S is an algebraic surface and is said to be of general type. The pluricanonical mappings of such a surface were studied by Šafarevič (1), Kodaira (6) and others; for example, Φ_{5K} is a birational mapping whose image is a normal surface with at worst rational double points. If $\kappa(S)=1$, then S is an elliptic surface, i. e., there exists a holomorphic mapping of S onto a curve such that almost all fibres are elliptic curves. The theory of elliptic surfaces was studied in detail by Kodaira (4). If S is a minimal surface with $\kappa(S)=0$, then S is one of the following: a complex torus, a $K3$ surface, a hyperelliptic surface, an Enriques surface or a non-algebraic elliptic surface of special kind. An algebraic surface S with $\kappa(S)=-\infty$ is a ruled surface (i. e. a surface birationally equivalent to a product of a projective line and a curve). The famous criterion of Castelnuovo asserts that S is rational if and only if $P_2(S)=q(S)=0$. Non-algebraic surfaces with $\kappa(S)=-\infty$ have been studied by Kodaira and Inoue. Thus the theory of surfaces is full of many interesting examples, affording much life to the general theory of complex manifolds, and in turn being enriched by it. For example, an arbitrary $K3$ surface, which is defined as a surface S with trivial K and $q(S)=0$, turns out to be a deformation of a non-singular quartic surface, say $x_0^4+x_1^4+x_2^4+x_3^4=0$, in $\boldsymbol{P}^3$, as shown by Kodaira.

3. It will be perhaps too early to "review" current developments in the present decade. Zariski, in the preface to his Collected Papers (3), says, "There are signs at the present moment of the pendulum swinging back from "schemes", "motives", and so on toward concrete but difficult unsolved questions concerning the old pedestrian concept of a projective variety (and even of algebraic surfaces)." Indeed, some famous problems have recently been solved. Clemens and Griffiths (2) have shown that a non-singular cubic 3-fold, which is well-known to be unirational, is not rational. There are several other examples of a similar kind. Pjateckii-Šapiro and Šafarevič (2) have obtained a Torelli theorem for the period mapping of polarized $K3$ surfaces, which was conjectured by Andreotti and Weil and whose local version was earlier proved by Kodaira and Tjurina. As we have already mentioned, Deligne (2) has succeeded in the proof of the Riemann-Weil hypothesis on the zeta function of a non-singular projective variety over a finite field. Also Deligne (1) has extended the Hodge theory to the case of (possibly open) non-singular algebraic varieties over the complex numbers, and the variations of Hodge structure, in its relation to period mappings, have been discussed by Griffiths (1) and others. Furthermore Iitaka has proposed the classification theory of higher dimensional varieties by means of their Kodaira dimensions, and we refer to Ueno (1) for the recent developments in this direction.

It goes without saying that there are many other important results in complex analysis and algebraic geometry which we have not mentioned in the above, partly because of space limitation but more essentially because of our lack of knowledge of the subjects.

In closing this introduction, we shall give some explanation of the articles in the present volume. Assuming the complete classification of surfaces, there are still many important problems left in the theory of surfaces. For example, one can raise such problems as (i) generalization to the characteristic p case, (ii) topological or differentiable structures of surfaces, (iii) structure of surfaces of some special interest (e. g. geometric, arithmetic, etc.), or (iv) applications in other fields (e. g. automorphic functions or number theory). In Part I of this book, the reader will find the following results. (i) Bombieri and Mumford extend the classification of algebraic surfaces to char p. Artin constructs smooth coverings of rational double points in the char p case. (ii) Kas determines the deformation type of certain elliptic surfaces. (iii) Among surfaces of general type, Horikawa investigates those with a pencil of curves of genus 2, while Miyaoka studies those with relatively small numerical invariants. Shioda and Inose determine the $K3$ surfaces with maximum Picard number. Inoue constructs new examples of (non-algebraic) surfaces of type VII_0. (iv) Hirzebruch and Zagier complete the classification of the Hilbert modular surfaces, and van der Geer and Van de Ven examine the minimality of some of these surfaces.

Of course, one can ask questions, similar to those posed above, for higher dimensional cases. In Part II of this volume : (i) Ueno studies certain fibre spaces from the viewpoint of the classification theory. Fujita determines the defining equations of certain polarized varieties, thus improving earlier results of Mumford. (ii) Iitaka and Sakai give a systematic approach to open varieties through the notion of Kodaira dimension. (iii) The papers of Nakamura and Namikawa are related to the moduli and the degeneration of polarized abelian varieties. (iv) Akao and Kato study the structure of certain non-Kähler complex manifolds, and Suwa determines compact manifolds which are quotients of a 3-dimensional Euclidean space by affine transformation groups.

We have mentioned in this introduction very little about the connection of algebraic geometry and number theory. In Part III of this book, the reader will find in Igusa's article an application of algebraic geometry to some arithmetical questions. Also, Baily investigates certain arithmetic groups and Eisenstein series. As we have seen above, some of the most basic results in complex analysis and algebraic geometry are based on the theory of differential equations. Part III also includes the papers of Andreotti and of Goldschmidt and Spencer in this direction. Kashiwara reports some results in microlocal calculus, which has recently been developed by himself, Sato and others. Sasakura discusses a new approach to the de Rham type theory of a real analytic variety with singularities. Finally Kawai investigates the effect of Steenrod's cohomological operations on algebraic cycles on a projective variety, thus generalizing a result of Atiyah and Hirzebruch.

References

Atiyah, M. F. and Singer, I. M. : (1) The index of elliptic operators on compact manifolds, Bull. Amer. Math. Soc., **69** (1963).

Borel, A. and Serre, J.-P. : (1) Le théorème de Riemann-Roch (d'après Grothendieck), Bull. Soc. Math. France, **86** (1958).

Cartan, H. : (1) Séminaire E. N. S., 1951–1952 ; (2) (with J.-P. Serre) Un théorème de finitude concernant les variétés analytiques compacts, C. R. Acad. Sci. Paris, **237** (1953).

Deligne, P. : (1) Théorie de Hodge I, II, Actes du Congrès Int. Math. Nice, 1970 and Publ. Math. IHES, **40** (1971) ; (2) La conjecture de Weil I, Publ. Math. IHES, **43** (1974).

de Rham, G. : (1) Variétés différentiables, Hermann, Paris, 1955.

Enriques, F. : (1) Le superificie algebriche, Nicola Zanichelli Editore, Bologna, 1949.

Grauert, H. : (1) Ein Theorem der analytischen Garbentheorie und die Modulräume komplexer Structuren, Publ. Math. IHES, **5** (1960).

Griffiths, P. A. : (1) Report on variations of Hodge structure, Bull. Amer. Math. Soc., **76** (1970) ; (2) (with H. Clemens) The intermediate Jacobian of the cubic threefold, Ann. of Math., **95** (1972).

Grothendieck, A. : (1) Fondements de Géométrie Algébrique, Extraits du Séminaire Bourbaki 1957–1962, Paris, 1962 ; (2) (with J. Dieudonné) Éléments de Géométrie Algébrique, Publ. Math. IHES, **4**, **8**, **11**, etc. (1960–) ; (3) (with others) Séminaire de Géométrie Algébrique, 1960–1968, Springer Lecture Notes, **224** (1971), etc. ; (4) (with others) Dix exposés sur la cohomologie de schémas, North-Holland, Amsterdam, 1968.

Hironaka, H. : (1) Resolution of singularities of an algebraic variety over a field of characteristic 0, Ann. of Math., **79** (1964) ; (2) Bimeromorphic smoothing of a complex analytic space, Preprint, Univ. of Warwick, 1971.

Hirzebruch, F. : (1) Topological methods in algebraic geometry, 3rd ed., Springer, Berlin-Heidelberg-New York, 1966.

Hodge, W. V. D. : (1) The theory and application of harmonic integrals, 2nd ed., Cambridge Univ. Press, Cambridge, 1952 ; (2) The topological invariants of algebraic varieties, Proc. Int. Congress Math., Cambridge, Mass., 1950.

Igusa, J. : (1) A fundamental inequality in the theory of Picard varieties, Proc. Nat. Acad. Sci. U. S. A., **41** (1955) ; (2) On some problems in abstract algebraic geometry, ibid.

Kodaira, K. : (0) Collected Works, I, II, III, Iwanami Shoten Publishers, Tokyo, and Princeton Univ. Press, 1975 ; esp. (1) On a differential-geometric method in the theory of analytic stacks, Proc. Nat. Acad. Sci. U. S. A., **39** (1953) ; (2) On Kähler varieties of restricted type (an intrinsic characterization of algebraic varieties), Ann. of Math., **60** (1954) ; (3) (with D. C. Spencer) On deformations of complex analytic structures, I, II, III, Ann. of Math., **67** (1958), **71** (1960) ; (4) On compact analytic surfaces I, II, III, Ann. of Math., **71** (1960), **77** (1963), **78** (1963) ; (5) On the structure of compact complex analytic surfaces, I, II, III, IV, Amer. J. Math., **86** (1964), **88** (1966), **90** (1968) ; (6) Pluricanonical systems on algebraic surfaces of general type, J. Math. Soc. Japan, 20 (1968).

Kuranishi, M. : (1) Deformation of compact complex manifolds, Montréal Univ. Press, 1971.

Lefschetz, S. : (1) L'analysis situs et la géométrie algébrique, Gauthier-Villars, Paris, 1924 ; (2) A page of mathematical autobiography, Bull. Amer. Math. Soc., **74** (1968).

Mumford, D. : (1) Geometric invariant theory, Springer, Berlin-Heidelberg-New York, 1965 ; (2) Lectures on curves on an algebraic surface, Princeton Univ. Press, 1966.

Šafarevič, I. R. : (1) (with others) Algebraic surfaces, Proc. Steklov Inst. Math., **75** (1965) ; (2) (with I. I. Pjateckii-Šapiro) A Torelli theorem for algebraic surfaces of type *K*3, Izv. Akad. Nauk SSSR, Ser. Math., **35** (1971).

Serre, J.-P. : (1) Un théorème de dualité, Comm. Math. Helv., **29** (1955) ; (2) Géométrie analytique et géométrie algébrique, Ann. Inst. Fourier, **6** (1955–56) ; (3) Faisceaux algébriques cohérents, Ann. of Math., **61** (1955) ; (4) Sur la topologie des variétés algébriques en characteristique *p*, Symp. of Alg. Top., Mexico, 1956.

Seshadri, C. S. : (1) Theory of moduli, AMS Proc. Symp. in Pure Math., **29** (1975).

Ueno, K. : (1) Classification theory of algebraic varieties and compact complex spaces, Springer Lecture Notes **439**, 1975.

Weil, A. : (1) Foundations of algebraic geometry, AMS Coll. Publ. **29**, 2nd ed., Providence, 1962 ; (2) Courbes algébriques et variétés abéliennes, 2nd ed., Hermann, Paris, 1971 ; (3) Number theory and algebraic geometry, Proc. Int. Congress Math., Cambridge, Mass., 1950 ; (4) Numbers of solutions of equations in finite fields, Bull. Amer. Math. Soc., **55** (1949) ; (5) Introduction à l'étude des variétés kähleriennes, Hermann, Paris, 1958.

Zariski, O. : (1) Algebraic surfaces, 2nd ed., Springer, Berlin-Heidelberg-New York, 1971 ; (2) The fundamental ideas of abstract algebraic geometry, Proc. Int. Congress Math., Cambridge, Mass., 1950 ; (3) Collected Papers, I, II, MIT Press, 1972–73.

Part I

Coverings of the Rational Double Points in Characteristic p

M. Artin[1)]

The rational double points of surfaces in characteristic zero are related to the finite subgroups G of SL_2 [6, 7]. Namely, if V denotes the affine plane with its linear G-action, then the variety $X=V/G$ has a singularity at the origin, which is the one corresponding to G. Let p be a prime integer. If p divides the order of G, this subgroup will degenerate when reduced modulo p, and the smooth reduction of V will usually not be compatible with an equisingular reduction of X. Nevertheless, it turns out that every rational double point in characteristic p has a finite (possibly ramified) covering by a smooth scheme. In this paper we prove the existence of such a covering by direct calculation, and we compute the local fundamental groups of the singularities.

1. Generalities on Coverings

We are interested in the local behavior of singularities and so we work with a scheme of the form $X=\operatorname{Spec} A$, where A is the henselization of the local ring of a normal algebraic surface over an algebraically closed field k. We could also work with complete local rings.

In general, U will denote the complement of the closed point of X : $U=X-x_0$. By *fundamental group* of X, we mean $\pi=\pi_1(U)$. This is the group which classifies finite étale coverings of U, or equivalently, normal, pure 2-dimensional schemes Y, finite over X, which are étale except above x_0.

Let us call a *covering* of X any finite surjective map $Y\to X$ such that Y is irreducible and normal, and let us call the covering *unramified* if it is étale above U, i.e., is unramified in codimension 1 on X.

Proposition (1. 1). *Let the solid arrows in the diagram be given coverings of X.*

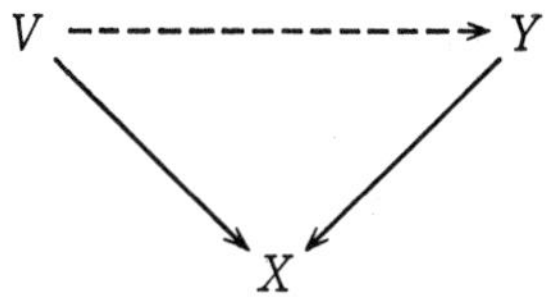

1) Supported by NSF.

Assume that V is smooth, i. e., $V \approx \mathrm{Spec}\ k\{x,y\}$, where $k\{x,y\}$ denotes the henselization of the polynomial ring, and that Y is unramified. Then a dotted arrow exists, i. e., V dominates Y.

Proof. This follows from purity of the branch locus: The scheme $V\times_X Y$ is étale over V except at the closed point, and therefore its normalization decomposes completely into a sum of copies of V. Each copy determines the graph of a map $V \to Y$.

As an immediate consequence, we have

Corollary (1. 2). (i) *If X admits a smooth covering $V\to X$, then the fundamental group π of X is finite.*

(ii) *If in addition V/X is totally ramified along some curve of X, then* $\pi=0$.

In characteristic zero, the converse of (1. 2i) is true. Mumford [9] proved that if $\pi=0$ then X is smooth[1]. If π is finite, then the universal cover $\tilde{U}$ of U is finite over U. The normalization V of X in $K(\tilde{U})$ is a singularity with trivial fundamental group, hence is smooth.

Mumford's theorem is easily seen to be false in characteristic $p\neq 0$. Some rational double points furnish examples (cf. sections 3–5). But it is natural to ask whether the converse of (1. 2) continues to hold:

Question (1. 3). *Suppose the fundamental group π of X is finite. Does there exist a covering $V\to X$ which is smooth?*

This would be a very beautiful fact, if true. Our calculations provide some slight positive evidence, since we exhibit smooth coverings for the rational double points. But even for these special singularities, we do not know a conceptual proof of their existence.

Now suppose that X is a rational double point [3]. Let $X' \xrightarrow{\varphi} X$ be the minimal resolution of the singularity of X. It is known [2, 2. 7] that rational double points X are characterized by the existence of a double differential ω whose divisor on X' is zero. This fact restricts the possible unramified coverings:

Proposition (1. 4). *An unramified covering Y of a rational double point is either smooth, or is a rational double point.*

Proof. Let $Y''\to Y$ be a resolution of Y. The differential ω is regular on X', and hence has no pole along any prime divisor of $K(X)$ centered at x_0. (In other words, ω is regular on every resolution of X.) Since every prime divisor of Y centered at the closed point lies over some prime divisor of X, it follows that ω is regular on Y''.

1) Actually, Mumford [9] works with the classical topology. However, he has extended his result to the algebraic context (unpublished).

Moreover, ω has no zeros on $Y-y_0$ since $Y-y_0$ is étale over U. Therefore the divisor K of ω on Y'' is supported on the exceptional curves of the map $Y''\to Y$. Let Z be the fundamental cycle on Y'' [3, p. 132]. Then since $K\geq 0$, $(Z\cdot K)\leq 0$. Also, $(Z^2)<0$. Hence $p(Z)=0$, and $(Z^2)=-1$ or -2. The proposition follows from [3, thm. 4].

Proposition(1. 5). *With the above notation, let $Y''\to Y$ be the minimal resolution of the singularity of Y, and let Y' be the normalization of X' in $K(Y)$. Then Y' dominates Y'' :*

$$\begin{array}{ccc} Y' & \longrightarrow & Y'' \\ & \searrow \qquad \swarrow & \\ & Y & \end{array}$$

and the curves in Y' which contract on Y'' are the ones ramified over X'. In particular, Y is smooth ($Y=Y''$) if and only if every curve is ramified.

Proof. The divisor of ω on Y'' is zero since it is concentrated on the closed fibre of Y''/Y and Y is a rational double point (or $Y=Y''$). Let D be a prime divisor of $K(Y)$ lying over a prime divisor C of $K(X)$. Then the order of zero of ω on D will be zero only if ω has order zero on C and the extension is not ramified on D. The prime divisors of X on which ω has order zero are those corresponding to the curves of X', and similarly, the prime divisors of Y on which ω has order zero correspond to the curves of Y''. The proposition now follows from Zariski's Main Theorem.

Example.

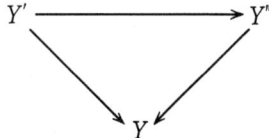

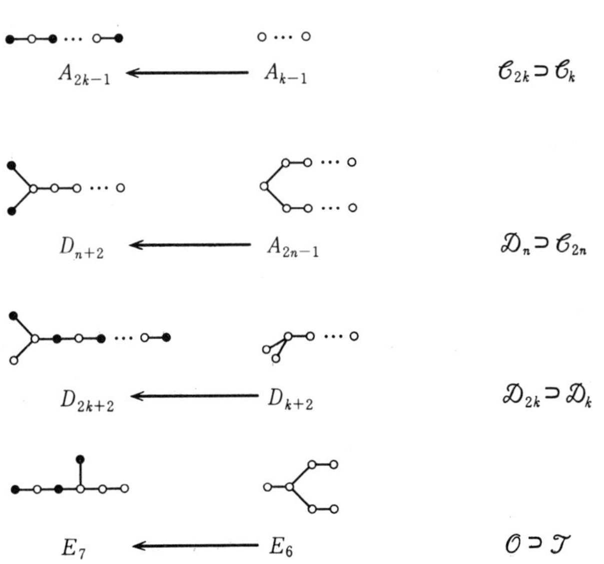

(1. 6) Double covers in characteristic zero

The unramified double covers $Y\to X$ of the rational double points in characteristic zero are listed here. The dark vertices of the Dynkin diagram stand for the ramified curves on X', whose inverse images are contracted on Y''. The corresponding subgroups of SL_2 are on the right. We use script letters to denote them, viz. $\mathcal{C}_n$=cyclic group of order n, $\mathcal{D}_n$=binary dihedral group of order $4n$, and $\mathcal{T}$, $\mathcal{O}$, $\mathcal{I}$ the binary tetrahedral, octahedral and icosahedral groups. The fact that the ramified curves must alternate with unramified ones is easily checked.

2. Reduction modulo p

Any rational double point in characteristic p can be obtained by specialization from one in characteristic zero. This is clear from the equations given in the next section. The main facts that we need about specialization of unramified coverings were proved in [4, Sect. 4], where the case of complete local rings was treated. The henselian case is similar, and can be deduced from [4] using standard algebraization techniques.

Let S=Spec R, where R is an unequal characteristic discrete valuation ring with algebraically closed residue field k. Let X/S be a henselian local scheme, essentially of finite type, which represents an equisingular family of two-dimensional singularities, in the sense of [10]. Suppose that the closed fibre X_0 is a rational double point. Then the generic geometric fibre X_η will have a single rational double point of the same type. Let $\tilde{X}_\eta$ denote the henselization at that point. It follows from [4] that

(**2. 1**) (i) *The unramified coverings of X_0 which extend to X are those which are tamely ramified along every prime divisor of X_0.*

(ii) *If $Y \to X$ is a cover such that Y_0 is unramified and reducible, then Y is reducible.*

(iii) *There is a natural 1–1 correspondence between unramified Galois covers of order prime to p of X_0 and of $\tilde{X}_\eta$.*

Let us return to the case of a single rational double point X in characteristic p, dropping the subscript 0. In this case, any unramified Galois covering Y of X whose order is divisible by p must be wildly ramified along some prime divisor centered at x_0. For, let G be the Galois group and let $H \subset G$ be a subgroup of order p. This subgroup corresponds to an intermediate covering $Y \to Z \to X$, and Y/Z is cyclic of order p. It is enough to show that the covering Y/Z is ramified on some prime divisor. Let $Z' \to Z$ be its minimal resolution, and let $C \subset Z'$ be the exceptional set. This is a simply connected union of rational curves, and therefore it has no étale cover. It follows that Z' has no étale cover either. Hence Y/Z is ramified on some component of C.

Corollary(**2. 2**). *Let X be a rational double point in characteristic p. The unramified Galois covers of X which lift to characteristic zero are those of order prime to p.*

Let us call the fundamental group π of X *tame* if it is not wildly ramified, or equivalently, if it has order prime to p. In this case, it will be the maximal quotient of order prime to p of the corresponding characteristic zero group, and so it is known.

The A_n singularities have tame fundamental groups. They can be described in all characteristics by the equations

$$xy+z^{n+1} = 0. \tag{2. 3}$$

In characteristic zero, A_{n-1} corresponds to the representation of the cyclic group $\mathcal{C}_n$ by the matrices

$$\begin{bmatrix} \zeta^i & 0 \\ 0 & \zeta^{n-i} \end{bmatrix} \qquad \zeta = e^{2\pi i/n}. \tag{2. 4}$$

If we let $\mathcal{C}_n$ act by (2. 4) on a vector space V with basis $\{v_1, v_2\}$, then the invariant functions are generated by

$$v_1^n = x, \qquad v_2^n = y, \qquad v_1v_2 = z. \tag{2. 5}$$

The equations (2. 5) define a covering of the A_n singularity (2. 3) by a smooth scheme V, uniformly over Spec $\boldsymbol{Z}$. When p divides n, this covering becomes inseparable modulo p, and the fundamental group of the characteristic p singularity reduces, therefore, to $\mathcal{C}_{\bar{n}}$, where $n=p^e\bar{n}$, and $p\nmid\bar{n}$. The inseparable phenomena can be explained in this case if we replace the cyclic group $\mathcal{C}_n$ by the group scheme μ_n of nth roots of unity, operating in the analogous way. Such an explanation will *not* be possible for the other rational double points.

One can begin the analysis of the other singularities by passing to a covering of degree prime to p, which will be a simpler rational double point. This reduces the problem to those types whose characteristic zero groups have no quotient of order prime to p, so that the fundamental group is completely wild. Excluding A_n, which was treated above, we are left with the following cases :

$$\begin{aligned} &D_n, \ p = 2 \\ &E_6, \ p = 3 \\ &E_7, \ p = 2 \\ &E_8, \ p = 2, 3, 5. \end{aligned}$$

(2. 6) The singularities having no non-trivial tame covering

Corollary (2. 7). *The fundamental group of a rational double point X in characteristic p is tame if $p\neq 2, 3, 5$.*

3. List of the Singularities and their Equations

Lipman [8] classified the E_8 singularities in all characteristics, and we need to extend his classification to the other rational double points[1]. Following tradition, we will omit the rather tedious verification of these results. We do not know an a priori reason for the fact that there are only finitely many singularities of each type.

In a family of singularities X_n^r, the index n is upper semi-continuous, while the co-index r is lower semi-continuous. The number to the right of the equation is the dimension of the space of deformations of the singularity, which can be used to check that the cases listed are all different.

For convenience, we include the nonsingular local scheme Spec $k\{x,y\}$ in our lists, and denote it by A_0.

A_n	:	$z^{n+1}+xy$		$n\geq 0$
D_{2n}^0	:	$z^2+x^2y+xy^n$	$4n$	$n\geq 2$
D_{2n}^r	:	$z^2+x^2y+xy^n+xy^{n-r}z$	$4n-2r$	$r=1,\cdots,n-1$
D_{2n+1}^0	:	$z^2+x^2y+y^nz$	$4n$	$n\geq 2$
D_{2n+1}^r	:	$z^2+x^2y+y^nz+xy^{n-r}z$	$4n-2r$	$r=1,\cdots,n-1$
E_6^0	:	$z^2+x^3+y^2z$	8	
E_6^1	:	$z^2+x^3+y^2z+xyz$	6	
E_7^0	:	$z^2+x^3+xy^3$	14	
E_7^1	:	$z^2+x^3+xy^3+x^2yz$	12	
E_7^2	:	$z^2+x^3+xy^3+y^3z$	10	
E_7^3	:	$z^2+x^3+xy^3+xyz$	8	
E_8^0	:	$z^2+x^3+y^5$	16	
E_8^1	:	$z^2+x^3+y^5+xy^3z$	14	
E_8^2	:	$z^2+x^3+y^5+xy^2z$	12	
E_8^3	:	$z^2+x^3+y^5+y^3z$	10	
E_8^4	:	$z^2+x^3+y^5+xyz$	8	

Rational double points in characteristic 2

A_n, D_n	:	classical forms.	
E_6^0	:	$z^2+x^3+y^4$	9
E_6^1	:	$z^2+x^3+y^4+x^2y^2$	7
E_7^0	:	$z^2+x^3+xy^3$	9
E_7^1	:	$z^2+x^3+xy^3+x^2y^2$	7

1) See also Arnold [1]. But note that in non-zero characteristics the classification of singularities does not lead to the same answers as the classification of germs of maps. They have continuous parameters. Moreover, if $p=2$, the classification of singularities depends on their dimension, which is 2 in our case.

$E_8^0 \quad : \quad z^2+x^3+y^5 \qquad 12$

$E_8^1 \quad : \quad z^2+x^3+y^5+x^2y^3 \qquad 10$

$E_8^2 \quad : \quad z^2+x^3+y^5+x^2y^2 \qquad 8$

Rational double points in characteristic 3

A_n, D_n, E_6, E_7 : classical forms.

$E_8^0 \quad : \quad z^2+x^3+y^5 \qquad 10$

$E_8^1 \quad : \quad z^2+x^3+y^5+xy^4 \qquad 8$

Rational double points in characteristic 5

Rational double points in characteristic $p>5$: classical forms only

4. The Computations in Characteristic 2

The singularities D_N^r, $N=2n$ or $2n+1$ can be analyzed by the substitution

$$(4.\ 1) \qquad \begin{aligned} u^2+y^{n-r}u+y &= 0 \qquad &\text{if } r>0, \text{ or} \\ u^2+y &= 0 \qquad &\text{if } r=0. \end{aligned}$$

Set

$$\bar{v} = z+xu,$$

so that

$$\begin{aligned} \bar{v}^2+xy^{n-r}\bar{v}+xy^n &= 0 \qquad &\text{if } N \text{ even, } r>0 \\ \bar{v}^2+xy^n &= 0 \qquad &\text{if } N \text{ even, } r=0 \\ \bar{v}^2+xy^{n-r}\bar{v}+y^nz &= 0 \qquad &\text{if } N \text{ odd, } r>0 \\ \bar{v}^2+y^nz &= 0 \qquad &\text{if } N \text{ odd, } r=0. \end{aligned}$$

Case 1. $N = 2n$, $2r \geq n$.

Set $\qquad \bar{v} = y^{n-r}v$, so that

$$v^2+xv+xy^{2r-n} = 0.$$

Substitution for y using (4. 1) leads to an equation of type $A_{8r-4n-1}$, if $2r>n$, and of type A_0 if $2r=n$. It is an unramified double cover having ramification on the resolution as indicated :

$$D_{2n}^r \xleftarrow{\ \deg 2\ } A_{8r-4n-1} \quad (2r\geq n).$$

Since the fundamental group of A_k is tame and the fundamental group π of D_N^r has no non-trivial tame quotient (2. 6), π is the dihedral group of order $2m$, where

$m=\overline{2r-n}$ is the greatest divisor of $2r-n$ prime to 2, or is 1 if $2r=n$.

Case 2. $N=2n+1,\ 2r\geq n$.
Set $v=y^{n-r}\bar{v}$, $v_1=v+y^{2r-n}u$, and $x_1=x+y^r$, to obtain an equation of type $A_{8r-4n+1}$, which is an unramified double cover of D_N^r:

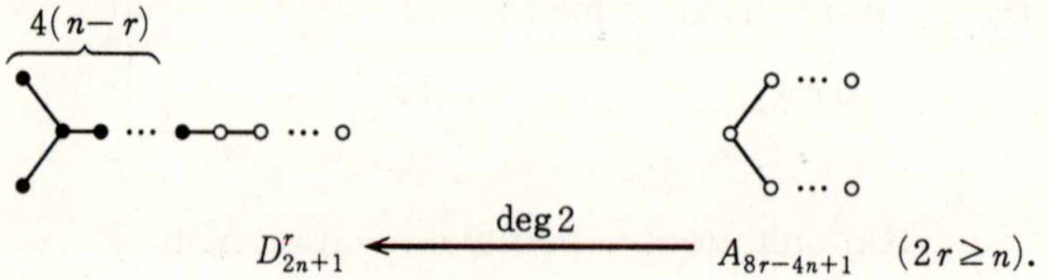

As in case 1, π is the dihedral group of order $2(4p-2n+1)$.

Case 3. $N=2n,\ 2r<n$.
This case leads to a ramified double cover of D_{2n}^r by A_0, and $\pi=0$.

Case 4. $N=2n+1,\ 2r<n$.
This leads to a ramified double cover of D_{2n+1}^r by A_1, and $\pi=0$.

The E_6^r singularities ($r=0,1$) have a tame cyclic cover of degree 3 defined by the equation

$$u^3=z.$$

The resulting cover is D_4^r, which has fundamental group 0 or $\mathscr{C}_2$ according as r is 0 or 1. In both cases, π is cyclic, and the double cover of E_6^1 is

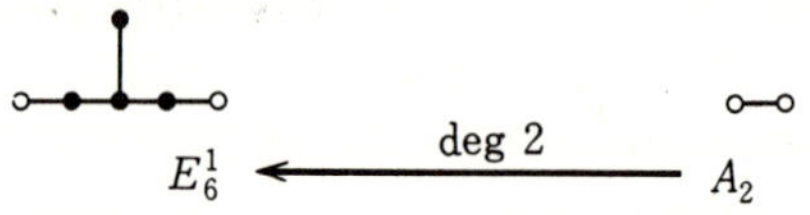

The singularity E_7^0 has the purely inseparable cover of degree 2 by A_0 defined by $u^2=x$. The cases E_7^r with $r=1,2$ can be treated together if the equation for E_7^2 is replaced by the equivalent one

$$z^2+x^3+xy^3+y^3z+x^2yz.$$

The substitution

$$u^2+xyu+x=0$$

defines a ramified double cover by A_0 if $r=0$, or D_5^0 if $r=1$. It follows that $\pi=0$ for E_7^r, and $r=0,1,2$.

The case E_7^3 has an unramified double cover by D_4^1, given by the substitution

$$u^2+yu+x=0:$$

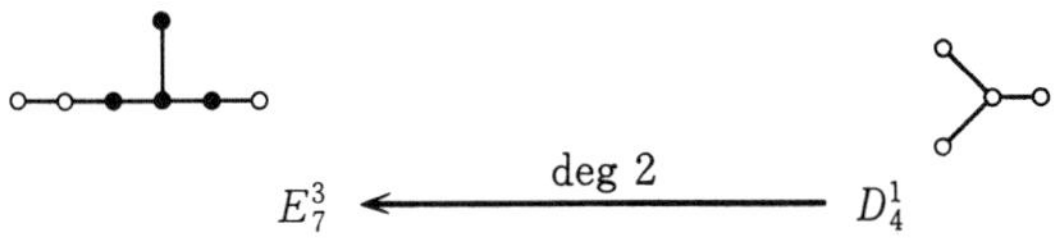

and the fundamental group is cyclic of order 4.

It remains to treat the E_8 singularities. The case E_8^0 has a purely inseparable cover by A_0 defined by $u^2=x$, E_8^1 has a ramified double cover by A_0, defined by

$$u^2+y^3u+x = 0,$$

and E_8^2 has an unramified double cover by A_0 defined by

$$u^2+y^2u+x = 0$$

(see [5]). The substitution $u^2=x$ defines a purely inseparable cover of E_8^3 by E_7^2, which has trivial fundamental group. Thus the fundamental group of E_8^3 is trivial, too.

The most interesting case is that of E_8^4, which has an unramified double cover by E_6^1, defined by the equation

$$u^2+yu+x = 0:$$

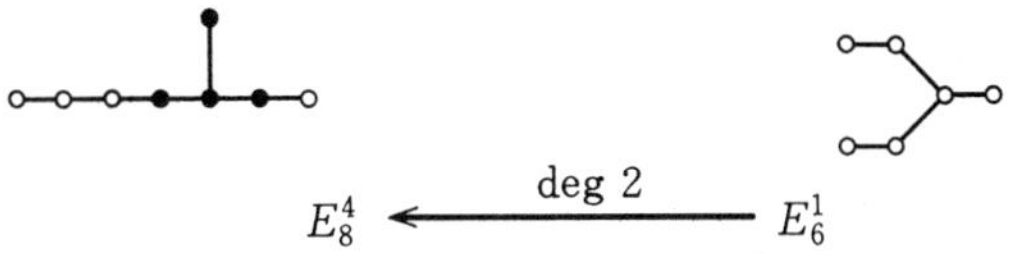

Thus its fundamental group π has order 12. It has no quotient of order 3 (2. 6), and therefore is either the dihedral group or the *metacyclic group* with generators σ, τ:

(4. 2) $$\sigma^3 = 1, \qquad \tau^4 = 1, \qquad \tau\sigma\tau^{-1} = \sigma^2.$$

The dihedral group is isomorphic to $\mathcal{S}_3 \oplus \mathcal{C}_2$, and so if this is π there must be a second double cover Y of $E_8^4=X$, whose fundamental group is $\mathcal{S}_3$. Consider the subset of the Dynkin diagram on which this second cover is ramified. We can contract the curves corresponding to this subset on the minimal resolution X' of X, to obtain a scheme $\bar{X}$ having finitely many rational double points, and such that the normalization $\bar{Y}$ of $\bar{X}$ in $K(Y)$ is unramified at these points. Moreover, $\bar{Y}$ will be the minimal resolution of Y, by (1. 5). Looking over the singularities having double covers by A_0 and whose Dynkin diagrams are subsets of E_8, we find the two possibilities D_4 and E_8. The second is ruled out because Y, having fundamental group $\mathcal{S}_3$, is not A_0. In the case D_4, the cover is by an E_6 singularity, as in the above figure. There is only one double cover of D_4^1. Thus in this case the tensor product of the two given covers of X would be unramified on X', and hence would split completely, contradicting the assumption that they are different. Therefore, π is the metacyclic group.

We collect the results together :

A_n	:	π is tame
$D_{2n}^r,\ 0 \leq 2r < n$	:	$\pi = 0$
$D_{2n}^r,\ n \leq 2r < 2n$	:	π is dihedral, of order $2m$, where m is the greatest divisor of $2r-n$ prime to 2, or is 1 if $2r=n$
$D_{2n+1}^r,\ 0 \leq 2r < n$	:	$\pi = 0$
$D_{2n+1}^r,\ n \leq 2r < 2n$	:	π is dihedral, of order $2(4r-2n+1)$
E_6^0	:	$\pi = \mathcal{C}_3$ is tame
E_6^1	:	$\pi = \mathcal{C}_6$
$E_7^r,\ r = 0, 1, 2$	:	$\pi = 0$
E_7^3	:	$\pi = \mathcal{C}_4$
$E_8^r,\ r = 0, 1$	:	$\pi = 0$
E_8^2	:	$\pi = \mathcal{C}_2$
E_8^3	:	$\pi = 0$
E_8^4	:	π is the metacyclic group (4. 2) of order 12.

The fundamental groups in characteristic 2

Note that the order of π is always smaller than the order of the corresponding group in characteristic zero. But for D_N^r, the prime factors do not always correspond. Therefore any relation between smooth coverings in characteristics 0 and p must be very subtle.

5. The Computations in Characteristics 3 and 5

We look for a p-cyclic covering of X using Artin-Schreier theory, and the étale cohomology sequence

$$(5.\ 1) \qquad 0 \longrightarrow \boldsymbol{Z}/p \longrightarrow \mathcal{O}^+ \xrightarrow{F-1} \mathcal{O}^+ \longrightarrow 0, \qquad (F = p\text{th power})$$

on the open set $U = X - x_0$. The cohomology sequence of (5. 1) shows that p-cyclic étale covers of U are given by elements $\bar{\alpha} \in H^1(U, \mathcal{O})$ such that $F\bar{\alpha} - \bar{\alpha} = 0$. We take as affine cover the open sets $U_0 = U - \{x=0\}$ and $U_1 = U - \{y=0\}$, and try the cohomology class defined by the cocycle

$$\alpha = x^{-1}y^{-1}z$$

on $U_0 \cap U_1$. In other words, we ask whether $\alpha^p - \alpha$ is a coboundary $\beta_0 - \beta_1$. This is so in the following cases :

$$E_6^1,\ p = 3: \quad \alpha^3 - \alpha = (x^{-3}yz) - (-y^{-3}z)$$
$$E_7^1,\ p = 3: \quad \alpha^3 - \alpha = (x^{-2}z) - (-y^{-3}z)$$
$$E_8^2,\ p = 3: \quad \alpha^3 - \alpha = (y^2x^{-3}z) - (-y^{-3}z)$$
$$E_8^1,\ p = 5: \quad \alpha^5 - \alpha = x^{-5}(y^5 + x^2y^3 + 2x^3 + 2xy^4)z - (-y^{-5}xz).$$

Thus these singularities have p-cyclic unramified covers defined by the equations

$$u^3-u+y^{-3}z = 0 \qquad (p = 3)$$
$$u^5-u+y^{-5}xz = 0 \qquad (p = 5)$$

or,

$$u^3-y^2u+z = 0 \qquad \text{if } p = 3$$

and

$$u^5-y^4u+xz = 0 \qquad \text{if } p = 5.$$

In the case of $X=E_8^1$, $p=5$, every curve of the minimal resolution X' must be ramified. For, the singularities defined by contracting the ramification curves must have a p-cyclic cover, and there are no wild covers in characteristic 5 except for E_8 (see (2. 6)). Thus the cover is nonsingular, by (1. 5) :

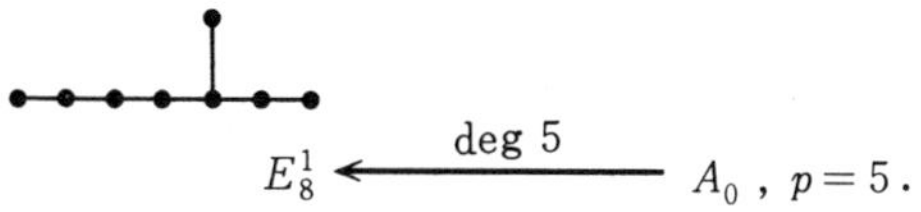

A similar argument applies to E_6^1 in characteristic 3. The other two cases must be computed, and they lead to

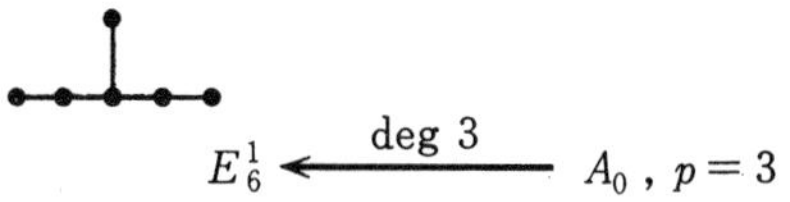

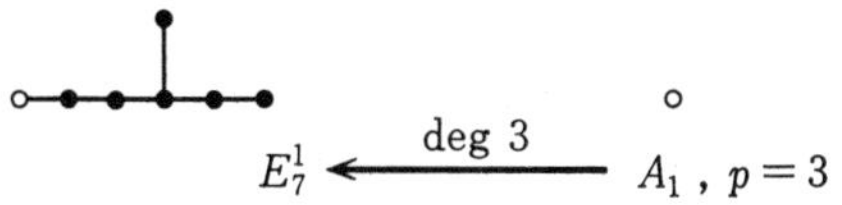

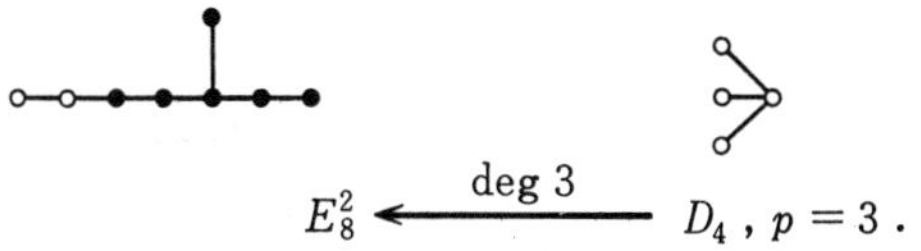

The fundamental group of E_8^2, $p=3$, is an extension of the cyclic group $\mathcal{C}_3$ by the quaternion group $\mathcal{D}_2$, and it has no quotient prime to 3. These properties characterize the binary tetrahedral group : $\pi=\mathcal{T}$. For, the extension splits, hence is determined by a non-trivial operation of $\mathcal{C}_3$ on $\mathcal{D}_2$, and there is essentially only one such operation. The fundamental groups are cyclic in the other cases considered above.

The only remaining singularity to consider in characteristic 5 is E_8^0, which has a purely inseparable cover by A_0, defined by $u^5=y$. In characteristic 3, the singularities E_6^0, E_8^0 have purely inseparable covers by A_0 defined by $u^3=y$, and E_7^0 has a tame double cover by E_6^0. Finally, the substitution

$$u^3 = x^2 + y^2$$

defines a purely inseparable cover of E_8^1 by E_6^0.

A_n, D_n	:	π is tame
E_6^0	:	$\pi = 0$
E_6^1	:	$\pi = \mathcal{C}_3$
E_7^0	:	$\pi = \mathcal{C}_2$ is tame
E_7^1	:	$\pi = \mathcal{C}_6$
E_8^r, $r = 0, 1$	:	$\pi = 0$
E_8^2	:	$\pi = \mathcal{T}$ is the binary tetrahedral group.

The fundamental groups in characteristic 3

A_n, D_n, E_6, E_7	:	π is tame
E_8^0	:	$\pi = 0$
E_8^1	:	$\pi = \mathcal{C}_5$

The fundamental groups in characteristic 5

References

[1] Arnold, V. I.: Normal forms for functions near degenerate critical points..., Funct. Anal. **6** (1974), 254–272.
[2] Artin, M.: Some numerical criteria for contractability..., Amer. J. Math. **84** (1962), 485–496.
[3] Artin, M.: On isolated rational singularities of surfaces, Amer. J. Math. **88** (1966), 129–136.
[4] Artin, M.: Lifting of two-dimensional singularities..., Amer. J. Math. **88** (1966), 747–762.
[5] Artin, M.: Wildly ramified $\boldsymbol{Z}/2$ actions in dimension two, Proc. Amer. Math. Soc. **52** (1975), 60–64.
[6] Brieskorn, E.: Rationale Singularitäten Komplexer Flächen, Invent. Math. **4** (1968), 336–358.
[7] du Val, P.: Homographies, quaternions, and rotations, Oxford 1964.
[8] Lipman, J.: Rational singularities..., Publ. Math. IHES **36** (1969), 195–280.
[9] Mumford, D.: The topology of normal singularities of an algebraic surface..., Publ. Math. IHES **9** (1961).
[10] Wahl, J.: Equisingular deformations of surface singularities, (to appear).

Massachusetts Institute
of Technology

(Received March 14, 1975)

Enriques' Classification of Surfaces in Char. p, II

E. Bombieri and D. Mumford

Introduction and Preliminary Reductions

The purpose of this paper is to carry further the extension of Enriques' classification of surfaces from the case of a char. 0 groundfield to the case of a char. p groundfield. The first part of this extension was made in the paper [10] of one of the present authors. The main results of that paper are as follows[1)] : let X be a non-singular complete algebraic surface without exceptional curves over a field k of any characteristic. We may divide such X's into 4 classes :

a) $\exists$ a curve C on X with $(K_X \cdot C) < 0$

b) $\forall$ curve C on X, $(K_X \cdot C) = 0$, or equivalently, for any $l \neq$ char. p, the fundamental class $[K_X] \in H^2_{\acute{e}t}(X, \boldsymbol{Q}_l)$ is zero.

c) $(K_X \cdot C) \geq 0$ for all curves C and $(K_X^2) = 0$ but $(K_X \cdot H) > 0$ for all ample divisors H.

d) $(K_X \cdot C) \geq 0$ for all C and $(K_X^2) > 0$, hence $(K_X \cdot H) > 0$ for all ample H. (Other cases are excluded by using the following well-known consequences of Hodge's Index Theorem : (1) $(K_X \cdot H) = 0$ for some ample H, $(K_X^2) \geq 0$ implies $(K_X \cdot C) = 0$ all C and (2) $(K_X \cdot C) \geq 0$ all curves C implies $(K_X^2) \geq 0$). Then in [10], it is proven that

(a) holds $\Leftrightarrow$ X is ruled, in which case $|nK_X| = \phi$, all n.

(b) holds $\Leftrightarrow$ either i) $2K_X \equiv 0$

or ii) $\exists\ \pi : X \to D$, D a curve, almost all fibres of π non-singular elliptic and hence $nK_X = \pi^*(\mathfrak{A})$, $\mathfrak{A}$ divisor on D of degree 0, $n \geq 1$ an integer.

(c) holds $\Leftrightarrow$ $\exists\ \pi : X \to D$ almost all fibres either non-singular elliptic or rational with one cusp, hence $nK_X = \pi^*(\mathfrak{A})$ where deg $(\mathfrak{A}) > 0$, $n \geq 1$.

(d) holds $\Leftrightarrow$ $|nK_X|$ is base-point free and defines a birational map from X to $\boldsymbol{P}^N$, for $n \gg 0$. Moreover, in this case $|2K_X| \neq \phi$.

Our *first goal* in this paper is to prove the following result, well known in char. 0 :

Theorem 1. *In cases* (b) *and* (c), *either* $|4K_X| \neq \phi$ *or* $|6K_X| \neq \phi$. *Therefore, in case* (b), *either* $4K_X \equiv 0$ *or* $6K_X \equiv 0$, *and in case* (c), *either* $4K_X$ *or* $6K_X$ *is represented by a*

1) The notation used is summarized below in "list of notations".

positive divisor.

In particular, this shows that the 4 cases above correspond to the classification of surfaces by Kodaira-dimension κ, i.e.,

$$\kappa = \text{tr. deg.}_k \bigoplus_{n=0}^{\infty} \Gamma(X, \mathcal{O}(nK_X)) - 1.$$

Then we see that:

In case (a), $\kappa = -1$
In case (b), $\kappa = 0$
In case (c), $\kappa = 1$
In case (d), $\kappa = 2$.

Thereafter, our *next goal* in this and a subsequent 3rd paper is the further analysis of all surfaces in case (b). It turns out that these can be divided into 4 types *by their Betti numbers*. This division into 4 types is based on a rather mysterious calculation that appears again and again in all work on the classification of surfaces. This calculation is as follows:

Assume $(K_X^2)=0$. Then by the Riemann-Roch theorem on X,

$$\text{(1)} \qquad 12(\dim H^0(\mathcal{O}_X) - \dim H^1(\mathcal{O}_X) + \dim H^2(\mathcal{O}_X)) = c_{2,X} = B_0 - B_1 + B_2 - B_3 + B_4$$

hence substituting $1 = B_0 = B_4 = \dim H^0(\mathcal{O}_X)$, we find

$$\text{(2)} \qquad 10 + 12\, p_g = 8 \dim H^1(\mathcal{O}_X) + 2(2 \dim H^1(\mathcal{O}_X) - B_1) + B_2.$$

Write $\Delta = 2 \dim H^1(\mathcal{O}_X) - B_1$. This is a "non-classical" term because when $\text{char}(k)=0$, then $\Delta=0$. In fact, we know that for *almost all* primes l:

$$\begin{aligned}(\boldsymbol{Z}/l\boldsymbol{Z})^{B_1} &\approx H^1_{\acute{e}t}(X, \boldsymbol{Z}/l\boldsymbol{Z}) \\ &\cong \{x \in \text{Pic}(X) \mid lx = 0\} \\ &\cong \{x \in \text{Pic}^0(X) \mid lx = 0\} \\ &\approx (\boldsymbol{Z}/l\boldsymbol{Z})^{2q}\end{aligned}$$

hence in any characteristic $B_1 = 2q$. On the other hand,

$$H^1(\mathcal{O}_X) \cong [\text{tangent space to Pic}(X) \text{ at } 0].$$

Thus if $\text{char}(k)=0$, $\text{Pic}(X)$, like any group scheme, is reduced, hence

$$\dim H^1(\mathcal{O}_X) = q$$

and $\Delta=0$. In general, we conclude that

$$\dim H^1(\mathcal{O}_X) \geq q$$

hence $\Delta \geq 0$, Δ even. We can say a bit more: if β_i are the Bockstein operators from $H^1(\mathcal{O}_X)$ to $H^2(\mathcal{O}_X)$, we know ([9], Lecture 27) that

$$\text{tangent space to } \text{Pic}^0_{\text{red}} \cong \bigcap_{i=1}^{\infty} \ker(\beta_i)$$

hence

$$\dim(\overbrace{\text{tang.sp.to Pic}^0}^{T_P})-\dim(\overbrace{\text{tang.sp.to Pic}^0_{\text{red}}}^{T_{P,\text{red}}}) = \dim H^1(\mathcal{O}_X)-\dim\bigcap_{i=1}^{\infty}\ker\beta_i$$
$$\leq \dim\bigcup_{i=1}^{\infty}\text{Im }\beta_i$$
$$\leq p_g.$$

Thus

$$\Delta = 2(\dim T_P-\dim T_{P,\text{red}}) \leq 2p_g.$$

Although it is not used in what follows, it is interesting at this point to consider what happens for arbitrary *analytic* surfaces over $\boldsymbol{C}$. The equation (2) is perfectly valid and in this case Kodaira ([4], p. 755, Th. 3) has shown that $\Delta=0$ or 1 according to the parity of B_1.

Now assume that the surface X has $\kappa=0$, i.e. : $(K_X\cdot C)=0$ for all curves C. Then not only is $(K_X^2)=0$ but either $p_g=0$ or $p_g=1$ and $K_X\equiv 0$. It is then easy to list *all* solutions to equation (2) :

Table of Possible Invariants for Surfaces with $\kappa=0$

B_2	B_1	c_2	$\chi(\mathcal{O}_X)$	$\dim H^1(\mathcal{O}_X)$	p_g	Δ
22	0	24	2	0	1	0
14	2	12	1	1	1	0
10	0	12	1	0	0	0
				1	1	2
6	4	0	0	2	1	0
2	2	0	0	1	0	0
				2	1	2

(B_2, B_1, c_2, $\chi(\mathcal{O}_X)$:) invariants under deformation

($\dim H^1(\mathcal{O}_X)$, p_g, Δ:) invariants which are in general only upper semi-continuous under deformation

Concerning these categories of surfaces, we shall prove in this paper the following results :

Theorem 5. *The surfaces with $\kappa=0$, $B_2=22$, known as K3-surfaces, have the following properties :*

i) *for all divisors D on X, $(D\cdot C)=0$ for all curves C implies $D\equiv 0$, hence*

$$\text{Pic}^0(X) = (0).$$

ii) *X has no connected étale coverings, i.e.,*

$$\pi_{1,\text{alg}}(X) = (e).$$

No surfaces with $\kappa=0$, $B_2=14$ exist.
Surfaces with $\kappa=0$, $B_2=10$, $p_g=1$ cannot exist if $\text{char}(k)\neq 2$.

Theorem 6. *All surfaces with* $\kappa=0$, $B_2=6$ *are abelian varieties.*

Moreover, the following is easy to see from the above table and the results of [15]:

Proposition. *If* X *is a surface with* $\kappa=0$, $B_2=2$, *then* $B_1=2$, *hence* $\mathrm{Alb}(X)$ *is an elliptic curve and the fibres of the canonical map*

$$\pi : X \to \mathrm{Alb}(X)$$

are either almost all non-singular elliptic curves, or almost all rational curves with ordinary cusps. The latter is only possible if $\mathrm{char}(k)=2$ *or* 3.

We call surfaces of this type *hyperelliptic* or *quasi-hyperelliptic* surfaces, depending on which type of fibre π has. In this paper, we shall also analyze hyperelliptic surfaces. However, the analysis of the case of quasi-hyperelliptic surfaces and the case of surfaces with $\kappa=0$, $B_2=10$, which we propose to call *Enriques surfaces* (regardless of whether $K\equiv 0$ or $K\not\equiv 0$!), we postpone to a 3rd part of the paper. Since Enriques surfaces in $\mathrm{char}(k)\neq 2$ are fairly easily seen to have the same behaviour as in char. 0, Part III of this paper will deal largely with the curious pathology of char. 2 and 3.

Finally, for use in § 2, we note that the analysis leading to the Table does not use completely the assumption $\kappa=0$: in fact, it really only uses $(K_X^2)=0$, $p_g\leq 1$. Thus the analysis also shows:

Corollary. *If* X *is a non-singular complete surface with* $(K_X^2)=p_g=0$, *then* X *belongs to one of the 2 following types:*

i) $B_1 = \dim H^1(\mathcal{O}_X) = 0$, *hence* $\mathrm{Pic}^0(X) = (0)$; $\chi(\mathcal{O}_X) = 1$; $B_2 = 10$

ii) $B_1 = 2$, $\dim H^1(\mathcal{O}_X) = 1$, *hence* $\mathrm{Pic}^0(X)$ *is a reduced elliptic curve* ; $\chi(\mathcal{O}_X) = 0$; $B_2 = 2$.

List of Notations

X usually a non-singular projective surface
Alb X = Albanese variety of X
Pic X = Picard scheme of X
$\mathrm{Pic}^0\, X$ = connected component of $0 \in \mathrm{Pic}(X)$
$q = \dim \mathrm{Pic}\, X = \dim \mathrm{Alb}\, X$, the "irregularity" of X
K_X = the canonical divisor class on X
B_i = i^{th} Betti number of X
$h^{p,q} = \dim H^q(X, \Omega^p)$
$p_g = h^{0,2} = h^{2,0}$, the geometric genus of X
$\omega_X = \Omega_X^2$, the sheaf of 2–forms, if X is smooth
= the dualizing sheaf of Grothendieck for general Cohen-Macauley surfaces.

1. K_X of Elliptic or Quasi-elliptic Surfaces

An elliptic or quasi-elliptic surface is a fibration $f: X \to B$ of a surface X over a non-singular curve B, with $f_* \mathcal{O}_X = \mathcal{O}_B$, with almost all fibres elliptic or rational with a cusp (by a result of Tate [15], the latter situation can occur only if char$(k)=2$ or 3). Note that since the function field $k(X)$ is separable over $k(B)$, almost all fibres are generically smooth. Also every fibre of f is a curve of canonical type[1]. At finitely many points $b_1, \cdots, b_r \in B$ the fibre $f^{-1}(b_\lambda)$ is multiple, i.e.,

$$f^{-1}(b_\lambda) = m_\lambda P_\lambda$$

with $m_\lambda \geqq 2$ and P_λ indecomposable of canonical type. We have

$$R^1 f_* \mathcal{O}_X = L \oplus T$$

where L is an invertible sheaf and T is supported precisely at the points $b \in B$ at which

$$\dim H^0(f^{-1}(b), \mathcal{O}_{f^{-1}(b)}) \geqq 2.$$

To see this, note that by E. G. A. III 7. 8, the sheaf $R^1 f_* \mathcal{O}_X$ is locally free at b if and only if $\mathcal{O}_X$ is cohomologically flat at b in dimension 0.
This suggests

Definition. The fibres of f over supp T are called *wild fibres.*

Noting that if C is indecomposable of canonical type then $\dim H^0(C, \mathcal{O}_C)=1$ (see Mumford [10], p. 332), we get

Proposition 3. *Every wild fibre is a multiple fibre.*

In the following, we consider only relatively minimal fibrations $f: X \to B$, i.e., no exceptional curve of the first kind is a component of a fibre.

Theorem 2. *Let $f: X \to B$ be a relatively minimal elliptic or quasi-elliptic fibration and let $R^1 f_* \mathcal{O}_X = L \oplus T$. Then*

$$\omega_X = f^*(L^{-1} \otimes \omega_B) \otimes \mathcal{O}(\textstyle\sum a_\lambda P_\lambda)$$

where

(i) *$m_\lambda P_\lambda$ are the multiple fibres*
(ii) $0 \leqq a_\lambda < m_\lambda$
(iii) *$a_\lambda = m_\lambda - 1$ if $m_\lambda P_\lambda$ is not wild*
(iv) $\deg(L^{-1} \otimes \omega_B) = 2p(B) - 2 + \chi(\mathcal{O}_X) + \text{length } T$

where $p(B)$ is the genus of B.

---

1) In the notation of [10], a curve $D = \sum n_i E_i$ is said to be of canonical type if $(K \cdot E_i) = (D \cdot E_i) = 0$ for all i.

Note that in the case char$(k)=0$ or in the complex analytic case there are no wild fibres, so that $a_i=m_i-1$; see Kodaira [4], p. 772, Th. 12.

Proof. For any non-multiple fibre $f^{-1}(y)$ we have

$$\mathcal{O}_{f^{-1}(y)}\otimes\omega_X\cong\omega_{f^{-1}(y)}\cong\mathcal{O}_{f^{-1}(y)},$$

hence if $y_1,\cdots,y_r$ are distinct general points of B the cohomology sequence of

$$0\to\omega_X\to\omega_X\otimes\mathcal{O}\big(\sum_{i=1}^r f^{-1}(y_i)\big)\to\bigoplus_{i=1}^r\mathcal{O}_{f^{-1}(y_i)}\to 0$$

yields

$$\dim\left|\omega_X\otimes\mathcal{O}\big(\sum_{i=1}^r f^{-1}(y_i)\big)\right|\geqq 0$$

for large enough r. If D is a divisor in the linear system above, we have

$$(D\cdot f^{-1}(y))=0$$

hence we can write

$$K_X\equiv(\text{sum of fibres})+\Delta$$

where $\Delta\geqq 0$ is contained in a union of fibres and does not contain fibres of f. Let Δ_0 be a connected component of Δ and let $C=f^{-1}(y)$ be the fibre containing Δ_0. Then Δ_0 is a rational submultiple of C, i.e., we have

$$C=mP,\qquad \Delta_0=aP$$

where P is indecomposable of canonical type and $0\leqq a<m$. This follows from

Lemma. *Let $D=\sum n_iC_i$ be an effective divisor on a surface X with each C_i irreducible. Assume that*

$$(C_i\cdot D)\leqq 0,\qquad \text{all } i$$

and that D is connected.

Then every divisor $Z=\sum m_iC_i$ satisfies $Z^2\leqq 0$ and equality holds if and only if $D^2=0$ and $Z=\lambda D$, $\lambda\in\boldsymbol{Q}$.

Proof. Write $x_i=m_i/n_i$. We have

$$\begin{aligned}Z^2&=\sum x_ix_jn_in_j(C_i\cdot C_j)\\&\leqq\sum x_i^2n_i^2(C_i\cdot C_i)+\sum_{i\neq j}\frac{1}{2}(x_i^2+x_j^2)n_in_j(C_i\cdot C_j)\\&=\sum x_i^2n_i(C_i\cdot D)\leqq 0.\end{aligned}$$

If equality holds everywhere, we have either $x_i=x_j$ or $(C_i\cdot C_j)=0$ for all i,j; since D is connected, x_i is constant, i.e., $m_i=\lambda n_i$, $\lambda\in\boldsymbol{Q}$. q. e. d.

Going back to the proof that $\Delta_0=aP$, if Δ_ν are the connected components of Δ, we have

$$0=K_X^2=\sum\Delta_\nu^2;$$

since each $\Delta_\nu^2\leqq 0$ by the previous lemma, we must have $\Delta_\nu^2=0$ and now the equality case of the lemma proves that Δ_ν is a rational multiple of the fibre containing it.

We have proved that

$$\omega_X = f^*\mathcal{O}_B(\mathfrak{A}) \otimes \mathcal{O}(\sum a_\lambda P_\lambda)$$

for some divisor $\mathfrak{A} \in \mathrm{div}(B)$ and integers a_λ with $0 \leqq a_\lambda < m_\lambda$. We deduce that

$$f_*(\omega_X) = \mathcal{O}_B(\mathfrak{A}).$$

Now the duality theorem for a map says that

$$\begin{aligned} f_*\omega_X &= \mathrm{Hom}(R^1 f_*\mathcal{O}_X, \omega_B) \\ &= L^{-1} \otimes \omega_B \end{aligned}$$

because the dual of the torsion sheaf is 0 ; this can be found in Deligne-Rapoport [2], pp. 19–20, formula (2. 2. 3). Hence

$$\omega_X = f^*(L^{-1} \otimes \omega_B) \otimes \mathcal{O}(\sum a_\lambda P_\lambda).$$

The spectral sequence of the map f yields

$$\begin{aligned} \chi(\mathcal{O}_X) &= \chi(\mathcal{O}_B) - \chi(R^1 f_*\mathcal{O}_X) \\ &= \chi(\mathcal{O}_B) - \chi(L) - \text{length } T \\ &= -\deg L - \text{length } T, \end{aligned}$$

by the Riemann-Roch theorem on the curve B, and since $\deg(\omega_B) = 2p(B) - 2$ we obtain (iv) of Theorem 2.

It remains to prove (iii), and this follows from

Proposition 4. *Let m_λ, P_λ, a_λ be as in Theorem 2 and let*

$$\nu_\lambda = \mathrm{order}(\mathcal{O}_{P_\lambda} \otimes \mathcal{J}_{P_\lambda}^{-1})$$

where $\mathcal{J}_{P_\lambda}$ is the sheaf of ideals of P_λ, be the order of the normal sheaf of P_λ in X. Then we have

i) *ν_λ divides m_λ and $a_\lambda + 1$,*

ii) $\dim H^0(P_\lambda, \mathcal{O}_{(\nu_\lambda+1)P_\lambda}) \geqq 2$, $\dim H^0(P_\lambda, \mathcal{O}_{\nu_\lambda P_\lambda}) = 1$,

iii) $\dim H^0(P_\lambda, \mathcal{O}_{rP_\lambda})$ *is non-decreasing with r.*

In particular, if $a_\lambda < m_\lambda - 1$ then $\nu_\lambda < m_\lambda$ and this is equivalent to the multiple fibre $m_\lambda P_\lambda$ being wild.

Proof. Let us write $m, P, a, \nu, \mathcal{J}$ for $m_\lambda, P_\lambda, a_\lambda, \nu_\lambda, \mathcal{J}_{P_\lambda}$. If $r \geqq s \geqq 1$, the restriction map $\mathcal{O}_{rP} \to \mathcal{O}_{sP}$ is surjective, hence $\dim H^1(P, \mathcal{O}_{rP})$ is non-decreasing with r. Since $\chi(\mathcal{O}_{rP}) = 0$, this proves that $\dim H^0(P, \mathcal{O}_{rP})$ is non-decreasing too.

We have an isomorphism

$$\mathcal{O}_P \otimes \mathcal{J}^\nu \cong \mathcal{O}_P$$

and via this isomorphism we get an exact sequence

$$0 \to \mathcal{O}_P \to \mathcal{O}_{(\nu+1)P} \xrightarrow{\text{res}} \mathcal{O}_{\nu P} \to 0$$

where res is the restriction. Since constants in $H^0(P, \mathcal{O}_{(\nu+1)P})$ are mapped into constants in $H^0(P, \mathcal{O}_{\nu P})$, the cohomology sequence shows that $\dim H^0(P, \mathcal{O}_{(\nu+1)P}) \geqq 2$. Finally, ν divides both m and $a+1$, because $\mathcal{O}_P \otimes \mathcal{J}^{-m} \cong \mathcal{O}_P$ (trivial) and

$$\mathcal{O}_P \otimes \mathcal{J}^{-a-1} \cong \omega_P \cong \mathcal{O}_P$$

(Mumford [10], p. 333). q. e. d.

It is shown in Raynaud [13], Prop. 6. 3. 5, that m_λ/ν_λ is a power of the characteristic p of k. In particular the multiplicity of a wild fibre is divisible by p, and wild fibres do not occur in char. 0.

Corollary. *If* $\dim H^1(X, \mathcal{O}_X) \leqq 1$ *we have either*

$$a_\lambda + 1 = m_\lambda \quad \text{or} \quad \nu_\lambda + a_\lambda + 1 = m_\lambda.$$

Proof. Since $\chi(\mathcal{O}_{(\nu+1)P}) = 0$ and $\dim H^0(P, \mathcal{O}_{(\nu+1)P}) \geqq 2$, using duality we find that

$$\dim H^0(P, \omega_{(\nu+1)P}) \geqq 2.$$

Now the cohomology sequence of

$$0 \to \omega_X \to \mathcal{I}^{-\nu-1} \otimes \omega_X \to \omega_{(\nu+1)P} \to 0$$

yields

$$\dim H^0(X, \mathcal{I}^{-\nu-1} \otimes \omega_X) > \dim H^0(X, \omega_X),$$

since we have $\dim H^1(X, \omega_X) = \dim H^1(X, \mathcal{O}_X) \leqq 1$ by hypothesis. This increase in dimension is possible only if $\nu + a + 1 \geqq m$, or $1 + (a+1)/\nu \geqq m/\nu$. Therefore $(a+1)/\nu = m/\nu$ or $m/\nu - 1$. q. e. d.

We conclude this section with a remark on hyperelliptic or quasi-hyperelliptic surfaces.

Proposition 5. *Let* $f: X \to E$, $E = \mathrm{Alb}(X)$ *be an hyperelliptic surface. Then every fibre of* f *is smooth.*

Moreover if $f: X \to E$ *is quasi-hyperelliptic then every fibre of* f *is a rational curve with a cusp, i.e., there are no reducible fibres.*

Proof. Since $p(E) = 1$, $\chi(\mathcal{O}_X) = 0$ and $K_X \sim 0$ ($\sim$ is numerical equivalence), Theorem 2 gives

$$(\text{length } T) f^{-1}(y) + \sum a_\lambda P_\lambda \sim 0$$

therefore there are no multiple fibres. Also since the Picard number is $\rho \leqq B_2 = 2$, there are no reducible fibres. In the elliptic case the smoothness of f follows by considering the differential $f^*(\omega)$, where $\omega \in \Gamma(\Omega^1_E)$. $f^*(\omega)$ will only be zero at the points where f is not smooth and since these are finite in number,

$$c_{2,X} = [\text{number of zeroes of } f^*(\omega) \text{ counted with multiplicity}].$$

But $c_{2,X} = 0$, so $f^*\omega$ has no zeroes, so f is smooth. In any elliptic or quasi-elliptic surface, every irreducible fibre is either a) non-singular elliptic, b) rational with a node, or c) rational with a cusp. In the quasi-elliptic case, the generic fibre is of type (c) and since such a curve cannot specialize to type (a) or type (b), every irreducible fibre is rational with a cusp.

q. e. d.

2. Proof of Theorem 1

We shall prove here that if $f: X \to B$ is elliptic or quasi-elliptic, $(K_X \cdot C) \geqq 0$ for all curves C and $K_X^2 = 0$, then :

$$(*) \qquad |4K_X| \neq \phi \quad \text{or} \quad |6K_X| \neq \phi.$$

In proving this result we may assume $p_g = 0$ and use Table 1 as a list of numerical invariants. Theorem 2 implies

$$p_g = \dim H^0(B, L^{-1} \otimes \omega_B)$$

and since $\chi(\mathcal{O}_X) \geqq 0$, the Riemann-Roch theorem on B shows that $p_g = 0$ implies $p(B) = 0$ or 1 and if $p(B) = 1$ we must also have $T = (0)$. So if $p(B) = 1$ there are no wild fibres and $a_\lambda = m_\lambda - 1$ in Theorem 2. If there is a multiple fibre, it is easily seen that $|2K_X| \neq \phi$. If there are no multiple fibres at all, then

$$\omega_X = f^*(L^{-1} \otimes \omega_B)$$

and deg $(L^{-1} \otimes \omega_B) = 0$, thus $K_X \sim 0$ and X is hyperelliptic or quasi-hyperelliptic.

Theorem 3. *If X is hyperelliptic or quasi-hyperelliptic, then there is a second structure $f: X \to \boldsymbol{P}^1$ of X as an elliptic surface over $\boldsymbol{P}^1$.*

Proof. By the results in [10], it is sufficient to show the existence of a curve C of canonical type, transversal to the Albanese fibration, $\varphi: X \to E$ with $E = \mathrm{Alb}(X)$. Let F_t be the fibre $\varphi^{-1}(t)$ of φ over $t \in E$. There exists a divisor D on X such that

$$(D^2) = 0, \qquad (D \cdot F_0) > 0,$$

for example some linear combination of an ample divisor and F_0 ; let

$$D_t = D + F_t - F_0.$$

There is a point $t \in E$ such that $|D_t| \neq \phi$. If not, use $\chi(\mathcal{O}(D_t)) = 0$ and the Riemann-Roch theorem to prove

$$\dim H^0(X, \mathcal{O}(D_t)) = \dim H^1(X, \mathcal{O}(D_t)) = 0$$

for all t. The cohomology sequence of

$$0 \to \mathcal{O}(D_t) \to \mathcal{O}(D + F_t) \xrightarrow{r_{F_0}} \mathcal{O}_{F_0} \otimes \mathcal{O}(D) \to 0$$

then gives an isomorphism

$$r_{F_0}: H^0(X, \mathcal{O}(D + F_t)) \simeq H^0(F_0, \mathcal{O}_{F_0} \otimes \mathcal{O}(D))$$

where r_{F_0} is the restriction. Since $(D \cdot F_0) > 0$, there is a non-trivial section $\sigma \in \Gamma(\mathcal{O}_{F_0} \otimes \mathcal{O}(D))$, and let $s_t = r_{F_0}^{-1}(\sigma)$. Clearly $X = \text{closure} \bigcup_{t \neq 0} \mathrm{div}(s_t)$ and $\mathrm{div}(s_t) \cap F_0$ has support in $\mathrm{div}(\sigma)$, for all $t \neq 0$. It follows that as $t \to 0$ we must have $\mathrm{div}(s_t) \to F_0 + C \equiv D + F_0$, and $C \in |D|$, proving our assertion.

We have found a curve $C > 0$ with $(C^2) = 0$ and $(C \cdot F_0) > 0$, and we claim that C is of canonical type. In fact, since $K_X \sim 0$ and $(C^2) = 0$, our assertion will follow from the fact that X has no irreducible curve Γ with $(\Gamma^2) = -2$. Such a curve Γ

cannot be transversal to the Albanese fibering because Γ is rational, and cannot be a component of a fibre, since every fibre is irreducible by Proposition 5.

q. e. d.

In view of Theorem 3, we have only to examine the case in which $p(B)=0$. Since B is rational, the canonical bundle formula becomes

$$K_X \equiv rf^{-1}(y)+\sum_\lambda a_\lambda P_\lambda$$

where

$$r=-2+\chi(\mathcal{O}_X)+\text{length } T.$$

If H is an ample divisor on X, since $(K_X\cdot H)\geqq 0$ we have

$$r+\sum_\lambda \frac{a_\lambda}{m_\lambda}\geqq 0.$$

Moreover

$$\dim|nK_X| = nr+\sum_\lambda\left[\frac{na_\lambda}{m_\lambda}\right].$$

It is now easy to see, using $\chi(\mathcal{O}_X)\geqq 0$ and Proposition 4, Corollary that we can have only the following cases :

(A) length $T=0$, so $a_\lambda=m_\lambda-1$, $\nu_\lambda=m_\lambda$.

If $\chi(\mathcal{O}_X)=0$, then there are at least 3 multiple fibres and we can have :

a) there are 4 or more multiple fibres, i. e., $m_\lambda\geq 2$, $1\leq\lambda\leq 4$, and then $|2K_X|\neq\phi$.

b) there are 3 multiple fibres with all multiplicities $m_\lambda\geq 3$. Then $|3K_X|\neq\phi$.

c) there are 3 multiple fibres with $m_1=2$, m_2, $m_3\geq 4$. Then $|4K_X|\neq\phi$.

d) there are 3 multiple fibres with $m_1=2$, $m_2=3$, $m_3\geq 6$. Then $|6K_X|\neq\phi$.

If $\chi(\mathcal{O}_X)=1$, then there are at least 2 multiple fibres, m_1, $m_2\geq 2$, and $|2K_X|\neq\phi$.

If $\chi(\mathcal{O}_X)\geq 2$, then $|K_X|\neq\phi$.

(B) length $T=1$. If $\chi(\mathcal{O}_X)=0$, then $|K_X|=\phi$, so $\dim H^1(\mathcal{O}_X)=1$ and Prop. 4, Cor. applies. So if $f^{-1}(P_1)$ is the wild fibre, then we have $a_1=m_1-1$ or $a_1=m_1-1-\nu_1$ where $\nu_1|$g. c. d. (m_1, a_1+1), while $a_\lambda=m_\lambda-1$, $\nu_\lambda=m_\lambda$ for $\lambda\geq 2$. Moreover there are at least 2 multiple fibres and we can have :

a′) there are 2 or more multiple fibres with $a_\lambda=m_\lambda-1$ and then $|2K_X|\neq\phi$.

b′) the wild fibre satisfies $m_1=3$, $a_1=1$, $\nu_1=1$ (hence char.$=3$) and the tame fibre satisfies $m_2\geq 3$. Then $|3K_X|\neq\phi$.

c_1') the wild fibre satisfies $m_1=4$, $a_1=1$, $\nu_1=2$ (hence char.$=2$) and the tame fibre satisfies $m_2\geq 4$. Then $|4K_X|\neq\phi$.

c_2') the wild fibre satisfies $m_1=\mu_1\nu_1$, where $\mu_1\geq 4$ (any positive char.). In this case, $a_1/m_1\geq 1/2$ and $|2K_X|\neq\phi$.

d_1') the wild fibre satisfies $m_1=2\nu_1$, $a_1=\nu_1-1$, $\nu_1\geq 3$ (hence char.$=2$) and the tame fibre satisfies $m_2\geq 3$. Then $|3K_X|\neq\phi$.

d_2') the wild fibre satisfies $m_1=3\nu_1$, $a_1=2\nu_1-1$, $\nu_2\geq 2$ (hence char.$=3$). In this case $|2K_X|\neq\phi$.

If $\chi(\mathcal{O}_X)\geq 1$, then $|K_X|\neq\phi$.
(C) length $T\geq 2$, then also $|K_X|\neq\phi$.

If we specialize to the case $\kappa=0$, then we easily get the following list of possible multiple fibres for elliptic or quasi-elliptic surfaces $f\colon X\to\boldsymbol{P}^1$ with K_X a torsion divisor :

		length T	$\chi(\mathcal{O}_X)$	a_λ/m_λ (*=wild fibre)	order K_X	char.
tame cases	i)	0	0	(1/2, 1/2, 1/2, 1/2)	2	
	ii)	0	0	(2/3, 2/3, 2/3)	3	
	iii)	0	0	(1/2, 3/4, 3/4)	4	
	iv)	0	0	(1/2, 2/3, 5/6)	6	
	v)	0	1	(1/2, 1/2)	2	
	vi)	0	2	none	1	
wild cases	vii)	1	0	(0/2*, 1/2, 1/2)	2	2
	viii)	1	0	(1/2*, 1/2)	2	2
	ix)	1	0	(1/3*, 2/3)	3	3
	x)	1	0	(1/4*, 3/4)	4	2
	xi)	1	0	(2/4*, 1/2)	2	2
	xii)	1	0	(2/6*, 2/3)	3	2
	xiii)	1	0	(3/6*, 1/2)	2	3
	xiv)	1	1	(0/2*)	1	2
	xv)	2	0	one or two wild fibres $0/p^r$	1	p

Note that each of the wild cases may be thought of as coming from the confluence of 2 tame fibres in one of the tame cases.

3. Analysis of Hyperelliptic Surfaces

In this section, we study more closely surfaces X such that :

a) $\kappa=0$
b) the Albanese mapping is $\pi: X\to E$, E elliptic
c) almost all fibres C_x of π are non-singular.

By the Table of the Introduction, it follows also that

d) $B_2=2$, $c_2=0$, $\chi(\mathcal{O}_X)=0$.

Moreover, by Proposition 5 it follows that

c′) *all* fibres C_x are non-singular elliptic.

By Theorem 3, § 2, we see :

e) There is a second elliptic pencil $\pi': X\to\boldsymbol{P}^1$ on X.

We want to compare π and π' and see the effect of 2 simultaneous elliptic fibrations! Let C'_y be the fibres of π'. Then all the C'_y are finite coverings of E :

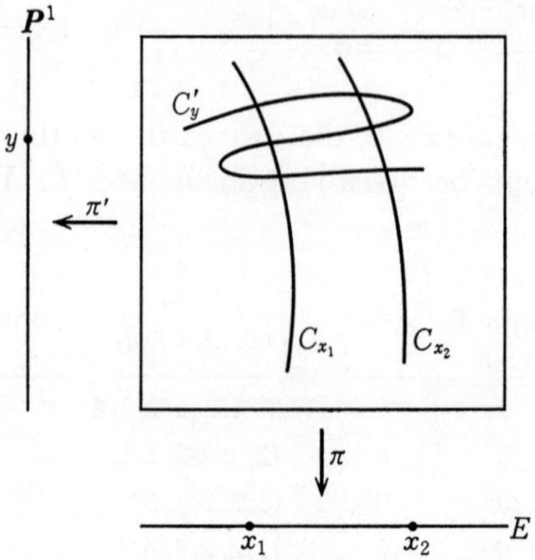

Hence all the C'_y are either non-singular elliptic or multiples of non-singular elliptic curves, and

$$p_y = \text{res } \pi : C'_y \to E$$

is an isogeny. Let $S = \{y \in \boldsymbol{P}^1 | C'_y \text{ multiple}\}$. p_y defines a pull-back on Pic^0:

$$\text{Pic}^0(C'_y) \xleftarrow{p_y^*} \text{Pic}^0(E).$$

Choosing a base point $x_0 \in E$, we can identify $\text{Pic}^0(E)$ with E by associating the sheaf $\mathcal{O}_E(x-x_0)$ with the point x. As usual, this makes E into an algebraic group with identity x_0. Now we cannot choose base points on each C'_y varying nicely with y unless $\pi' : X \to \boldsymbol{P}^1$ has a section. However, we can instead note that $\text{Pic}^0(C'_y)$ *acts* canonically on C'_y by translations : i.e., the sheaf L of degree 0 maps $u \in C'_y$ to the unique point v such that $L(u) \cong \mathcal{O}_{C'_y}(v)$. Then via the maps p_y^*, we find that E is acting by translations simultaneously on *all* the curves C'_y. If we stick to the non-multiple curves, it follows easily that this is an algebraic action of E:

$$\sigma_0 : E \times \pi'^{-1}(\boldsymbol{P}^1 - S) \to \pi'^{-1}(\boldsymbol{P}^1 - S).$$

But since X is a minimal model, any automorphism of the Zariski-open set $\pi'^{-1}(\boldsymbol{P}^1 - S)$ extends to an automorphism of X so we actually get an action :

$$\sigma : E \times X \to X.$$

To relate this action to π, say $x \in E$, $u \in C'_y$. Then x takes u to v where

$$\pi^*(\mathcal{O}_E(x-x_0)) \otimes \mathcal{O}_{C'_y}(u) \cong \mathcal{O}_{C'_y}(v).$$

Let $n = (C'_y \cdot C_x) =$ (degree of res $\pi : C'_y \to E$). Then taking $\text{Norm}_{C'_y/E}$ of the 2 sides of the above isomorphism :

$$\mathcal{O}_E(nx - nx_0 + \pi u) \cong \mathcal{O}_E(\pi v),$$

hence we get a commutative diagram

$$(*)\qquad \begin{array}{ccc} X & \xrightarrow{\text{action of } x} & X \\ {\scriptstyle\pi}\Big\downarrow & & \Big\downarrow{\scriptstyle\pi} \\ E & \xrightarrow[\text{by } n(x-x_0)]{\text{translation}} & E \end{array} \quad ; \quad \begin{array}{ccc} E\times X & \xrightarrow{\sigma} & X \\ {\scriptstyle 1\times\pi}\Big\downarrow & & \Big\downarrow{\scriptstyle\pi} \\ E\times E & \longrightarrow & E \\ x,y & \longrightarrow & nx+y \end{array}$$

We can now use this action of E to describe the whole surface X as follows : let $E_0 = C_{x_0}$ be the fibre over x_0, and let $A_n = \text{Ker } (n_E : E \to E)$ considered as a subgroup scheme of E. Then by $(*)$ the action of A_n on X preserves the fibres of π, hence A_n acts on E_0, and give this action the name α :

$$\alpha : A_n \to \text{Aut}(E_0) = \text{group scheme of automorphisms of } E_0.$$

Then by restriction of the action σ of E, we get a morphism :

$$\tau : E \times E_0 \to X$$

which by $(*)$ fits into a diagram :

$$\begin{array}{ccc} E\times E_0 & \xrightarrow{\quad\tau\quad} & X \\ & {\scriptstyle n_E\cdot p_1}\searrow \qquad \swarrow{\scriptstyle\pi} & \\ & E & \end{array}$$

Note that

$$\tau(x, y) = \tau(x', y') \Leftrightarrow \sigma(x-x', y) = y'$$
$$\Leftrightarrow x-x' \in A_n \text{ and } \alpha(x-x')(y) = y'$$

hence it follows that $X \cong \text{quotient } (E \times E_0/A_n)$, via the action

$$x(u, v) = (u+x, \alpha(x)(v)), \quad x \in A_n, \ u \in E, \ v \in E_0.$$

If we replace E by $E_1 = E/\text{Ker } \alpha$, this proves :

Theorem 4. *Every hyperelliptic surface X is of the form :*

$$X = E_1 \times E_0/A, \qquad E_1, E_0 \textit{ elliptic curves}$$

where A is a finite subgroupscheme of E_1, and A acts by

$$k(u, v) = (u+k, \alpha(k)(v))$$

for some injective homomorphism

$$\alpha : A \to \text{Aut}(E_0).$$

Moreover, the 2 elliptic fibrations on X are given by :

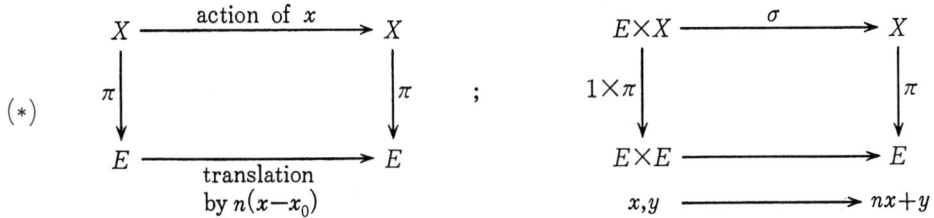

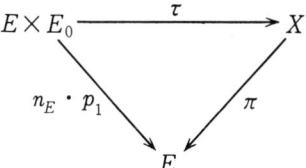

$$\begin{array}{ccc} E_1 \times E_0/A & & E_1 \times E_0/A \\ \downarrow & & \downarrow \\ E_1/A & & E_0/\alpha(A) \\ \text{(elliptic)} & & \wr\| \\ & & \boldsymbol{P}^1 \end{array}$$

This theorem can easily be used a) to classify such X's and b) to compute the order of K_X in Pic (X). We use the fact that choosing a base point $0 \in E_0$, $\mathrm{Aut}(E_0)$ becomes a semi-direct product :

$$\mathrm{Aut}(E_0) = \underset{\substack{\nearrow \\ \text{normal subgroup} \\ \text{of translations}}}{E_0} \cdot \underset{\substack{\nwarrow \\ \text{finite, discrete group} \\ \text{of autos, fixing } 0}}{\mathrm{Aut}(E_0, 0)}$$

Note that $\alpha(A) \not\subset E_0$, or else $E_0/\alpha(A)$ would be elliptic instead of rational as required. Moreover, from the tables in Lang [5], Appendix 1, we find :

$$\begin{aligned} \mathrm{Aut}(E_0, 0) &= \{1_E, -1_E\} \cong \boldsymbol{Z}/2\boldsymbol{Z} && \text{if } j(E_0) \neq 0, 12^3 \\ &\cong \boldsymbol{Z}/4\boldsymbol{Z} && \text{if } j(E_0) = 12^3, \text{ char} \neq 2, 3 \\ &\cong \boldsymbol{Z}/6\boldsymbol{Z} && \text{if } j(E_0) = 0, \text{ char} \neq 2, 3 \end{aligned}$$

$\cong$ semi-direct product $\boldsymbol{Z}/4\boldsymbol{Z} \cdot \boldsymbol{Z}/3\boldsymbol{Z}$, $\boldsymbol{Z}/3\boldsymbol{Z}$ normal, $i \in \boldsymbol{Z}/4\boldsymbol{Z}$ acting by mult. by $(-1)^{2i}$

if $j(E_0) = 0$, char $= 3$

$\cong$ semi-direct product (Quat. gp. of order 8) $\cdot\, \boldsymbol{Z}/3\boldsymbol{Z}$, Quat. gp. normal, $\boldsymbol{Z}/3\boldsymbol{Z}$ permuting cyclically $i, j, k \in$ Quat. gp.

if $j(E_0) = 0$, char $= 2$

The important point here is that since A is commutative, so is $\alpha(A)$ and now even in the last 2 nasty cases, the maximal *abelian* subgroups are still $\boldsymbol{Z}/4\boldsymbol{Z}$ and $\boldsymbol{Z}/6\boldsymbol{Z}$, which in all cases are cyclic.

Let $k \in A$ be such that

$$\mathrm{Im}\, \alpha(k) \in \mathrm{Aut}(E_0)/E_0$$

generates

$$\mathrm{Im}\, \alpha(A) \subset \mathrm{Aut}(E_0)/E_0.$$

Then $\alpha(k) \notin E_0$, hence it has some fixed point. Replacing 0 by this fixed point, it follows that $\alpha(A)$ itself is a direct product :

$$\alpha(A) = \underset{\substack{\nearrow \\ \text{finite gp. scheme of} \\ \text{translations } A_0 \subset E_0}}{A_0} \cdot \underset{\substack{\nwarrow \\ \text{cyclic gp. generated by } k, \\ n = 2, 3, 4 \text{ or } 6}}{\boldsymbol{Z}/n\boldsymbol{Z}}$$

Since A_0 and k must commute, $A_0 \subset$ (fix pt. set F of k). Again referring to Lang to check the fix point sets, we find :

a) $n = 2$, (so $k = -1_E$), then $F = \mathrm{Ker}\; 2_{E_0}$

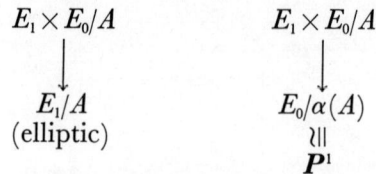

b) $n=3$, then $\# F = 3$ so $F \cong \boldsymbol{Z}/3\boldsymbol{Z}$ if char $\neq 3$
$F \cong \alpha_3$ if char $= 3$ (because E_0 is supersingular !)

c) $n=4$, then $\# F = 2$ so $F \cong \boldsymbol{Z}/2\boldsymbol{Z}$ if char $\neq 2$
$F \cong \alpha_2$ if char $= 2$ (because E_0 is supersingular !)

d) $n=6$, then $F=(e)$

We can now mechanically compile a list of all possible $\alpha(K)$' s, hence all possible X' s:

a1) $E_1 \times E_0/(\boldsymbol{Z}/2\boldsymbol{Z})$;	action $(x,y) \mapsto (x+a, -y)$
a2) $E_1 \times E_0/(\boldsymbol{Z}/2\boldsymbol{Z})^2$;	action $(x,y) \mapsto (x+a, -y)$, $(x+b, y+c)$ (here char $\neq 2$)
a3) $E_1 \times E_0/(\boldsymbol{Z}/2\boldsymbol{Z}) \cdot \mu_2$;	action $(x,y) \mapsto (x+a, -y)$, μ_2 acts by transl. on both factors.
b1) $E_1 \times E_0/(\boldsymbol{Z}/3\boldsymbol{Z})$;	action $(x,y) \mapsto (x+a, \omega y)$ where $j(E_0)=0$, $\omega: E_0 \to E_0$ an automorphism of order 3
b2) $E_1 \times E_0/(\boldsymbol{Z}/3\boldsymbol{Z})^2$;	action $(x,y) \mapsto (x+a, \omega y)$, $(x+b, y+c)$, E_0, ω as before and $\omega c = c$, order $c=3$ (here char $\neq 3$)
c1) $E_1 \times E_0/(\boldsymbol{Z}/4\boldsymbol{Z})$;	action $(x,y) \mapsto (x+a, iy)$, where $j(E_0)=12^3$, $i: E_0 \to E_0$ an automorphism of order 4
c2) $E_1 \times E_0/(\boldsymbol{Z}/2\boldsymbol{Z}) \cdot (\boldsymbol{Z}/4\boldsymbol{Z})$;	action $(x,y) \mapsto (x+a, iy)$, $(x+b, y+c)$, E_0, i as before and $ic = c$, order $c=2$ (here char $\neq 2$)
d) $E_1 \times E_0/\boldsymbol{Z}/6\boldsymbol{Z}$;	action $(x,y) \mapsto (x+a, -\omega y)$, E_0, ω as in b.

The list obtained here coincides with the classical list in characteristic 0 (see Bagnera and DeFranchis [1], Enriques and Severi [3], pp. 283–392, Šafarevič [14], p. 181). Note here that the requirements $A_0 \subset E_0$ *and* $A \subset E_1$ eliminate the possibilities $n=2$, $A_0 = \text{Ker } 2_{E_0}$ and $n=3$ or 4, $A_0 = \alpha_3$ or α_2. Striking features of this list are the missing cases. From a moduli point of view, even in case a1), one may ask what happens if we start with such an X in characteristic 0 and specialize to characteristic 2 in such a way that the point a goes to $0 \in E_1$. One would hope for instance that the moduli spaces of these X's were proper over $\boldsymbol{Z}[j(E_0), j(E_1)]$ but this is not true. The answer seems to be that the X's become quasi-hyperelliptic ! This is an interesting point to investigate.

The order of K_X is easily obtained, since if ω is the 2-form on $E_1 \times E_0$ with no zeros or poles, then

$$\text{order of } K_X = \text{least } n \text{ such that } A \text{ acts trivially on } \omega^{\otimes n}$$

and we find

$$\begin{aligned} \text{order of } K_X &= 2, 3, 4, 6 \text{ in cases a), b), c), d)} \\ &\quad \text{and char}(k) \neq 2, 3 \\ &= 1, 3, 1, 3 \text{ in cases a), b), c), d)} \\ &\quad \text{and char}(k) = 2 \\ &= 2, 1, 4, 2 \text{ in cases a), b), c), d)} \\ &\quad \text{and char}(k) = 3 \end{aligned}$$

It is interesting to check exactly which wild multiple fibres (in the sense of § 1) occur here for $\pi' : X \to \boldsymbol{P}^1$. One can check that we get the following cases in the list of § 2 :

case	char. $\neq 2,3$	char. 3	char. 2
a	(i)	(i)	(xv)-one or two fibres 0/2
b	(ii)	(xv)-one fibre 0/3	(ii)
c	(iii)	(iii)	(xv)-one fibre 0/4
d	(iv)	(xiii)	(xii)

4. Proof of Theorem 5

First of all, let X be a $K3$-surface, i.e., $K_X \equiv 0$, $B_2=22$, $B_1=0$, $\chi(\mathcal{O}_X)=2$, $H^1(\mathcal{O}_X) = (0)$ (cf. Table in Introduction). Then

i) if $\pi : Y \to X$ were a connected étale covering of degree d, one would have $K_Y \equiv \pi^* K_X \equiv 0$, hence Y would be a surface in the Table too. But

$$c_{2,Y} = \pi^{-1}(c_{2,X})$$

hence

$$\deg c_{2,Y} = 24d > 24$$

and there are no such surfaces in the table.

ii) Since $H^{-}(\mathcal{O}_X)$ is isomorphic to the tangent space to $\mathrm{Pic}(X)$, it follows that Pic^0_X is a finite discrete group. Let $L = \mathcal{O}_X(D)$ represent a point of Pic^0_X. Then $(D^2) = (D \cdot K_X) = 0$, so $\chi(L) = \chi(\mathcal{O}_X) = 2$. Therefore $H^0(L) \neq (0)$ or $H^2(L) \neq (0)$ But by Serre duality $H^2(L)$ is dual to $H^0(L^{-1})$. Thus L or L^{-1} is represented by an effective divisor E, but since it is in Pic^0, $E=0$. So finally $L \cong \mathcal{O}_X$ and $\mathrm{Pic}^0_X = (0)$.

Secondly, let X be a surface with $K_X \equiv 0$, $B_2=14$, $B_1=2$, $\chi(\mathcal{O}_X)=1$, $\dim H^1(\mathcal{O}_X) = 1$. Since $B_1>0$, X has a positive dimensional Picard variety. This means that X does indeed support invertible sheaves $L = \mathcal{O}_X(D)$ such that D is numerically equivalent to zero but $D \not\equiv 0$. Then $\chi(L) = \chi(\mathcal{O}_X) = 1$, so $H^0(L) \neq (0)$ or $H^2(L) \neq (0)$. As above, Serre duality shows that $H^2(L) \neq (0) \Rightarrow H^0(L^{-1}) \neq (0)$, so L or L^{-1} is represented by an effective divisor E. E numerically equivalent to 0 implies $E=0$, so $L \cong \mathcal{O}_X$ contrary to our assumption.

Alternatively, we could argue that because $B_1>0$, X has connected cyclic étale coverings $\pi : Y \to X$ of every order d prime to the characteristic. As in (i) above, $c_{2,Y} = 12d$ and if $d>2$, no such Y appears in our table.

Arguments of the above type, using μ_p or α_p-coverings of X (cf. Mumford [11]) do not quite seem to be strong enough to prove that if X is a $K3$-surface, then $H^0(X, \Omega^1_X) = (0)$. It remains a very intriguing open question[1)] whether or not

1) (added in proof) Rudakov and Šafarevič have just settled this. They show that Ω^1_X has no sections when X is a $K3$-surface. Moreover, P. Deligne has used their result to prove that all $K3$-surfaces lift to char. 0.

$H^0(X, \Omega^1_X)$ is (0) for every $K3$-surface of char. p.

Thirdly, let X be a surface with $K_X \equiv 0$, $B_2=10$, $B_1=0$, $\chi(\mathcal{O}_X)=1$, $\dim H^1(\mathcal{O}_X)=1$. Let $\{a_{ij}\} \in Z^1(\mathcal{O}_X)$ be a non-trivial cocycle and consider the $\boldsymbol{G}_a$-bundle

$$\pi : W \to X$$

defined locally as $\boldsymbol{A}^1 \times U_i$, coordinate z_i on $\boldsymbol{A}^1$, and glued by

$$z_i = z_j + a_{ij}.$$

If ω is a non-zero 2-form on X with no zeroes or poles,

$$\eta = dz_i \wedge \omega$$

is a non-zero 3-form on W with no zeroes or poles, i.e:, $K_W \equiv 0$. Now since $H^1(\mathcal{O}_X)$ is 1-dimensional, there is a constant $\lambda \in k$ such that $\{a_{ij}^p\}$, $\{\lambda a_{ij}\}$ are cohomologous :

$$a_{ij}^p = \lambda a_{ij} + b_i - b_j$$

Consider the global function f on W defined locally by

$$f = z_i^p - \lambda z_i - b_i.$$

Let Y be the 2-dimensional scheme $f=0$. If $\lambda \neq 0$, Y is étale over X, hence non-singular. If $\lambda=0$, still $b_i \notin \mathcal{O}_X^p$ (or else $a_{ij}=b_i^{1/p}-b_j^{1/p}$ is cohomologous to zero), so Y is a reduced Gorenstein surface. Since $K_W \equiv 0$ and Y has trivial normal sheaf in W, in both cases $\omega_Y \cong \mathcal{O}_Y$. Thus

$$\chi(\mathcal{O}_Y) \leq \dim H^0(\mathcal{O}_Y) + \dim H^2(\mathcal{O}_Y) = \dim H^0(\mathcal{O}_Y) + \dim H^0(\omega_Y) = 2.$$

On the other hand,

$$\operatorname{res} \pi : Y \to X$$

is finite and flat and $(\operatorname{res} \pi)_* \mathcal{O}_Y$ is filtered by the subsheaves :

$$\mathcal{O}_X \subset [\mathcal{O}_X \oplus \mathcal{O}_X \cdot z_i] \subset [\mathcal{O}_X \oplus \mathcal{O}_X \cdot z_i \oplus \mathcal{O}_X \cdot z_i^2] \subset \cdots \subset (\operatorname{res} \pi)_* \mathcal{O}_Y.$$

The quotients here are all isomorphic to $\mathcal{O}_X$, thus

$$\chi(\mathcal{O}_Y) = p \cdot \chi(\mathcal{O}_X) = p.$$

Thus $p \leq 2$ as asserted.

5. Analysis of the Case Leading to Abelian Surfaces

In this section, we prove Theorem 6, that a surface X with $K_X \equiv 0$ and $B_2=6$ is an abelian surface. As we see from the table in § 1, the surface X also has the properties :

a) $\dim H^1(\mathcal{O}_X) = 2$, $\dim H^2(\mathcal{O}_X) = 1$, $\chi(\mathcal{O}_X) = 0$,

b) $c_{2,X} = 0$, $B_1 = 4$, $q = 2$.

In particular, $\mathrm{Pic}^0 X$ is reduced and 2-dimensional and its dual $\mathrm{Alb}\, X$ is 2-dimensional. Let

$$\phi : X \to \mathrm{Alb}\, X$$

be the Albanese mapping. First of all, we can see that ϕ is surjective as follows : if not, since $\phi(X)$ generates $\mathrm{Alb}\, X$, $\phi(X)$ is a curve of genus $g \geq 2$. Consider the diagram :

$$\begin{array}{ccc} X'=X\times_{\mathrm{Alb}}\mathrm{Alb} & \xrightarrow{\ \psi\ } & \mathrm{Alb}\,X \\ \Big\downarrow & & \Big\downarrow n \\ X & \xrightarrow[\ \phi\ ]{} & \mathrm{Alb}\,X \end{array}$$

where n denotes multiplication by n and $p \nmid n$. Then $\psi(X')$ is an étale covering of $\phi(X)$ of degree n^{2g}. Also $\psi(X')$ is connected because $\psi(X')=n^{-1}(\phi(X))$ and $\phi(X)$ is an ample curve[1] on Alb X. Therefore, $\psi(X')$ has genus $g'>2$. Therefore, Alb X' can be mapped onto $\mathrm{Jac}(\psi(X'))$ which is an abelian variety of dimension>2 : i.e., $q(X')>2$. But X' is an étale cover of X. So $K_{X'}\equiv 0$ and looking in the Table, we see that no such surface X' exists.

Therefore, ϕ is surjective, and hence of finite degree. If ϕ were separable, e.g., if char.$=0$, then we could quickly finish up as follows :

Let ω be the translation-invariant 2-form on Alb X. Then ω has no zeroes or poles and because ϕ is separable, $\phi^*\omega\neq 0$. But $\phi^*\omega$ has zeroes at all points where ϕ is not étale, and $K_X=(\phi^*\omega)$. Since $K_X\equiv 0$, $\phi^*\omega$ has no zeroes, hence ϕ is everywhere étale. But then by the Theorem of § 18 [12], X itself is an abelian surface. Unfortunately if ϕ is inseparable, this argument breaks down. However, when we are in characteristic p, we can use another trick and reduce the Theorem to the case where the ground field k is finite! In fact X lies in a smooth and proper algebraic family of surfaces defined over a finite field and all members of this family have the same invariants (e.g., because by the table in § 1, these surfaces are also characterized by saying $K_X\sim 0$ in étale cohomology and $q=2$). Therefore, if we prove that the surfaces in this family over closed points of the base are abelian, it follows that all are abelian (cf. Theorem 6. 14, [8]).

Now assume the ground field k is finite. We follow a line of argument similar to that in Tate [16]. Consider the infinite sequence of surfaces :

$$\begin{array}{ccc} X_l=X\times_{\mathrm{Alb}}\mathrm{Alb} & \xrightarrow{\ \phi_l\ } & \mathrm{Alb}\,X \\ \Big\downarrow & & \Big\downarrow \\ X & \xrightarrow{\ \phi\ } & \mathrm{Alb}\,X \end{array}$$

for all $l\geq 2$ with $p \nmid l$. Note that $\deg \phi_l=\deg \phi$ for all : call this degree d. Note that $X_l \to X$ is étale and hence X_l is a surface of the same type as X (in fact, $K_{X_l}\equiv 0$ and $q(X_l)\geq 2$, hence by table I, $q(X_l)=2$). We can deduce quickly that ϕ and hence ϕ_l are all finite morphisms : in fact, if not, let $E\subset X$ be a curve such that $\phi(E)$ is a point $e\in\mathrm{Alb}\,X$. Then considering the Stein factorization $X\to Y\to\mathrm{Alb}$ of ϕ, we see that E can be blown down in a birational map $X\to Y$, hence $(E^2)<0$. Now

1) A suitable multiple of an ample curve C on any surface Y is a hyperplane section of Y for some projective embedding and all hyperplane sections of varieties of dimension>1 are connected.

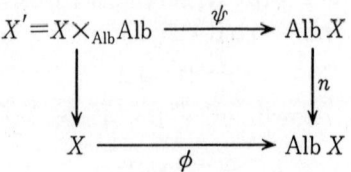

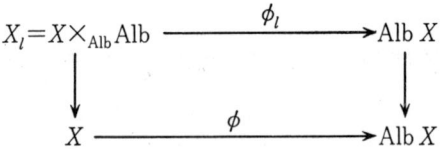

for each l, $l^{-1}(e)$ consists of l^4 points $e_i \in \text{Alb } X$, and $\phi^{-1}(e_i)$ contains a curve E_i that is contracted by ϕ_l. These curves are disjoint since $\phi_l(E_i)=e_i \neq e_j=\phi_l(E_j)$. Thus X_l has l^4 disjoint curves E_i with $(E_i^2)<0$; thus $B_2(X_l)\geq l^4$. But for all surfaces of the same type as X, $B_2=6$. This is a contradiction if $l>1$, hence ϕ is finite.

Next, fix L_0, an ample sheaf on Alb X. It follows that $L_l=\phi_l^*(L_0)$ is ample on X_l, with Hilbert polynomial

$$\chi(L_l^{\otimes n}) = d\cdot\chi(L_0^{\otimes n})$$

independent of l. By the Main Theorem of Matsusaka-Mumford [7], there is also a number N independent of l such that $L_l^{\otimes N}$ is very ample for all l. Therefore the infinite set of k-varieties X_l can all be embedded in a fixed $\boldsymbol{P}^M$ with fixed degree. Since there are only finitely many k-varieties of this degree (as k is finite), it follows that all the pairs $(X_l, L_l^{\otimes N})$ are isomorphic to finitely many of them !

Now consider the facts—

a) for any variety X and ample sheaf L, the group of automorphisms f of X such that f^*L is numerically equivalent to L is an algebraic group ; esp. it has only finitely many components (Matsusaka [6]),

b) The group A_l of translations by points of order l acts on X_l since by definition, it is the fibre product $X\times_{\text{Alb}}(\text{Alb}, l)$; moreover each $g\in A_l$ carries L_l into a sheaf algebraically equivalent to L_l.

Let $(X_l, L_l^{\otimes N})$ be isomorphic to infinitely many other $(X_{l'}, L_{l'}^{\otimes N})$'s. Then $A_{l'}$ acts on X_l. Let $G_l\subset \text{Aut}(X_l)$ be the group of automorphisms f such that f^*L_l is numerically equivalent to L_l. Then $A_{l'}\subset G_l$ which implies that the order of G_l is infinite, hence G_l^0 (the connected component) is positive dimensional. But if G_l^0 contains a non-trivial *linear* subgroup, then when this acts on X_l, it would follow that X_l was a ruled surface : since $K_{X_l}\equiv 0$, this is absurd. Therefore G_l^0 is an abelian variety. On the other hand, $A_{l'}\cong(\boldsymbol{Z}/l'\boldsymbol{Z})^4$, and subgroups of fixed bounded index in $A_{l'}$ are inside G_l^0. Therefore dim $G_l^0\geq 2$. It follows that X_l consists in only one orbit under G_l^0, hence X_l is a coset space G_l^0/H, hence X_l itself is an abelian variety. Finally X itself is now caught in the middle between 2 abelian varieties :

$$X_l \xrightarrow{\text{étale}} X \longrightarrow \text{Alb } X.$$

With a suitable origin, $X_l\rightarrow \text{Alb } X$ is then a homomorphism, hence if K is its kernel, we find :

$$X_l\times_{\text{Alb}X}X_l = \{(x, x+k)\mid x\in X_l, k\in K\}.$$

But $X_l\times_X X_l\subset X_l\times_{\text{Alb}X}X_l$ and $X_l\times_X X_l$ is (i) étale over X_l, and (ii) the graph of an equivalence relation on X_l. (i) implies that

$$X_l\times_X X_{l'} = \{(x, x+k)\mid x\in X_l, k\in K'\}$$

for some subset $K'\subset K$, and (ii) implies that K' is a subgroup. It follows that $X\cong X_l/K'$, hence X is also an abelian variety.

References

[1] Bagnera, G. and De Franchis, M. : Le nombre ρ de Picard pour les surfaces hyperelliptiques, Rend. Circ. Mat. Palermo **30** (1910).
[2] Deligne, P. and Rapoport, M. : Les schémas de modules de courbes elliptiques, in Springer Lecture Notes **349** (1973).
[3] Enriques, F. and Severi, F. : Mémoire sur les surfaces hyperelliptiques, Acta Math. **32** (1909).
[4] Kodaira, K. : On the structure of compact complex analytic surfaces I, Amer. J. Math. **86** (1964).
[5] Lang, S. : Elliptic Functions, Addison-Wesley, 1973.
[6] Matsusaka, T. : Polarized varieties and fields of moduli, Amer. J. Math. **80** (1958).
[7] Matsusaka, T. and Mumford, D. : Two fundamental theorems on deformations of polarized varieties, Amer. J. Math. **86** (1964).
[8] Mumford, D. : Geometric Invariant Theory, Springer-Verlag, 1965.
[9] —— : Lectures on curves on surfaces, Princeton Univ. Press, 1966.
[10] —— : Enriques' classification of surfaces I, in Global Analysis, Princeton Univ. Press, 1969.
[11] —— : Pathologies III, Amer. J. Math. **89** (1967).
[12] —— : Abelian Varieties, Tata Studies in Math., Oxford Univ. Press, 1970.
[13] Raynaud, M. : Spécialisation du foncteur de Picard, Publ. Math. IHES **38**, p. 27.
[14] Šafarevič, I.et al : Algebraic Surfaces, Proc. Steklov Inst. Math. **75** (1965).
[15] Tate, J. : Genus change in purely inseparable extensions of function fields, Proc. AMS **3** (1952), p. 400.
[16] —— : Endomorphisms of abelian varieties over finite fields, Inv. Math. **2** (1966).

Scuola Normale Superiore, Pisa
Harvard University

(Received January 14, 1976)

Classification of Hilbert Modular Surfaces

F. Hirzebruch and D. Zagier

I. Introduction and Statement of Results

1. 1 In the paper [6] non-singular models for the Hilbert modular surfaces were constructed. In [9] it was investigated how these algebraic surfaces fit into the Enriques-Kodaira rough classification of surfaces ([11], [12]). But this was only done for the surfaces $Y(p)$ belonging to a real quadratic field of prime discriminant. We shall solve the corresponding problem for real quadratic fields of arbitrary discriminant. We shall use the notation of [6] and [9] and refer to these papers very often.

1. 2 Let K be the real quadratic field of discriminant D and $\mathfrak{o}$ its ring of integers. The Hilbert modular group $G = \boldsymbol{SL}_2(\mathfrak{o})/\{1, -1\}$ acts on $\mathfrak{H}\times\mathfrak{H}$ where $\mathfrak{H}$ is the upper half plane. The complex space $\mathfrak{H}^2/G$ can be compactified by finitely many cusps. This gives a compact normal complex space of dimension 2 denoted by $\overline{\mathfrak{H}^2/G}$ which has finitely many singularities (resulting from the cusps and the elliptic fixed points of G). If one resolves these singularities in the canonical minimal way, one gets a non-singular algebraic surface $Y(D)$. Thus for any discriminant D of a real quadratic field (i. e. $D\equiv 1 \bmod 4$ or $D\equiv 0 \bmod 4$, where $D\geqq 5$ and D or $D/4$ respectively is square free) an algebraic surface $Y(D)$ is defined. (Here we have changed the notation of [6] § 4. 5. There $Y(D)$ was called $Y(d)$ where d is the square free part of D.)

1. 3 The rough classification of algebraic surfaces without exceptional curves was recalled in [9] (Chap. I, Theorem ROC). Since the surface $Y(D)$ is regular (see [1] Part I, [2] or [9] Prop. II. 4), it is either *rational* or admits a unique minimal model which is a $K3$-*surface*, an *honestly elliptic surface* (fibred over the projective line) or a surface of *general type*. Thus there are four distinct possibilities, and we wish to decide for every D which of these four cases happens. It was proved recently that $Y(D)$ is simply-connected ([17] and A. Kas, unpublished). Therefore, the Enriques surface (which is an honestly elliptic surface) cannot occur as minimal model of any $Y(D)$ and the class (rational, blown-up $K3$ surface, blown-up honestly elliptic surface, general type) of $Y(D)$ can be characterized by the Kodaira dimension $\kappa(Y(D))$ (defined as the maximal dimension of the images of $Y(D)$

under the pluricanonical mappings). *Thus* $Y(D)$ *is*

rational	*if and only if*	$\kappa(Y(D)) = -1,$
a blown-up K3 surface	*if and only if*	$\kappa(Y(D)) = 0,$
a blown-up honestly elliptic surface	*if and only if*	$\kappa(Y(D)) = 1,$
of general type	*if and only if*	$\kappa(Y(D)) = 2.$

In Chap. II we shall recall the formulas for the arithmetic genus of $Y(D)$. Since $Y(D)$ is regular, we have $\chi(Y(D))=1+p_g\geqq 1$. It is easy to see that $\chi(Y(D))$ tends to ∞ for $D\to\infty$ and certain estimates (Chap. IV) and explicit calculations will show that

$$\chi(Y(D)) = 1 \Leftrightarrow D = 5, 8, 12, 13, 17, 21, 24, 28, 33, 60. \tag{1}$$

It was proved in [6] § 4. 5 that $Y(D)$ is rational for these values of D. Since the arithmetic genus of any rational surface equals 1, the ten values of D given in (1) are exactly the values for which $Y(D)$ is rational.

The following result was proved in [9]. For convenience we express it in terms of the Kodaira dimension.

If p *is a prime congruent to* 1 mod 4, *then*

$$\begin{aligned} \kappa(Y(p)) &= -1 \Leftrightarrow \chi(Y(p)) = 1 \Leftrightarrow p = 5, 13, 17,\\ \kappa(Y(p)) &= 0 \Leftrightarrow \chi(Y(p)) = 2 \Leftrightarrow p = 29, 37, 41,\\ \kappa(Y(p)) &= 1 \Leftrightarrow \chi(Y(p)) = 3 \Leftrightarrow p = 53, 61, 73,\\ \kappa(Y(p)) &= 2 \Leftrightarrow \chi(Y(p)) \geqq 4 \Leftrightarrow p > 73. \end{aligned} \tag{2}$$

To generalize such results to any discriminant we have to calculate $c_1^2(Y(D))$ which equals $K\cdot K$ where K is a canonical divisor of $Y(D)$. Namely, if $Y(D)$ *is not rational and* $c_1^2(Y(D))>0$, *then* $Y(D)$ *is of general type*. This follows from the rough classification theorem : For the unique minimal model $Y_{\min}(D)$ of $Y(D)$ we have

$$c_1^2(Y_{\min}(D)) \geqq c_1^2(Y(D)) > 0.$$

Therefore, $Y_{\min}(D)$ cannot be a $K3$-surface or an honestly elliptic surface, because for such a surface $c_1^2=K^2=0$. Since $c_1^2(Y(D))$ tends to ∞ for $D\to\infty$, we can reduce the classification to a finite list. This requires certain estimates. In Chap. IV we will prove :

Theorem 1. *If* $D>285$, *then* $c_1^2(Y(D))>0$.

(The proof depends on computer calculations.) There are exactly 50 discriminants with $c_1^2(Y(D))\leqq 0$; they are listed in Chap. IV. We consider these cases by hand and can settle all of them using the methods of [9] (in particular, Proposition I. 8 and I. 9). Many cases are already covered by (1) and (2) above. The result is the following theorem (Chap. V).

Theorem 2. *The Hilbert modular surface* $Y(D)$ *is*

rational	*for*	$D = 5, 8, 12, 13, 17, 21, 24, 28, 33, 60,$
blown-up K3	*for*	$D = 29, 37, 40, 41, 44, 56, 57, 69, 105,$

blown-up honestly elliptic for $D = 53, 61, 65, 73, 76, 77, 85, 88, 92, 93, 120, 140, 165,$
of general type otherwise (i. e. $D = 89$ or $D \geqq 97$, but $D \neq 105, 120, 140, 165$).

1. 4 The Hilbert modular group G belonging to $\boldsymbol{Q}(\sqrt{D})$ acts also on $\mathfrak{H} \times \mathfrak{H}^-$ where $\mathfrak{H}^-$ is the lower half plane. Compactifying $(\mathfrak{H} \times \mathfrak{H}^-)/G$ and resolving all singularities of the compactification $\overline{(\mathfrak{H} \times \mathfrak{H}^-)/G}$ in the minimal way lead to an algebraic surface $Y_-(D)$. Here we get

Theorem 3. *The Hilbert modular surface* $Y_-(D)$ *is*
rational *for* $D = 5, 8, 12, 13, 17,$
blown-up $K3$ *for* $D = 21, 24, 28, 29, 33, 37, 40, 41,$
blown-up honestly elliptic for $D = 44, 53, 57, 61, 65, 73, 85,$
of general type otherwise (i.e. $D = 56, 60, 69, 76, 77$ or $D \geqq 88$).

1. 5 Let $\mathfrak{b}$ be an ideal in the ring $\mathfrak{o}$ of integers of K. We introduce the group $\boldsymbol{SL}_2(\mathfrak{o}, \mathfrak{b})$ consisting of all matrices $\begin{bmatrix} \alpha & \beta \\ \gamma & \delta \end{bmatrix} \in \boldsymbol{SL}_2(K)$ such that $\alpha, \delta \in \mathfrak{o}$ and $\beta \in \mathfrak{b}^{-1}$, $\gamma \in \mathfrak{b}$. The actions of $\boldsymbol{SL}_2(\mathfrak{o}, \mathfrak{b})$ and $\boldsymbol{SL}_2(\mathfrak{o})$ on $\mathfrak{H}^2$ are equivalent (i. e. the groups are conjugate in $\boldsymbol{GL}_2^+(K)$) if $\mathfrak{b} = \lambda \mathfrak{a}^2$ where λ is a totally positive element of K and $\mathfrak{a}$ an ideal in $\mathfrak{o}$ (see [6], 3. 7 (40)). The action of $\boldsymbol{SL}_2(\mathfrak{o})$ on $\mathfrak{H} \times \mathfrak{H}^-$ and the action of $\boldsymbol{SL}_2(\mathfrak{o}, \mathfrak{b})$ on $\mathfrak{H}^2$ are equivalent if $\mathfrak{b} = (\Delta)$ where Δ is an element of $\mathfrak{o}$ of negative norm. The following four conditions on the field K are equivalent:

i) There exists an element Δ of negative norm and an ideal $\mathfrak{a}$ in $\mathfrak{o}$ with $(\Delta) = \mathfrak{a}^2$.
ii) The number -1 is the norm of an element of K.
iii) The discriminant D is a sum of two natural square numbers.
iv) The discriminant D has no prime factor $\equiv 3 \bmod 4$.

If one of these conditions is satisfied, then the actions of $\boldsymbol{SL}_2(\mathfrak{o})$ *on* $\mathfrak{H}^2$ *and* $\mathfrak{H} \times \mathfrak{H}^-$ *are equivalent* under an isomorphism of $\mathfrak{H}^2$ and $\mathfrak{H} \times \mathfrak{H}^-$ given by an element of $\boldsymbol{GL}_2(K)$ whose determinant is positive but has negative norm. The converse is also true (compare 2. 2). For this whole section 1.5 we refer the reader to Hammond [3].

For the group $\boldsymbol{SL}_2(\mathfrak{o}, \mathfrak{b})$ we consider $\mathfrak{H}^2/\boldsymbol{SL}_2(\mathfrak{o},\mathfrak{b})$, its compactification $\overline{\mathfrak{H}^2/\boldsymbol{SL}_2(\mathfrak{o},\mathfrak{b})}$ and the algebraic surface $Y(D, \mathfrak{b})$ obtained by resolving all singularities (cusps and quotient singularities) of $\overline{\mathfrak{H}^2/\boldsymbol{SL}_2(\mathfrak{o}, \mathfrak{b})}$ in the minimal way. The surface $Y(D,\mathfrak{b})$ is also simply-connected [17]. The surfaces $Y_-(D)$ and $Y(D, \mathfrak{b})$ are isomorphic if $\mathfrak{b} = (\Delta)$, where Δ is an element of $\mathfrak{o}$ of negative norm. *The surfaces* $Y(D)$ *and* $Y_-(D)$ *are isomorphic if one of the above conditions* i)–iv) *is satisfied.*

1. 6 We consider the involution T on $\overline{\mathfrak{H}^2/G}$ induced by $(z_1, z_2) \mapsto (z_2, z_1)$ and study the minimal resolution of $\overline{(\mathfrak{H}^2/G)}/T$. Here we cannot calculate the invariants c_1^2 and χ, because we do not have complete information on the fixed points of $G \cup G \cdot T$ in general. However, if $D = p$ is a prime, the fixed points are known [16].

The question, for which primes the surface $\overline{(\mathfrak{H}^2/G)}/T$ is rational, was completely answered in [6] ; there are 24 such primes, the largest being 317. For $p>17$ we define in Chap. II a certain non-singular model $Y_T(p)$ of this surface and give its numerical invariants ; this is needed to determine how the surface fits into the rough classification scheme. In particular, we need to estimate $c_1^2(Y_T(p))$. In Chap. IV, we prove

Theorem 4. $c_1^2(Y_T(p))>0$ *for* $p>821$.

This reduces the classification question to a finite list. In fact, all cases can be settled here, too, but for this we must refer to [8]. The result was announced in [7].

1. 7 For the surfaces $Y(p)$ (and also for the $Y_T(p)$) the arithmetic genus χ surprisingly determines the Kodaira dimension :

$$\kappa = \min\,[2, \chi-2].$$

For arbitrary D this is no longer true :
for $D=85$, 140 and 165 we have $\chi(Y(D))=4$, but the surface is nevertheless a blown-up honestly elliptic surface ($\kappa=1$). In all cases studied up to now, however,

$\chi = 1 \Leftrightarrow \kappa = -1$ (rational)
$\chi = 2 \Leftrightarrow \kappa = 0$ (blown-up $K3$)
$\chi = 3 \Rightarrow \kappa = 1$ (blown-up honestly elliptic)
$\chi \geqq 5 \Rightarrow \kappa = 2$ (general type).

It would be interesting to know whether this is an accident or whether there is some general property of simply-connected algebraic surfaces, valid for all Hilbert modular surfaces, which ensures, for example, that the surface is rational if $\chi=1$ and that it is $K3$ if it is minimal and $\chi=2$. It is known that there exist simply-connected algebraic surfaces with arithmetic genus one which are not rational (I. V. Dolgacev, Dokl. **7**(1966)).

II. Numerical Invariants of Hilbert Modular Surfaces

2. 1 The basic term for the calculation of invariants of Hilbert modular surfaces is the volume of $\mathfrak{H}^2/G$ with respect to the normalized Euler volume form ([6] § 1 (5))

(1) $$\omega = (2\pi)^{-2} y_1^{-2} y_2^{-2}\, dx_1 \wedge dy_1 \wedge dx_2 \wedge dy_2.$$

If $K=\boldsymbol{Q}(\sqrt{D})$ is the underlying field and G the Hilbert modular group, then

(2) $$\int_{\mathfrak{H}^2/G} \omega = 2\zeta_K(-1),$$

where $\zeta_K(s)$ is the ζ-function of K. We have

(3) $$\zeta_K(-1) = \frac{1}{60} \sum_{\substack{|k|<\sqrt{D} \\ k^2 \equiv D \bmod 4}} \sigma_1\left(\frac{D-k^2}{4}\right)$$

(compare [6], § 1 (11), (12)). Here $\sigma_1(n)$ is the sum of the divisors of n.

Let $a_r(G)$ be the number of points in $\mathfrak{H}^2/G$ for which the corresponding points in $\mathfrak{H}^2$ have isotropy groups of order r. Then the Euler number of $\mathfrak{H}^2/G$ is given by the formula ([6] § 1 (21))

$$e(\mathfrak{H}^2/G) = \int_{\mathfrak{H}^2/G} \omega + \sum_{r \geq 2} a_r(G) \frac{r-1}{r}. \tag{4}$$

We restrict ourselves to discriminants ≥ 13. Thus we exclude $D=5, 8, 12$. *Then the $a_r(G)$ vanish for $r>3$.*

We write

$$a_3(G) = a_3^+(G) + a_3^-(G). \tag{5}$$

Here $a_3^+(G)$ is the number of quotient singularities of $\mathfrak{H}^2/G$ of type $(3 ; 1, 1)$ whereas $a_3^-(G)$ is the number of quotient singularities of type $(3 ; 1, -1)$. Compare [6] § 3 (13). We have complete information [15] on $a_2(G)$, $a_3^+(G)$, $a_3^-(G)$. We will state the result in terms of the discriminant D. It is convenient to introduce also the square free part d of D:

$$D = d \quad \text{if} \quad d \equiv 1 \mod 4$$
$$D = 4d \quad \text{if} \quad d \equiv 2 \mod 4 \text{ or } d \equiv 3 \mod 4.$$

By $h(-N)$ we denote the class number of the imaginary quadratic number field of discriminant $-N$.

We have

$$a_2(G) = \begin{cases} h(-4d) & \text{if} \quad d \equiv 1 \mod 4 \\ 3h(-4d) & \text{if} \quad d \equiv 2 \mod 4 \\ 10h(-d) & \text{if} \quad d \equiv 3 \mod 8 \\ 4h(-d) & \text{if} \quad d \equiv 7 \mod 8 \end{cases} \tag{6}$$

$$a_3^+(G) = \begin{cases} \frac{1}{2}h(-3D) & \text{if} \quad D \not\equiv 0 \mod 3 \\ 4h(-D/3) & \text{if} \quad D \equiv 3 \mod 9 \\ 3h(-D/3) & \text{if} \quad D \equiv 6 \mod 9 \end{cases} \tag{7}$$

$$a_3^-(G) = \begin{cases} \frac{1}{2}h(-3D) & \text{if} \quad D \not\equiv 0 \mod 3 \\ h(-D/3) & \text{if} \quad D \equiv 3 \mod 9 \\ 0 & \text{if} \quad D \equiv 6 \mod 9. \end{cases} \tag{8}$$

The Euler number of $\mathfrak{H}^2/G$ is now calculable. It is not difficult to write a computer program for $\zeta_K(-1)$ as given by formula (3), for the class numbers $h(-N)$ and finally for the Euler number of $\mathfrak{H}^2/G$.

(9) *For* $D \geq 13$, $\quad e(\mathfrak{H}^2/G) = 2\zeta_K(-1) + \frac{1}{2} a_2(G) + \frac{2}{3} a_3(G).$

The second important invariant of the 4-dimensional rational homology manifold $\mathfrak{H}^2/G$ is the signature. It has no volume contribution. In the formula for sign $\mathfrak{H}^2/G$ only contributions from the quotient singularities of order 3 and from the cusps enter (compare [6] § 3 (43), (44)).

(10) *For* $D \geqq 13$, *we have* $\quad \operatorname{sign} \mathfrak{H}^2/G = 4w - \frac{2}{9}a_3^+(G) + \frac{2}{9}a_3^-(G).$

Here w is the total parabolic contribution in the sense of Shimizu. According to [4] Theorem 2. 1, it can be expressed in the following form :

(11) $$w = -4\sum h(D_1)h(D_2)u(D_1)^{-1}u(D_2)^{-1}$$

with the summation taken over all decompositions $D=D_1 \cdot D_2$ *with* $D_1<D_2$ *in which* D_1, D_2 *are discriminants of imaginary quadratic fields.* $u(D_1)$ *and* $u(D_2)$ *respectively are the orders of the groups of units in the corresponding fields.* The parabolic contribution vanishes if and only if D is the sum of two squares, i.e. D contains no prime$\equiv 3$ mod 4. Otherwise it is negative ([4] Corollary 2. 2). In particular, it vanishes if the fundamental unit of $K=\boldsymbol{Q}(\sqrt{D})$ has negative norm. (Compare 1. 5.)

As was shown in [6] § 3.6, the arithmetic genus $\chi(Y(D))$ of the non-singular model $Y(D)$ of the compactification $\overline{\mathfrak{H}^2/G}$ of $\mathfrak{H}^2/G$ can be calculated in terms of the topological invariants of the non-compact rational homology manifold $\mathfrak{H}^2/G$. We have

(12) $$\chi(Y(D)) = \frac{1}{4}(e(\mathfrak{H}^2/G) + \operatorname{sign}(\mathfrak{H}^2/G)).$$

2. 2 We now consider the action of G on $\mathfrak{H}\times\mathfrak{H}^-$. The rational homology manifold $(\mathfrak{H}\times\mathfrak{H}^-)/G$ admits an orientation reversing homeomorphism onto $\mathfrak{H}^2/G$. Therefore

(13) $$\begin{aligned} e(\mathfrak{H}^2/G) &= e((\mathfrak{H}\times\mathfrak{H}^-)/G), \\ \operatorname{sign}(\mathfrak{H}^2/G) &= -\operatorname{sign}((\mathfrak{H}\times\mathfrak{H}^-)/G). \end{aligned}$$

For the non-singular model $Y_-(D)$ mentioned in the introduction we have

(14) $$\chi(Y_-(D)) = \frac{1}{4}(e(\mathfrak{H}^2/G) - \operatorname{sign}(\mathfrak{H}^2/G)).$$

The formulas (7), (8), (10) imply that $\operatorname{sign}(\mathfrak{H}^2/G)$ is always non-positive. Therefore,

(15) $$\chi(Y_-(D)) \geqq \chi(Y(D)).$$

If we exclude $D=12$, then sign $(\mathfrak{H}^2/G)=0$ if and only if D is not divisible by a prime$\equiv 3$ mod 4. For $D=12$ the signature vanishes ([6] § 3. 9). As Hammond showed (see 1. 5), the signature of $\mathfrak{H}^2/G$ vanishes ($D=12$ again excluded) if and only if the actions of G on $\mathfrak{H}^2$ and on $\mathfrak{H}\times\mathfrak{H}^-$ are equivalent, one direction of this equivalence being clear by the second formula of (13).

2. 3 A "cusp" is described by a pair (M, V) where M is a complete $\boldsymbol{Z}$-module in the real quadratic field K and V a subgroup of finite index in the (infinite cyclic) group of all totally positive units ε with $\varepsilon M=M$. To such a pair (M, V) we associate in a topological way ([6] § 3) a rational number $\delta(M, V)$. Then

(16) $$4w(M, V) = \delta(M, V),$$

where $w(M, V)$ is the Shimizu number of a cusp given by evaluating the L-function

$L(M, V, s)$ for $s=1$ (see [6] 3. 5). The sum of the $w(M, V)$ for all cusps of the Hilbert modular group of the real quadratic field with discriminant D is the number w given in (11). We wish to recall here the expression ([6] 3. 2 Theorem) for $\delta(M, V)$ using the continued fraction describing the resolution of the "cusp singularity" of type (M, V):

There exists a totally positive number α in K such that

$$\alpha M = \mathbf{Z}w_0 + \mathbf{Z}\cdot 1,$$

where w_0 is reduced, i.e.

$$0 < w_0' < 1 < w_0.$$

Here $x \mapsto x'$ denotes the non-trivial automorphism of K. The number w_0 has a purely periodic continued fraction development

$$w_0 = b_0 - \cfrac{1}{b_1 - \cfrac{\ddots}{\ \ - \cfrac{1}{b_{r-1} - \cfrac{1}{b_0 - \ddots}}}} \qquad (b_i \in \mathbf{Z},\ b_i \geqq 2)$$

where $((b_0, \ldots, b_{r-1}))$ is the primitive period which (up to cyclic permutations) depends only on the strict equivalence class of the module and conversely determines this strict equivalence class. (We recall that by definition the modules M and $\tilde{M}$ are strictly equivalent if and only if there exists an element β of K of positive norm such that $\tilde{M} = \beta M$. They are called equivalent if there exists an element β of K such that $\tilde{M} = \beta M$.) We define

$$\delta(M) = -\frac{1}{3}\sum_{i=0}^{r-1}(b_i - 3) \tag{17}$$

and

$$l(M) = r. \tag{18}$$

Thus $l(M)$ is the length of the period which we shall also call the length of the module.

If γ is an element of K with negative norm, then

$$\begin{aligned} \delta(\gamma M) &= -\delta(M), \\ 3\delta(M) &= l(M) - l(\gamma M). \end{aligned} \tag{19}$$

In particular, $\delta(M)=0$ if there exists a unit ε of K with negative norm such that $\varepsilon M = M$.

To prove (19) we observe that w_0 (see above) admits an ordinary continued fraction

$$w_0 = c_0 + \cfrac{1}{c_1 + \cfrac{1}{c_2 + \ddots}} \qquad (c_i \in \mathbf{Z},\ c_i \geqq 1 \text{ for } i > 0) \tag{20}$$

which is not necessarily purely periodic.

We denote the shortest period of even length by

(21) $$(a_1, \cdots, a_{2s}).$$

Thus it is either the primitive period or twice the primitive period, the latter if and only if the primitive period has odd length. The period $(a_1, \cdots, a_{2s})$ (up to cyclic permutations) depends only on the equivalence class of M and also determines this equivalence class.

A period (21) determines two periods in the sense of continued fractions with minus signs, namely

(22) $$((\underbrace{2, \cdots, 2}_{a_1-1}, a_2+2, \underbrace{2, \cdots, 2}_{a_3-1}, a_4+2, \cdots, \underbrace{2, \cdots, 2}_{a_{2s-1}-1}, a_{2s}+2))$$

and

(23) $$((\underbrace{2, \cdots, 2}_{a_2-1}, a_3+2, \underbrace{2, \cdots, 2}_{a_4-1}, a_5+2, \cdots, \underbrace{2, \cdots, 2}_{a_{2s}-1}, a_1+2))$$

(compare [6] 2. 5 (19) and 3. 10). These two periods coincide (up to cyclic permutation) if and only if the period (21) is twice the primitive period of (20), i. e. if the primitive period has odd length. The periods (22), (23) determine the strict equivalence classes contained in the equivalence class of M. There is only one such equivalence class if and only if the primitive period of (20) is odd because this happens if and only if (22) and (23) coincide. Therefore the primitive period of (20) is odd if and only if there exists a unit ε of negative norm with $\varepsilon M = M$. The formulas (19) are an easy consequence of (22), (23). We also observe that $\delta(M)$ is up to sign the alternating sum of the a_i.

If we have a cusp of type (M, V), then V is a subgroup of finite index in the infinite cyclic group U_M^+ of all totally positive units ε with $\varepsilon M = M$, and we have

(24) $$\begin{aligned} \delta(M, V) &= [U_M^+ : V] \cdot \delta(M), \\ l(M, V) &= [U_M^+ : V] \cdot l(M). \end{aligned}$$

For the cusps of the Hilbert modular group the modules M are always strictly equivalent to ideals in the ring $\mathfrak{o}$ of all integers of K. The strict equivalence classes mentioned above correspond to narrow ideal classes, the equivalence classes to ordinary ideal classes. Let C^+ be the group of narrow and C the group of ordinary ideal classes of $\mathfrak{o}$. Then $\mathfrak{a} \mapsto \mathfrak{a}^{-2}$ (where $\mathfrak{a}$ is an ideal in $\mathfrak{o}$) induces homomorphisms $Sq : C \to C^+$ and $Sq : C^+ \to C^+$ (see [6] 3. 7 (42)). There are h cusps for the Hilbert modular group $\boldsymbol{SL}_2(\mathfrak{o})/\{1, -1\}$ where h equals $|C|$ and is the class number of K. These cusps are of type $(\mathfrak{a}^{-2}, U^2)$ where U denotes the group of units of $\mathfrak{o}$. Let U^+ be the group of all totally positive units; then $U^+ = U^2$ if and only if there exists a unit of negative norm, otherwise $[U^+ : U^2] = 2$. In the first case $|C| = |C^+| = h$, in the latter $|C^+| = 2 \cdot |C| = 2h$. Let $\varepsilon_0 \in U$ be the fundamental unit ($\varepsilon_0 > 1$). Then

(25) $$\begin{aligned} \delta(\mathfrak{a}^{-2}, U^2) &= 2\delta(\mathfrak{a}^{-2}) & &\text{if } N(\varepsilon_0) = 1, \\ \delta(\mathfrak{a}^{-2}, U^2) &= \delta(\mathfrak{a}^{-2}) = 0 & &\text{if } N(\varepsilon_0) = -1, \\ l(\mathfrak{a}^{-2}, U^2) &= 2l(\mathfrak{a}^{-2}) & &\text{if } N(\varepsilon_0) = 1, \\ l(\mathfrak{a}^{-2}, U^2) &= l(\mathfrak{a}^{-2}) & &\text{if } N(\varepsilon_0) = -1. \end{aligned}$$

The numbers δ and l depend only on the strict module *class*. Therefore, δ and l can be regarded as functions on C^+. For the total parabolic contribution we have in view of (16) and (25)

$$(26)\qquad \begin{aligned} w &= \frac{1}{2}\sum_{\mathfrak{a}\in C}\delta(Sq(\mathfrak{a})), \quad (Sq\colon C\to C^+) \\ &= \frac{1}{4}\sum_{\mathfrak{a}\in C^+}\delta(Sq(\mathfrak{a})), \quad (Sq\colon C^+\to C^+) \end{aligned}$$

which because of (11) is a relation between continued fractions and class numbers of imaginary quadratic fields. (Compare [6] 3. 10 (55)).

The pair (M, V) determines a singularity whose minimal resolution is cyclic. The number of curves in this resolution equals $l(M, V)$ (see [6] 2. 5 Theorem). The Hilbert modular surface $\mathfrak{H}^2/G$ for the field K of discriminant D is compactified by h points. They are singularities in the compactification $\overline{\mathfrak{H}^2/G}$ which when resolved minimally give rise to h cycles of curves. The number of all these curves will be denoted by $l_0(D)$. We have

$$(27)\qquad l_0(D) = \begin{cases} \sum_{\mathfrak{a}\in C} l(Sq(\mathfrak{a})) & \text{if } N\varepsilon_0 = -1 \\ 2\sum_{\mathfrak{a}\in C} l(Sq(\mathfrak{a})) & \text{if } N\varepsilon_0 = 1 \end{cases}$$

or equivalently

$$(28)\qquad l_0(D) = \sum_{\mathfrak{a}\in C^+} l(Sq(\mathfrak{a})) \qquad (Sq : C^+\to C^+).$$

The Hilbert modular surface $(\mathfrak{H}\times\mathfrak{H}^-)/G$ is also compactified by h points. These cusps are of type $(\gamma\mathfrak{a}^{-2}, U^2)$ where γ is an element of K of negative norm. We denote by $l_0^-(D)$ the number of curves needed to resolve all these cusp singularities minimally. Then

$$l_0^-(D) = \sum_{\mathfrak{a}\in C^+} l(\gamma Sq(\mathfrak{a}))$$

and by (19) and (26)

$$(29)\qquad l_0(D) - l_0^-(D) = 12w.$$

2. 4 Let $Y(D)$ be the surface obtained from $\overline{\mathfrak{H}^2/G}$ by minimal resolutions of all the singular points (see Chap. I). If we assume $D\geqq 13$, we have only quotient singularities of order 2 or 3. Those of order 2 are resolved by one curve; those of order 3 by one or two curves depending on whether the type is (3 ; 1, 1) or (3 ; 1, -1). As in [9] (Proposition II. 2 and (7)) we conclude

$$(30)\qquad \begin{aligned} e(Y(D)) &= e(\mathfrak{H}^2/G) + a_2(G) + a_3^+(G) + 2a_3^-(G) + l_0(D) \\ &= 2\zeta_K(-1) + \frac{3}{2}a_2(G) + \frac{5}{3}a_3^+(G) + \frac{8}{3}a_3^-(G) + l_0(D) \end{aligned}$$

for $D \geqq 13$.

Noether's formula states that $c_1^2(Y(D)) + e(Y(D)) = 12\chi(Y(D))$. Using (9), (10), (12), (29), (30) we obtain

$$(31') \qquad c_1^2(Y(D)) = 4\zeta_K(-1) - l_0^-(D) - \frac{a_3^+(G)}{3}.$$

If we consider the action of G on $\mathfrak{H}\times\mathfrak{H}^-$ instead of $\mathfrak{H}\times\mathfrak{H}$, then $a_3^+(G)$, $a_3^-(G)$ interchange their role. The same is true for $l_0(D)$, $l_0^-(D)$. This implies

$$(32) \qquad c_1^2(Y_-(D)) = 4\zeta_K(-1) - l_0(D) - \frac{a_3^-(G)}{3}.$$

We have

$$(33) \qquad c_1^2(Y_-(D)) \geqq c_1^2(Y(D)), \qquad e(Y_-(D)) \geqq e(Y(D))$$

and in fact

$$\begin{aligned} c_1^2(Y_-(D)) - c_1^2(Y(D)) &= l_0^-(D) - l_0(D) && \text{if } D \not\equiv 0(3) \\ &= l_0^-(D) - l_0(D) + h\left(-\frac{D}{3}\right) && \text{if } D \equiv 0(3) \\ e(Y_-(D)) - e(Y(D)) &= l_0^-(D) - l_0(D) && \text{if } D \not\equiv 0(3) \\ &= l_0^-(D) - l_0(D) + 3h\left(-\frac{D}{3}\right) && \text{if } D \equiv 0(3). \end{aligned}$$

The corresponding inequality for the arithmetic genus was mentioned before (15).

2. 5 As mentioned in Chapter I, the surfaces $Y(D)/T$ will be investigated for prime discriminants in a later paper [8]. However the necessary estimates for c_1^2 will be done in this paper.

Let p be a prime $\equiv 1$ mod 4. The surface $Y(p)$ has some exceptional curves which can be blown down to give a surface $Y^0(p)$. We always assume $p>17$ to ensure that $Y(p)$ is not rational and exceptional curves do not meet. (For details see [6] § 5 and [9]). The involution $(z_1, z_2) \mapsto (z_2, z_1)$ induces an involution T on $Y^0(p)$ which has no isolated fixed points. The fixed point set is a non-singular curve F_p^0. We have

$$(34) \qquad e(Y^0(p)/T) = \frac{1}{2}(e(Y^0(p)) + e(F_p^0)).$$

The Euler number $e(F_p^0)$ is given by a classical formula. Namely, the curve F_p^0 is the compact non-singular model of $\mathfrak{H}/\Gamma_0^*(p)$ where $\Gamma_0^*(p)$ is the normal extension of $\Gamma_0(p)$ by the element $\begin{bmatrix} 0 & -1/\sqrt{p} \\ \sqrt{p} & 0 \end{bmatrix}$. This element induces an involution on $\overline{\mathfrak{H}/\Gamma_0(p)}$ which has $h(-4p)$ fixed points according to Fricke (loc. cit in [6]). Put $\varepsilon=1$ if $p\equiv 1$ mod 3 and $\varepsilon=0$ if $p\equiv 2$ mod 3. Then $\Gamma_0(p)$ has 2ε fixed points of order 3 and 2 fixed points of order 2 and two cusps. Therefore

$$e(\overline{\mathfrak{H}/\Gamma_0(p)}) = -\frac{p+1}{6} + \frac{4}{3}\varepsilon + 3$$

and

$$(35) \qquad e(F_p^0) = \frac{1}{2}\left(-\frac{p+1}{6} + \frac{4}{3}\varepsilon + 3 + h(-4p)\right).$$

Put $\delta=1$ if $p\equiv 1 \bmod 8$ and $\delta=0$ if $p\equiv 5 \bmod 8$. Then $Y^0(p)$ was obtained from $Y(p)$ by blowing down $4+2\delta+\varepsilon$ curves. By (30) we get

$$e(Y^0(p)) = 2\zeta_K(-1)+\frac{3}{2}h(-4p)+\frac{13}{6}h(-3p)+l_0(p)-4-2\delta-\varepsilon \tag{36}$$

and by (34) and (35)

$$e(Y^0(p)/T) = \zeta_K(-1)+h(-4p)+\frac{13}{12}h(-3p)+\frac{1}{2}l_0(p) -\frac{p+1}{24}-\frac{5}{4}-\frac{\varepsilon}{6}-\delta.$$

For the arithmetic genera of $Y^0(p)$ and $Y^0(p)/T$ we have the following formulas (cf. [6] 5. 6 (20), (21))

$$\chi(Y^0(p)) = \frac{1}{2}\zeta_K(-1)+\frac{h(-4p)}{8}+\frac{1}{6}h(-3p) \tag{37}$$

$$\chi(Y^0(p)/T) = \frac{1}{2}\left(\chi(Y^0(p))-\frac{p+1}{24}+\frac{\varepsilon}{3}+\frac{5}{4}+\frac{\delta}{2}\right). \tag{38}$$

By Noether's formula

$$c_1^2(Y^0(p)/T) = 12\chi(Y^0(p)/T)-e(Y^0(p)/T)$$

which yields

$$c_1^2(Y^0(p)/T) = 2\zeta_K(-1)-\frac{h(-4p)}{4}-\frac{1}{12}h(-3p)-\frac{1}{2}l_0(p) -\frac{5p}{24}+\frac{13}{6}\varepsilon+4\delta+8+\frac{13}{24}. \tag{39}$$

Since $K=\boldsymbol{Q}(\sqrt{p})$ has a unit of negative norm, $l_0(p)$ and $l_0^-(p)$ coincide. The class number $h(p)$ is odd. Thus $Sq: C\to C$ is bijective and $l_0(p)$ equals the number of *all* reduced quadratic irrationalities of discriminant p which was denoted in [9] by $l(p)$. In [6] it was shown that many curves on $Y^0(p)/T$ can be blown down. The "tail" of the resolution of the principal cusp (see [6] 5. 8) admits $\left[\frac{\sqrt{p}-1}{2}\right]$ blow-downs (for $p>17$). If we use the basic configuration of curves on $Y^0(p)$ (see [6] 5. 4 (8)) we get on $Y^0(p)/T$ exceptional curves which come from the $h(-3p)/2$ "crosses" and the $h(-4p)/2$ curves of self-intersection number -2 on $Y^0(p)$. (The "crosses" were denoted in [9] p. 18 by C_i, C_i', the (-2)-curves by D_i.) We have $T(C_i)=C_i'$ and $T(D_i)=D_i$. The images of C_i and D_i are the exceptional curves in $Y^0(p)/T$ we are looking for.

The surface obtained from $Y^0(p)/T$ by these blow-downs will be denoted by $Y_T(p)$. We have (for $p>17$)

$$c_1^2(Y_T(p)) = 2\zeta_K(-1)+\frac{h(-4p)}{4}+\frac{5}{12}h(-3p)-\frac{1}{2}l_0(p)-\frac{5p}{24} +\left[\frac{\sqrt{p}-1}{2}\right]+\frac{13}{6}\varepsilon+4\delta+8+\frac{13}{24}. \tag{40}$$

III. The Hurwitz-Maaß Extension of the Hilbert Modular Group, Skew-Hermitian Curves on $Y(D)$

3. 1 Let $K=\boldsymbol{Q}(\sqrt{D})$ be as before a real quadratic field and $\mathfrak{o}$ its ring of integers. We consider the matrices $\begin{bmatrix} a & b \\ c & d \end{bmatrix}$ with entries in $\mathfrak{o}$ such that $ad-bc$ is a totally positive unit of $\mathfrak{o}$. These matrices constitute a group which we divide by its center $\left\{\begin{bmatrix} a & 0 \\ 0 & a \end{bmatrix} \middle| a\in U\right\}$ where U is again the group of all units of $\mathfrak{o}$ (cf. introduction). We get the extended Hilbert modular group G_e. We have $G_e/G \simeq U^+/U^2$. It is a group of order 1 or 2.

Now we take the matrices $\begin{bmatrix} a & b \\ c & d \end{bmatrix}$ with entries in $\mathfrak{o}$ such that $w=ad-bc$ is totally positive and $a/\sqrt{w}$, $b/\sqrt{w}$, $c/\sqrt{w}$, $d/\sqrt{w}$ are algebraic integers not necessarily in $\mathfrak{o}$. The group of all these matrices has to be divided by its center $\left\{\begin{bmatrix} a & 0 \\ 0 & a \end{bmatrix} \middle| a\in \mathfrak{o}\right\}$. We get a group G_m which is a normal extension of G. It was introduced and studied by Hurwitz [10] § 3 and Maaß [13]. Obviously, the square of every element of G_m/G is the identity element. If we associate to $\begin{bmatrix} a & b \\ c & d \end{bmatrix}$ the ideal $(\sqrt{w})$ of $\mathfrak{o}$ (consisting of all elements $x\in\mathfrak{o}$ such that $x/\sqrt{w}$ is an algebraic integer) we get a homomorphism π : $G_m/G\rightarrow C$ which maps G_m/G onto the kernel of $Sq : C\rightarrow C^+$. The group G_e/G is the kernel of π. Thus $[G_m : G]$ equals the order of the kernel of $Sq : C^+\rightarrow C^+$ which is 2^{t-1} where t is the number of primes dividing the discriminant D.

We remark that every line and column of $\begin{bmatrix} a & b \\ c & d \end{bmatrix}$ generates the ideal $(\sqrt{w})$ in $\mathfrak{o}$.

3. 2 The group G_m/G operates on $\mathfrak{H}^2/G$ and also on the compactification $\overline{\mathfrak{H}^2/G}$. The cusps (considered as singular points of $\overline{\mathfrak{H}^2/G}$) are in one-to-one correspondence with C if one associates to a point $m/n\in\boldsymbol{P}_1(K)$ with m, $n\in\mathfrak{o}$ the ideal (m, n), see [6] 3. 7. Then $g\in G_m/G$ operates on C by multiplication with $\pi(g)$. This is easy to check. Two cusps represented by ideals $\mathfrak{a}$, $\mathfrak{b}$ are in the same orbit of the G_m/G-action if and only if $\mathfrak{a}^{-2}$, $\mathfrak{b}^{-2}$ represent the same element of C^+. This is true if and only if they have the same cycle of curves in their resolution. The group G_m/G operates also on $Y(D)$. The subgroup G_e/G keeps the cusps invariant, but is on each cycle the identity or the "covering translation" of order 2 depending on whether $|G_e/G| =|U^+/U^2|$ equals 1 or 2. The latter case is true if and only if there is no unit of negative norm. In this case the resolution cycle is twice the primitive cycle belonging to the module $\mathfrak{a}^{-2}$. Thus the group G_m/G of order 2^{t-1} operates freely on the union of the h cuspidal cycles of curves of $Y(D)$ (where $h=|C|$). *Each primitive cycle belonging to an element in the image of $Sq : C\rightarrow C^+$ occurs in $Y(D)$ exactly 2^{t-1} times (a twofold cycle counts as twice the primitive cycle).* The union of these 2^{t-1} primitive cycles is an orbit of the G_m/G action on $Y(D)$.

3. 3 We shall discuss the curves on the Hilbert modular surfaces defined by skew-hermitian matrices. By a *skew-hermitian matrix* we mean a matrix of the form

$$\begin{bmatrix} a_1\sqrt{D} & \lambda \\ -\lambda' & a_2\sqrt{D} \end{bmatrix} \quad \text{where } \lambda \in \mathfrak{o} \text{ and } a_1, a_2 \in \boldsymbol{Z}. \tag{1}$$

Its determinant is

$$N = a_1 a_2 D + \lambda\lambda'. \tag{2}$$

The matrix (1) is called primitive if there is no natural number >1 dividing a_1, a_2, λ. For a given natural number N the curve F_N in $\mathfrak{H}^2/G$ is defined to be the set of all points of $\mathfrak{H}^2/G$ which have representatives $(z_1, z_2) \in \mathfrak{H}^2$ for which there exists a primitive skew-hermitian matrix of determinant N such that

$$a_1\sqrt{D} z_1 z_2 - \lambda' z_1 + \lambda z_2 + a_2\sqrt{D} = 0. \tag{3}$$

It can be shown that F_N defines a curve in $\overline{\mathfrak{H}^2/G}$ and in $Y(D)$, also to be denoted by F_N. The curve F_N is not necessarily irreducible. By (2) *the curve F_N is non-empty if and only if the residue class of N modulo D can be represented by a norm in* $\mathfrak{o}$. If N is prime to D, this condition can be expressed in terms of values of "genus characters" of N; see [5] Satz 141. The group G_m/G operates on F_N and on the set of its irreducible components. The component of F_N defined by (3) passes through a cusp if and only if there exists an element $x \in K \cup \infty = \boldsymbol{P}_1(K)$ such that

$$a_1\sqrt{D} xx' - \lambda' x + \lambda x' + a_2\sqrt{D} = 0. \tag{4}$$

Since the matrix (1) can be diagonalized over the field K, this holds if and only if N is a norm in K. This is a condition only on N, so either all components of F_N pass through a cusp or none of them do.

The reduced quadratic irrationalities of discriminant D are of the form

$$w = \frac{M + \sqrt{D}}{2N}$$

where M and N are natural numbers, $0 < w' < 1 < w$, and $M^2 - D \equiv 0 \pmod{4N}$. There are only finitely many. Their continued fractions are *purely periodic*. Thus the reduced quadratic irrationalities of discriminant D are arranged in cycles which correspond bijectively to the elements of C^+ (see [6] 2. 6 and 4. 1 (5)). For a given cycle we index the reduced quadratic irrationalities as $w_k = \dfrac{M_k + \sqrt{D}}{2N_k}$ where k runs through $\boldsymbol{Z}/l\boldsymbol{Z}$ with l being the length of the cycle. We illustrate such a cycle as follows:

(5)

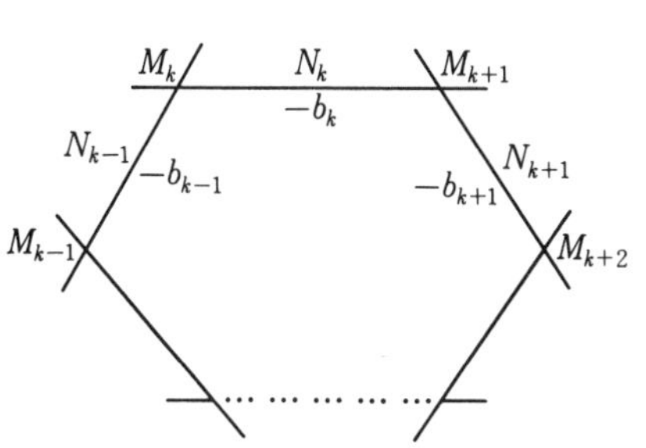

$$\left(w_k = b_k - \frac{1}{w_{k+1}}, \quad b_k = (M_k + M_{k+1})/2N_k\right).$$

The k-th line of (5) represents the rational curve S_k of self-intersection number $-b_k$ in the resolution of the corresponding cyclic singularity. To the kth corner of (5) we associate the quadratic form $p^2N_{k-1}+pqM_k+q^2N_k$ of discriminant D. In the resolution of the cusps of $\overline{\mathfrak{H}^2/G}$ we have exactly the cycles associated to the squares in C^+, i.e. to the image of $Sq: C^+ \to C^+$. In $Y(D)$ there are 2^{t-1} cycles of curves (see 3. 2) belonging to a given element of $Sq(C^+)$; in each cycle we have a local coordinate system (u_k, v_k) centered at the kth corner of the cycle. (The curve S_k is given by $v_k=0$ and S_{k-1} by $u_k=0$.) To relate this coordinate system as in [6] 2. 3 (11) to the coordinates (z_1, z_2) of $\mathfrak{H}^2$ by equations

$$2\pi i z_1 = A_{k-1} \log u_k + A_k \log v_k$$
$$2\pi i z_2 = A'_{k-1} \log u_k + A'_k \log v_k$$

we must transform the cusp to ∞. This is done by an isomorphism between $Y(D)$ and $Y(D, \mathfrak{b}_0^2)$. (see 1. 5) where $\mathfrak{b}_0$ is an ideal in $\mathfrak{o}$ such that $\mathfrak{b}_0^{-1}$ represents the element of $Sq(C^+)$ corresponding to the cycle (compare [6] 3. 7). Two such isomorphisms differ by an element of G_m/G. For integers $p, q \geqq 0$ (not both 0) we consider the local curve $u_k^q = v_k^p$. It has (p, q) branches

(6) $$u_k^{q/(p,q)} = \zeta v_k^{p/(p,q)} \quad \text{with} \quad \zeta^{(p,q)} = 1$$

of which $\varphi((p, q))$ are primitive, i.e. ζ is a (p, q)-th primitive root of unity.

As can be checked, the $\varphi((p, q))$ primitive branches (6) belong to F_N, where

(7) $$N = p^2N_{k-1} + pqM_k + q^2N_k.$$

We identify the triples $(k|0, 1)$ and $(k+1|1, 0)$. For any triple $(k|p, q)$ belonging to an element of $Sq(C^+)$ we have 2^{t-1} local curves $u_k^p = v_k^q$ in $Y(D)$. They are transformed to each other under G_m/G. It is not difficult to prove the following lemma.

Lemma. *For given N the union of all the primitive branches* (6) *satisfying* (7) *(restricted to a sufficiently small neighborhood of all the resolved cusps of $Y(D)$) equals the intersection of F_N with this neighborhood.*

The equation (3) defines a curve in $\mathfrak{H}^2$ which is the graph of the fractional linear transformation

(8) $$z_1 \to z_2 = \frac{\lambda' z_1 - a_2\sqrt{D}}{a_1\sqrt{D}\, z_1 + \lambda}$$

from $\mathfrak{H}$ to $\mathfrak{H}$. Thus the curve (3) can be identified in a specific way with $\mathfrak{H}$. Then the irreducible component of F_N (given by (3)) has $\overline{\mathfrak{H}/\Gamma}$ as its non-singular model where Γ is the subgroup of the Hilbert modular group G consisting of all elements of G which map the curve (3) to itself. The non-singular compact curve $\overline{\mathfrak{H}/\Gamma}$ is obtained from $\mathfrak{H}/\Gamma$ by "adding" finitely many cusps. Their number will be denoted by $\sigma(\Gamma)$. The non-singular model of F_N is a disjoint union of finitely many curves $\overline{\mathfrak{H}/\Gamma_j}$. The sum of the $\sigma(\Gamma_j)$ is by definition the number $\sigma(F_N)$ of cusps of F_N. For given

$\mathfrak{B} \in C^+$ ($\mathfrak{B}^{-1}$ representing a cycle (5)) the triples $(k|p, q)$, where $(k|0, 1)$ is to be identified with $(k+1|1, 0)$, are in one-to-one correspondence with the (integral) ideals $\mathfrak{b} \in \mathfrak{B}$ (see [6] 4. 1). We have $\mathfrak{b}\mathfrak{b}' = (N)$ with N as in (7) and $(p, q) = n(\mathfrak{b})$ where $n(\mathfrak{b})$ is the greatest natural number such that $\mathfrak{b}/n(\mathfrak{b})$ is an integral ideal. The set of all ideals $\mathfrak{b} \subset \mathfrak{o}$ which belong to an ideal class $\mathfrak{B} \in Sq(C^+)$ is the principal genus $\mathfrak{P}$. By the lemma we have

$$(9) \qquad \sigma(F_N) = 2^{t-1} \sum_{\substack{\mathfrak{b} \in \mathfrak{P} \\ \mathfrak{b}\mathfrak{b}' = (N)}} \varphi(n(\mathfrak{b})).$$

The curve F_N has a cusp ($\sigma(F_N) \geqq 1$) if and only if N is a norm in K (see (4)). Thus (9) is in agreement with the well-known fact that a natural number is a norm in K if and only if it is the norm of an ideal in the principal genus. If N is a norm in K, then the sum in (9) can be taken over all integral ideals $\mathfrak{b}$ with $\mathfrak{b}\mathfrak{b}' = (N)$. They are automatically in the principal genus. In some cases (9) gives information on the number of components of F_N.

First we need a definition. N is called *admissible* if it is the norm of an ideal $\mathfrak{b}$ in the principal genus which is primitive, i.e. $n(\mathfrak{b}) = 1$. This happens if and only if N is a norm in K and every prime factor of N decomposes or ramifies in $\mathfrak{o}$, the ramifying prime factors having exponent 1 in N.

Proposition. *If N is admissible and not divisible by the square free part d of D, then F_N has 2^{t-1-r} components where r is the number of primes dividing (D, N). If N is admissible and divisible by d, then F_N has 2^{t-r} (thus 1 or 2) components. The group G_m/G operates transitively on the set of components.*

We indicate the proof. If $(p, q) = 1$ then (6) can be represented by the "diagonal" in $\mathfrak{H}^2/\boldsymbol{SL}_2(\mathfrak{o}, \mathfrak{b})$ where $\mathfrak{b}$ is the primitive ideal with norm N corresponding to $(k|p, q)$. Compare [6] 4. 1. Therefore, in this case, the non-singular model of the component of F_N represented by (6) is $\overline{\mathfrak{H}/\Gamma}$ where $\Gamma = \Gamma_0(N)/\{1, -1\}$ or where Γ is a certain extension of index 2 of $\Gamma_0(N)/\{1, -1\}$. The latter case happens if and only if N is divisible by d. As is well-known, the cusps of $\overline{\mathfrak{H}/\Gamma}$ can be represented by rational numbers a/c with $(a, c) = 1$, $c > 0$ and $c|N$. For any divisor c of N we have $\varphi((c, N/c))$ cusps. If $d|N$ and Γ is an extension of index 2 of $\Gamma_0(N)/\{1, -1\}$, then a cusp with denominator c is identified with a cusp of denominator $cd/(c, d)^2$. The given equation (6) from which we started is a description of the embedding of $\overline{\mathfrak{H}/\Gamma}$ in $Y(D)$ near the cusp of $\overline{\mathfrak{H}/\Gamma}$ given by $c = N$. For a given divisor c of N it can be shown that $\overline{\mathfrak{H}/\Gamma}$ near a cusp with denominator c is imbedded in $Y(D)$ by an equation (6) where $(k|p, q)$ corresponds to the ideal $\tilde{\mathfrak{b}} = (\mathfrak{b} \cdot c)/(\mathfrak{b}, c)^2$ which has norm N and for which $(p, q) = n(\tilde{\mathfrak{b}}) = (c, N/c)$. All ideals with norm N are obtained in this way. As can be checked, we get for given c for the various cusps with denominator c all the $\varphi((p, q))$ primitive roots of unity in (6). We conclude that all components of F_N are equivalent under G_m/G. The number of cusps of $\overline{\mathfrak{H}/\Gamma_0(N)}$ equals

$$\sigma(\Gamma_0(N)) = \sum_{c|N} \varphi((c, N/c)).$$

Formula (9) now implies the proposition if N is not divisible by d. If N is divisible by d, then all components have $\frac{1}{2}\sigma(\Gamma_0(N))$ cusps. Again (9) implies the result.

3. 4 Suppose we have two different skew-hermitian curves in $\mathfrak{H}^2$, one given by (3) with determinant N and the second one by

$$b_1\sqrt{D}z_1z_2-\mu' z_1+\mu z_2+b_2\sqrt{D}=0$$

with determinant M. They intersect in $\mathfrak{H}^2$ if and only if the matrix

$$B=\begin{bmatrix}\mu & b_2\sqrt{D}\\ -b_1\sqrt{D} & \mu'\end{bmatrix}\begin{bmatrix}\lambda' & -a_2\sqrt{D}\\ a_1\sqrt{D} & \lambda\end{bmatrix}$$

has a fixed point in $\mathfrak{H}$ (compare (8)) which happens if and only if

$$4NM-\operatorname{tr}(B)^2>0.$$

It is easy to check that $\operatorname{tr}(B)^2-4NM$ is divisible by D and its quotient by D is a discriminant (i.e. $\equiv 0$ or 1 mod 4). *Therefore, if the two curves intersect in* $\mathfrak{H}^2$, *then the following condition holds.*

(10) There exists $x\in\mathbf{Z}$ such that $|x|<\sqrt{4NM}$ and $4NM-x^2\equiv 0 \bmod D$ with $(x^2-4NM)/D\equiv 0$ or 1 mod 4.

If (10) *is not satisfied for* $N\neq M$, *then* F_N *and* F_M *do not intersect in* $\mathfrak{H}^2/G$.

Lemma. *If* (10) *is not satisfied for* $M=N$, *then two different components of* F_N *do not intersect in* $\mathfrak{H}^2/G$ *and moreover* F_N *is non-singular in* $Y(D)$ *outside the resolved cusps.*

Proof. Assume that (10) is not satisfied for $M=N$. If a component of F_N is given by (8) with $\mathfrak{H}/\Gamma$ as its non-singular model, then the isotropy group of the Hilbert modular group G at a point x of $\mathfrak{H}^2$ satisfying (8) is contained in Γ. It also follows that there is only one skew-hermitian curve of determinant N in $\mathfrak{H}^2$ passing through x. If the isotropy group of G at x is trivial, then F_N is non-singular in the point of $Y(D)$ represented by x. If the isotropy group is of order r, then it is of type $(r;1,1)$. This follows from (8). (For $D>12$ we have $r=2$ or 3; see [15].) The curve F_N passes in $Y(D)$ transversally through the curve of self-intersection number $-r$ which gives the resolution of the quotient singularity. (Condition (10) and the lemma were suggested to us by P. Hahnel and H.-P. Kraft.)

The necessary and sufficient condition that F_N be non-singular in the neighborhood of a resolved cusp given by a cycle (5) is that for all p, q satisfying (7) one of the exponents $p/(p,q)$ or $q/(p,q)$ in (6) be equal to 1. Thus:

If (10) *is not satisfied for* $M=N$ *and if in the lemma in* 3. 3 *all pairs* p, q *are such that* $p|q$ *or* $q|p$, *then* F_N *is non-singular in* $Y(D)$.

In particular F_1 is non-singular in $Y(D)$ and has 2^{t-1} components.

3. 5 If N is a prime, then the curve F_N is non-empty if and only if N is a norm in K, and N is a norm in K if and only if the t characters χ_i $(i=1,\cdots,t)$ do not take a value -1 at N. Here we define the χ_i as follows. We write D as product of prime discriminants

$$D = \prod_{i=1}^{t} D_i,$$

for example $60=(-3)\cdot(-4)\cdot 5$. Then

$$\chi_i(N) = \left(\frac{D_i}{N}\right) \qquad \text{for } N \text{ odd}$$

and

$$\chi_i(2) = \left(\frac{D_i}{2}\right) = \begin{cases} 0 & \text{for } D_i \equiv 0 \quad (4) \\ 1 & \text{for } D_i \equiv 1 \quad (8) \\ -1 & \text{for } D_i \equiv 5 \quad (8). \end{cases}$$

Using the proposition in 3. 3, we get

Proposition. *If N is a prime and $D \neq N$, $4N$, then the number of components of F_N equals*

(11) $$\frac{1}{2}\prod_{i=1}^{t}(1+\chi_i(N)).$$

3. 6 Let N be a prime. We wish to study the curve F_{N^2} in $Y(D)$. If N decomposes in $\mathfrak{o}$, i.e. $(D/N)=1$, then N^2 is admissible and we have the proposition in 3. 3; the curve F_{N^2} has 2^{t-1} components. By (9), F_{N^2} has $2^{t-1}(N-1)$ cusps if $(D/N)\neq 1$ and $2^{t-1}(N+1)$ cusps if $(D/N)=1$.

If $\mathfrak{o}=\boldsymbol{Z}w_0+\boldsymbol{Z}$ where w_0 is reduced, then one of the local coordinate systems for the cusp at ∞ is given by

(12) $$\begin{aligned} 2\pi i z_1 &= w_0 \log u_0 + \log v_0 \\ 2\pi i z_2 &= w_0' \log u_0 + \log v_0. \end{aligned}$$

There are $N-1$ cusps of F_{N^2} corresponding to

(13) $$u_0 = \zeta \quad \text{where} \quad \zeta^N = 1, \quad \zeta \neq 1$$

(compare (6), (7); we have $N_0=1, p=0, q=N$, and the N in (7) has to be replaced here by N^2) or to skew-hermitian forms

(14) $$Nz_1 - Nz_2 = r(w_0 - w_0') \text{ where } (r, N) = 1, \ (w_0 - w_0' = \sqrt{D}).$$

The curve S_0 (given by $v_0=0$) intersects the $N-1$ branches (13) of F_{N^2} transversally.

The component of F_{N^2} given by (14) has $\overline{\mathfrak{H}/\Gamma}$ as model where the subgroup Γ of G carrying (14) to itself has to be determined. The result is independent of r. We list it and give also the number of cusps $\sigma(\Gamma)$ which is well-known for the groups in question:

(15) $$\text{If}\left(\frac{D}{N}\right) = -1, \text{ then } \Gamma = \Gamma'(N)/\{1, -1\} \text{ and } \sigma(\Gamma) = N-1.$$

Here $\Gamma'(N)$ is defined as follows. Consider the multiplicative group of the field $\boldsymbol{F}_{N^2}$

as subgroup of $\boldsymbol{GL}_2(\boldsymbol{F}_N)$, take the intersection with $\boldsymbol{SL}_2(\boldsymbol{F}_N)$. Its inverse image in $\boldsymbol{SL}_2(\boldsymbol{Z})$ is $\Gamma'(N)$.

$$\text{(16)}\quad \begin{array}{ll} \text{If } N|D,\ N\neq 2 \text{ and } D\neq N, 4N, \text{ then } \Gamma=\Gamma_1(N)/\{1,-1\} & \\ & \text{and } \sigma(\Gamma)=N-1. \\ \text{If } D \text{ even, } N=2,\ D\neq 8, \text{ then } \Gamma=\Gamma_1(2)/\{1,-1\}=\Gamma_0(2)/\{1,-1\} & \\ & \text{and } \sigma(\Gamma)=2. \\ \text{If } D=N \text{ or } D=4N\ (N\neq 2), \text{ then } \Gamma=\Gamma_1^*(N)/\{1,-1\} & \\ & \text{and } \sigma(\Gamma)=\dfrac{N-1}{2}. \\ \text{If } D=8,\ N=2, \text{ then } \Gamma=\Gamma_1^*(2)/\{1,-1\} & \text{and } \sigma(\Gamma)=1. \end{array}$$

The group $\Gamma_1(N)$ consists of those matrices in $\boldsymbol{SL}_2(\boldsymbol{Z})$ which are of the form $\pm\begin{bmatrix}1 & *\\ 0 & 1\end{bmatrix}$ modulo N and $\Gamma_1^*(N)$ is an extension of index 2 of $\Gamma_1(N)$. The proof of (15) and (16) is carried out by applying the method of [6] p. 270 to equation (14). To bring Γ into the above form one must conjugate in $\boldsymbol{GL}_2(K)$. Using (9), (15), (16) (and the Proposition in 3. 3 for the case $(D/N)=1$) we get

Proposition. *If N is an odd prime, then the curve F_{N^2} has 2^{t-1} components, except in the case $D=N$ or $D=4N$ where it has 2 or 4 components respectively. If $N=2$, then F_4 has 2^{t-1} components if D is odd or if $D=8$. If D is even $(D\neq 8)$ then F_4 has 2^{t-2} components.*

Remark. For $N\nmid D$ the skew-hermitian curves (14) all belong to the same component of F_{N^2} and G_m/G operates transitively on the set of components. If $N|D$ $(N\neq 2)$, then two skew-hermitian curves (14) belong to the same component if and only if the two values of (r/N) are both equal to $+1$ or both equal to -1. In [7] § 3 it was stated that the curve $F_N((N/p)\neq -1)$ on $Y(p)$ (p prime) is irreducible. This has to be corrected as pointed out by Hammond. It will be shown in a forthcoming dissertation by Hans-Georg Franke (Bonn) that F_N is irreducible if $N\not\equiv 0\ (p^2)$. If $N\equiv 0\ (p^2)$ then F_N has exactly two components.

3. 7 An exceptional curve on an algebraic surface is a non-singular rational curve of self-intersection number -1. If the surface is regular and not rational, then any two exceptional curves are disjoint and can be blown down simultaneously. *In this section we assume that $Y(D)$ is not rational.* Thus we exclude 10 discriminants (Chap. I (1)). How many exceptional curves can be found on $Y(D)$ using skew-hermitian curves ?

For a discrete subgroup Γ of $\boldsymbol{PL}_2^+(\boldsymbol{R})$ with $\mathfrak{H}/\Gamma$ of finite volume the number

$$c_1(\Gamma)=2e(\overline{\mathfrak{H}/\Gamma})-\sum_{r\geq 2}a_r(\Gamma)-\sigma(\Gamma) \qquad (17)$$

was introduced in [6] 4. 3. We recall that e denotes the Euler number, $a_r(\Gamma)$ the number of Γ-equivalence classes of fixed points of order r of Γ and $\sigma(\Gamma)$ the number of cusps. If a component E of a skew-hermitian curve in $Y(D)$ has the non-singular model $\overline{\mathfrak{H}/\Gamma}$, then

$$c_1 \cdot E \geqq c_1(\Gamma)$$

where $c_1 \cdot E$ denotes the value of the first Chern class of $Y(D)$ on E. Since $Y(D)$ is not rational, $c_1 \cdot E \geqq 1$ implies that $c_1 \cdot E = 1$ and E is an exceptional curve (see [6] 4. 4 Corollary I). The curve F_1 has 2^{t-1} components (see the Proposition in 3. 3). Each component passes through a quotient singularity of order 2 and one of order 3 on $\mathfrak{H}^2/G$ and is on $Y(D)$ an exceptional curve which gives rise to a configuration

(18)

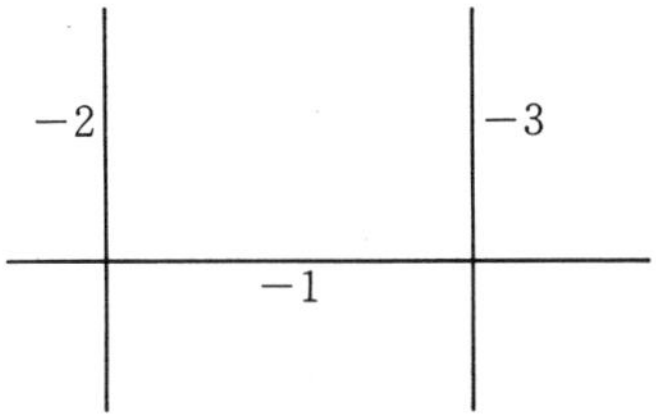

of non singular rational curves. These configurations are disjoint to each other. *Using F_1 we have found on $Y(D)$ in this way $3 \cdot 2^{t-1}$ curves which can be blown down.*

The groups Γ which occur for the components of F_2, F_3 are $\Gamma_0(2)$ and $\Gamma_0(3)$ respectively (to be divided by $\{1, -1\}$). For the components of F_4 we have $\Gamma = \Gamma_0(4)$ or $\Gamma = \Gamma'(2)$ if D is odd (to be divided by $\{1, -1\}$). (These groups were treated in [6] 5. 5 if D is a prime.) If D is even, the group for F_4 is $\Gamma_0(2)/\{1, -1\}$. If $3|D$ $(D \neq 12)$, then the components of F_9 have the group $\Gamma_1(3) = \Gamma_0(3)$ (to be divided by $\{1, -1\}$). For these groups Γ (namely $\Gamma_0(2)$, $\Gamma_0(3)$, $\Gamma'(2)$, $\Gamma_0(4)$, always divided by $\{1, -1\}$) the value of $c_1(\Gamma)$ equals 1. Since $Y(D)$ is supposed to be not rational, *all components of F_2, F_3, F_4 and* (if $3|D$) *F_9 give exceptional curves.* Each component of F_2 passes through a quotient singularity of order 2 on $\mathfrak{H}^2/G$ and gives on $Y(D)$ a configuration.

(19)

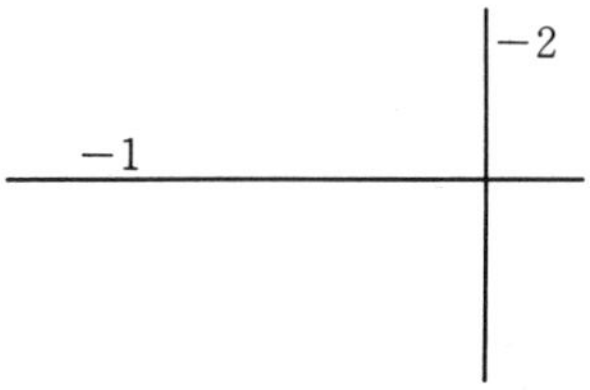

Every component of F_2 gives two curves which can be blown down. The curve F_4 has 2^{t-1} or 2^{t-2} components. In the latter case we have a configuration (19) for each component because the group is $\Gamma_0(2)$ (see (16)). Therefore F_4 gives always 2^{t-1} curves to blow down, *F_1 and F_4 together give 2^{t+1} curves to blow down.* For F_3 and F_9 the corresponding group has no fixed point of order 2. There is no configuration (19). No additional blow-downs occur in this way. For $D = 105$ a special situation occurs. We have

$$\frac{4\cdot9\cdot9-3^2}{105}=3.$$

Thus condition (10) is satisfied. In fact, it can be checked that the 4 components of F_9 meet in quotient singularities of order 3 on $\mathfrak{H}^2/G$ and this leads on $Y(105)$ to a configuration like this

(20)

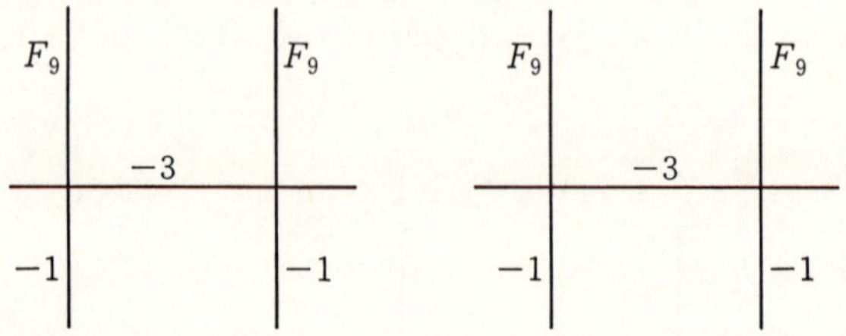

$D=105$

which gives two extra curves to blow down. In fact, for $D=105$ there are six quotient singularities of order 3 on $\mathfrak{H}^2/G$, all of type $(3\,;1,1)$. Four of them lie on the 4 components of F_1. The two others give rise to the two curves of self-intersection number -3 in (20). Two intersecting components of F_9 never occur for other D (with $3|D$) as can be checked by condition (10).

By the propositions in 3. 5 and 3. 6 we know the number of components of F_2, F_3, F_4, F_9, hence we can collect the information on exceptional curves in the following theorem.

Theorem. *Suppose $Y(D)$ is not rational. Then $\beta(D)$ curves on $Y(D)$ can be blown down where*

$$\beta(D)=2^{t+1}+\prod_{i=1}^{t}(1+\chi_i(2))+\frac{1}{2}\prod_{i=1}^{t}(1+\chi_i(3)) +2^{t-1}\left(1-\left(\frac{D}{3}\right)^2\right)+\begin{cases}2 & \text{for } D=105\\ 0 & \text{for } D\neq105.\end{cases} \tag{21}$$

We call $Y^0(D)$ the surface obtained from $Y(D)$ by blowing down these $\beta(D)$ curves. We define $Y^0(D)$ only if $Y(D)$ is not rational. Clearly

$$c_1^2(Y^0(D))=c_1^2(Y(D))+\beta(D). \tag{22}$$

We conjecture that $Y^0(D)$ is the minimal model. For D equal to a prime, this was conjectured in [9]. In fact, van der Geer and van de Ven have checked the conjecture for several prime values of D where $Y^0(D)$ is of general type. When $Y^0(D)$ is not of general type, then the conjecture holds because $c_1^2(Y^0(D))=0$ as we shall see.

3. 8 For the surface $Y_-(D)$ introduced in Chapter I similar considerations hold. We have a curve F_N given by all primitive equations (3) with $a_1a_2D+\lambda\lambda'=-N$. This curve passes through a cusp if and only if $-N$ is a norm in K. If -1 is a norm in K, then $Y(D)$ and $Y_-(D)$ are isomorphic. In this case (provided $Y(D)$

is not rational) we can blow down $\beta(D)$ curves on $Y_-(D)$; the resulting surface $Y^0_-(D)$ is then isomorphic to $Y^0(D)$. If condition (10) is not satisfied for N, M $(N \neq M)$, then F_N and F_M do not meet on $(\mathfrak{H}\times\mathfrak{H}^-)/G$. If condition (10) is not satisfied for $M=N$, then F_N is non-singular on $Y_-(D)$ outside the resolved cusps. The lemma in 3. 3 holds in the same way except that one has to take the triples $(k|p,q)$ belonging to an element of $(\sqrt{D})Sq(C^+)$ where $(\sqrt{D})$ denotes here the element of C^+ represented by the ideal $(\sqrt{D})$. The natural number N is called *admissible* for $Y_-(D)$ if $-N$ is a norm in K and all primes dividing N decompose or ramify in $\mathfrak{o}$, but where a prime which ramifies occurs in N only with exponent 1. The number of components of F_N (N admissible for $Y_-(D)$) is as in the proposition in 3. 3. If -1 is not a norm in K, then F_1, F_4, F_9 are empty on $Y_-(D)$, so we can only blow down F_2 and F_3, and this gives (*if $Y_-(D)$ is not rational*)

$$\beta_-(D) = \prod_{i=1}^{t}(1+\chi_i(-2)) + \frac{1}{2}\prod_{i=1}^{t}(1+\chi_i(-3)) \tag{23}$$

blow-downs. (Note that $\chi_i(-N) = (\text{sign } D_i)\ \chi_i(N)$.) Again we conjecture that the surface $Y^0_-(D)$ obtained by these blow-downs is minimal.

IV. Estimates of the Numerical Invariants

4. 1 The purpose of this chapter is to prove the facts

$$\begin{aligned}
&\chi(Y(D)) = 1 \Leftrightarrow D = 5, 8, 12, 13, 17, 21, 24, 28, 33, 60,\\
&\chi(Y_-(D)) = 1 \Leftrightarrow D = 5, 8, 12, 13, 17,\\
&c_1^2(Y(D)) \leq 0 \Rightarrow D \leq 285,\\
&c_1^2(Y_-(D)) \leq 0 \Rightarrow D \leq 136,\\
&c_1^2(Y_T(p)) \leq 0 \Rightarrow p \leq 821 \qquad (p \equiv 1 (\text{mod } 4) \text{ prime})
\end{aligned}$$

(compare Chapter I), thus reducing the problem of classifying all Hilbert modular surfaces to the consideration of a finite list. Since all of the invariants have been calculated (by computer) up to at least $D=1500$, it will suffice to prove

(1) $D > 1500 \Rightarrow \chi(Y(D)) > 1,\ \chi(Y_-(D)) > 1,\ c_1^2(Y(D)) > 0,\ c_1^2(Y_-(D)) > 0,$
(2) $p > 1500 \Rightarrow c_1^2(Y_T(p)) > 0.$

There are precisely 50 discriminants for which the four inequalities of (1) are not all satisfied ; complete numerical data on these discriminants is given in section 4. 5.

4. 2 As explained in 2. 1, the dominant term in the formula for all of these numerical invariants is

$$\zeta_K(-1) = \frac{1}{60}\sum_{\substack{k^2<D\\ k^2\equiv D (\text{mod } 4)}} \sigma_1\left(\frac{D-k^2}{4}\right). \tag{3}$$

From $\sigma_1(n) \geq n+1$ we deduce easily that

$$\zeta_K(-1) > D^{3/2}/360\ ; \tag{4}$$

this result can also be obtained by writing

$$\zeta_K(2) = \zeta(2)\prod\left(1-\left(\frac{D}{p}\right)p^{-2}\right)^{-1} > \zeta(2)\prod(1+p^{-2})^{-1} = \zeta(4) = \frac{\pi^4}{90}$$

and applying the functional equation of $\zeta_K(s)$.

From 2. 4 (31) and 2. 1 (9), (10), (12) we have

$$c_1^2(Y(D)) = 4\zeta_K(-1) - l_0^-(D) - \frac{1}{3}a_3^+(G)$$

and

$$\chi(Y(D)) = \frac{1}{2}\zeta_K(-1) + \frac{1}{8}a_2(G) + \frac{1}{9}a_3^+(G) + \frac{2}{9}a_3^-(G) + w\,;$$

moreover, by 2. 3 (29),

$$w = \frac{1}{12}(l_0(D) - l_0^-(D)) > -\frac{1}{12}l_0^-(D),$$

and hence

$$\chi(Y(D)) > \frac{1}{2}\zeta_K(-1) + w > \frac{1}{6}\zeta_K(-1) + \frac{1}{12}(4\zeta_K(-1) - l_0^-(D))$$

$$> \frac{1}{2160}D^{3/2} + \frac{1}{12}c_1^2(Y(D)) > \frac{c_1^2(Y(D))}{12} + 1 \qquad (D > 200).$$

Hence the inequality $\chi(Y(D))>1$ in (1) will follow once we have proved that $c_1^2(Y(D))$ is positive; since (by 2. 2 (15) and 2. 4 (33)) the values of χ and c_1^2 for $Y_-(D)$ are at least as large as for $Y(D)$, the remaining two inequalities in (1) will also follow. Thus to prove (1) we have to show

(5) $$4\zeta_K(-1) - l_0^-(D) - a_3^+(G)/3 > 0 \qquad (D > 1500),$$

while for (2) the inequality

(6) $$2\zeta_K(-1) - \frac{1}{2}l_0(p) - \frac{5p}{24} + \frac{\sqrt{p}}{2} > 0 \qquad (p > 1500)$$

will certainly suffice (equation 2. 5 (40)).

We introduce a new invariant

(7) $$l(D) = \sum_{\mathfrak{a}\in C^+} l(\mathfrak{a})$$

(notation as in 2. 3). If $D=p$ is a prime, then $Sq : C^+ \to C^+$ is an isomorphism and

(8) $$l_0(p) = l_0^-(p) = l(p)\,;$$

if, however, D has t distinct prime factors, then Sq has a kernel of order 2^{t-1} and so

(9) $$l_0^-(D) = \sum_{\mathfrak{a}\in C^+} l(\gamma Sq(\mathfrak{a})) = 2^{t-1}\sum_{\substack{\mathfrak{b}\in C^+\\ \gamma\mathfrak{b}\in \mathrm{Im}(Sq)}} l(\mathfrak{b}) \leq 2^{t-1}l(D)$$

The advantage of working with $l(D)$ rather than $l_0^-(D)$ is that it can be evaluated by a formula analogous to formula (3) for $\zeta_K(-1)$. Indeed, $l(D)$ is the sum of the lengths of all cycles occurring as the primitive period of the continued fraction of some quadratic irrationality w of discriminant D (the discriminant of w is defined as b^2-4ac, where $aw^2+bw+c=0$, $(a, b, c)=1$). This is simply the number of *reduced* quadratic irrationalities w of discriminant D (i. e. w satisfying $w>1>w'>0$), since, as discussed in 2. 3, such w have purely periodic continued fractions, and a cycle $((b_0,\cdots,b_{r-1}))$ of length r gives rise to precisely r reduced numbers

$$b_i-\cfrac{1}{b_{i+1}-\ddots}\qquad (i=0, 1,\cdots, r-1).$$

If $aw^2+bw+c=0$, $b^2-4ac=D$, then the condition $(a, b, c)=1$ is automatically satisfied since D is the discriminant of a quadratic field. Therefore

$$l(D)=\#\left\{(a, b, c)\in \boldsymbol{Z}^3 \mid b^2-4ac=D,\ \frac{-b+\sqrt{D}}{2a}>1>\frac{-b-\sqrt{D}}{2a}>0\right\}.$$

The inequalities are equivalent to

$$a>0,\qquad |-b-2a|<\sqrt{D},\qquad -b>\sqrt{D}\,;$$

therefore replacing b by $k=-b-2a$ gives

$$l(D)=\#\{(a, k)\in \boldsymbol{Z}^2 \mid a>0,\ k^2<D,\ k^2\equiv D \pmod{4a},\ k+2a>\sqrt{D}\}.$$

We claim that this is precisely half of

$$\#\{(a, k)\in \boldsymbol{Z}^2 \mid a>0,\ k^2<D,\ k^2\equiv D \pmod{4a}\}.$$

Indeed, $(a, k)\mapsto(a', k')=((D-k^2)/4a, -k)$ is an involution on this latter set with

$$\frac{2a'+k'-\sqrt{D}}{2a+k-\sqrt{D}}=-\frac{k+\sqrt{D}}{2a}<0,$$

so precisely half of the elements (a, k) satisfy $2a+k>\sqrt{D}$. Therefore

$$l(D)=\frac{1}{2}\sum_{\substack{k^2<D\\ k^2\equiv D \pmod 4}}\ \sum_{\substack{a>0\\ a\mid \frac{D-k^2}{4}}}1=\frac{1}{2}\sum_{\substack{k^2<D\\ k^2\equiv D \pmod 4}}\sigma_0\!\left(\frac{D-k^2}{4}\right). \tag{10}$$

This formula will be the basis for our estimates of c_1^2.

4. 3 In this section we prove the estimate (6) ; this case is easier than estimate (5) for composite D because of (8). We will prove (for all D, prime or composite) that

$$2\zeta_K(-1)-\frac{1}{2}l(D)\geq\frac{D+15}{180}\sqrt{D-200}-3.6\qquad (D\geq 730)\,; \tag{11}$$

since the right-hand side is $>\dfrac{5D}{24}-\dfrac{\sqrt{D}}{2}$ for $D>1500$, this implies (6).

By (3) and (10), the left-hand side of (11) equals

$$\sum_{\substack{k^2<D\\ k^2\equiv D \pmod 4}}\phi\!\left(\frac{D-k^2}{4}\right) \tag{12}$$

with

$$\phi(n)=\frac{1}{30}\sigma_1(n)-\frac{1}{4}\sigma_0(n).$$

We have

$$(13)\qquad \begin{aligned} \psi(n) &\geq -0.6 && \text{for } n<50, \\ \psi(n) &\geq \frac{n-14}{30} && \text{for } n\geq 50. \end{aligned}$$

Indeed, for $n\leq 56$ we can check this by hand, while, for $n>56\frac{1}{4}=\left(7\frac{1}{2}\right)^2$,

$$\begin{aligned} \psi(n) &= \frac{1}{30}\sum_{d|n}\left(d-7\frac{1}{2}\right) \\ &= \frac{1}{30}\left[\left(n-7\frac{1}{2}\right)+\left(1-7\frac{1}{2}\right)+\frac{1}{2}\sum_{\substack{d|n \\ 1<d<n}}\left(d+\frac{n}{d}-15\right)\right] \\ &\geq \frac{n-14}{30} \end{aligned}$$

(each term $d+n/d-15$ is ≥ 0). For $D>729$ there are at most 4 values of k (two positive and two negative) for which $k\equiv D \pmod 2$ and $0<(D-k^2)/4<50$ (because the interval $(\sqrt{D-200}, \sqrt{D})$ has length<4), so the first line of (13) is used at most four times in (12); the second estimate in (13) now gives

$$\begin{aligned} \sum_{\substack{k^2<D \\ k^2\equiv D \pmod 4}} \psi\left(\frac{D-k^2}{4}\right) &\geq -2.4+\sum_{\substack{k^2<D-200 \\ k^2\equiv D \pmod 4}}\left(\frac{D-k^2}{120}-\frac{14}{30}\right) \\ &\geq -3.6+\sqrt{D-200}\left(\frac{D-201}{180}+\frac{200}{120}-\frac{14}{30}\right), \end{aligned}$$

where we have used the easy estimates (valid for any positive A and integer D)

$$\sum_{\substack{k^2<A \\ k\equiv D \pmod 2}} 1 \geq \sqrt{A}-1, \quad \sum_{\substack{k^2<A \\ k\equiv D \pmod 2}} (A-k^2) \geq \frac{2\sqrt{A}}{3}(A-1)$$

with $A=D-200$. This proves the inequality (11).

4. 4 We now want to prove the estimate (5). The number $l_0^-(D)$ in that equation will be estimated using (9) and (10); for the number $a_3^+(G)$, given exactly by 2. 1 (7), we use the estimate

$$h(-N)<\frac{\sqrt{N}}{2\pi}(2+\log N) \qquad (N>4)$$

(cf. [14]) to obtain

$$\frac{1}{3}a_3^+(G)<0.13\, D^{1/2}(\log D+1).$$

The formula to be proved then becomes

$$(14)\quad 4\zeta_K(-1)-0.13\sqrt{D}(\log D+1)>2^{t-2}\sum_{\substack{k^2<D \\ k^2\equiv D \pmod 4}}\sigma_0\left(\frac{D-k^2}{4}\right) \qquad (D>1500).$$

Because of the factor 2^{t-2}, the method of 4. 3 does not work here and we must have recourse to far cruder estimates. We would like to thank Henri Cohen, who suggested the method for estimating the right-hand side of (14) and carried out the necessary computer calculations.

Lemma. *Set* $\varepsilon = \log 2/\log 11 = 0.289064826\cdots$. *Then for all n*

$$\sigma_0(n) < 5.1039782\, n^{\varepsilon}. \tag{15}$$

Proof. The function $\sigma_0(n)/n^{\varepsilon}$ is multiplicative and $(a+1)/p^{a\varepsilon} \leq 1$ for $p \geq 11$, $a \geq 1$ by the choice of ε. Hence

$$\frac{\sigma_0(n)}{n^{\varepsilon}} \leq \prod_{p} \max_{a \geq 0}\left(\frac{a+1}{p^{a\varepsilon}}\right) = \frac{5}{2^{4\varepsilon}} \cdot \frac{3}{3^{2\varepsilon}} \cdot \frac{2}{5^{\varepsilon}} \cdot \frac{2}{7^{\varepsilon}} = 5.103978196\cdots$$

If we now estimate $\zeta_K(-1)$ by (4), and the right-hand side of (14) by the product of the number of terms in the sum with the estimate of the individual terms given by equation (15), we find as a sufficient condition for (14) the inequality

$$\frac{D^{3/2}}{90} - 0.13\, D^{1/2}(\log D + 1) > 2^{t-2}(\sqrt{D}+1) \cdot 5.1039782 \cdot (D/4)^{\varepsilon} \tag{16}$$

with $\varepsilon = 0.289064826\cdots$ as before. A desk calculator computation now shows that (16) holds if

$$\begin{array}{lllll} & t \leq 3 & \textit{and} & D > & 9{,}000 \\ \textit{or} & t = 4 & \textit{and} & D > & 23{,}000 \\ \textit{or} & t = 5 & \textit{and} & D > & 60{,}000 \\ \textit{or} & t = 6 & \textit{and} & D > & 157{,}000 \\ \textit{or} & t = 7 & \textit{and} & D > & 420{,}000. \end{array} \tag{17}$$

But the smallest discriminant with $t=7$ is $4.\ 3.\ 5.\ 7.\ 11.\ 13.\ 17 = 1{,}021{,}020 > 420{,}000$, so (17) implies that (16) holds for all D with $t=7$. A similar argument holds for any $t \geq 7$, since a D with $t \geq 7$ distinct prime factors is greater than

$$4.\ 3.\ 5.\ 7.\ 11.\ 13.\ 16^{t-6} = \frac{60060}{65536}(2^{t-2})^4,$$

so $2^{t-2} < \sqrt[4]{1.1D}$, more than sufficient to prove (16) for $D > 420{,}000$. Therefore (17) implies that (16) (and hence (14)) holds for all $D > 157{,}000$, and a computer calculation showed that (14) holds for all D up to this point.

4. 5 As already stated, the calculation of the various invariants for $D < 1500$ showed that $c_1^2(Y(D)) \leq 0$ for just 50 discriminants, the largest being $D = 285$. We have tabulated all numerical invariants of $Y(D)$ and $Y_-(D)$ for these discriminants. The following notation is used :

Topological Invariants :

$Z = 6\zeta_K(-1)$ (this is an integer for $D > 8$)

$l_0, l_0^- = l_0(D), l_0^-(D)$ (§§ 2. 3, 2. 4)

$a_2, a_3^+, a_3^- = a_2(G), a_3^+(G), a_3^-(G)$ (§ 2. 1 ; for $D = 5$, 8 and 12 there are also fixed points of order 5, 4 and 6 respectively)

$e = e(\mathfrak{H}^2/G) = e(\mathfrak{H} \times \mathfrak{H}^-/G)$ (2. 1 (9), 2. 2 (13))

$\tau = -\text{sign}\,(\mathfrak{H}^2/G) = \text{sign}\,(\mathfrak{H} \times \mathfrak{H}^-/G)$ (2. 1 (10), 2. 2 (13))

Invariants of $Y(D)$:

$\chi = \frac{1}{4}(e-\tau) = \chi(Y(D))$ (2. 1 (12))

$c = c_1^2(Y(D))$ (2. 4 (31))

$c^0 = c_1^2(Y^0(D)) = c+\beta(D)$ (3. 7 (22) ; c^0 is not listed if $Y(D)$ is rational, since $Y^0(D)$ was not defined in this case)

Invariants of $Y_-(D)$ (not given if D is a sum of two squares since then $Y_-(D)$ is isomorphic to $Y(D)$) :

$\chi_- = \frac{1}{4}(e+\tau) = \chi(Y_-(D))$ (2. 2 (14))

$c_- = c_1^2(Y_-(D))$ (2. 4 (32))

$c_-^0 = c_1^2(Y_-^0(D)) = c_-+\beta_-(D)$ (§ 3. 8).

D	Z	l_0	l_0^-	a_2	a_3^+	a_3^-	e	τ	χ	c	c^0	χ_-	c_-	c_-^0
5	1/5	1	1	2	1	1	4	0	1	−2	•			
8	1/2	2	2	2	1	1	4	0	1	−3	•			
12	1	2	4	3	2	0	4	0	1	−4	•	1	−4	•
13	1	3	3	2	2	2	4	0	1	−3	•			
17	2	5	5	4	1	1	4	0	1	−4	•			
21	2	2	6	4	4	1	6	2	1	−6	•	2	−1	0
24	3	4	8	6	3	0	6	2	1	−7	•	2	−2	0
28	4	4	10	4	2	2	6	2	1	−8	•	2	−2	0
29	3	5	5	6	3	3	8	0	2	−4	0			
33	6	8	12	4	3	0	6	2	1	−9	•	2	−4	0
37	5	7	7	2	4	4	8	0	2	−5	0			
40	7	12	12	6	2	2	8	0	2	−8	0			
41	8	11	11	8	1	1	8	0	2	−6	0			
44	7	6	12	10	2	2	10	2	2	−8	0	3	−2	0
53	7	7	7	6	5	5	12	0	3	−4	0			
56	10	4	16	12	2	2	12	4	2	−10	0	4	2	2
57	14	14	18	4	4	1	10	2	2	−10	0	3	−5	0
60	12	4	24	8	6	0	12	8	1	−18	•	5	4	4
61	11	11	11	6	4	4	12	0	3	−5	0			
65	16	18	18	8	2	2	12	0	3	−8	0			
69	12	4	16	8	9	0	14	6	2	−11	0	5	4	4
73	22	21	21	4	2	2	12	0	3	−7	0			
76	19	14	20	10	2	2	14	2	3	−8	0	4	−2	2
77	12	2	14	8	6	6	16	4	3	−8	0	5	4	4
85	18	18	18	4	6	6	16	0	4	−8	0			
88	23	12	24	6	4	4	16	4	3	−10	0	5	2	4
89	26	21	21	12	1	1	16	0	4	−4	2			
92	20	4	22	12	4	4	18	6	3	−10	0	6	8	8
93	18	6	18	4	12	3	18	6	3	−10	0	6	5	6
97	34	27	27	4	2	2	16	0	4	−5	2			
101	19	11	11	14	5	5	20	0	5	0	4			
104	25	20	20	18	2	2	20	0	5	−4	4			
105	36	12	44	8	6	0	20	12	2	−22	0	8	12	12
109	27	17	17	6	6	6	20	0	5	−1	4			
113	36	23	23	8	3	3	20	0	5	0	6			
120	34	8	40	12	8	2	24	12	3	−20	0	9	14	14
124	40	16	34	12	2	2	22	6	4	−8	2	7	10	12

D	Z	l_0	l_0^-	a_2	a_3^+	a_3^-	e	τ	χ	c	c^0	χ_-	c_-	c_-^0
129	50	30	34	12	4	1	26	2	6	-2	8	7	3	8
133	34	12	24	4	8	8	24	4	5	-4	4	7	8	10
136	46	32	32	12	2	2	24	0	6	-2	8			
140	38	4	40	20	4	4	28	12	4	-16	0	10	20	20
141	36	8	28	8	15	0	26	10	4	-9	2	9	16	16
156	52	16	48	16	8	2	32	12	5	-16	4	11	18	20
161	64	14	50	16	2	2	32	12	5	-8	4	11	28	28
165	44	4	44	8	16	4	32	16	4	-20	0	12	24	24
168	54	8	48	12	12	0	32	16	4	-16	4	12	28	28
184	74	16	52	12	4	4	36	12	6	-4	8	12	32	32
204	78	28	56	20	12	0	44	12	8	-8	12	14	24	28
220	92	16	64	16	4	4	44	16	7	-4	12	15	44	44
285	96	4	60	16	24	0	56	24	8	-4	16	20	60	60

V. The Rough Classification of Hilbert Modular Surfaces

5. 1 In this chapter we prove Theorems 2 and 3 of the introduction (Chap. I). Our basic reference for the classification is the joint paper [9] with Van de Ven. In the proposition below we state the main classification principle. A (-2)-curve is a non-singular rational curve with self-intersection number -2. An elliptic configuration on an algebraic surface X is a finite set of irreducible curves on X having the same genera and intersection numbers as the configurations occurring as fibres (without exceptional curves) in an elliptic fibration of some surface ([11] Part II). We give a complete list of the elliptic configurations:

A non-singular curve E of genus 1 with $EE=0$; a rational curve E having exactly one singular point (a cusp or a double point) with $EE=0$; a configuration of (-2)-curves with one of the following intersection diagrams

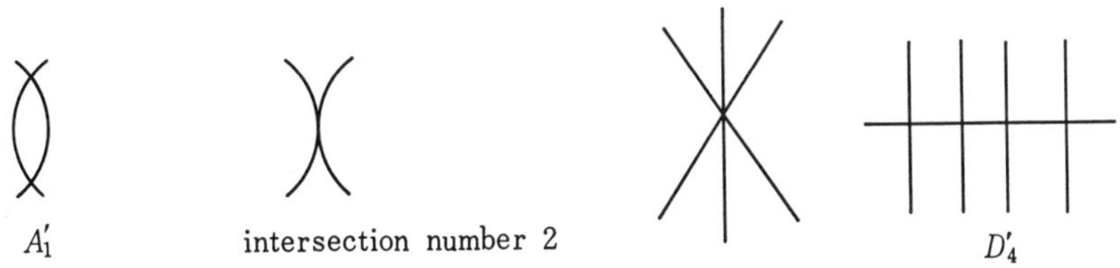

A_1' intersection number 2 D_4'

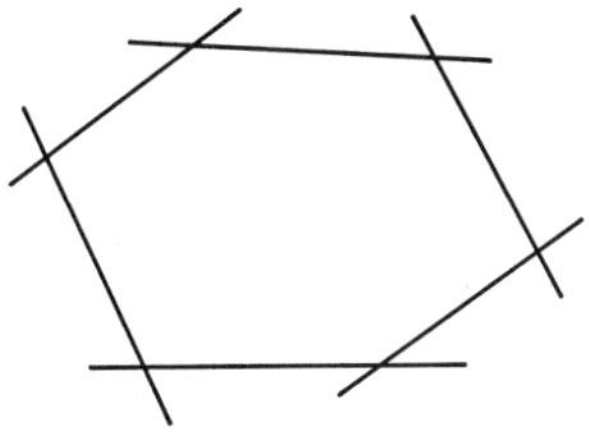

A_{k-1}' (cycle of any length $k \geqq 3$)

or of the diagrams, better indicated by their dual graphs (a dot indicates a (-2)-curve and a line a transversal intersection):

D'_{k-1} (k curves, $k-1 \geqq 5$)

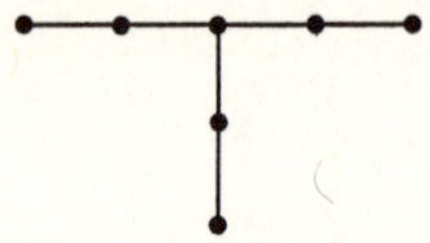

E'_6

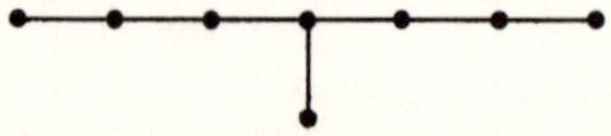

E'_7

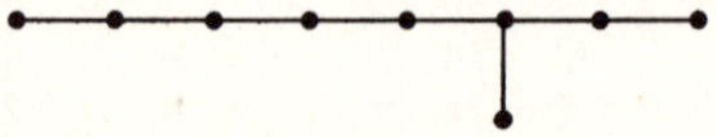

E'_8

Proposition. *Let X be a simply-connected non-rational algebraic surface. If X contains an elliptic configuration, then X is a blown-up $K3$-surface or a blown-up honestly elliptic surface. If X contains an elliptic configuration which intersects a (-2)-curve on X not belonging to the configuration, then X is a blown-up $K3$-surface. If X is simply-connected, not rational, and not a blown-up $K3$-surface and E is an irreducible curve on X such that $c_1 \cdot E \geqq 0$ (where $c_1 \in H^2(X, \boldsymbol{Z})$ is the first Chern class), then either $c_1 \cdot E = 1$ and E is an exceptional curve or $c_1 \cdot E = 0$ and E is either a (-2)-curve or a curve of genus 1 or 0 with $EE=0$.*

The proof is obtained as in [9] (compare Proposition I. 9). For the second part of the proposition we use [9] (Propositions I. 1 and I. 5) and in particular the fact that $c_1 \cdot E \geqq 2$ implies the rationality of X. When passing to the minimal model X' of X, a certain configuration L of rational curves on X is blown down. If $c_1 \cdot E \geq 0$, then either E belongs to L and is an exceptional curve or a (-2)-curve on X, or E and L are disjoint, $c_1 \cdot E = 0$, and E is a component of the unique elliptic fibration of X' or a (-2)-curve on the surface X' of general type.

5. 2 The results and the tables of the preceding chapter have shown that $c_1^2(Y(D)) > 0$ except for 50 discriminants. The arithmetic genus equals 1 for 10 discriminants (5, 8, 12, 13, 17, 21, 24, 28, 33, 60) ; they are among those 50. The corresponding 10 surfaces $Y(D)$ are known to be rational ([6] 4. 5 Theorem). For the remaining 40 discriminants we calculated $c_1^2(Y^0(D))$ (table in Chap. IV) using 3. 7 (21), (22) and obtained $c_1^2(Y^0(D)) > 0$ (which implies general type !) except for the 22 discriminants

(1) 29, 37, 40, 41, 44, 53, 56, 57, 61, 65, 69
73, 76, 77, 85, 88, 92, 93, 105, 120, 140, 165,

for which we get $c_1^2(Y^0(D)) = 0$. These 22 have to be investigated by hand.

The surface $Y_-(D)$ has arithmetic genus 1 for 5 discriminants (5, 8, 12, 13, 17). These surfaces are rational. Namely, except for $D=12$ they are isomorphic to $Y(D)$, and for $D=12$ it was shown in [6] 4. 5 that $Y_-(D)$ is rational. For $D \neq 5$, 8, 12, 13, 17 (i.e. $D > 17$) there are 23 discriminants for which $c_1^2(Y_-(D)) \leqq 0$. For these we consider $Y_-^0(D)$ (see 3. 8) and obtain $c_1^2(Y_-^0(D)) > 0$ except for 15 discriminants (see table in Chap. IV)

(2) 21, 24, 28, 29, 33, 37, 40, 41
44, 53, 57, 61, 65, 73, 85

for which we get $c_1^2(Y_-^0(D))=0$. These 15 surfaces have to be investigated by hand. All other $Y_-(D)$ are rational (5 cases) or of general type.

5. 3 The components of the curves F_N in $Y(D)$ or $Y_-(D)$ all have the same non-singular model if N is admissible (3. 3 and 3. 8). This model is $\overline{\mathfrak{H}/\Gamma_0(N)}$ if N is not divisible by the square free part d of D.

The values of $c_1(\Gamma_0(N)/\{1, -1\})$ (see 3. 7 (17)) are denoted by $c_1(N)$ and were listed in [6] 4. 3 for the case that the genus of $\overline{\mathfrak{H}/\Gamma_0(N)}$ is 0 or 1.

Let c_1 be the first Chern class of $Y(D)$ or $Y_-(D)$ respectively. Then

$$(3) \qquad c_1 \cdot E \geqq c_1(N) + \Sigma\left(\frac{p+q}{(p, q)} - 1\right)$$

for any component E of F_N (N admissible, $N \not\equiv 0 \bmod d$) where the sum is over all the branches of F_N belonging to E near the cusps (see 3. 3 (6)). For (3) compare [6] 4. 5 (34). The sum in (3) equals the intersection number of E with the Chern divisors of the cusps minus $\sigma(\Gamma_0(N))$.

For $N=5, 6, 7, 8, 9$ the curve $\overline{\mathfrak{H}/\Gamma_0(N)}$ is rational and $c_1(N)=0$. In these cases $c_1 \cdot E \geqq 0$ for all components E of F_N. If $c_1 \cdot E=0$ and E is non-singular, then E is a (-2)-curve.

5. 4 In this section we settle the rough classification of the surfaces $Y_-(D)$ using the proposition in 5. 1.

The principal cusp of $Y_-(D)$ has the resolution cycle belonging to the strict ideal class of $(\sqrt{D})$. We consider the reduced quadratic irrationalities

$$(4) \quad w_{(b)} = \frac{b+2+\sqrt{D}}{b+\sqrt{D}}, \qquad b \in \mathbf{Z},\ b \equiv D \bmod 2,\ -\sqrt{D} < b < \sqrt{D}-2.$$

The module $\mathbf{Z}w_{(b)}+\mathbf{Z}$ is strictly equivalent to the ideal $(\sqrt{D})$. We have $w_{(b)} = 2 - \dfrac{1}{w_{(b-2)}}$ for $|b| < \sqrt{D}-2$. Furthermore $w_{(b)}$ is of the form $\dfrac{1}{2N}(M+\sqrt{D})$ with $N = \dfrac{1}{4}(D-b^2)$. Hence we have on $Y_-(D)$ a configuration

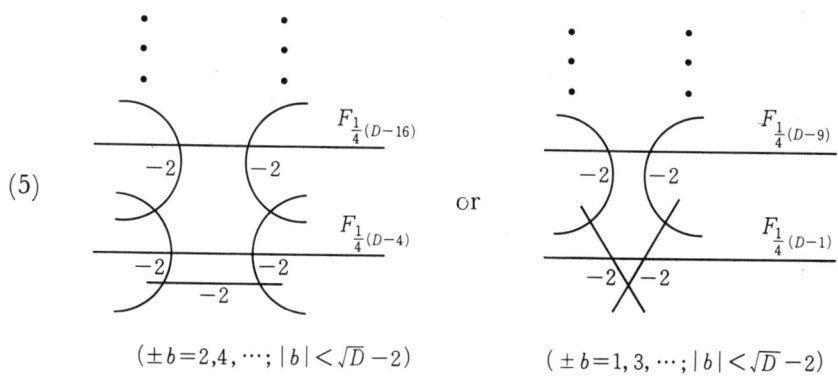

(5)

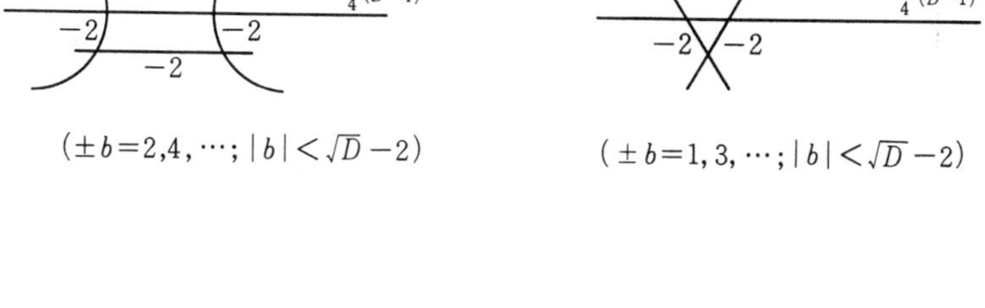

$(\pm b=2,4,\cdots;\ |b|<\sqrt{D}-2)$ $\qquad$ $(\pm b=1, 3, \cdots;\ |b|<\sqrt{D}-2)$

depending on whether D is even or odd. The (-2)-curves belong to the resolution of the cusp. The symmetry in the configutation comes from the canonical involution on $Y_-(D)$ induced by $(z_1, z_2) \mapsto (-z_2, -z_1)$. Two b's differing only up to sign give the same component of $F_{(D-b^2)/4}$. It is carried to itself by the involution ([6] § 4. 5).

For the rest of this section 5. 4 we suppose that $D>17$ so that $Y_-(D)$ is *not rational*. Then we can use (5) for the rough classification as follows.

If in (5) *any of the* $F_{(D-b^2)/4}$ (*with* $|b|<\sqrt{D}-2$) *is* F_5, F_6, F_7, F_8 *or* F_9, *then* $Y_-(D)$ *is a blown-up* K3 *or an honestly elliptic surface.*

Namely, let E be the component drawn in (5) of such an $F_{(D-b^2)/4}$. Then $c_1 \cdot E \geqq 0$, but E is not an exceptional curve, because blowing it down would give by (5) two intersecting exceptional curves which is not possible on a non-rational regular surface.. The proposition in 5. 1 and diagram (5) now show: If $Y_-(D)$ is not a blown-up $K3$-surface, then E is a (-2)-curve (in fact it has to belong to the largest $|b|$ in (5)) and E and the (-2)-curves of the resolved cusp indicated in (5) give a cyclic elliptic configuration proving that the surface is blown-up honestly elliptic.

Let us first consider the values D in (2) for which $\chi(Y_-(D)) \geqq 3$. These are certainly not blown-up $K3$-surfaces. For $D=44$, 53, 57, 61, 85 we get on $Y_-(D)$ a cycle of non-singular rational curves of self-intersection number -2 using F_7, F_7, F_8, F_9, F_9 respectively. For $D=65$ we have a configuration

(6)

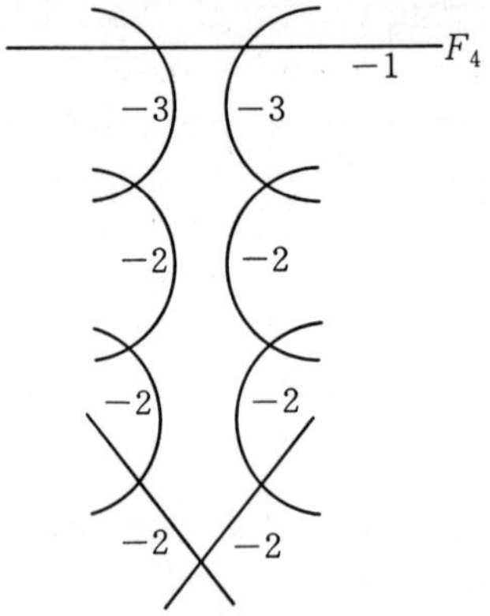

where the F_4 belongs to $w_{(-7)}=(-5+\sqrt{65})/(-7+\sqrt{65})$ and $[w_{(-7)}+1]=3$. Blowing down the component of F_4 drawn in (6) we get again a cyclic elliptic configuration. For $D=73$ see [9] (for prime discriminants D the surfaces $Y(D)$ and $Y_-(D)$ are isomorphic). In all cases we have a cyclic elliptic configuration. The surfaces are blown-up honestly elliptic.

Now we study the eight values in (2) for which $\chi(Y_-(D))=2$. These are 21, 24, 28, 29, 33, 37, 40, 41.

We wish to prove that the corresponding surfaces are blown-up $K3$-surfaces. We may assume that the components of F_5, F_6, F_7, F_8, F_9 occurring in (5) are (-2)-curves, since if they are not, the surfaces are certainly blown-up $K3$-surfaces (Proposition 5. 1). For $D=21$, 24, 29 the curve F_5 occurs in configuration (5). Every component of F_5 passes through a curve A of self-intersection number -2 coming from the resolution of a quotient singularity of order 2, because $\Gamma_0(5)$ has fixed points of order 2 in $\mathfrak{H}$. Thus (5) leads to an elliptic configuration which intersects A, so the surface is a blown-up $K3$-surface. For $D=33$, 37, 40 we have in (5) two different values of b for which $(D-b^2)/4=6$, 7, 8, 9. Hence the surfaces are blown-up $K3$-surfaces.

For $D=28$ we have a configuration

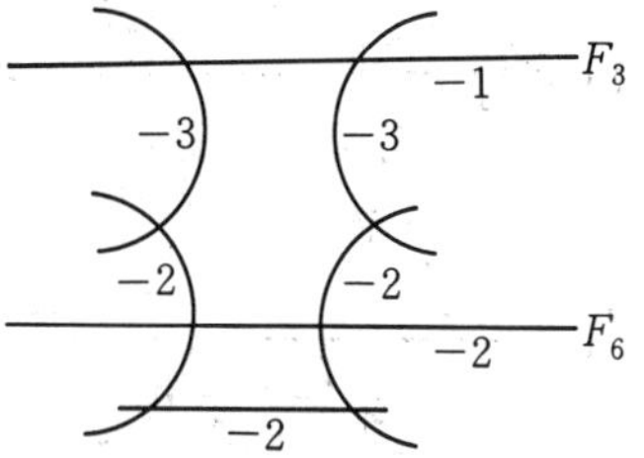

which proves by blowing down F_3 that $Y_-(28)$ is blown-up $K3$. For $D=41$ the same argument works : The curve F_8 occurs and one has to blow down F_4.

Theorem 3 in Chapter I is now completely proved.

5. 5 In this section and in the following one we shall do the rough classification of the surfaces $Y(D)$ and prove Theorem 2 of Chap. I. Since $Y(D)$ is equivalent to $Y_-(D)$ if D is not divisible by a prime$\equiv 3 \bmod 4$, it remains to study the following 13 discriminants from the list (1)

$$\begin{aligned} &D = 44, 56, 57, 69, 105 && (\chi(Y(D)) = 2) \\ &D = 76, 77, 88, 92, 93, 120, 140, 165 && (\chi(Y(D)) \geqq 3). \end{aligned} \tag{7}$$

For these 13 discriminants we indicate the resolution of the cusps with the notation of 3. 3 (5). The reader should consult these diagrams, which are printed at the end of the paper, during the course of the proofs.

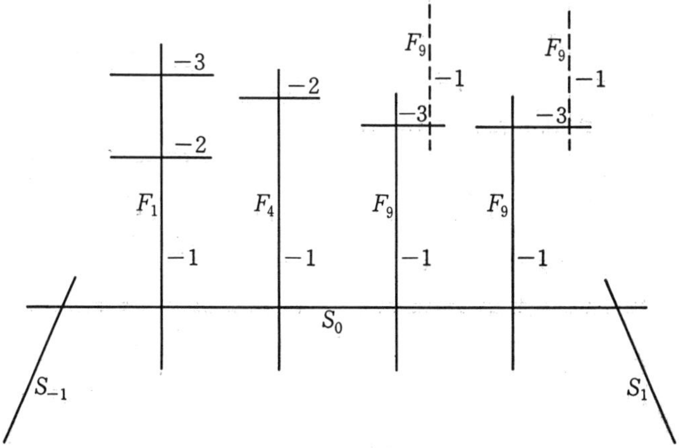

For $D=44, 56, 57, 69, 105$ we consider the curve S_0 in the resolution of the principal cusp (see 3. 6).

The exceptional curves F_1, F_4 and F_9 (for $3|D$) give on $Y(D)$ the configuration in the previous page. The curve of self-intersection -2 intersecting F_4 exists if and only if D is even. The exceptional curves F_9 exist if and only if $3|D$. The dotted components of F_9 intersect the (-3)-curves if and only if $D=105$ (see 3. 7). (It has to be checked for $D=105$ that the two components of F_9 intersecting S_0 do not meet in $\mathfrak{H}^2/G$.) We have $S_0 \cdot S_0 = -b_0 = -[w_0]-1$ and $c_1 \cdot S_0 = -b_0+2$. On passing to $Y^0(D)$ the exceptional curves in the above diagram are blown down successively and we get on $Y^0(D)$ (whose first Chern class we denote by $\tilde{c}_1$) an image curve $\tilde{S}_0$ which has exactly one singular point (a cusp) and for which

$$\tilde{c}_1 \cdot \tilde{S}_0 = -b_0+7+1(\text{if } D \text{ is even})+2(\text{if } 3|D)+2(\text{if } D=105).$$

For $D=44, 56, 57, 69, 105$ the values of b_0 are 8, 8, 9, 9, 11 and we get $\tilde{c}_1 \cdot \tilde{S}_0=0$ and hence $\tilde{S}_0 \cdot \tilde{S}_0=0$. Thus the single curve $\tilde{S}_0$ is an elliptic configuration.

The curve S_1 has for $D=44, 57$ the self-intersection number -2. For $D=69, 105$ the curve S_1 has self-intersection number -3 and intersects F_3, F_4 respectively; therefore in $Y^0(D)$ the image curve $\tilde{S}_1$ has self-intersection number -2. For $D=56$ we have $S_1 \cdot S_1=-4$, but the curve S_1 meets the exceptional curve F_2 which by 3. 7 (19) leads to two blow-downs, so the image curve $\tilde{S}_1$ on $Y^0(56)$ has the self-intersection number -2. Thus by the proposition in 5. 1 (and because $c_1^2(Y^0(D))=0$), the surfaces $Y^0(D)$ are $K3$-surfaces for $D=44, 56, 57, 69, 105$.

5. 6 We now study the discriminants in the second line of (7). In all cases we shall find an elliptic configuration on $Y(D)$ which proves that $Y(D)$ is blown-up honestly elliptic and finishes the proof of Theorem 2 in Chapter I.

For $D=76$ consider the curve F_6. It has one component which meets four curves of the resolution of the cusp. This gives rise to an elliptic configuration:

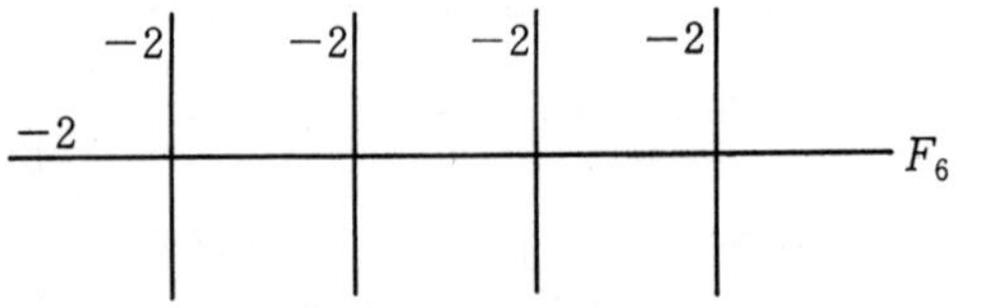

(The proposition in 5. 1 implies that F_6 is a (-2)-curve.)

For $D=77$ the irreducible curve F_{11} passes through the two corners of the resolution of the cusp. The genus of F_{11} is 1. We have $c_1(11)=-2$ and $c_1 \cdot F_{11} \geqq 0$ by (3). By the proposition in 5. 1 the curve F_{11} is an elliptic configuration ($c_1 \cdot F_{11} = F_{11} \cdot F_{11}=0$).

For $D=88$ the two components of F_9 together with 6 curves of the resolved cusp give a cyclic elliptic configuration of length 8.

For $D=92$ the two components of F_{13} pass through the four corners of the resolution of the cusp. We have $c_1(13)=-2$, but (3) implies that $c_1 \cdot E \geqq 0$ for each component E of F_{13}. Since E is a rational curve and meets two (-2)-curves (coming from two quotient singularities of order 2), it follows from the proposition in 5. 1 that E is a (-2)-curve. The curve F_9 was considered in 3. 6. Here 9 is not admissible ; the group Γ in 3. 6 (15) is $\Gamma'(3)/\{1, -1\}$ for which $a_2(\Gamma)=2$, $a_3(\Gamma)=0$, $\sigma(\Gamma)=2$, $e(\overline{\mathfrak{H}/\Gamma})=2$ and $c_1(\Gamma)=0$. The curve F_9 has two components. It follows as before that each component of F_9 is a (-2)-curve. As can be checked, F_9 and F_{13} intersect in $\mathfrak{H}^2/G$ in quotient singularities of order 2. (Condition 3. 3 (10) is satisfied : $(4\cdot 9\cdot 13-10^2)/92=4$.) We have on $Y(D)$ the following elliptic configurations :

where the "vertical" curves come from the resolution of quotient singularities of order 2.

For $D=93$ the two components of F_7 together with 4 curves of the resolved cusp give a cyclic elliptic configuration of length 6.

For $D=120$ the curve F_6 has one component. It passes through both cusps and gives rise to the elliptic configuration :

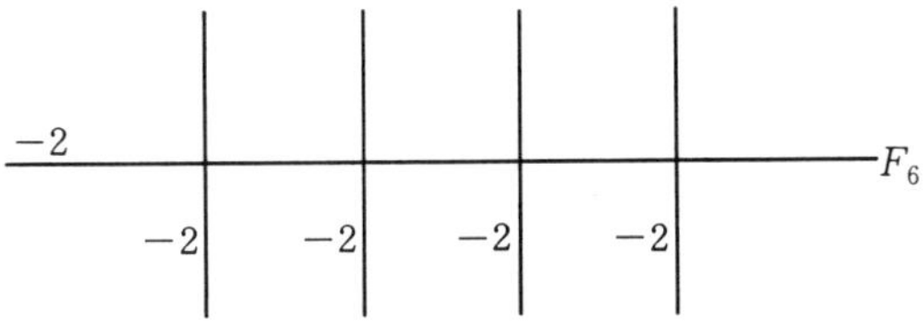

For $D=140$ the curve F_{14} passes through the four corners of the 2 resolved cusps. It has one component. The genus of F_{14} is 1. We have $c_1(14)=-4$, and $c_1\cdot F_{14}=0$ by (3). By proposition 5. 1 the curve F_{14} is an elliptic configuration.

For $D=165$ the same argument works with F_{15}.

$D=44$

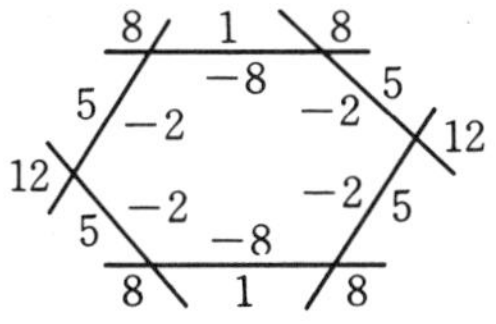

$D=56$

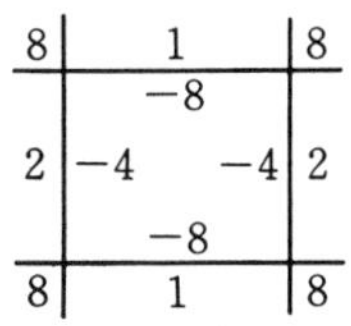

$D=57$

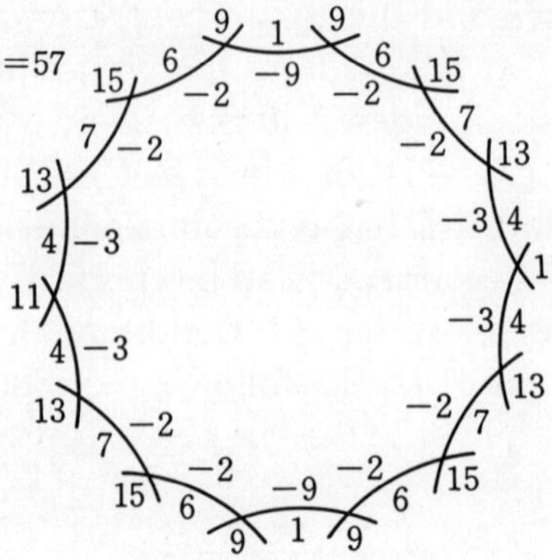

$D=69$

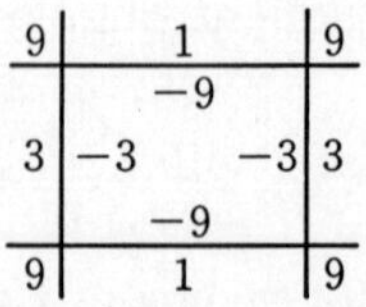

$D=77$

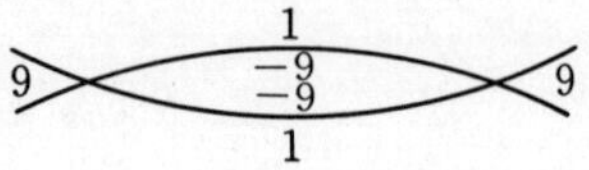

$D=76$

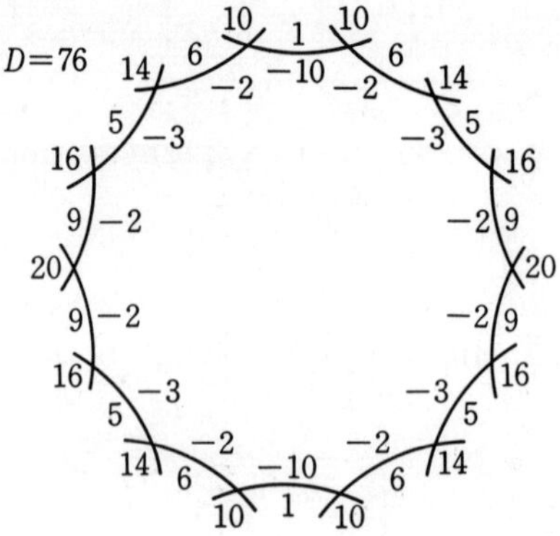

$D=88$

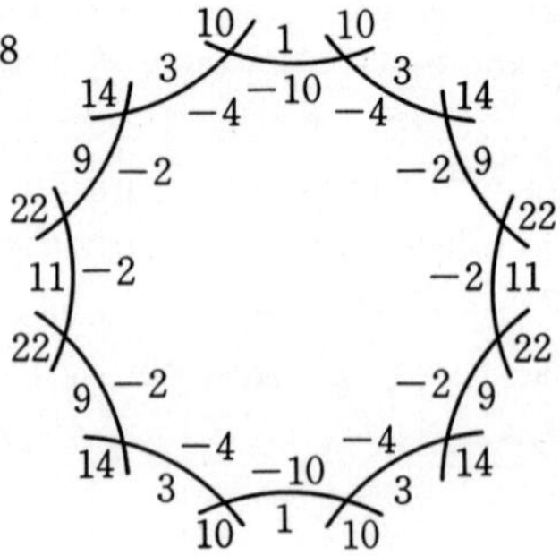

$D=92$

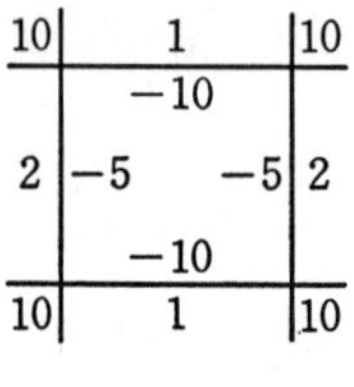

$D=93$

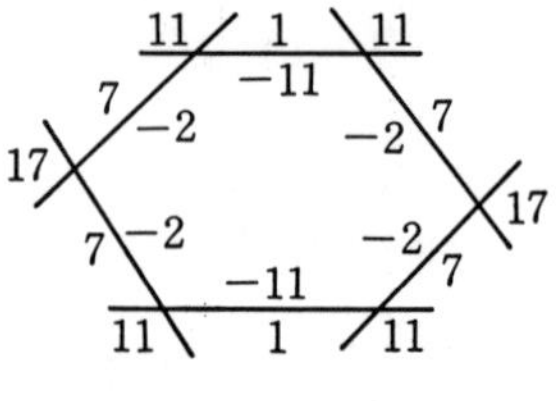

$D=105$
(two cusps)

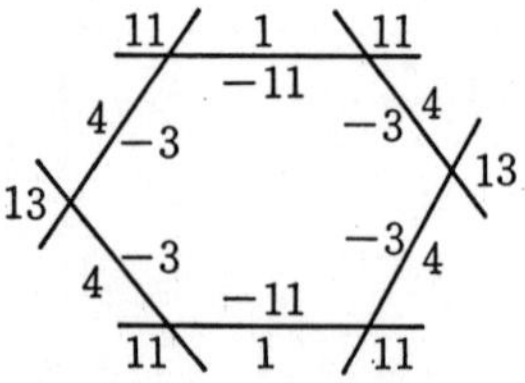

$D=120$
(two cusps)

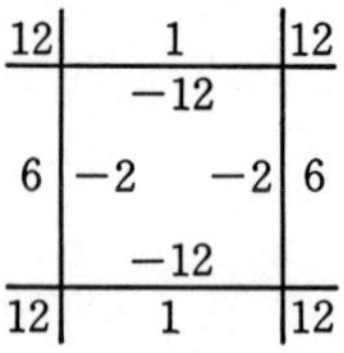

$D=140$
(two cusps)

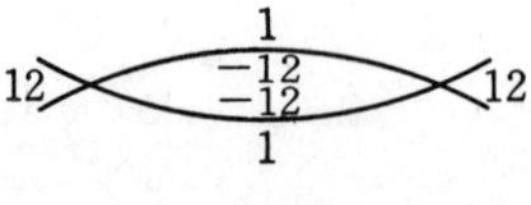

$D=165$
(two cusps)

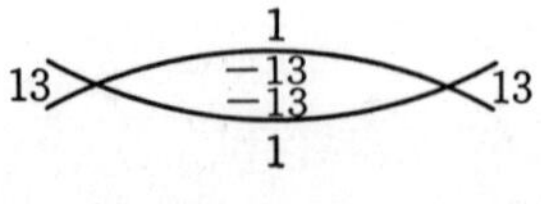

References

[1] Freitag, E. : Über die Struktur der Funktionenkörper zu hyperabelschen Gruppen I. II. J. Reine Angew. Math. (Crelle) **247** (1971), 97–117, **254** (1972), 1–16.

[2] Hammond, W. F. : The Hilbert modular surface of a real quadratic field. Math. Ann. **200** (1973), 25–45.

[3] Hammond, W. F. : The two actions of Hilbert's modular group (to appear).

[4] Hammond, W. F. and Hirzebruch, F. : *L*-Series, modular imbeddings and signatures. Math. Ann. **204** (1973), 263–270.

[5] Hecke, E. : Vorlesungen über die Theorie der algebraischen Zahlen. Akademische Verlagsgesellschaft, Leipzig 1923.

[6] Hirzebruch, F. : Hilbert modular surfaces. Enseignement Math. **29** (1973), 183–281.

[7] Hirzebruch, F. : Kurven auf den Hilbertschen Modulflächen und Klassenzahlrelationen, in Classification of Algebraic Varieties and Compact Complex Manifolds. Lecture Notes in Mathematics, Springer-Verlag, Berlin-Heidelberg-New York (1974), 75–93.

[8] Hirzebruch, F. : Rough classification of the algebraic surfaces arising from the symmetric Hilbert modular group (in preparation).

[9] Hirzebruch, F. and Van de Ven A. : Hilbert modular surfaces and the classification of algebraic surfaces. Inventiones math. **23** (1974), 1–29.

[10] Hurwitz, A. : Die unimodularen Substitutionen in einem algebraischen Zahlkörper. Nachr. Akad. Wiss. Göttingen, Math. Physik Klasse, 332–356 (1895). (Mathematische Werke Bd. II, Birkhäuser Verlag, Basel und Stuttgart 1963, 244–268.)

[11] Kodaira, K. : On compact complex analytic surfaces I, II, III. Ann. of Math. **71** (1960) 111–152, **77** (1963) 563–626, **78** (1963), 1–40.

[12] Kodaira, K. : On the structure of compact complex analytic surfaces I, II, III, IV. Amer. J. Math. **86** (1964), 751–798, **88** (1966), 682–721, **90** (1968), 55–83, **90** (1968), 1048–1066.

[13] Maass, H. : Über die Erweiterungsfähigkeit der Hilbertschen Modulgruppe, Math. Z. **51** (1948), 255–261.

[14] Newman, M. : Bounds for Class Numbers, in Theory of Numbers, Proceedings of Symposia in Pure Mathematics, Vol. VIII, A. M. S., Providence, 1965, 70–77.

[15] Prestel, A. : Die elliptischen Fixpunkte der Hilbertschen Modulgruppen. Math. Ann. **177** (1968), 181–209.

[16] Prestel, A. : Die Fixpunkte der symmetrischen Hilbertschen Modulgruppe zu einem reellquadratischen Zahlkörper mit Primzahldiskriminante. Math. Ann. **200** (1973), 123–139.

[17] Švarčman, O. V. : Simply-connectedness of the factor space of the Hilbert modular group (in Russian), Functional Analysis and its Applications (2) **8** (1974), 99–100.

SFB Theoretische Mathematik
Mathematisches Institut der
Universität Bonn

(Received October 11, 1975)

On Algebraic Surfaces with Pencils of Curves of Genus 2

E. Horikawa[1)]

Introduction

Inspired by Kodaira's work on elliptic surfaces, several authors have studied pencils of curves of genus 2 on compact complex analytic surfaces. We understand that they have established a "local theory" on such pencils. We refer the reader to [7] for a brief account of results and references. We do not use the results of our predecessors in the following.

In this paper we shall study pencils of curves of genus 2 from a little more global point of view. We are more interested in surfaces S which carry these pencils rather than in the pencils themselves. We note that these surfaces are projective algebraic.

Our main results are as follows. Let $g: S \to \Delta$ be a surjective holomorphic map onto a non-singular complete curve Δ whose general fibre C is an irreducible non-singular curve of genus 2. We let K denote the canonical bundle of S. Then, for a sufficiently ample divisor $\mathfrak{m}$ on Δ, the linear system $|K+g^*\mathfrak{m}|$ determines a rational map $\Phi: S \to W^{\natural}$ of degree 2 onto a surface $W^{\natural}$ which is a $\boldsymbol{P}^1$-bundle over Δ. Let $\varpi: \check{S} \to S$ be a composition of quadric transformations such that $\Phi \circ \varpi$ is everywhere defined. We define the branch locus $B^{\natural}$ of Φ to be that of $\Phi \circ \varpi$, which is independent of the choice of ϖ. The singularities of $B^{\natural}$ are classified into six types (see Lemma 6). We can calculate numerical characters of S in terms of $B^{\natural}$ (see Theorems 2 and 3). More accurately these characters are determined by specifying a line bundle $F^{\natural}$ which satisfies $2F^{\natural}=[B^{\natural}]$.

Our approach is summarized as follows. We first define a rational map $f: S \to W$ of degree 2 onto a $\boldsymbol{P}^1$-bundle W over Δ, which induces a double covering $C \to \boldsymbol{P}^1$. This rational map, in turn, determines its branch locus B on W. By applying "elementary transformations" to W, we normalize the singularities of B. By this process, W is transformed into another $\boldsymbol{P}^1$-bundle $W^{\natural}$ over Δ, which turns out to be the image of the rational map Φ associated with $|K+g^*\mathfrak{m}|$ for sufficiently ample $\mathfrak{m}$. If $\Delta=\boldsymbol{P}^1$, $W^{\natural}$ and Φ can be identified respectively with W and f as defined at the beginning. In the final § 5 we shall make a remark on pencils of curves of genus 2 with base points.

1) Supported in part by NSF grant MPS72 05055 A02.

Corresponding results are valid over any algebraically closed field of characteristic $\neq 2$.

This paper is an outgrowth of afterthoughts about [3], § 2. In fact, the whole paper depends only on several lemmas there and other more or less standard results. In addition, the surfaces of type II studied in [3] serve as clarifying examples for the present study.

1. A Fundamental Construction

Let $g: S \rightarrow \Delta$ be a surjective holomorphic map of a compact complex analytic surface S onto a non-singular complete algebraic curve Δ of genus π. We assume that a general fibre of g is an irreducible non-singular curve of genus 2. By results of Kodaira [5], this implies that S is projective algebraic. We also assume that S contains no exceptional curve (of the first kind) in any fibre of g.

Let K denote the canonical bundle of S. For any divisor $\mathfrak{m}$ on Δ, we set

$$a(\mathfrak{m}) = \dim H^0(S, \mathcal{O}(K+g^*\mathfrak{m})).$$

Let p be a point on Δ. Then $b(\mathfrak{m}, p) = a(\mathfrak{m}) - a(\mathfrak{m}-p)$ is lower semi-continuous in p. Hence, if p is a generic point of Δ, $b(\mathfrak{m}, p)$ does not depend on the choice of p, and we denote it by $b(\mathfrak{m})$.

Lemma 1. *For sufficiently ample* $\mathfrak{m}$, *we have*

$$a(\mathfrak{m}) = p_a + 2 \deg \mathfrak{m} + \pi,$$

where p_a *denotes the arithmetic genus of* S.

Proof. By the Riemann-Roch theorem and the Serre duality, we have

$$a(\mathfrak{m}) = p_a + 1 + \deg \mathfrak{m} + \dim H^1(S, \mathcal{O}(-g^*\mathfrak{m})).$$

Hence it suffices to prove

$$\dim H^1(S, \mathcal{O}(-g^*\mathfrak{m})) = \deg \mathfrak{m} - 1 + \pi$$

for sufficiently ample $\mathfrak{m}$. We assume that $\mathfrak{m} = \sum_i p_i$ is a sum of distinct points and that $C_i = g^{-1}(p_i)$ are irreducible and non-singular. In view of the exact sequence

$$0 \rightarrow \mathcal{O}(-g^*\mathfrak{m}) \rightarrow \mathcal{O} \rightarrow \bigoplus_i \mathcal{O}_{C_i} \rightarrow 0,$$

we need to prove

$$\dim \operatorname{Ker}(H^1(S, \mathcal{O}_S) \rightarrow \bigoplus_i H^1(C_i, \mathcal{O}_{C_i})) = \pi.$$

For this purpose we use the exact sequence

$$0 \rightarrow H^1(\Delta, \mathcal{O}_\Delta) \rightarrow H^1(S, \mathcal{O}_S) \rightarrow H^0(\Delta, R^1 g_* \mathcal{O}_S).$$

This implies that the above kernel is at least π-dimensional. Furthermore we have $H^0(\Delta, (R^1 g_* \mathcal{O}_S) \otimes \mathcal{O}(-\mathfrak{m})) = 0$ for sufficiently ample $\mathfrak{m}$. It follows that any element of $H^1(S, \mathcal{O}_S)$ which vanishes on $\sum C_i$ is in the image of $H^1(\Delta, \mathcal{O}_\Delta)$.

The following lemma is an immediate consequence of Lemma 1.

Lemma 2. *For sufficiently ample* $\mathfrak{m}$, *we have* $b(\mathfrak{m})=2$.

Let p be a "variable" point on Δ and $C=g^{-1}(p)$. We let K_C denote the restriction to C of the canonical bundle K of S. If C is non-singular, K_C coincides with the canonical bundle of C.

Let Θ_1 denote the set of those divisors $\mathfrak{m}$ on Δ such that $b(\mathfrak{m})\geqq 1$, and let $\mathfrak{m}_1$ be a minimal element in Θ_1. Then we take a divisor $\mathfrak{m}_2$ such that the images of $H^0(S, \mathcal{O}(K+g^*\mathfrak{m}_1))$ and $H^0(S, \mathcal{O}(K+g^*\mathfrak{m}_2))$ generate $H^0(C, \mathcal{O}(K_C))$ for any general fibre C. Such a divisor exists by Lemma 2. We further assume that $\mathfrak{m}_2$ is minimal among those which have the above property. We set $\mathfrak{d}=\mathfrak{m}_2-\mathfrak{m}_1$.

By our choice of $\mathfrak{m}_1$ and $\mathfrak{m}_2$, we can find $\zeta_i \in H^0(S, \mathcal{O}(K+g^*\mathfrak{m}_i))$, $i=1, 2$, such that their restrictions to C form a basis of $H^0(C, \mathcal{O}(K_C))$. The pair (ζ_1, ζ_2) defines a generically surjective rational map

$$f: S \to \boldsymbol{P}(\mathcal{O}(-\mathfrak{m}_1) \oplus \mathcal{O}(-\mathfrak{m}_2))$$

of degree 2. Here $\boldsymbol{P}(\mathcal{O}(-\mathfrak{m}_1)\oplus\mathcal{O}(-\mathfrak{m}_2))$ denotes the $\boldsymbol{P}^1$-bundle over Δ associated with $\mathcal{O}(-\mathfrak{m}_1)\oplus\mathcal{O}(-\mathfrak{m}_2)$. This $\boldsymbol{P}^1$-bundle is identified with $\boldsymbol{P}(\mathcal{O}(\mathfrak{d})\oplus\mathcal{O})$ and we shall denote it by $\Sigma_\mathfrak{d}$.

Let $\varpi : \tilde{S}\to S$ be a composition of a finite number of quadric transformations such that $f \circ \varpi$ extends to a holomorphic map $\tilde{f}: \tilde{S} \to \Sigma_\mathfrak{d}$. We let $q: \Sigma_\mathfrak{d}\to\Delta$ denote the natural projection and take a section Δ_0 of q whose normal bundle is $[-\mathfrak{d}]$ via an obvious identification. This section will be called the 0-section of $\Sigma_\mathfrak{d}$. Let R denote the ramification divisor of $\tilde{f}$, i.e., if $\tilde{f}$ is given locally by $(z_1, z_2)\to(w_1, w_2)$, R is the divisor of the zeroes of the Jacobian determinant $\partial(w_1, w_2)/\partial(z_1, z_2)$. We define the branch locus B of $\tilde{f}$ to be the direct image $\tilde{f}_*R$ of R as a divisor. We note that B has no multiple component.

Lemma 3. *There exists a line bundle* F *on* $\Sigma_\mathfrak{d}$ *such that* $2F=[B]$, *where* $[B]$ *denotes the line bundle associated with the divisor* B.

Proof. Clearly B is linearly equivalent to $6\Delta_0+q^*\theta$ with some divisor θ on Δ. Let Z be a very ample divisor on $\Sigma_\mathfrak{d}$ such that q induces a biholomorphic map $Z\to\Delta$. We assume that Z intersects B transversally at a finite number of points. Then $\tilde{Z}=\tilde{f}^{-1}(Z)$ is an irreducible non-singular curve and the intersection cycle BZ coincides with the branch locus of $\tilde{Z}\to Z$. From this we infer that the line bundle $[\theta]$ is divisible by 2. q. e. d.

Let F be a line bundle on $\Sigma_\mathfrak{d}$ which satisfies $2F=[B]$. We take a covering $\{U_i\}$ of $\Sigma_\mathfrak{d}$ by sufficiently small open subsets and assume that F is defined by a system of transition functions $\{f_{ij}\}$ on the nerve of the covering $\{U_i\}$. Then we can take the equations $b_i=0$ of B on U_i which satisfy $b_i=f_{ij}^2 b_j$ on $U_i \cap U_j$. Let w_i denote fibre coordinates on F over U_i. We define a subvariety S' of F by the equations $w_i^2=b_i$ over U_i. We call S' the double covering of $\Sigma_\mathfrak{d}$ in F with branch locus B. Since B has no multiple component, S' is a normal complex space. We note that S' depends on the choice of F.

Lemma 4. *We can find a line bundle F on Σ_b satisfying $2F=[B]$ such that there exists a birational holomorphic map from $\tilde{S}$ onto the double covering S' of Σ_b in F with branch locus B.*

Proof. We set $W=\Sigma_b$. Let $\tilde{S}\to S'\to W$ be the Stein decomposition of $\tilde{f}$ and let $f' : S'\to W$ be the induced holomorphic map. Then S' is normal and $\tilde{f}_*\mathcal{O}_{\tilde{S}}=f'_*\mathcal{O}_{S'}$. Since f' is flat ([2], IV, (6. 1. 5)), $f'_*\mathcal{O}_{S'}$ is locally free of rank 2 ([2], IV, (2. 1. 12)). By [2], 0_I, (5. 5. 5), the cokernel of the natural injection $\mathcal{O}_W\to f'_*\mathcal{O}_{S'}$ is invertible. Hence we obtain an exact sequence

$$0\to\mathcal{O}_W\to f'_*\mathcal{O}_{S'}\to\mathcal{O}(-F)\to 0$$

with some line bundle F on W.

Let $\{U_i\}$ be a covering of W by sufficiently small open sets U_i and let $\{1, w_i\}$ be a basis of $f'_*\mathcal{O}_{S'}$ over each U_i. We may assume that w_i^2 is in $\Gamma(U_i, \mathcal{O}_W)$ and we denote it by b_i. Then the branch locus B is defined by the equations $b_i=0$. It follows that F satisfies $2F=[B]$ and that S' is nothing but the double covering of W in F with branch locus B. q. e. d.

2. Elementary Transformations

Let S' denote the double covering of Σ_b in F with branch locus B as in Lemma 4. In this section we shall apply elementary transformations to Σ_b in order to study the singularities of B. We set $W=\Sigma_b$.

Let x be a singular point of B of multiplicity m. Here the multiplicity means the sum of the multiplicities of irreducible components of B at x. Let $p : W_1\to W$ be the quadric transformation with center at x and let $L=p^{-1}(x)$. We set $B_1=p^*B-2[m/2]L$ where $[m/2]$ denotes the greatest integer not exceeding $m/2$. Let S_1 be the double covering of W_1 in $F_1=p^*F-[m/2]\,[L]$ with branch locus B_1. Then we have a natural birational holomorphic map $S_1\to S'$ which is compatible with p (see [3], § 2). Let M be the proper transform of the fibre of $W\to\Delta$ through x. Since M is an exceptional curve on W_1, we can contract M into a non-singular point. Let $q : W_1\to W_2$ be the contraction of M and set $B_2=q_*B_1$. We let n denote the multiplicity of B_2 at $y=q(M)$. In view of the equality $MB_1=6-2[m/2]$, n is equal to $7-2[m/2]$ or $6-2[m/2]$ according as M is a component of B_1 or not. In either case we have $B_1=q^*B_2-2[n/2]M$. Let F_2 be the unique line bundle over W_2 which satisfies $q^*F_2=F_1+[n/2]\,[M]$. Then F_2 satisfies $2F_2=[B_2]$. Therefore we can consider the double covering S_2 of W_2 in F_2 with branch locus B_2. Again we have a birational holomorphic map $S_1\to S_2$ which is compatible with q.

Thus, from the birational point of view, we can replace the triple (W, B, F) by (W_2, B_2, F_2) as above. It is well known that W_2 is a $\boldsymbol{P}^1$-bundle over Δ. We shall call this operation *the elementary transformation at x*.

Lemma 5. *After a finite number of elementary transformations, the triple (Σ_b, B, F) is transformed into another triple (W', B', F') such that B' has no singular point of multiplicity* *greater than* 3.

Proof. If B has a singular point x of multiplicity $\geqq 4$, we apply the elementary transformation at x. Retaining our previous notation, y is of multiplicity $\leqq 3$ on B_2. Therefore our assertion follows from the theorem on resolution of singularities of curves on surfaces. q. e. d.

In view of [3], Lemma 5, we need to study "infinitely near triple points" of B'. Let s_1 be a triple point of B' and let $W_1' \to W'$ be the quadric transformation with center at s_1. If the proper transform B_1' of B' has a triple point over s_1, let us call it s_2. We note that s_2 is unique if it exists. Next let $W_2' \to W_1'$ be the quadric transformation with center at s_2, and look for a triple point over s_2 on the proper transform of B_1'. We continue this process until we obtain a maximal sequence s_1, $s_2, \cdots, s_k$ of triple points over s_1. In this situation we call s_1 a *k-fold triple point*. If $k=1$, we call it a *simple triple point*, while if $k \geqq 2$, we refer to the collection $(s_1, s_2, \cdots, s_k)$ as *infinitely near triple points*.

Lemma 6. *After a finite number of elementary transformations, the triple* (W', B', F') *is transformed into another triple* $(W^\natural, B^\natural, F^\natural)$ *such that* $B^\natural$ *has only the following singularities* (0), (I_k), (II_k), (III_k), (IV_k) *and* (V) *with* $k \geqq 1$.

(0) *A double point or a simple triple point.*

(I_k) *A fibre* Γ *plus two triple points on it (hence these are quadruple points of* $B^\natural$*). Each of these triple points is* $(2k-1)$ *or* $2k$*-fold.*

(II_k) *Two triple points on a fibre, each of which is* $2k$ *or* $(2k+1)$*-fold.*

(III_k) *A fibre* Γ *plus a* $(4k-2)$ *or* $(4k-1)$*-fold triple point on it which has a contact of order* 6 *with* Γ.

(IV_k) *A* $4k$ *or* $(4k+1)$*-fold triple point* x *which has a contact of order* 6 *with the fibre through* x.

(V) *A fibre* Γ *plus a quadruple point* x *on* Γ, *which, after a quadric transformation with center at* x, *results in a double point on the proper transform of* Γ.

Proof. Let x be an m-fold triple point of B' with $m \geqq 2$ and let Γ be the fibre of $W' \to \Delta$ through x. Let $p: W_1' \to W'$ be the quadric transformation with center at x, $L = p^{-1}(x)$, $B_1 = p^* B' - 2L$ and let M be the proper transform of Γ. By the assumption $m \geqq 2$, the proper transform B_1^0 of B' has an $(m-1)$-fold triple point x_1 on L. Let $q: W_1 \to W_2$ be the contraction of M and $B_2 = q_* B_1$.

We first assume that Γ is a component of B'. Then $C = B' - \Gamma$ has a double point at x and the proper transform C_1 of C has a double point at the intersection of L and M. Since we have $B_1 = C_1 + L + M$, B_2 contains $q(L)$ as a component. Moreover $C_2 = q_* C_1$ has a quadruple point at $q(M)$. Thus B_2 has a singularity of type (V).

Next we shall study the case in which Γ is not a component of B'. This case will be divided into two subcases according as x_1 is on M or not.

First assume that x_1 is on M. This implies that B' has a contact of order 6 with Γ at x. Hence, if $m \equiv 0$ or 1 mod 4, B' itself has a singularity of type (IV_k) with $k = [m/4]$. While, if $m \equiv 2$ or 3 mod 4, B_2 contains $q(L)$ as a component and $B_2^0 =$

$B_2-q(L)$ has an m-fold triple point x_2 on $q(L)$. Furthermore, since $B_1^0=B_1-L$ has a contact of order 3 with L at x_1, B_2^0 has a contact of order 6 with $q(L)$ at x_2. Thus B_2 has a singularity of type (III_k) with $k=[m/4]$.

Next we assume that x_1 is not on M. This implies that B' has a contact of order 3 with Γ at x. If B' has another triple point on Γ, let us call it y. We may assume that y is n-fold with $n \leqq m$. If B' has no triple point on Γ other than x, we set $n=0$.

We apply an elementary transformation at x. Then the resulting curve B_2 contains $q(L)$ as a component and $B_2^0=B_2-q(L)$ has two triple points at $q(x_1)$ and $q(M)$. The former one is $(m-1)$-fold and the latter is $(n+1)$-fold. If we perform another elementary transformation at $q(x_1)$, we obtain a curve B_4 with two triple points x_4 and y_4. One is $(m-2)$-fold and the other is $(n+2)$-fold. We note that in this case the fibre through x_4 and y_4 is not a component of B_4.

If $m+n$ is even we obtain singularities of type (I_k) or (II_k) after a finite number of elementary transformations.

We now assume that $m+n$ is odd. In this case the above argument shows that we obtain B_6 of the form either B_6^0 or $B_6^0+\Gamma$ where B_6^0 has $(l+1)$-fold and l-fold triple points on a fibre Γ. We distinguish cases according as l is even or odd.

Assume that l is even. If, in addition, Γ is not a component of B_6, then B_6 has singularities of type (II_k) with $k=l/2$. If Γ is a component of B_6, we apply an elementary transformation at the $(l+1)$-fold triple point of B_6^0. Then we obtain the singularities of type (II_k) with $k=l/2$ again.

If l is odd, a similar argument works. This time we obtain singularities of type (I_k) with $k=(l+1)/2$. This completes the proof of Lemma 6.

Definition. We say that a fibre Γ is a singular fibre of type (I_k), (II_k), (III_k), (IV_k) or (V), if $B^{\natural}$ has the corresponding singularity or singularities on Γ.

3. Canonical Resolution

Let $W^{\natural}$, $B^{\natural}$ and $F^{\natural}$ be as in Lemma 6 and let $S^{\natural}$ be the double covering of $W^{\natural}$ in $F^{\natural}$ with branch locus $B^{\natural}$. The purpose of this section is to study the canonical resolution of singularities of $S^{\natural}$.

If $B^{\natural}$ has a k-fold triple point with $k \geqq 2$ or a quadruple point x, let $p_1: W_1 \to W^{\natural}$ be the quadric transformation with center at x. We set $L_1=p_1^{-1}(x)$, $B_1=p_1^*B^{\natural}-2[m_1/2]L_1$ and $F_1=p_1^*F^{\natural}-[m_1/2][L_1]$, where m_1 denotes the multiplicity of x. Let S_1 be the double covering of W_1 in F_1 with branch locus B_1. If B_1 has a k-fold triple point with $k \geqq 2$ or a quadruple point x_1, we apply the same transformation as above at x_1. Then after a finite number of these transformations we obtain a surface $W^{\flat}$, a curve $B^{\flat}$ and a line bundle $F^{\flat}$ on $W^{\flat}$ such that $2F^{\flat}=[B^{\flat}]$ and such that $B^{\flat}$ has no infinitely near triple points nor quadruple points. Let $S^{\flat}$ be the double covering of $W^{\flat}$ in $F^{\flat}$ with branch locus $B^{\flat}$. Then $S^{\flat}$ has only rational double points as its singularities ([3], Lemma 5) and is birationally equivalent to $S^{\natural}$, hence to the original S.

Lemma 7. *The inverse image on $W^\flat$ of each singular fibre of type* (I_k), (II_k), (III_k), (IV_k) *and* (V) *has the following configuration.*

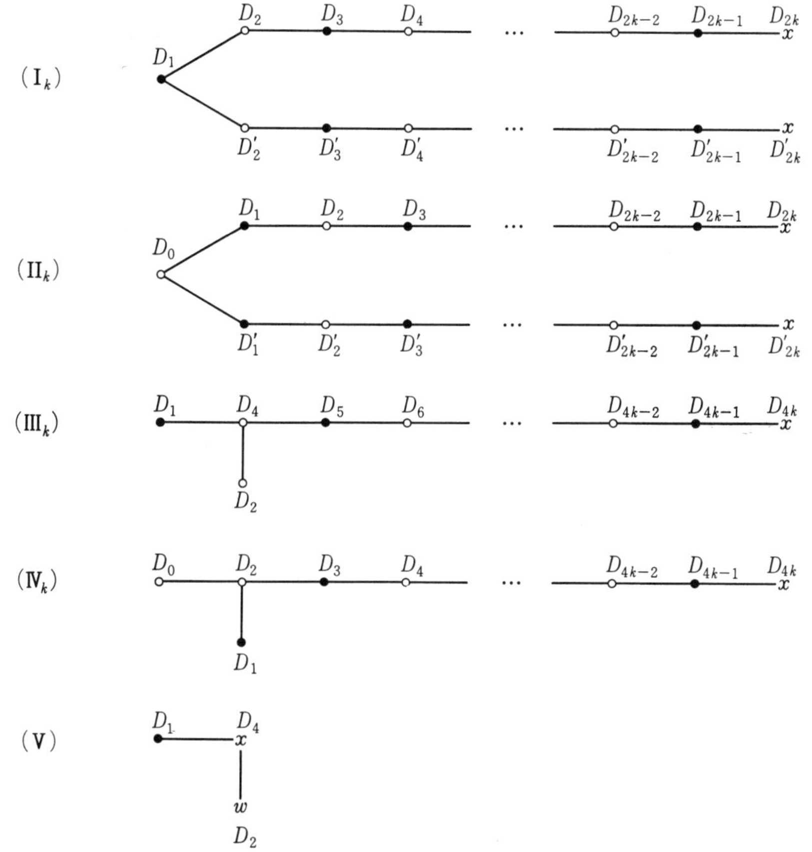

Here •, ∘ or w denotes a non-singular rational curve with self-intersection number -2. • *is a component of $B^\flat$ which is disjoint from other components of $B^\flat$ and ∘ is not a component of $B^\flat$, while a curve denoted by w is a component of $B^\flat$ which intersects with another component of $B^\flat$. Finally x denotes a non-singular rational curve with self-intersection number* -1. *The left end ∘ or • is the proper transform of the corresponding singular fibre. Two curves intersect transversally at a point if and only if they are connected by a segment in the diagram.*

The proof of Lemma 7 is a standard calculation of quadric transforms of curves on surfaces and is left to the reader.

Let $\check{S}\to S^\flat$ be the minimal resolution of singularities of $S^\flat$. $\check{S}$ is called *the canonical resolution* of singularities of $S^\flat$. We let $\check{g}$ denote the projection $\check{S}\to\Delta$. Since there is no exceptional curve in any fibre of $g: S\to\Delta$, there exists a unique birational holomorphic map $\check{S}\to S$ which is compatible with $\check{g}$ and g (see [8]). The following lemma is an immediate consequence of Lemma 7.

Lemma 8. *The inverse image on $\check{S}$ of each singular fibre of type* (I_k), (II_k), (III_k), (IV_k) *or* (V) *has the following configuration.*

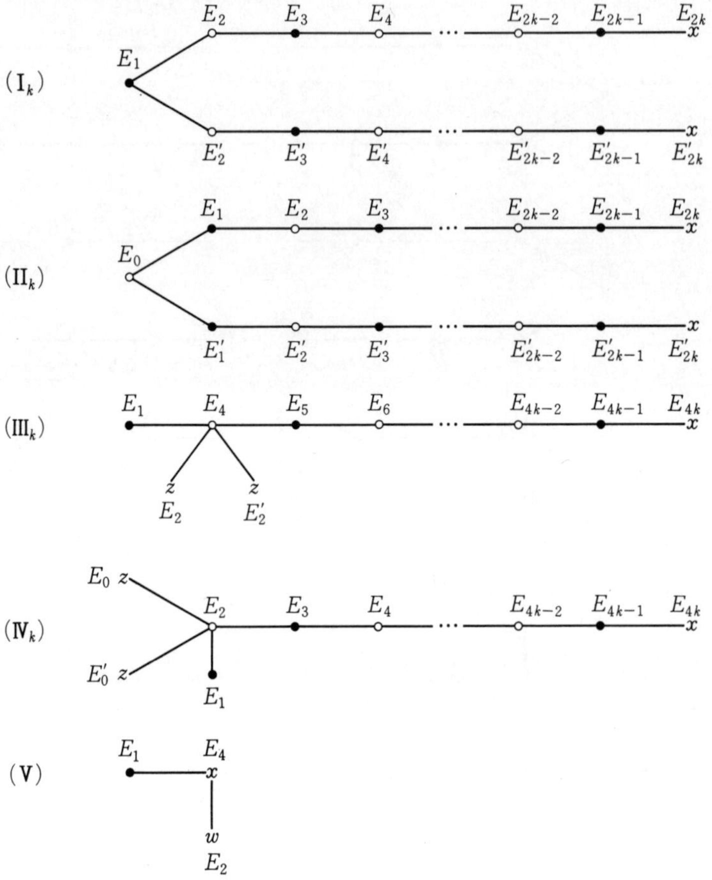

Here $\bullet$, $\circ$ *and* z *denote non-singular rational curves with self-intersection numbers* -1, -4 *and* -2, *respectively. The symbols* x *and* w *denote effective divisors, one of the components of* w *being a non-singular rational curve with self-intersection number* -2.

Let $\psi: W^{\natural}\to\Delta$ be the natural projection and let $\Delta_0^{\natural}$ be a section of ψ with normal bundle $[-\mathfrak{d}^{\natural}]$. We denote by σ the composition of quadric transformations $W^{\flat}\to W^{\natural}$ as above.

Lemma 9. *Let* $\mathfrak{k}$ *be the canonical bundle of* Δ *and write* $F^{\natural}$ *in the form*

$$3[\Delta_0^{\natural}]+\psi^*(-\mathfrak{k}+\mathfrak{f}+[\mathfrak{d}^{\natural}])$$

with some line bundle $\mathfrak{f}$ *on* Δ. *Then the canonical bundle* $\check{K}$ *of* $\check{S}$ *is induced by*

$$\sigma^*([\Delta_0^{\natural}]+\psi^*\mathfrak{f})-[\textstyle\sum \mathcal{G}_i],$$

where the summation extends over all singular fibres Γ_i *except those of type* (0). *We have* $\mathcal{G}_i=\sum_{\alpha\geqq 2}[\alpha/2](D_\alpha+D'_\alpha)$ *for singular fibres of types* (I_k) *and* (II_k), *and* $\mathcal{G}_i=\sum_{\alpha\geqq 2}[\alpha/2]D_\alpha$ *for singular fibres of types* (III_k), (IV_k) *and* (V).

Proof. Let $\mathfrak{K}$ denote the canonical bundle of $W^\flat$. Then $\check{K}$ is induced by $\mathfrak{K}+F^\flat$ (see [3], § 2). Our assertion follows from this fact (cf. Proofs of [3], Lemmas 5 and 6).

q. e. d.

Let $\{\Gamma_1, \Gamma_2, \cdots, \Gamma_N\}$ be the set of all singular fibres except those of type (0) and let $p_i=\psi(\Gamma_i)$. We set $\beta_i=k$ if Γ_i is of type (I_k), (II_k), (III_k) or (IV_k), and $\beta_i=1$ if Γ_i is of type (V). Then we obtain a divisor $\sum\beta_i p_i$ on Δ, and we denote it by $\mathfrak{c}$.

Lemma 10. *The canonical bundle* $\check{K}$ *of* $\check{S}$ *is induced by*

$$\sigma^*([\Delta_0]+\psi^*(\mathfrak{f}-[\mathfrak{c}]))+[\sum\mathscr{F}_i]$$

where the summation runs over all singular fibres except those of type (0). *For each singular fibre* Γ_i, $\mathscr{F}_i$ *is an effective divisor which is a linear combination of* D_α *and* D'_α. *However* D_α *or* D'_α *with maximum index does not appear in* $\mathscr{F}_i$. *Furthermore, if* Γ_i *is of type* (V), D_2 *does not appear in* $\mathscr{F}_i$.

Proof. For each singular fibre Γ_i, the total transform of Γ_i on $W^\flat$ is as follows.

(I_k) $D_1+\sum_{\alpha\geqq 2}(D_\alpha+D'_\alpha)$,

(II_k) $D_0+\sum_{\alpha\geqq 1}(D_\alpha+D'_\alpha)$,

(III_k) $D_1+D_2+2\sum_{\alpha\geqq 4}D_\alpha$,

(IV_k) $D_0+D_1+2\sum_{\alpha\geqq 2}D_\alpha$,

(V) $D_1+D_2+2D_4$.

The assertion of Lemma 10 follows from Lemma 9 and the definition of $\mathfrak{c}$.

q. e. d.

4. Return to the Original Surface

We return to the original pencil $g: S\rightarrow\Delta$.

Theorem 1. *Let* $\mathfrak{m}$ *be a sufficiently ample divisor on* Δ. *Then the linear system* $|K+g^*\mathfrak{m}|$ *defines a rational map of degree* 2 *onto the* $\boldsymbol{P}^1$*-bundle* $W^\sharp$ *over* Δ. *Its branch locus* $B^\sharp$ *has only those singularilies of types* (0), (I_k), (II_k), (III_k), (IV_k) *and* (V) *which are listed in Lemma* 6. *If* $\Delta=\boldsymbol{P}^1$, $W^\sharp$ *is biholomorphically equivalent to* $\Sigma_{\mathfrak{d}}$ *as defined in* § 1 *and* $B^\sharp$ *can be identified with* B.

Proof. Let $\Delta_0^\sharp$ be a section of $\psi: W^\sharp\rightarrow\Delta$ with normal bundle $[-\mathfrak{d}^\sharp]$. We write $F^\sharp=3[\Delta_0^\sharp]+\psi^*(-\mathfrak{k}+\mathfrak{f}+[\mathfrak{d}^\sharp])$ as in Lemma 9. Then, by [3], Lemma 6, the arithmetic genus p_a of $\check{S}$ or S is given by

$$p_a=\deg(2\mathfrak{f}-\mathfrak{d}^\sharp-2\mathfrak{c})-3\pi+2. \tag{1}$$

By Lemma 1, we have

$$a(\mathfrak{m})=\deg(2\mathfrak{f}-\mathfrak{d}^\sharp-2\mathfrak{c}+2\mathfrak{m})-2\pi+2$$

for sufficiently ample $\mathfrak{m}$. Hence we have

$$a(\mathfrak{m}) = \dim H^0(W^\natural, \mathcal{O}(\Delta_0^\natural + \psi^*(\mathfrak{f}-\mathfrak{c}+\mathfrak{m}))),$$

provided that $\mathfrak{m}$ is sufficiently ample. In view of Lemma 10, this implies that there is a natural isomorphism

$$(2) \qquad H^0(\check{S}, \mathcal{O}(\check{K}+\check{g}^*\mathfrak{m})) \cong H^0(W^\natural, \mathcal{O}(\Delta_0^\natural + \psi^*(\mathfrak{f}-\mathfrak{c}+\mathfrak{m}))).$$

This proves our first assertion and the second assertion follows from Lemma 6.

Before proving the last assertion, we prepare with the following

Lemma 11. *There is a natural isomorphism* (2) *for any* $\mathfrak{m}$.

Proof. Let $\sum\check{\mathcal{F}}_i$ denote the pull-back of $\sum\mathcal{F}_i$ in Lemma 10. From what we have seen above, it follows that $\sum\check{\mathcal{F}}_i$ is the fixed part of $|\check{K}+\check{g}^*\mathfrak{m}|$ for sufficiently ample $\mathfrak{m}$. Therefore $\sum\check{\mathcal{F}}_i$ is contained in the fixed part of $|\check{K}+\check{g}^*\mathfrak{m}|$ for any $\mathfrak{m}$. Hence it suffices to prove the following

Lemma 12. *For any line bundle* $\mathfrak{l}$ *on* Δ, *we have a natural isomorphism*

$$H^0(\check{S}, \mathcal{O}(\check{f}^*(\Delta_0^\natural+\psi^*\mathfrak{l}))) \cong H^0(W^\natural, \mathcal{O}(\Delta_0^\natural+\psi^*\mathfrak{l})),$$

where $\check{f}: \check{S} \to W^\natural$ *denotes the natural projection.*

Proof. We note that $\check{S}$ is a double covering of a surface $\check{W}$ in a line bundle $\check{F}$ with a branch locus $\check{B}$. Here $\check{W}$ is obtained from $W^\natural$ by a finite number of quadric transformations. Let $\pi: V \to \check{W}$ be the completion of $\check{F}$ as a $\boldsymbol{P}^1$-bundle. π admits two sections $\check{W}_0$ and $\check{W}_\infty$ such that $[\check{W}_0] = [\check{W}_\infty] + \pi^*\check{F}$. As a divisor of V, $\check{S}$ is linearly equivalent to $2\check{W}_0$.

Let L denote the pull-back on $\check{W}$ of the line bundle $[\Delta_0^\natural]+\psi^*\mathfrak{l}$. We use the following exact sequences on V:

$$0 \to \mathcal{O}(\pi^*L-\check{S}) \to \mathcal{O}(\pi^*L) \to \mathcal{O}_{\check{S}}(\pi^*L) \to 0,$$
$$0 \to \mathcal{O}(\pi^*L-\check{S}) \to \mathcal{O}(\pi^*L-\check{W}_0) \to \mathcal{O}_{\check{W}}(L-\check{F}) \to 0,$$
$$0 \to \mathcal{O}(\pi^*L-\check{W}_0) \to \mathcal{O}(\pi^*L) \to \mathcal{O}_{\check{W}}(L) \to 0.$$

The last two yield an isomorphism

$$H^0(\check{W}, \mathcal{O}_{\check{W}}(L-\check{F})) \cong H^1(V, \mathcal{O}(\pi^*L-\check{S})).$$

In view of the first exact sequence, it is sufficient for our purpose to prove $H^0(\check{W}, \mathcal{O}_{\check{W}}(L-\check{F}))=0$. But this follows from $\check{\Gamma}(L-\check{F}) = -2 < 0$, where $\check{\Gamma}$ denotes a fibre of the projection $\check{W} \to \Delta$. This completes the proof of Lemma 12, hence that of Lemma 11.

We now prove the last assertion of Theorem 1. If $\Delta = \boldsymbol{P}^1$, we may assume that $W^\natural = \Sigma_{\mathfrak{d}^\natural}$. Then we have, by Lemma 11,

$$a(\mathfrak{m}) = \dim H^0(\Delta, \mathcal{O}(\mathfrak{f}-\mathfrak{c}+\mathfrak{m})) + \dim H^0(\Delta, \mathcal{O}(\mathfrak{f}-\mathfrak{c}+\mathfrak{m}-\mathfrak{d}^\natural)).$$

This implies that $\mathfrak{d}$ is linearly equivalent to $\mathfrak{d}^\natural$ or $-\mathfrak{d}^\natural$. Hence $W^\natural$ can be identified with $\Sigma_{\mathfrak{d}}$. q. e. d.

Lemma 11, combined with (1), has as a consequence the following theorem.

Theorem 2. *With the same notations as above, the geometric genus* p_g *and the irregularity* q *of* S *are given by*

$$p_g = \dim H^0(W^\natural, \mathcal{O}(\Delta_0^\natural + \psi^*(\mathfrak{f}-\mathfrak{c}))),$$
$$q = \pi + \dim H^1(W^\natural, \mathcal{O}(\Delta_0^\natural + \psi^*(\mathfrak{f}-\mathfrak{c}))).$$

Our next purpose is to study the relation between $\check{S}$ and S. We note that $\check{S}$ contains the exceptional curves E_α and E'_α, $\alpha=1, 3, 5, \cdots$, over each singular fibre (see Lemma 8).

Lemma 13. *S is obtained from $\check{S}$ by contracting the exceptional curves E_α and E'_α with odd α.*

Proof. First we note the following fact. Let E be an exceptional curve on $\check{S}$ which is contained in a fibre of $\check{g} : \check{S} \to \Delta$. Then E is a fixed component of $|\check{K} + \check{g}^*\mathfrak{m}|$ for any $\mathfrak{m}$.

Let $\bar{S}$ be the surface obtained from $\check{S}$ by contracting E_α and E'_α with odd α and let $\bar{g} : \bar{S} \to \Delta$ be the natural projection. In order to prove our assertion, it suffices to show that there is no exceptional curve in any fibre of $\bar{g}$ (see [8]). But, as we have noted above, such an exceptional curve is the image of a fixed component of $|\check{K} + \check{g}^*\mathfrak{m}|$. Hence it is the image of a component of $\sum \check{\mathcal{F}}_i$ (see the proof of Lemma 11). This is impossible by Lemmas 8 and 10. q. e. d.

This leads us to the following

Theorem 3. *We have*

$$K^2 = 2p_a - 4 + 6\pi + \sum_k \{(2k-1)(\nu(\mathrm{I}_k) + \nu(\mathrm{III}_k)) + 2k(\nu(\mathrm{II}_k) + \nu(\mathrm{IV}_k))\} + \nu(\mathrm{V}),$$

where $\nu(*)$ denotes the number of singular fibres of type $(*)$.

Proof. By [3], Lemma 6, we have

$$\check{K}^2 = 2\deg(2\mathfrak{f} - \mathfrak{d}^\natural - 2\mathfrak{c}).$$

Combining this with (1), we obtain $\check{K}^2 = 2p_a - 4 + 6\pi$. Now the assertion is an immediate consequence of Lemma 13. q. e. d.

We conclude this section with the following theorem.

Theorem 4. *Assume that K is ample. Then the branch locus $B^\natural$ has no singularities except those of type* (I_1).

Proof. From Lemma 8 and the above observations, it follows that a singular fibre of any other type (including that of type (0)) results in a non-singular rational curve with self-intersection -2. q. e. d.

5. A Remark on Base Points

In this section we consider the case in which S contains an exceptional curve E

which is mapped onto Δ(hence we have $\pi=0$). In other words, if $S\to S_0$ is the contraction of E, S_0 carries a pencil of curves of genus 2 with a base point. As far as algebraic surfaces of general type are concerned, we have the following

Theorem 5. *Let S be a minimal algebraic surface of general type which contains a linear pencil $|C|$ of curves of genus 2 with $C^2>0$. Then S has the following numerical characters: $p_g\leqq 2$, $q=0$ and $c_1^2=1$.*

Proof. We borrow an argument from Kodaira [6]. Let K denote the canonical bundle of S. Then we have $KC+C^2=2$. Since both KC and C^2 are positive, they are equal to 1. Hence we have $C(K-C)=0$. Then, by Hodge's index theorem, $(K-C)^2=K^2-1$ is non-positive. Since K^2 is positive, we conclude that $K^2=1$. This implies that $q=0$ and $p_g\leqq 2$ ([1], Theorems 9 and 11 or [6]). q. e. d.

The surfaces with $p_g=2$, $q=0$ and $c_1^2=1$ are studied by Kodaira in [6]. They admit holomorphic maps of degree 2 onto a quadric cone in $\boldsymbol{P}^3$. The inverse images of lines through the vertex of the cone form a pencil of curves of genus 2 which has one base point (see [4], for details).

References

[1] Bombieri, E.: Canonical models of surfaces of general type, Publ. Math. IHES **42** (1973), 171–219.
[2] Grothendieck, A.: Élément de Géométrie Algébriques, Publ. Math. IHES **4**, **8**, ···1960 ff.
[3] Horikawa, E.: On deformations of quintic surfaces, Invent. Math., **31** (1975), 43–85.
[4] ——: Algebraic surfaces of general type with small c_1^2, II, Invent. Math., **37**(1976),121–155.
[5] Kodaira, K.: On compact complex analytic surfaces I, Ann. of Math. **71** (1960), 111–152.
[6] ——: Pluricanonical systems on algebraic surfaces of general type, II, unpublished.
[7] Namikawa, Y.: Studies on degeneration, in Classification of Algebraic Varieties and Compact Complex Manifolds. Lecture Notes in Math. **412**, Springer, Berlin, 1974.
[8] Shafarevich, I. R.: Lectures on Minimal Models and Birational Transformations of Two Dimensional Schemes, Tata Inst. Fund. Res., Bombay, 1966.

Department of Mathematics
University of Tokyo

(Received November 5, 1975)

New Surfaces with No Meromorphic Functions, II

M. Inoue[1)]

0. Introduction

In our previous note [3], we have constructed a kind of (compact complex) surface S satisfying

$$(\alpha) \qquad b_1(S) = 1, \qquad b_2(S) \neq 0$$

and

$$(\beta) \qquad S \text{ is minimal.}$$

As for the significance of these conditions we refer to Kodaira [4], [5] or [2], [3].

In this note, we shall construct other such surfaces and study their properties. They have close connections with the cusps of Hilbert modular surfaces. We shall use some methods devised by Hirzebruch in [1] to resolve the cusp singularities of Hilbert modular surfaces. In particular, we shall refer to [1] for some results on continued fractions and real quadratic number fields.

Notations. As usual, we denote by $\boldsymbol{Z}$, $\boldsymbol{Q}$, $\boldsymbol{R}$, $\boldsymbol{R}^+$, $\boldsymbol{C}$ and $\boldsymbol{C}^*$ the ring of rational integers, the field of rational numbers, the field of real numbers, the set of positive real numbers, the field of complex numbers and the multiplicative group of non-zero complex numbers, respectively. We denote by $[-\infty, \infty]$, $[0, \infty]$ and $[-\infty, 0]$ the intervals which are mapped homeomorphically onto $[-1, 1]$, $[0, 1]$ and $[-1, 0]$, respectively, under the principal value of $\dfrac{2}{\pi}\tan^{-1}$.

1. Algebraic Preliminaries

For a real quadratic irrational number x, we denote by x' the conjugate of x, by $M(x)$ the $\boldsymbol{Z}$-module generated by 1 and x, and by $\boldsymbol{Q}(x)$ the real quadratic number field obtained by adjoining x to $\boldsymbol{Q}$. Let

$$U(x) = \{\alpha \in \boldsymbol{Q}(x) \mid \alpha > 0, \alpha \cdot M(x) = M(x)\},$$
$$U^+(x) = \{\alpha \in U(x) \mid \alpha \cdot \alpha' > 0\}.$$

1) Supported in part by the Sakkôkai Foundation.

Then $U(x)$ and $U^+(x)$ are infinite cyclic groups and

$$[U(x) : U^+(x)] = 1 \quad \text{or} \quad 2.$$

x has a unique development as an infinite (modified) continued fraction which is periodic from a certain point on:

$$x = [[e_0, e_1, \cdots, e_{t-1}, \overline{e_t, \cdots, e_{t+r-1}}]]$$

where $e_i \in \boldsymbol{Z}$, $e_i \geq 2$ for $i \geq 1$ and $e_j \geq 3$ for at least one $j \geq t$. x is purely periodic (namely, $t=0$) if and only if $x > 1 > x' > 0$ (see [1]). Throughout this note we shall fix a real quadratic irrational number ω satisfying

$$\omega > 1 > \omega' > 0.$$

ω is developed uniquely into a purely periodic continued fraction:

$$\omega = [[\overline{n_0, n_1, \cdots, n_{r-1}}]] \tag{1}$$

where r is the *smallest* period and

$$n_i \geq 2, \qquad 0 \leq i \leq r-1,$$
$$n_j \geq 3 \qquad \text{for at least one } j.$$

Let M_1 and M_2 be $\boldsymbol{Z}$-modules of rank 2 in $\boldsymbol{Q}(\omega)$. We say that M_1 and M_2 are *strictly equivalent* if there exists $\delta \in \boldsymbol{Q}(\omega)$ such that $\delta > 0$, $\delta' > 0$ and $\delta \cdot M_1 = M_2$. For the following two lemmas we refer to § 2 of [1].

Lemma 1. *For any $\boldsymbol{Z}$-module M of rank 2 in $\boldsymbol{Q}(\omega)$, there exists $\eta \in \boldsymbol{Q}(\omega)$ such that M and $M(\eta)$ are strictly equivalent and*

$$\eta > 1 > \eta' > 0. \tag{2}$$

Lemma 2. *Let η_1, η_2 be irrational numbers in $\boldsymbol{Q}(\omega)$ satisfying the condition (2), and let*

$$\eta_1 = [[\overline{m_0, m_1, \cdots, m_{s-1}}]]$$

be the development of η_1 as a purely periodic continued fraction. Then $M(\eta_1)$ and $M(\eta_2)$ are strictly equivalent if and only if

$$\eta_2 = [[\overline{m_t, m_{t+1}, \cdots, m_{t+s-1}}]]$$

for some $t \in \boldsymbol{Z}$, $0 \leq t \leq s-1$.

The following proposition plays an essential role in our construction of surfaces.

Proposition (1.1). (Hirzebruch). i) *For ω as above, there exist ω^*, $\beta \in \boldsymbol{Q}(\omega)$ such that $\beta > 0$, $\beta' < 0$, $\beta \cdot M(\omega) = M(\omega^*)$ and $\omega^* > 1 > (\omega^*)' > 0$.*

ii) *$U^+(\omega) = U^+(\omega^*)$, $U(\omega) = U(\omega^*)$.*

iii) *If there exists another γ such that $\gamma > 0$, $\gamma' < 0$ and $\gamma \cdot M(\omega) = M(\omega^*)$, then $\beta \cdot \gamma^{-1}$ belongs to $U^+(\omega)$.*

iv) *Let*

$$\omega^* = [[\overline{m_0, m_1, \cdots, m_{s-1}}]] \tag{3}$$

be the development of ω^* *as a purely periodic continued fraction, where* s *is the smallest period. If there exist* $\eta, \gamma \in \boldsymbol{Q}(\omega)$ *such that* $\gamma>0$, $\gamma'<0$, $\gamma\cdot M(\omega)=M(\eta)$ *and* $\eta>1>\eta'>0$, *then* $\eta=[[\overline{m_t, m_{t+1}, \cdots, m_{t+s-1}}]]$ *for some* $t\in \boldsymbol{Z}$, $0\leq t\leq s-1$.

Proof. ii), iii) and iv) are easily derived from the definitions and Lemma 2. We shall show i). Take $\varepsilon\in\boldsymbol{Q}(\omega)$ such that $\varepsilon>0$, $\varepsilon'<0$. We set $M=\varepsilon\cdot M(\omega)$. Then by Lemma 1 there exist ω^*, $\delta\in\boldsymbol{Q}(\omega)$ such that $\delta>0$, $\delta'>0$, $\delta\cdot M=M(\omega^*)$ and $\omega^*>1>(\omega^*)'>0$. We set $\beta=\delta\cdot\varepsilon$. Then $\beta>0$ and

$$\beta\cdot M(\omega)=\delta\cdot M=M(\omega^*),\qquad \beta'=(\delta\cdot\varepsilon)'=\delta'\cdot\varepsilon'<0,\qquad \text{q. e. d.}$$

We take ω^* and β as in Proposition (1. 1). We take a generator α of the infinite cyclic group $U^+(\omega)$ such that $\alpha>1$, and define *integral* matrices N, N^* and B by

$$\begin{aligned}(\omega, 1)\cdot N&=\alpha\cdot(\omega, 1),\\ (\omega^*, 1)\cdot N^*&=\alpha\cdot(\omega^*, 1),\\ (\omega^*, 1)\cdot B&=\beta\cdot(\omega, 1).\end{aligned}\tag{4}$$

Then we have

$$\det N=\det N^*=\alpha\cdot\alpha'=1,\qquad \det B=-1\tag{5}$$

and

$$B\cdot N=N^*\cdot B.\tag{6}$$

Remark. i) ω^*, β and B are obtained more explicitly as follows: Let $1/\omega=[[e_0, e_1, \cdots, e_{t-1}, \overline{e_t, \cdots, e_{t+s-1}}]]$ be the development of $1/\omega$ as a continued fraction, where s is the smallest period. We define $\eta_1, \eta_2, \cdots, \eta_t\in\boldsymbol{Q}(\omega)$ inductively by

$$1/\omega=e_0-1/\eta_1,\qquad \eta_{i-1}=e_{i-1}-1/\eta_i,\qquad 2\leq i\leq t.$$

Then $\eta_i>0$ and $\eta_t=[[\overline{e_t, \cdots, e_{t+s-1}}]]$. We set

$$\omega^*=\eta_t,\qquad \beta=(\eta_1\cdot\eta_2\cdot\cdots\cdot\eta_t)/\omega,$$
$$B=\begin{bmatrix}e_{t-1}, & 1\\ -1, & 0\end{bmatrix}\cdot\cdots\cdot\begin{bmatrix}e_0, & 1\\ -1, & 0\end{bmatrix}\cdot\begin{bmatrix}0, & 1\\ 1, & 0\end{bmatrix}.$$

Then we can easily prove that

$$\beta>0,\qquad (\omega^*, 1)\cdot B=\beta\cdot(\omega, 1),\qquad \det B=-1.$$

(Compare (9) and (10) below.) Since $\det B=-1$, $\omega-\omega'>0$ and $\omega^*-(\omega^*)'>0$, we obtain

$$\beta\cdot M(\omega)=M(\omega^*),\qquad \beta'<0.$$

ii) If $[U(\omega):U^+(\omega)]=2$, there exists $\alpha_0\in U(\omega)$ such that $\alpha_0>1$ and $\alpha_0'<0$. In this case, we may take ω itself as ω^* and α_0 as β. Then we have

$$r=s,\qquad n_i=m_i \text{ for all } i,$$
$$N=N^*,\ B^2=N.$$

For any integers i, k we define

$$\begin{aligned}n_i&=n_j, & i&\equiv j \text{ (modulo } r), & 0&\leq j\leq r-1,\\ m_k&=m_l, & k&\equiv l \text{ (modulo } s), & 0&\leq l\leq s-1,\end{aligned}$$

where n_j, m_l are the natural numbers in (1), (3). Then we know that

(7) $$n_i,\ m_k \geq 2 \quad \text{for any } i,\ k,$$

(8) $$n_j,\ m_l \geq 3 \quad \text{for at least one } j,\ l.$$

We define ω_i, $\omega_k^* \in \boldsymbol{Q}(\omega)$ inductively by

(9) $$\begin{aligned} \omega_0 &= \omega, & \omega_{i-1} &= n_{i-1} - 1/\omega_i, \\ \omega_0^* &= \omega^*, & \omega_{k-1}^* &= m_{k-1} - 1/\omega_k^*. \end{aligned}$$

Then ω_i, $\omega_k^* > 0$ and

$$\omega_i = \omega_{i+r} \text{ for any } i, \qquad \omega_k^* = \omega_{k+s}^* \text{ for any } k.$$

We set

$$a_i = \begin{cases} (\omega_1 \cdot \cdots \cdot \omega_i)^{-1} & \text{for } i \geq 1 \\ 1 & \text{for } i = 0 \\ \omega_0 \cdot \omega_{-1} \cdot \cdots \cdot \omega_{i+1} & \text{for } i \leq -1, \end{cases}$$

$$b_k = \begin{cases} (\omega_1^* \cdot \cdots \cdot \omega_k^*)^{-1} & \text{for } k \geq 1 \\ 1 & \text{for } k = 0 \\ \omega_0^* \cdot \omega_{-1}^* \cdot \cdots \cdot \omega_{k+1}^* & \text{for } k \leq -1. \end{cases}$$

Then (9) implies

(10) $$\begin{aligned} \begin{bmatrix} a_{i-1}, & a_i \\ a'_{i-1}, & a'_i \end{bmatrix} \cdot \begin{bmatrix} n_{i-1}, & 1 \\ -1, & 0 \end{bmatrix} &= \begin{bmatrix} a_{i-2}, & a_{i-1} \\ a'_{i-2}, & a'_{i-1} \end{bmatrix}, \\ \begin{bmatrix} b_{k-1}, & b_k \\ b'_{k-1}, & b'_k \end{bmatrix} \cdot \begin{bmatrix} m_{k-1}, & 1 \\ -1, & 0 \end{bmatrix} &= \begin{bmatrix} b_{k-2}, & b_{k-1} \\ b'_{k-2}, & b'_{k-1} \end{bmatrix}. \end{aligned}$$

For a proof of the following lemma, we refer to (2. 5) of [1].

Lemma 3. $$1/\alpha = a_r = b_s.$$

Proposition (1. 2).

$$N = \begin{bmatrix} n_{r-1}, & 1 \\ -1, & 0 \end{bmatrix} \cdot \begin{bmatrix} n_{r-2}, & 1 \\ -1, & 0 \end{bmatrix} \cdot \cdots \cdot \begin{bmatrix} n_0, & 1 \\ -1, & 0 \end{bmatrix},$$

$$N^* = \begin{bmatrix} m_{s-1}, & 1 \\ -1, & 0 \end{bmatrix} \cdot \begin{bmatrix} m_{s-2}, & 1 \\ -1, & 0 \end{bmatrix} \cdot \cdots \cdot \begin{bmatrix} m_0, & 1 \\ -1, & 0 \end{bmatrix}.$$

Proof. From (10) it follows that

$$\begin{bmatrix} a_{r-1}, & a_r \\ a'_{r-1}, & a'_r \end{bmatrix} \cdot \begin{bmatrix} n_{r-1}, & 1 \\ -1, & 0 \end{bmatrix} \cdot \cdots \cdot \begin{bmatrix} n_0, & 1 \\ -1, & 0 \end{bmatrix} = \begin{bmatrix} a_{-1}, & a_0 \\ a'_{-1}, & a'_0 \end{bmatrix}.$$

By Lemma 3 and the definition of a_i, we obtain

$$\begin{aligned} a_{r-1} &= \omega/\alpha, & a_r &= 1/\alpha, & a_{-1} &= \omega, & a_0 &= 1, \\ a'_{r-1} &= \omega'/\alpha', & a'_r &= 1/\alpha', & a'_{-1} &= \omega', & a'_0 &= 1. \end{aligned}$$

Hence

$$\begin{bmatrix} \omega, & 1 \\ \omega', & 1 \end{bmatrix} \cdot \begin{bmatrix} n_{r-1}, & 1 \\ -1, & 0 \end{bmatrix} \cdot \cdots \cdot \begin{bmatrix} n_0, & 1 \\ -1, & 0 \end{bmatrix} = \begin{bmatrix} \alpha, & 0 \\ 0, & \alpha' \end{bmatrix} \cdot \begin{bmatrix} \omega, & 1 \\ \omega', & 1 \end{bmatrix},$$

while (4) implies

$$\begin{bmatrix}\omega, & 1\\ \omega', & 1\end{bmatrix}\cdot N = \begin{bmatrix}\alpha, & 0\\ 0, & \alpha'\end{bmatrix}\cdot\begin{bmatrix}\omega, & 1\\ \omega', & 1\end{bmatrix}.$$

Since $\omega-\omega' \neq 0$, we obtain

$$N = \begin{bmatrix}n_{r-1}, & 1\\ -1, & 0\end{bmatrix}\cdot\dots\cdot\begin{bmatrix}n_0, & 1\\ -1, & 0\end{bmatrix}.$$

Similarly we obtain the second equality, q. e. d.

Example. Let $\omega=(3+\sqrt{7})/2$. Then we may choose ω^* and β as follows: $\omega^*=(5+\sqrt{7})/3$, $\beta=(\sqrt{7}-1)/3$. In this case we obtain

$$\omega = [[\overline{3, 6}]], \quad r = 2, \quad n_0 = 3, \quad n_1 = 6,$$
$$\omega^* = [[\overline{3, 3, 2, 2, 2}]], \quad s = 5, \quad m_0 = m_1 = 3, \quad m_2 = m_3 = m_4 = 2,$$
$$\alpha = 8+3\sqrt{7},$$
$$N = \begin{bmatrix}17, & 6\\ -3, & -1\end{bmatrix}, \quad N^* = \begin{bmatrix}23, & 9\\ -18, & -7\end{bmatrix}, \quad B = \begin{bmatrix}1, & 1\\ -1, & -2\end{bmatrix}.$$

2. Construction I

We take two series of infinitely many copies of $\boldsymbol{C}^2$:

$$V_i = \{(u_i, v_i) \in \boldsymbol{C}^2\}, \qquad i \in \boldsymbol{Z},$$
$$W_k = \{(z_k, w_k) \in \boldsymbol{C}^2\}, \qquad k \in \boldsymbol{Z}.$$

We identify $(u_i, v_i) \in V_i$ with $(u_{i-1}, v_{i-1}) \in V_{i-1}$ if and only if

$$\text{(11)} \qquad \begin{aligned} v_i &= v_{i-1}^{n_{i-1}}\cdot u_{i-1},\\ u_i &= 1/v_{i-1}, \qquad u_i \neq 0, \qquad v_{i-1} \neq 0,\end{aligned}$$

and form their union,

$$\mathscr{V} = \bigcup_{i\in \boldsymbol{Z}} V_i.$$

Similarly we form the union of W_k,

$$\mathscr{W} = \bigcup_{k\in \boldsymbol{Z}} W_k,$$

identifying $(z_k, w_k) \in W_k$ with $(z_{k-1}, w_{k-1}) \in W_{k-1}$ if and only if

$$\text{(12)} \qquad \begin{aligned} w_k &= w_{k-1}^{m_{k-1}}\cdot z_{k-1},\\ z_k &= 1/w_{k-1}, \qquad z_k \neq 0, \qquad w_{k-1} \neq 0.\end{aligned}$$

Proposition (2. 1). *$\mathscr{V}$ and $\mathscr{W}$ are Hausdorff spaces with countable bases and, hence, are complex manifolds with $\{V_i\}$ and $\{W_k\}$ as coordinate neighbourhoods.*
(For a proof of this proposition, see (2. 2) of [1].)

Let $\tilde{C}$ be the subvariety of $\mathscr{V}$ defined by

$$\tilde{C} \cap V_i = \{(u_i, v_i) \mid u_i\cdot v_i = 0\} \text{ for any } i \in \boldsymbol{Z}.$$

Then $\tilde{C}$ consists of infinitely many irreducible components $\tilde{C}_i$, $i \in \boldsymbol{Z}$, where $\tilde{C}_i$ is a non-singular rational curve and $\tilde{C}_i \cap V_i = \{(0, v_i)\}$. $\tilde{C}_i$ intersects $\tilde{C}_{i-1}$ *transversally* at the origin P_i of V_i and

$$(\tilde{C}_i)^2 = -n_i, \ \tilde{C}_i \cap \tilde{C}_j = \phi \text{ for } i, j \text{ with } i-j \neq \pm 1, 0.$$

Similarly, let $\tilde{D}$ be the subvariety of $\mathcal{W}$ defined by

$$\tilde{D} \cap W_k = \{(z_k, w_k) \mid z_k \cdot w_k = 0\} \text{ for any } k \in \boldsymbol{Z}.$$

Then $\tilde{D}$ consists of infinitely many irreducible components $\tilde{D}_k$, $k \in \boldsymbol{Z}$, where $\tilde{D}_k$ is a non-singular rational curve and $\tilde{D}_k \cap W_k = \{(0, w_k)\}$. $\tilde{D}_k$ intersects $\tilde{D}_{k-1}$ *transversally* at the origin Q_k of W_k and

$$(\tilde{D}_k)^2 = -m_k, \ \tilde{D}_k \cap \tilde{D}_l = \phi \text{ for } k, l \text{ with } k-l \neq \pm 1, 0.$$

Clearly we have

$$\mathcal{V} - \tilde{C} = \{(u_0, v_0) \in V_0 \mid u_0 \cdot v_0 \neq 0\},$$
$$\mathcal{W} - \tilde{D} = \{(z_0, w_0) \in W_0 \mid z_0 \cdot w_0 \neq 0\}.$$

Proposition (2. 2).

$$|u_i|^{a_i} \cdot |v_i|^{a_{i-1}} = |u_j|^{a_j} \cdot |v_j|^{a_{j-1}}$$
$$|u_i|^{a_i'} \cdot |v_i|^{a_{i-1}'} = |u_j|^{a_j'} \cdot |v_j|^{a_{j-1}'} \quad \textit{on } V_i \cap V_j,$$

and

$$|z_k|^{b_k} \cdot |w_k|^{b_{k-1}} = |z_l|^{b_l} \cdot |w_l|^{b_{l-1}}$$
$$|z_k|^{b_k'} \cdot |w_k|^{b_{k-1}'} = |z_l|^{b_l'} \cdot |w_l|^{b_{l-1}'} \quad \textit{on } W_k \cap W_l.$$

Proof. Each equality is derived from (10), (11) and (12). We shall show only the first one. Without loss of generality we may assume $i > j$. Let

$$\begin{bmatrix} e, & f \\ g, & h \end{bmatrix} = \begin{bmatrix} n_{i-1}, & 1 \\ -1, & 0 \end{bmatrix} \cdot \begin{bmatrix} n_{i-2}, & 1 \\ -1, & 0 \end{bmatrix} \cdots\cdots \begin{bmatrix} n_{j-1}, & 1 \\ -1, & 0 \end{bmatrix}.$$

Then (10) implies

$$(a_{i-1}, a_i) \cdot \begin{bmatrix} e, & f \\ g, & h \end{bmatrix} = (a_{j-1}, a_j).$$

Hence

$$a_j = h \cdot a_i + f \cdot a_{i-1}, \qquad a_{j-1} = g \cdot a_i + e \cdot a_{i-1}.$$

By (11) we have

$$v_i = v_j^e \cdot u_j^f \qquad \text{and} \qquad u_i = v_j^g \cdot u_j^h \quad \text{on } V_i \cap V_j.$$

Therefore

$$\begin{aligned} |u_i|^{a_i} \cdot |v_i|^{a_{i-1}} &= |v_j^g \cdot u_j^h|^{a_i} \cdot |v_j^e \cdot u_j^f|^{a_{i-1}} \\ &= |u_j|^{h \cdot a_i + f \cdot a_{i-1}} \cdot |v_j|^{g \cdot a_i + e \cdot a_{i-1}} \\ &= |u_j|^{a_j} \cdot |v_j|^{a_{j-1}}, \end{aligned}$$

q. e. d.

We define

$$p = |u_i|^{a_i} \cdot |v_i|^{a_{i-1}}, \qquad q = |u_i|^{a_i'} \cdot |v_i|^{a_{i-1}'} \quad \text{on } V_i$$
$$r = |z_k|^{b_k} \cdot |w_k|^{b_{k-1}}, \qquad s = |z_k|^{b_k'} \cdot |w_k|^{b_{k-1}'} \quad \text{on } W_k.$$

From Proposition (2. 2) we infer that p, q and r, s are non-negative *continuous* functions respectively on $\mathcal{V}$ and on $\mathcal{W}$. Moreover we have

$$\tilde{C} = \{P \in \mathcal{V} \mid p(P) = 0\} = \{P \in \mathcal{V} \mid q(P) = 0\},$$
$$\tilde{D} = \{Q \in \mathcal{W} \mid r(Q) = 0\} = \{Q \in \mathcal{W} \mid s(Q) = 0\}.$$

Let $a, b, c, d \in \boldsymbol{Z}$ be the components of the matrix B defined by (4) :

$$B = \begin{bmatrix} a, & b \\ c, & d \end{bmatrix}.$$

Proposition (2. 3). *For any* $(u_0, v_0) \in \mathcal{V} - \tilde{C} = \{(u_0, v_0) \in V_0 \mid u_0 \cdot v_0 \neq 0\}$,

$$r(v_0^c \cdot u_0^d,\ v_0^a \cdot u_0^b) = p(u_0, v_0)^{\beta},$$
$$s(v_0^c \cdot u_0^d,\ v_0^a \cdot u_0^b) = q(u_0, v_0)^{\beta'}.$$

Proof. From the definition of r we have

$$\begin{aligned} r(v_0^c \cdot u_0^d,\ v_0^a \cdot u_0^b) &= |v_0^c \cdot u_0^d|^{b_0} \cdot |v_0^a \cdot u_0^b|^{b_{-1}} \\ &= |v_0^c \cdot u_0^d| \cdot |v_0^a \cdot u_0^b|^{\omega^*} = |u_0|^{b\omega^* + d} \cdot |v_0|^{a\omega^* + c}, \end{aligned}$$

while (4) implies

$$\begin{aligned} b\omega^* + d &= \beta = \beta \cdot a_0, \\ a\omega^* + c &= \beta\omega = \beta \cdot a_{-1}. \end{aligned}$$

Hence

$$\begin{aligned} r(v_0^c \cdot u_0^d,\ v_0^a \cdot u_0^b) &= |u_0|^{\beta a_0} \cdot |v_0|^{\beta a_{-1}} \\ &= (|u_0|^{a_0} \cdot |v_0|^{a_{-1}})^{\beta} = p(u_0, v_0)^{\beta}. \end{aligned}$$

Similarly the second equality follows from (4) and the definition of s, q. e. d.

We form the union $\mathcal{V} \cup \mathcal{W}$ of $\mathcal{V}$ and $\mathcal{W}$ identifying $(u_0, v_0) \in \mathcal{V} - \tilde{C}$ with $(z_0, w_0) \in \mathcal{W} - \tilde{D}$ if and only if

$$\begin{aligned} w_0 &= v_0^a \cdot u_0^b \\ z_0 &= v_0^c \cdot u_0^d, \qquad u_0, v_0, w_0, z_0 \neq 0. \end{aligned} \tag{13}$$

Proposition (2. 4). $\mathcal{V} \cup \mathcal{W}$ *is a Hausdorff space and, hence, is a complex manifold with* $\{V_i, W_k\}$ *as coordinate neighbourhoods.*

Proof. Let Γ be the graph of the identification (13) in $\mathcal{V} \times \mathcal{W}$. To prove this proposition it suffices to show that Γ is closed in $\mathcal{V} \times \mathcal{W}$. We set

$$\Delta = \{(P, Q) \in \mathcal{V} \times \mathcal{W} \mid s(Q) = q(P)^{\beta'},\ q(P) \neq 0\}.$$

Since $\beta' < 0$ and q, s are continuous, Δ is closed in $\mathcal{V} \times \mathcal{W}$. On the other hand, Proposition (2. 3) implies that Γ is a closed subset of Δ. Thus Γ is closed in $\mathcal{V} \times \mathcal{W}$, q. e. d.

We set

$$\rho(P) = \begin{cases} p(P) & \text{for } P \in \mathcal{V}, \\ r(P)^{1/\beta} & \text{for } P \in \mathcal{W}, \end{cases}$$

$$\sigma(P) = \begin{cases} q(P) & \text{for } P \in \mathcal{V}, \\ s(P)^{1/\beta'} & \text{for } P \in \mathcal{W}, \end{cases}$$

where we define $s(P)^{1/\beta'}$ to be ∞ for $P\in\tilde{D}$. Since $\beta'<0$, Proposition (2. 3) implies that ρ and σ are *continuous* mappings of $\mathcal{V}\cup\mathcal{W}$ onto $[0, \infty)$ and onto $[0, \infty]$, respectively. Moreover we have

$$(14)\qquad \begin{aligned} \tilde{C}\cup\tilde{D} &= \{P\in\mathcal{V}\cup\mathcal{W} \mid \rho(P)=0\},\\ \tilde{C} &= \{P\in\mathcal{V}\cup\mathcal{W} \mid \sigma(P)=0\},\\ \tilde{D} &= \{P\in\mathcal{V}\cup\mathcal{W} \mid \sigma(P)=\infty\}. \end{aligned}$$

We note that $\tilde{C}\cap\tilde{D}=\phi$ and

$$\mathcal{V}\cup\mathcal{W}-\tilde{C}-\tilde{D} = \{(u_0, v_0)\in V_0 \mid u_0\cdot v_0\neq 0\}.$$

3. Construction II

We introduce an analytic automorphism g of $\mathcal{V}\cup\mathcal{W}$ as follows:

$$g \text{ sends } (u_i, v_i) \text{ of } V_i \text{ to } (u_i, v_i) \text{ of } V_{i-r},$$
$$g \text{ sends } (z_k, w_k) \text{ of } W_k \text{ to } (z_k, w_k) \text{ of } W_{k-s},$$

where r and s are the smallest periods in (1) and (3). (That g is well-defined is easily derived from (6) and Proposition (1. 2).) Clearly we have

$$(15)\qquad \begin{aligned} g(\tilde{C}_i) &= \tilde{C}_{i-r}, & g(P_i) &= P_{i-r},\\ g(\tilde{D}_k) &= \tilde{D}_{k-s}, & g(Q_k) &= Q_{k-s}. \end{aligned}$$

Proposition (3. 1). *For any* $P\in\mathcal{V}\cup\mathcal{W}$,

$$\rho(g(P)) = \rho(P)^{\alpha},$$
$$\sigma(g(P)) = \sigma(P)^{1/\alpha},$$

where the second equality means $\infty=\infty$ *whenever* $P\in\tilde{D}$.

Proof. These equalities are trivial if $P\in\tilde{C}\cup\tilde{D}$. Hence we may assume $P\in\mathcal{V}\cup\mathcal{W}-\tilde{C}-\tilde{D}=\{(u_0, v_0)\in V_0 \mid u_0\cdot v_0\neq 0\}$. Then $g(P)$ also belongs to $\mathcal{V}\cup\mathcal{W}-\tilde{C}-\tilde{D}$. We write

$$N=\begin{bmatrix} e, & f\\ g, & h\end{bmatrix}.$$

Then Proposition (1. 2) implies

$$g(P) = (v_0^g\cdot u_0^h,\ v_0^e\cdot u_0^f) \quad \text{for} \quad P=(u_0, v_0).$$

Hence by (4) and by the definition of ρ, we obtain

$$\begin{aligned} \rho(g(P)) = p(g(P)) &= |v_0^g\cdot u_0^h|\cdot|v_0^e\cdot u_0^f|^{\omega}\\ &= |u_0|^{f\omega+h}\cdot|v_0|^{e\omega+g} = |u_0|^{\alpha}\cdot|v_0|^{\alpha\omega}\\ &= p(P)^{\alpha} = \rho(P)^{\alpha}. \end{aligned}$$

Similarly we obtain the second equality, q. e. d.

We define an *open* submanifold $\mathcal{D}$ of $\mathcal{V}\cup\mathcal{W}$ as follows:

$$\mathcal{D} = \{P\in\mathcal{V}\cup\mathcal{W} \mid \rho(P)<1\}.$$

By (14) and Proposition (3. 1) we have

$$\mathcal{D} \supset \tilde{C}, \tilde{D},$$
$$g(\mathcal{D}) = \mathcal{D}.$$

Moreover we can easily show that $\mathcal{D}$ is *connected* and *simply connected*. We denote also by the same g the restriction of g to $\mathcal{D}$. We set

$$\tilde{F}(P) = \begin{cases} \log \rho(P) \cdot \log \sigma(P) & \text{for } P \in \mathcal{D} - \tilde{C} - \tilde{D}, \\ +\infty & \text{for } P \in \tilde{C}, \\ -\infty & \text{for } P \in \tilde{D}. \end{cases}$$

From (14) and Proposition (3. 1) we infer that $\tilde{F}$ is a *continuous* mapping of $\mathcal{D}$ onto $[-\infty, \infty]$ and

$$\text{(16)} \qquad \begin{aligned} &\tilde{F}(g(P)) = \tilde{F}(P) \quad \text{for any} \quad P \in \mathcal{D}, \\ &\tilde{F}^{-1}(+\infty) = \tilde{C}, \qquad \tilde{F}^{-1}(-\infty) = \tilde{D}. \end{aligned}$$

Proposition (3. 2). *g generates a properly discontinuous group $\langle g \rangle$ of analytic automorphisms of $\mathcal{D}$ without fixed points.*

Proof. In view of (15) and Proposition (3. 1), it is clear that no automorphism from $\langle g \rangle$ has fixed points in $\mathcal{D}$.

To prove that $\langle g \rangle$ is a properly discontinuous group we have to show that for any $P_1, P_2 \in \mathcal{D}$ there exist neighbourhoods U_1 of P_1 and U_2 of P_2 such that $g^n(U_1) \cap U_2 \neq \phi$ only for finitely many $n \in \boldsymbol{Z}$. By Proposition (3. 1), this is clear if both P_1 and P_2 belong to $\mathcal{D} - \tilde{C} - \tilde{D}$.

If $P_1 \in \tilde{C}$ and $P_2 \in \mathcal{D} - \tilde{C} - \tilde{D}$, we set

$$U_1 = \{P \in \mathcal{D} \mid \tilde{F}(P) > \tilde{F}(P_2) + 1\},$$
$$U_2 = \{P \in \mathcal{D} \mid \tilde{F}(P) < \tilde{F}(P_2) + 1\}.$$

Since $\tilde{F}$ is continuous and $\tilde{F}(g(P)) = \tilde{F}(P)$, U_1 and U_2 are respectively neighbourhoods of P_1 and of P_2, and

$$g^n(U_1) \cap U_2 = \phi \quad \text{for any} \quad n \in \boldsymbol{Z}.$$

Similarly we can show the existence of U_1, U_2 when $P_1 \in \tilde{D}$ and $P_2 \in \mathcal{D} - \tilde{C} - \tilde{D}$.

If $P_1 \in \tilde{C}$ and $P_2 \in \tilde{D}$, we set

$$U_1 = \{P \in \mathcal{D} \mid \tilde{F}(P) > 0\},$$
$$U_2 = \{P \in \mathcal{D} \mid \tilde{F}(P) < 0\}.$$

The U_1 and U_2 are the neighbourhoods we need.

If both of P_1 and P_2 belong to $\tilde{C}$ or belong to $\tilde{D}$, the existence of such neighbourhoods is assured by Hirzebruch in [1, (2. 3)], q. e. d.

We define S_ω to be the quotient space of $\mathcal{D}$ by $\langle g \rangle$:

$$S_\omega = \mathcal{D} / \langle g \rangle .$$

S_ω is a *complex manifold of dimension* 2. Let μ denote the canonical projection of $\mathcal{D}$ onto S_ω. We set

$$C = \mu(\tilde{C}), \qquad C_i = \mu(\tilde{C}_i) \qquad \text{for} \quad i = 0, 1, \cdots, r-1,$$
$$D = \mu(\tilde{D}), \qquad D_k = \mu(\tilde{D}_k) \qquad \text{for} \quad k = 0, 1, \cdots, s-1.$$

By (15), C and D are *compact* subvarieties of S_ω consisting of irreducible components $C_0, C_1, \cdots, C_{r-1}$ and $D_0, D_1, \cdots, D_{s-1}$, respectively. When $r \geq 2$, C is a *cycle consisting of non-singular rational curves* $C_0, C_1, \cdots, C_{r-1}$, where C_i intersects C_j transversally at exactly one point if and only if $j \equiv i \pm 1$ (modulo r) for $r \geq 3$, and C_1 intersects C_2 transversally at exactly two points for $r=2$. Moreover, from (11) we deduce that the self-intersection number C_i^2 is equal to $-n_i$, where n_i satisfies (7) and (8). When $r=1$, C is a *rational curve with one ordinary double point* and the self-intersection number C^2 is equal to $-n_0+2$, where $n_0 \geq 3$. We have a similar situation for D. Moreover we know that $\tilde{C} \cap \tilde{D} = \phi$ and $\tilde{C}$, $\tilde{D}$ are invariant under g. Thus we obtain

$$(17) \qquad \begin{cases} C^2 = -(n_0+n_1+\cdots+n_{r-1}-2r), \\ D^2 = -(m_0+m_1+\cdots+m_{s-1}-2s), \end{cases}$$

$$(18) \qquad C \cdot D = 0,$$

and

Proposition (3. 3). *Let $[C_i \cdot C_j]$ and $[D_k \cdot D_l]$ denote the matrices of the intersection numbers of $C_1, C_2, \cdots, C_{r-1}$ and of $D_1, D_2, \cdots, D_{s-1}$, respectively. Then $[C_i \cdot C_j]$ and $[D_k \cdot D_l]$ are negative definite.*

Remark. From iii) and iv) of Proposition (1. 1) we can easily derive that S_ω does not depend on the choice of ω^* and β and, moreover, depends only on the strict equivalence class of $M(\omega)$.

4. Topological Properties

Let (ξ, ζ) be the coordinates on $\boldsymbol{H} \times \boldsymbol{C}$ where $\boldsymbol{H}$ is the upper half of the complex plane. Let G be the group of analytic automorphisms of $\boldsymbol{H} \times \boldsymbol{C}$ generated by

$$(19) \qquad \begin{aligned} g_0 &: (\xi, \zeta) \rightarrow \left(\alpha \cdot \xi, \frac{1}{\alpha} \cdot \zeta\right), \\ g_1 &: (\xi, \zeta) \rightarrow (\xi + \omega, \zeta + \omega'), \\ g_2 &: (\xi, \zeta) \rightarrow (\xi + 1, \zeta + 1). \end{aligned}$$

We denote by Γ the subgroup generated by g_1 and g_2. Then G is a properly discontinuous group of analytic automorphisms of $\boldsymbol{H} \times \boldsymbol{C}$ without fixed points and Γ is a normal subgroup of G. We set

$$(20) \qquad \begin{aligned} 2\pi\sqrt{-1}\, \xi &= \omega \log v_0 + \log u_0 \qquad \text{and} \\ 2\pi\sqrt{-1}\, \zeta &= \omega' \log v_0 + \log u_0 \quad \text{for} \quad (u_0, v_0) \in \mathcal{D} - \tilde{C} - \tilde{D}. \end{aligned}$$

Then (ξ, ζ) is well-defined modulo Γ and

$$(g^*\xi, g^*\zeta) \equiv \left(\alpha \cdot \xi, \frac{1}{\alpha} \cdot \zeta\right) \quad \text{modulo } \Gamma$$

(compare with the proof of Proposition (3. 1)). Thus (ξ, ζ) induces an analytic isomorphism ϕ of $S_\omega - C - D$ onto $\boldsymbol{H} \times \boldsymbol{C}/G$.

Let (y, x_1, x_2) be the coordinates on $\boldsymbol{R}^+ \times \boldsymbol{R} \times \boldsymbol{R}$ and let $\mathcal{G}$ be the group of homeomorphisms of $\boldsymbol{R}^+ \times \boldsymbol{R} \times \boldsymbol{R}$ generated by

$$\begin{aligned} h_0 &: (y, x_1, x_2) \to \left(\alpha y, \alpha x_1, \frac{1}{\alpha} x_2\right), \\ h_1 &: (y, x_1, x_2) \to (y, x_1+\omega, x_2+\omega'), \\ h_2 &: (y, x_1, x_2) \to (y, x_1+1, x_2+1). \end{aligned} \tag{21}$$

$\mathcal{G}$ is a properly discontinuous group of homeomorphisms of $\boldsymbol{R}^+ \times \boldsymbol{R} \times \boldsymbol{R}$ without fixed points. Let $\mathcal{T} = \boldsymbol{R}^+ \times \boldsymbol{R} \times \boldsymbol{R}/\mathcal{G}$. $\mathcal{T}$ is a compact real 3-dimensional manifold, which is a real 2-torus bundle over a circle.

By (19) and (21), (Im $\xi \cdot$ Im ζ, Im ξ, Re ξ, Re ζ) induces a homeomorphism ψ of $\boldsymbol{H} \times \boldsymbol{C}/G$ onto $\boldsymbol{R} \times \mathcal{T}$. Composing ψ with ϕ, we obtain a *homeomorphism* Φ of $S_\omega - C - D$ onto $\boldsymbol{R} \times \mathcal{T}$. We write $\Phi = (\Phi_0, \Phi_1)$ where $\Phi_0(P) \in \boldsymbol{R}$ and $\Phi_1(P) \in \mathcal{T}$ for $P \in S_\omega - C - D$. By (16) $\tilde{F}$ induces a continuous mapping F of S_ω onto $[-\infty, \infty]$ such that $F^{-1}(\infty) = C$ and $F^{-1}(-\infty) = D$. (20) implies that $F = \Phi_0$ on $S_\omega - C - D$. Hence $F^{-1}(\tau)$ is homeomorphic to $\mathcal{T}$ for any $\tau \in \boldsymbol{R}$. Let $\mathcal{S}$ denote the space obtained from $[-\infty, \infty] \times \mathcal{T}$ by pinching $\infty \times \mathcal{T}$ to a point P_∞ and $-\infty \times \mathcal{T}$ to a point $P_{-\infty}$. Then Φ is extendable to a *continuous* mapping $\tilde{\Phi}$ of S_ω onto $\mathcal{S}$ such that $\tilde{\Phi}(C) = P_\infty$ and $\tilde{\Phi}(D) = P_{-\infty}$. We set

$$\begin{aligned} S_\omega^+ &= F^{-1}([0, \infty]), & \mathcal{S}^+ &= \tilde{\Phi}(S_\omega^+), \\ S_\omega^- &= F^{-1}([-\infty, 0]), & \mathcal{S}^- &= \tilde{\Phi}(S_\omega^-). \end{aligned}$$

$\mathcal{S}^+$ and $\mathcal{S}^-$ are *cones* over $\mathcal{T}$ with vertices P_∞ and $P_{-\infty}$, respectively. $S_\omega^+ \cap S_\omega^-$ is homeomorphic to $\mathcal{T}$. $\tilde{\Phi}$ maps $S_\omega^+ - C$ and $S_\omega^- - D$ homeomorphically onto $\mathcal{S}^+ - P_\infty$ and onto $\mathcal{S}^- - P_{-\infty}$, respectively.

For a pair (X, A) of topological spaces, we denote by $H_\nu(X, A)$ the ν-th homology group of (X, A) with coefficients in $\boldsymbol{Z}$. Since $\mathcal{S}^+$, $\mathcal{S}^-$ are contractible to points P_∞, $P_{-\infty}$, we have

$$H_\nu(\mathcal{S}^+, P_\infty) = H_\nu(\mathcal{S}^-, P_{-\infty}) = 0 \quad \text{for any} \quad \nu.$$

Since C, D are strong deformation retracts of their tubular neighbourhoods, $\tilde{\Phi}$ induces the isomorphisms :

$$\begin{aligned} H_\nu(S_\omega^+, C) &\simeq H_\nu(\mathcal{S}^+, P_\infty), \\ H_\nu(S_\omega^-, D) &\simeq H_\nu(\mathcal{S}^-, P_{-\infty}) \qquad \text{for any} \quad \nu. \end{aligned}$$

Hence we obtain

$$H_\nu(S_\omega^+, C) = H_\nu(S_\omega^-, D) = 0 \qquad \text{for any} \quad \nu.$$

From this and the exact sequences of homology groups we derive

$$H_\nu(S_\omega^+) \cong H_\nu(C), \quad H_\nu(S_\omega^-) \cong H_\nu(D) \qquad \text{for any} \quad \nu.$$

Thus we obtain

$$H_\nu(S_\omega^+) \cong \begin{cases} \boldsymbol{Z} & \text{for} \quad \nu = 0, 1 \\ \boldsymbol{Z}^r & \text{for} \quad \nu = 2 \\ 0 & \text{for} \quad \nu \geq 3, \end{cases} \tag{22}$$

$$(23)\qquad H_\nu(S^-_\omega) \cong \begin{cases} \boldsymbol{Z} & \text{for } \nu = 0, 1 \\ \boldsymbol{Z}^s & \text{for } \nu = 2 \\ 0 & \text{for } \nu \geq 3. \end{cases}$$

From (21) we easily derive

$$(24)\qquad H_\nu(\mathcal{T}) \cong \begin{cases} \boldsymbol{Z} & \text{for } \nu = 0, 2, 3 \\ \boldsymbol{Z} \oplus \boldsymbol{Z}_e \oplus \boldsymbol{Z}_f & \text{for } \nu = 1, \end{cases}$$

where e, f are elementary divisors of the matrix $N-I$.

Proposition (4. 1). i) *S_ω is a compact complex surface.*

ii) $\pi_1(S_\omega) \cong \boldsymbol{Z}$.

iii) $H_\nu(S_\omega) \cong \begin{cases} \boldsymbol{Z} & \textit{for } \nu = 0, 1, 3, 4 \\ \boldsymbol{Z}^r \oplus \boldsymbol{Z}^s & \textit{for } \nu = 2. \end{cases}$

Proof. Since $\{S^+_\omega, S^-_\omega\}$ is an excisive couple, we have the Mayer-Vietoris sequence:

$$(25)\qquad \cdots \to H_\nu(S^+_\omega \cap S^-_\omega) \to H_\nu(S^+_\omega) \oplus H_\nu(S^-_\omega) \to H_\nu(S_\omega) \to \cdots$$

where $S^+_\omega \cap S^-_\omega$ is homeomorphic to $\mathcal{T}$.

i) By (22), (23), (24) and (25), we obtain

$$H_4(S_\omega) \cong H_3(\mathcal{T}) \cong \boldsymbol{Z},$$

while $H_n(X)=0$ for any real n-dimensional *non-compact* manifold X. Hence S_ω is compact.

ii) is trivial because $\mathcal{D}$ is simply connected.

iii) By (22), (23), (24), (25) and ii), we obtain

$$0 \to H_3(S_\omega) \to \boldsymbol{Z} \to \boldsymbol{Z}^r \oplus \boldsymbol{Z}^s \to H_2(S_\omega)$$
$$\to \boldsymbol{Z} \oplus \boldsymbol{Z}_e \oplus \boldsymbol{Z}_f \to \boldsymbol{Z} \oplus \boldsymbol{Z} \to \boldsymbol{Z} \to 0.$$

Since $H_\nu(S_\omega)$ has no torsion part, iii) follows from this exact sequence,

q. e. d.

We denote by $b_\nu(S)$ and $c_2(S)$ the ν-th Betti number and the Euler number of a surface S.

Proposition (4. 2).

$$b_\nu(S_\omega) = 1 \quad \textit{for} \quad \nu = 0, 1, 3, 4,$$
$$b_2(S_\omega) = c_2(S_\omega) = r+s.$$

We define b^+ and b^- to be respectively the numbers of positive and of negative eigen-values of the bilinear form of the intersection pairing on $H_2(S_\omega)$. By (18) and Propositions (3. 3) and (4.2), we have

Proposition (4. 3). i) *$\{C_0, C_1, \cdots, C_{r-1}, D_0, D_1, \cdots, D_{s-1}\}$ is a Betti base of 2-cycles on S_ω with respect to rational coefficients.*

ii) $b^+ = 0$ *and* $b^- = r+s$.

5. Analytical Properties

We denote by K the canonical line bundle of S_ω, by p_g the geometric genus, by q the irregularity and by $[E]$ the complex line bundle determined by a divisor E on S_ω.

Proposition (5. 1). $K=[-C-D]$.

Proof. By (11), (12) and (13), the formula

$$\begin{aligned}\tilde{\Omega} &= (dv_i\wedge du_i)/v_i\cdot u_i = (dv_j\wedge du_j)/v_j\cdot u_j \\ &= (dz_k\wedge dw_k)/z_k\cdot w_k = (dz_l\wedge dw_l)/z_l\cdot w_l\end{aligned}$$

defines a meromorphic 2-form on $\mathcal{D}$ with simple poles on $\tilde{C}$ and $\tilde{D}$. Since $\tilde{\Omega}$ is invariant under g, $\tilde{\Omega}$ induces a meromorphic 2-form Ω on S_ω whose divisor (Ω) is equal to $-C-D$. Hence we obtain $K=[(\Omega)]=[-C-D]$, q. e. d.

We have the following two well-known formulae :

$$\begin{aligned}K^2-2\cdot c_2(S_\omega) &= 3(b^+-b^-), \\ K^2+c_2(S_\omega) &= 12(p_g-q+1)\end{aligned}$$

(the index theorem of Hirzebruch and the formula of Noether, for which we refer to Kodaira [4]). Combining these formulae with Propositions (4. 2), (4. 3) and (5. 1), we obtain

Proposition (5. 2). $p_g=0$, $q=1$ *and* $K^2=-(r+s)$.

From iv) of Proposition (1. 1) it follows that s and $m_0+\cdots+m_{s-1}$ in (3) do not depend on the choice of ω^* and, hence, are uniquely determined by ω.

Proposition (5. 3). i) *Let n_i, r and m_k, s be the natural numbers in* (1) *and* (3). *Then*

$$n_0+\cdots+n_{r-1}+m_0+\cdots+m_{s-1}=3(r+s).$$

ii) If $[U(\omega):U^+(\omega)]=2$, *then*

$$n_0+n_1+\cdots+n_{r-1}=3r.$$

Proof. i) Combining (17), (18) with Proposition (5. 1), we obtain

$$K^2=-(n_0+\cdots+n_{r-1}-2r)-(m_0+\cdots+m_{s-1}-2s),$$

while by Proposition (5. 2) we have $K^2=-(r+s)$. Thus we obtain i).

ii) By remark ii) in § 1, $r=s$ and we may assume $n_i=m_i$ for any i. Hence ii) follows from i), q. e. d.

Let E be a divisor on S_ω. Since $b^+=0$ and $H_2(S_\omega)$ has no torsion part, we obtain that $E^2\leq 0$ and $E^2=0$ if and only if E is homologous to zero with respect to integral coefficients. Moreover, we know that E is homologous to zero if and only if

$[E]$ is a *flat* line bundle (see Kodaira [5, § 10]).

Proposition (5. 4). *S_ω contains no irreducible curves other than $C_0, \cdots, C_{r-1}, D_0, \cdots, D_{s-1}$. In particular, S_ω is minimal and there are no meromorphic functions on S_ω other than constants.*

Proof. Take an irreducible curve E on S_ω and denote by $\pi(E)$ the virtual genus of E:

$$\pi(E) = \frac{1}{2}(E^2+K\cdot E)+1.$$

We assume that $E \neq C_i, D_k$ for all i, k. Since $\pi(E) \geq 0$, $K\cdot E \leq 0$ and $E^2 \leq 0$ where $E^2=0$ if and only if E is homologous to zero, we obtain three cases:

i) $E^2 = -2$, $K\cdot E = 0$ and $\pi(E) = 0$,

ii) $E^2 = -1$, $K\cdot E = -1$, and $\pi(E) = 0$,

iii) $E^2 = 0$, $K\cdot E = 0$ and $\pi(E) = 1$.

In case i), E is a non-singular rational curve and is contained in $S_\omega - C - D$, while $S_\omega - C - D \cong \boldsymbol{H} \times \boldsymbol{C}/G$. This is a contradiction.

In case ii), E is a non-singular rational curve and intersects one and only one, say C_j, of $C_0, \cdots, C_{r-1}, D_0, \cdots, D_{s-1}$. Hence E is contained in $\mu(\mathscr{D} - \bigcup_{i \neq j} \tilde{C}_i - \tilde{D})$, while $\mathscr{D} - \bigcup_{i \neq j} \tilde{C}_i - \tilde{D}$ is an open domain in $V_j = \boldsymbol{C}^2$. This is a contradiction.

In case iii), $[E]$ is a flat line bundle on S_ω. Hence there exists a holomorphic function f on $\mathscr{D}$ such that

$$\mu(\{P \in \mathscr{D} \mid f(P) = 0\}) = E,$$
$$g^*f = \lambda\cdot f \quad \text{for some } \lambda \in \boldsymbol{C}^*, \ \lambda \neq 1.$$

Since each irreducible component of $\tilde{C}$ and of $\tilde{D}$ is a non-singular rational curve, f is constant along $\tilde{C}$ and $\tilde{D}$, while $\tilde{C}$ and $\tilde{D}$ are invariant under g and $\lambda \neq 1$. Thus f vanishes on $\tilde{C}$ and $\tilde{D}$. This evidently contradicts our assumption.

By (7) and (8), C_i and D_k are not exceptional curves of the first kind. Hence S_ω is minimal, q. e. d.

Propositions (4. 2) and (5. 4) imply that S_ω *satisfies the conditions* (α) *and* (β) stated in the introduction.

Example. Take $\omega = (3+\sqrt{7})/2$ (see the example in § 1). Then $b_2(S_\omega) = 7$ and S_ω contains exactly seven non-singular rational curves $C_0, C_1, D_0, D_1, D_2, D_3, D_4$ with

$$C_0^2 = -3, \quad C_1^2 = -6, \quad D_0^2 = D_1^2 = -3, \quad D_2^2 = D_3^2 = D_4^2 = -2.$$

We note that

$$3(r+s) = 21, \quad n_0 + \cdots + n_{r-1} + m_0 + \cdots + m_{s-1} = 21.$$

6. Special Cases with $[U(\omega) : U^+(\omega)] = 2$

We assume $[U(\omega) : U^+(\omega)] = 2$. Then, by remark ii) in § 1, $r = s$ and we may assume $n_i=m_i$ for all i. In this case S_ω admits an analytic involution ι induced by an automorphism $\tilde{\iota}$ on $\mathcal{D}$ defined as follows:

$$\tilde{\iota} \text{ sends } (u_i, v_i) \text{ of } V_i \text{ to } (u_i, v_i) \text{ of } W_i,$$
$$\tilde{\iota} \text{ sends } (z_k, w_k) \text{ of } W_k \text{ to } (z_k, w_k) \text{ of } V_{k-r}.$$

We easily see that ι has no fixed points and $\iota(C_i)=D_i$. Let $\hat{S}_\omega$ denote the quotient surface of S_ω by $\langle\iota\rangle$. Then

$$\pi_1(\hat{S}_\omega) \cong \boldsymbol{Z},$$
$$b_\nu(\hat{S}_\omega) = \begin{cases} 1 & \text{for } \nu = 0, 1, 3, 4, \\ r & \text{for } \nu = 2. \end{cases}$$

We set

$$\hat{C} = (C \cup D)/\langle\iota\rangle, \ \hat{C}_i = (C_i \cup D_i)/\langle\iota\rangle \qquad \text{for} \quad i = 0, 1, \cdots, r-1.$$

When $r \geq 2$, $\hat{C}_i$ is a non-singular rational curve with $(\hat{C}_i)^2=-n_i$. When $r=1$, $\hat{C}=\hat{C}_0$ is a rational curve with one ordinary double point and $(\hat{C})^2=-n_0+2$. By Proposition (5. 4), $\hat{S}_\omega$ contains no irreducible curves other than $\hat{C}_0, \hat{C}_1, \cdots, \hat{C}_{r-1}$. Thus $\hat{S}_\omega$ also satisfies the conditions (α) and (β).

Example. Take $\omega=(3+\sqrt{5})/2$. Then $[U(\omega) : U^+(\omega)]=2$ and

$\alpha_0 =$ a generator of $U(\omega) = (1+\sqrt{5})/2$, $\alpha = \alpha_0^2 = (3+\sqrt{5})/2$, $\omega = [[\overline{3}]]$, $r = 1$.

In this case, $b_2(\hat{S}_\omega)=1$ and $\hat{S}_\omega$ contains exactly one curve $\hat{C}$. $\hat{C}$ is a rational curve with one ordinary double point and $(\hat{C})^2=-1$.

7. Remarks and Acknowledgements

The author is heartily grateful to the mathematicians who have communicated to him the following:

i) The author constructed S_ω first only for ω with $[U(\omega) : U^+(\omega)]=2$. Afterwards F. Hirzebruch has communicated to the author that one can construct S_ω for *any* ω satisfying $\omega>1>\omega'>0$ with the aid of Proposition (1. 1).

ii) The second part of Proposition (5. 3) is obtained by Hirzebruch as a consequence of a theorem on the invariant δ (see [1, § 3]). T. Shintani has communicated that he also has obtained both parts of this proposition as an application of his deep results on special values of zeta functions of totally real algebraic number fields (see [7]).

iii) D. Mumford, I. Nakamura, Y. Namikawa and T. Oda have communicated that one can construct S_ω by the methods of toroidal embeddings. Moreover Oda has pointed out that these methods are applicable to the study of degenerations of S_ω to rational surfaces with singularities (compare Kodaira [6]).

References

[1] Hirzebruch, F. : Hilbert modular surfaces, L'Enseignement Math., t. XIX (1973), 183–282.
[2] Inoue, M. : On surfaces of class VII_0, Inventiones Math., **24** (1974), 269–310.
[3] Inoue, M. : New surfaces with no meromorphic functions, Proceedings of the International Congress of Math., Vancouver (1974).
[4] Kodaira, K. : On the structure of compact complex analytic surfaces, I, Amer. J. Math., **86** (1964), 751–798.
[5] Kodaira, K. : On the structure of compact complex analytic surfaces, II, Amer. J. Math., **88** (1966), 682–721.
[6] Kodaira, K. : On the structure of compact complex analytic surfaces, III, Amer. J. Math., **90** (1968), 55–83.
[7] Shintani, T. : On evaluation of zeta functions of totally real algebraic number fields at non-positive integral places, J. Fac. Sci. Univ. Tokyo, Sec. IA, 23 (1976), 393–417.

Department of Mathematics
Aoyamagakuin University

(Received December 21, 1976)

On the Deformation Types of Regular Elliptic Surfaces

A. Kas

The purpose of this paper is to prove the following theorem :

Theorem. *Let S_1 and S_2 be elliptic surfaces over $\boldsymbol{P}^1$ with no multiple fibres, with at least one singular fibre, and with no exceptional curve contained in a fibre. Then S_1 and S_2 are deformations of one another if and only if S_1 and S_2 have the same geometric genus.*

In particular the theorem implies that S_1 and S_2 are diffeomorphic. As a corollary we show that any elliptic surface satisfying the conditions of the theorem is simply connected.

Let T_n = the set of pairs of polynomials $(g_2(u), g_3(u))$ with deg $g_2(u) \leq 4n$, deg $g_3(u) \leq 6n$ and satisfying the conditions :

(i) $D(u) = 4g_2(u)^3 - 27g_3(u)^2$ is not identically zero ;

(ii) the polynomials $g_2(u), g_2'(u), \cdots, g_2^{(3)}(u), g_3(u), g_3'(u), \cdots, g_3^{(5)}(u)$ have no common zero ;

(iii) either deg $g_2(u) > 4n-4$ or deg $g_3(u) > 6n-6$.

It is clear that T_n may be identified with a Zariski open subset of $\boldsymbol{C}^{10n+2}$.

Let $\boldsymbol{P}^2$ denote a projective plane on which a system of homogeneous coordinates $(x : y : z)$ is fixed. We take two copies $W_0 = \boldsymbol{P}^2 \times \boldsymbol{C}_0$ and $W_1 = \boldsymbol{P}^2 \times \boldsymbol{C}_1$ and form their union $W = W_0 \cup W_1$ by identifying $(x : y : z) \times u \in W_0$ with $(x_1 : y_1 : z_1) \times u_1 \in W_1$ if and only if :

$$uu_1 = 1, \qquad u^{2n}x_1 = x, \qquad u^{3n}y_1 = y, \qquad z_1 = z.$$

Let $\mathcal{B}_n$ be the subvariety of $T_n \times W$ defined by the equations :

$$y^2z = x^3 - g_2(u)xz^2 - g_3(u)z^3, \qquad \text{in } T_n \times W_0$$
$$y_1^2z_1 = x_1^3 - u_1^{4n}g_2(1/u_1)x_1z_1^2 - u_1^{6n}g_3(1/u_1)z_1^3, \text{ in } T_n \times W_1.$$

Let $\Psi : \mathcal{B}_n \rightarrow \boldsymbol{P}^1$ be defined by :

$$\Psi : (x : y : z) \times u \rightarrow u, \qquad \Psi : (x_1 : y_1 : z_1) \times u_1 \rightarrow u_1$$

where $\boldsymbol{P}^1 = \boldsymbol{C}_0 \cup \boldsymbol{C}_1$ identifying $u \in \boldsymbol{C}_0$ with $u_1 \in \boldsymbol{C}_1$ if and only if $uu_1 = 1$. Let $\pi : \mathcal{B}_n \rightarrow T_n$ be the mapping induced by the projection $T_n \times W \rightarrow T_n$. For each point $(g_2, g_3) \in T_n$, let $B_{(g_2, g_3)} = \pi^{-1}(g_2, g_3)$. Then $B_{(g_2, g_3)}$ together with the mapping $\Psi | B_{(g_2, g_3)} : B_{(g_2, g_3)} \rightarrow \boldsymbol{P}^1$ is a basic elliptic surface (i.e. an elliptic surface with a section) with a section defined by:

$$x = z = 0, \text{ in } B_{(g_2, g_3)} \cap W_0$$
$$x_1 = z_1 = 0, \text{ in } B_{(g_2, g_3)} \cap W_1.$$

$B_{(g_2,g_3)}$ may have a finite number of isolated singular points.

The following results are proved in [2]. (1) The only singularities, if any, of $B_{(g_2,g_3)}$ are rational double points. (2) The minimal resolution of $B_{(g_2,g_3)}$ is a basic elliptic surface with no exceptional curve contained in a fibre and with $p_g=n-1$. (3) If B is any basic elliptic surface over $\boldsymbol{P}^1$ with $q=0$ and with no exceptional curve contained in a fibre, then B is the minimal resolution of some $B_{(g_2,g_3)}$ with $(g_2, g_3) \in T_n$ for $n=p_g+1$.

Thus $\pi : \mathcal{B}_n \to T_n$ is a complex analytic family of surfaces with rational double points parametrized by the connected space T_n. It follows from well known results on the simultaneous resolution of surfaces with rational double points ([1]) that if B_1 and B_2 are basic elliptic surfaces over $\boldsymbol{P}^1$ with $q=0$ and with no exceptional curves in their fibres, then B_1 and B_2 are deformations of one another if and only if B_1 and B_2 have the same geometric genus. We will prove that any elliptic surface over $\boldsymbol{P}^1$ satisfying the assumptions of the theorem, may be deformed to a basic elliptic surface.

We recall certain results of Kodaira ([3]). Let $\Psi : B \to \boldsymbol{P}^1$ be a basic elliptic surface with no exceptional curve contained in a fibre and with a section $\sigma : \boldsymbol{P}^1 \to B$. Let $B_0^\sharp$ be the open surface obtained by deleting from B (1) those components of fibres which do not meet the section; and (2) all singular points of fibres of B. Then $B_0^\sharp \to \boldsymbol{P}^1$ is an analytic family of abelian Lie groups such that the identity element of the group $\Psi^{-1}(u) \cap B_0^\sharp$ is $\sigma(u)$. Let $u \to \phi(u)$ be a section of $B_0^\sharp$ over some open subset $U \subset \boldsymbol{P}^1$. Then the automorphism of $\Psi^{-1}(u) \cap B_0^\sharp$ defined by $z \to z+\phi(\Psi(z))$ has a unique extension to an automorphism of $\Psi^{-1}(U)$. Let $\mathcal{O}(B_0^\sharp)$ denote the sheaf of germs of sections of $B_0^\sharp$ over $\boldsymbol{P}^1$, and $\eta \in H^1(\boldsymbol{P}^1, \mathcal{O}(B_0^\sharp))$. An elliptic surface B^η is defined as follows: choose a representative 1-cocyle (η_{ij}) for η with respect to some covering $\{E_i\}$ of $\boldsymbol{P}^1$. Then $B^\eta = \bigcup_i \Psi^{-1}(E_i)$ where we identify z_i in $\Psi^{-1}(E_i)$ with z_j in $\Psi^{-1}(E_j)$ if and only if $\Psi(z_i)=\Psi(z_j)$ and $z_i=L_{ij}(z_j)$ where $L_{ij} : \Psi^{-1}(E_i \cap E_j) \to \Psi^{-1}(E_i \cap E_j)$ is the unique extension of the automorphism $z \to z+\eta_{ij}(\Psi(z))$. Every elliptic surface with no multiple fibres and with no exceptional curve contained in a fibre is of the form B^η.

Let $\mathfrak{f}$ be the normal bundle of the section $\sigma(\boldsymbol{P}^1)$ in B. $\mathfrak{f}$ is the line bundle on $\sigma(\boldsymbol{P}^1)$ of degree $-(p_g+1)$. The stalk of $\mathfrak{f}$ at $u \in \boldsymbol{P}^1$ may be identified with the Lie algebra of $B_0^\sharp \cap \Psi^{-1}(u)$. The exponential map determines a surjective homomorphism of sheaves $e : \mathcal{O}(\mathfrak{f}) \to \mathcal{O}(B_0^\sharp)$. The kernel G of the mapping e is called the homological invariant of B. There is also a more explicit description of G. Let Δ' be the punctured sphere obtained by deleting from $\boldsymbol{P}^1$ those points which correspond to singular fibres of B. Then $\Psi^{-1}(\Delta')$ is differentiably a fibre bundle of elliptic curves. This bundle is determined by its monodromy, i.e. by a representation $R : \pi_1(\Delta') \to \boldsymbol{SL}(2, \boldsymbol{Z})$. Let G' be the locally constant sheaf over Δ' with fibre

$\boldsymbol{Z}\oplus\boldsymbol{Z}$ determined by the monodromy R. G' may be extended to a sheaf G on $\boldsymbol{P}^1$ by defining $\Gamma(U, G)=\Gamma(U\cap\Delta', G')$ for each open set $U\subset\boldsymbol{P}^1$ (i.e., $G=i_*G'$ where $i: \Delta'\to\boldsymbol{P}^1$ is the inclusion mapping). We have an exact sequence of sheaves:

$$0\to G\to\mathcal{O}(\mathfrak{f})\xrightarrow{e}\mathcal{O}(B_0^\sharp)\to 0.$$

The collection of the complex analytic surfaces $\{B^{e(\theta)}\}$, $\theta\in H^1(\boldsymbol{P}^1, \mathcal{O}(\mathfrak{f}))$, forms a complex analytic family. Let $c: H^1(\boldsymbol{P}^1, \mathcal{O}(B_0^\sharp))\to H^2(\boldsymbol{P}^1, G)$ be the coboundary map of the above exact sequence. Then the surface B^η is a deformation of the basic elliptic surface $B(=B^0)$ if $c(\eta)=0$. The above results are in [3], Th. 9. 1, page 603, Th. 9. 2, page 609, Th. 10. 1, page 623, Th. 11. 1, page 1, Th. 11. 2, page 2, Th. 11. 3, 11. 4 on page 5.

There is a duality theorem for the sheaf G (cf. [5]) which implies that $H^2(\boldsymbol{P}^1, G)\cong H_0(\boldsymbol{P}^1, G)=\boldsymbol{Z}\oplus\boldsymbol{Z}/M$ where $M\subset\boldsymbol{Z}\oplus\boldsymbol{Z}$ is the submodule generated by the elements $(m, n)R_\gamma-(m, n)$, $\gamma\in\pi_1(\Delta')$. Let C_a be a singular fibre and let $\gamma\in\pi_1(\Delta')$ be represented by a simple closed loop around a in $\boldsymbol{P}^1$. Then it follows from what we have said that $H^2(\boldsymbol{P}^1, G)=0$ if det $(R_\gamma-1)=\pm1$. From the list of singular fibres and their monodromies ([3], page 604), we conclude that $H^2(\boldsymbol{P}^1, G)=0$ if B contains a singular fibre of type II (a rational curve with a cusp). It follows from [2] that if $B=B_{(g_2,g_3)}$, then $H^2(\boldsymbol{P}^1, G)=0$ if $g_2(u)$ and $g_3(u)$ have a common zero which is a simple zero of $g_3(u)$.

Lemma 1. *If every singular fibre of B is of type* I_1, *then* $H^2(\boldsymbol{P}^1, G)=0$.

Proof. According to [2], every singular fibre of $B=B_{(g_2,g_3)}$ is of type I_1 if and only if $D(u)=4g_2(u)^3-27g_3(u)^2$ has no multiple zeros and is of degree $\geq 12n-1$. Thus for all t in a dense Zariski open subset $V\subset T_n$, the elliptic surface B_t $(t=(g_2, g_3))$ has only singular fibres of type I_1. Moreover it is clear from the definition of G that $H^2(\boldsymbol{P}^1, G_t)$ is constant for $t\in V$ where G_t is the homological invariant of B_t. Let $t_0\in T_n$ be such that B_{t_0} contains a singular fibre of type II. Then $H^2(\boldsymbol{P}^1, G_{t_0})=\boldsymbol{Z}\oplus\boldsymbol{Z}/M_0=0$. It is clear that for t in a sufficiently small neighborhood of t_0, we have $M_0\subset M_t$; thus $H^2(\boldsymbol{P}^1, G_t)=0$ (compare [4], Lemma 3, page 781). Since V is dense in T_n, the lemma is proved.

Now let $B=B_{t_0}$, $t_0\in T_n$, be a basic elliptic surface and let $\eta\in H^1(\boldsymbol{P}^1, \mathcal{O}(B_0^\sharp))$.

Lemma 2. *There exists a neighborhood U of t_0 in T_n, a covering $\{E_i\}$ of $\boldsymbol{P}^1$ and an analytic family $\{\eta_{ij}(t)\}$ of 1-cocycles with coefficients in $\mathcal{O}(B_{0t}^\sharp)$ with respect to covering $\{E_i\}$ such that the $\eta_{ij}(t)$ depend holomorphically on $t\in U$ and η is the cohomology class of $(\eta_{ij}(t_0))$.*

Proof. Let $\gamma=c(\eta)\in H^2(\boldsymbol{P}^1, G_{t_0})$. It is clear that we can find a neighborhood U of t_0 in T_n, a finite open covering $\{E_i\}$ of $\boldsymbol{P}^1$ and a family of 2-cocycles $(\gamma_{ijk}(t))$ depending continuously on t such that $(\gamma_{ijk}(t_0))$ is a representative of γ. Under the inclusion mapping $0\to G\to\mathcal{O}(\mathfrak{f})$, the $(\gamma_{ijk}(t))$ may be regarded as a 2-cocycle in $\mathcal{O}(\mathfrak{f})$ depending holomorphically on t. Thus we may write $\gamma_{ijk}(t)=\theta_{jk}(t)-\theta_{ik}(t)+\theta_{ij}(t)$ where $(\theta_{ij}(t))$ is a 1-cocycle with coefficients in $\mathcal{O}(\mathfrak{f})$ depending holomorphically on

$t \in U$. Finally set $\eta_{ij}(t) = e(\theta_{ij}(t))$.

Finally to prove the theorem, let $S = B^\eta$ be an elliptic surface satisfying the hypothesis of the theorem. Then $B = B_{t_0}$ for some $t_0 \in T_n$, $n = p_g(S) + 1$. Choose U and $(\eta_{ij}(t))$, $t \in U$, as in Lemma 2, and let $\eta(t) \in H^1(\boldsymbol{P}^1, \mathcal{O}(B^\sharp_{0t}))$ be the cohomology class of $(\eta_{ij}(t))$. Then it is clear that the analytic surfaces $\{B_t^{\eta(t)}\}_{t \in U}$ form an analytic family of surfaces. Moreover, for some $t \in U$, we have $H^2(\boldsymbol{P}^1, G_t) = 0$, and therefore $B_t^{\eta(t)}$ may be deformed to a basic elliptic surface. We have already shown that all basic elliptic surfaces of the same geometric genus are deformations of one another. This completes the proof of the theorem.

Corollary. *Let S be an elliptic surface over* $\boldsymbol{P}^1$ *with at least one singular fibre such that S has no multiple fibres. Then S is simply connected.*

Proof. S is obtained by blowing up an elliptic surface S_0 which has no exceptional curves in a fibre. It suffices to prove that S_0 is simply connected. By our theorem, it suffices to exhibit a simply connected elliptic surface satisfying the hypothesis of the theorem with any given value of p_g. We claim in fact that any elliptic surface B with no multiple fibres which has a simply connected fibre is itself simply connected. Let $C_{a_1}, \cdots, C_{a_r}$ be the singular fibres of B and let C be a non-singular fibre. Let $B' = B - \bigcup_{i=1}^{r} C_{a_i}$. It is clear that $\pi_1(B')$ maps surjectively onto $\pi_1(B)$. Since B' is a differentiable fibre bundle over $\boldsymbol{P}^1 - \bigcup_{i=1}^{r} a_i$, it follows that $\pi_1(B')$ is generated by elements $\gamma_1, \gamma_2, \alpha_1, \cdots, \alpha_r$ where γ_1, γ_2 generate $\pi_1(C)$ and α_i is the lifting of a simple loop around a_i in $\boldsymbol{P}^1$. However, the existence of a local section of B at a_i implies that α_i may be chosen so that $\alpha_i \sim 0$ in B. Thus $\pi_1(B)$ is generated by γ_1 and γ_2. The loops γ_1 and γ_2 may be "moved" to a simply connected fibre where they are contractible. Thus $\pi_1(B) = \{1\}$. Now if $g_2(u)$ and $g_3(u)$ are polynomials with a common zero, then the minimal resolution of $B_{(g_2, g_3)}$ contains a simply connected fibre [2]. We can give a more explicit construction. Let Δ be a hyperelliptic curve with a canonical involution $\sigma : \Delta \to \Delta$ where σ has $2g+2$ fixed points, $g = \text{genus } \Delta$. Let E be an elliptic curve, $E = \boldsymbol{C}/L$, with involution $\tau : E \to E$, $\tau(z) = -z$. The surface $\Delta \times E/\sigma \times \tau$ has $8g+8$ ordinary double points. Let S be a minimal resolution and let $\Psi : S \to \boldsymbol{P}^1$ be induced by $\Delta \times E/\sigma \times \tau \to \Delta/\sigma = \boldsymbol{P}^1$. Then S is a basic elliptic surface with $2g+2$ singular fibres of type I_0^* (simply connected). It is not hard to verify that $p_g(S) = g(\Delta)$.

References

[1] Brieskorn, E.: Singular elements of semi-simple algebraic groups, Proc. Inter. Congr. Math., Nice 1970, Gauthier-Villars, Paris (1971).
[2] Kas, A.: Weierstrass normal forms and invariants of elliptic surfaces, to appear.
[3] Kodaira, K.: On compact analytic surfaces, II–III, Ann. of Math., **77** (1963), 563–626 and

78, 1–40.
[4] Kodaira, K. : On the structure of compact complex analytic surfaces, I, Amer. J. Math., **86** (1964), 751–798.
[5] Shioda, T. : On elliptic modular surfaces, J. Math. Soc. Japan, **24**(1972), 20–59.

Department of Mathematics
Oregon State University

(Received November 5, 1975)

On Numerical Campedelli Surfaces

Y. Miyaoka

1. Statement of Main Results

In this paper, a *surface* will mean a compact complex manifold of dimension 2. We denote by $|mK_S|$ $(m \in \boldsymbol{N})$ the pluricanonical system on a surface S, and by Φ_{mK_S} the associated rational map (the pluricanonical map), assuming that $|mK_S|$ is not empty. A surface S is called *of general type* if, for a large number m, $\Phi_{mK_S}(S)$ is a variety of dimension 2 in the projective space $\boldsymbol{P}^N$ ($N = \dim |mK_S|$). We shall study a certain class of surfaces of general type.

Definition. (1) A minimal surface of general type is called a *numerical Campedelli surface* if S satisfies the following numerical conditions :

$$\begin{cases} p_g(S) = \dim H^0(S, \mathcal{O}_S(K_S)) = 0, \\ q(S) = \dim H^1(S, \mathcal{O}_S) = 0, \\ K_S^2 = 2. \end{cases}$$

(2) A numerical Campedelli surface is called a *Campedelli surface* if its fundamental group $\pi_1(S)$ is isomorphic to $\boldsymbol{Z}/(2) \oplus \boldsymbol{Z}/(2) \oplus \boldsymbol{Z}/(2)$ (cf. [2]).

Then we have the following results.

Theorem A. *The tricanonical system* $|3K_S|$ *for a numerical Campedelli surface* S *is free from base points and fixed components.*

Remark. Except for numerical Campedelli surfaces, we can enumerate all surfaces for which the tricanonical maps are not birational (see Bombieri [1] and Miyaoka [4])[1].

Theorem B. *For a Campedelli surface* S, *the universal covering* $\tilde{S}$ *of* S *is birational to a complete intersection of type* (2, 2, 2, 2) *in* $\boldsymbol{P}^6$.

Remark. It is an interesting but, in general, a very difficult problem to determine the complex structures on a given underlying differentiable manifold. However, in the case of Campedelli surfaces, the problem is rather easy and the complex

1) (added in proof) Recently Bombieri has proved that the tricanonical maps of numerical Campedelli surfaces are birational.

structures are completely determined.

2. Preliminaries

If S is a surface of general type, the following results are well-known.

Theorem 1. (Mumford [5]). *If m is sufficiently large, Φ_{mK_S} is a birational morphism and $\Phi_{mK_S}(S) \cong X = \mathrm{Proj} \bigoplus_{r=0}^{\infty} H^0(S, \mathcal{O}_S(rK_S))$. X is a normal variety with only a finite number of rational double points as singularities. If S is a minimal surface, then S is the minimal resolution of X.* (X is called the canonical model of S.)

Theorem 2. (Mumford [5]). *Assume that S is minimal. Then we have*

$$H^1(S, \mathcal{O}_S(mK_S)) = 0, \quad \text{for } m \neq 0, 1, \quad m \in \mathbf{Z}.$$

Theorem 3. (Riemann-Roch Theorem for pluricanonical systems). *Letting $\bar{c}_1^2$ be the self intersection number for the canonical divisor of the minimal model of S, we have*

$$\dim H^0(S, \mathcal{O}_S(mK_S)) = \chi(\mathcal{O}_S) + (\bar{c}_1^2/2)m(m-1), \quad \text{for } m > 1,$$

where $\chi(\mathcal{O}_S)$ denotes the Euler characteristic of the structure sheaf $\mathcal{O}_S$ of S.

Theorem 4. (Iitaka [3]). *The m-genera $P_m(S) = \dim H^0(S, \mathcal{O}_S(mK_S))$ are deformation-invariants.*

As an immediate corollary to Theorems 3 and 4, we obtain the following

Theorem 5. (Deformation Invariance of the Minimality). *If S is minimal, then any deformation of S is also a minimal surface of general type.*

Especially, we have

Theorem 5′. *A deformation of a numerical Campedelli [resp. Campedelli] surface is also a numerical Campedelli [resp. Campedelli] surface.*

3. Proof of Theorem A

For the proof we need the following results.

Definition. An effective divisor D on a surface F is called 1-*connected* if

$$D_1 D_2 > 0,$$

for any non-trivial decomposition $D = D_1 + D_2$, $D_i > 0$.

Theorem 6. (Ramanujam vanishing theorem [6]). *If an effective divisor D on a* *regular surface* F (i.e. $q(F)=0$) *is* 1-*connected, then* $H^1(F, \mathcal{O}(-D))=0$.

Theorem 7. (Bombieri [1]). *Let F be a minimal surface of general type and P a point on F. Let $p: \tilde{F} \to F$ denote a quadric transformation at P and E the exceptional curve over P. If an effective divisor D is numerically equivalent to $2p^*K_F-2E$, then D is* 1-*connected unless $K_F^2=1$.*

Now we proceed to the proof of Theorem A. Let S be a numerical Campedelli surface, $p: \tilde{S} \to S$, the quadric transformation at a point P, and E, the associated exceptional curve. Let us consider the following natural exact sequence of sheaves :

$$0 \to \mathcal{O}_{\tilde{S}}(3p^*K_S-E) \to \mathcal{O}_{\tilde{S}}(3p^*K_S) \to \mathcal{O}_E \to 0.$$

Then it is obvious that $|3K_S|$ is free from base point at P if and only if $H^1(\tilde{S}, \mathcal{O}(3p^*(K_S-E))=0$. By the Serre duality, we have

$$\dim H^1(\tilde{S}, \mathcal{O}(3p^*K_S-E)) = \dim H^1(\tilde{S}, \mathcal{O}(2E-2p^*K_S)).$$

Hence Theorem 7 yields the vanishing of the cohomology group under the condition that $|2p^*K_S-2E| \neq \phi$. Now assume that $|2p^*K_S-2E|=\phi$. Since $\dim H^0(S, \mathcal{O}(2K_S))=3$, this implies that the rational map Φ_{2K_S} associated with the bicanonical system $|2K_S|$ is a *local isomorphism* at P. Therefore there exists an effective divisor $D \in |2p^*K_S-E|$ such that D is irreducible in a neighbourhood of E and that the unique irreducible component D_0 which *simply intersects* E satisfies $D_0^2 \geq 0$. Now we shall take the following exact sequence of cohomology groups :

$$\begin{aligned} 0 &\to H^0(\tilde{S}, \mathcal{O}(2E-2p^*K_S)) \to H^0(\tilde{S}, \mathcal{O}(E)) \to H^0(D, \mathcal{O}_D(E)) \\ &\to H^1(\tilde{S}, \mathcal{O}(2E-2p^*K_S)) \to H^1(\tilde{S}, \mathcal{O}(E)). \end{aligned}$$

Note that $H^0(\tilde{S}, \mathcal{O}(2E-2p^*K_S))=0$ and that

$$\begin{aligned} \dim H^1(\tilde{S}, \mathcal{O}(E)) &= \dim H^1(\tilde{S}, \mathcal{O}(p^*K_S)) = \dim H^1(S, \mathcal{O}(K_S)) \\ &= q(S)=0. \end{aligned}$$

Hence, for the proof of Theorem A, it is sufficient to show the equality

$$\dim H^0(D, \mathcal{O}(E)) = \dim H^0(\tilde{S}, \mathcal{O}(E)) = 1.$$

On the other hand we have the following natural commutative diagram

$$\begin{array}{ccccccc} 0 \to & H^0(D, \mathcal{O}) & \to & H^0(D, \mathcal{O}(E)) & \xrightarrow{r} & H^0(D\cdot E, \mathcal{O}) & \\ & \downarrow & & \downarrow & & \downarrow & \text{identity} \\ 0 \to & H^0(D_0, \mathcal{O}) & \to & H^0(D_0, \mathcal{O}(E)) & \xrightarrow{r} & H^0(D\cdot E, \mathcal{O}) & \end{array}$$

of which the rows are exact. But it is obvious that the virtual genus of D_0 is not 0. Since the degree of the divisor E on D_0 is 1, the restriction map r is the zero-map. This implies that

$$\dim H^0(D, \mathcal{O}(E)) = \dim H^0(D, \mathcal{O}).$$

Moreover we have $\dim H^0(D, \mathcal{O})=1$. In fact, there exists the following natural exact sequence

$$0 \to H^0(\tilde{S}, \mathcal{O}(E-2p^*K_S)) \to H^0(\tilde{S}, \mathcal{O}) \to H^0(D, \mathcal{O})$$
$$\to H^1(\tilde{S}, \mathcal{O}(E-2p^*K_S)),$$

where $\dim H^1(\tilde{S}, \mathcal{O}(E-2p^*K_S)) = \dim H^1(\tilde{S}, \mathcal{O}(3p^*K_S)) = \dim H^1(S, \mathcal{O}(3K_S)) = 0$. Thus $\dim H^0(D, \mathcal{O}) = \dim H^0(\tilde{S}, \mathcal{O}) = 1$ and the assertion is proved.

4. The Structure of Campedelli Surfaces

In this section we shall study Campedelli surfaces.

If S is a Campedelli surface, the universal covering $\bar{S}$ of S has the following numerical characters :

$$\begin{cases} \chi(\bar{S}, \mathcal{O}_{\bar{S}}) = 8\chi(S, \mathcal{O}_S) = 8, \\ q(\bar{S}) = 0, \\ p_g(\bar{S}) = \chi(\bar{S}, \mathcal{O}_{\bar{S}}) - q(\bar{S}) - 1 = 7, \\ K_{\bar{S}}^2 = 8K_S^2 = 16. \end{cases}$$

The fundamental group G of S acts on $\bar{S}$ as the covering transformation group of the unramified covering $e : \bar{S} \to S$, and G operates naturally on the vector space $H^0(\bar{S}, \mathcal{O}(K_{\bar{S}}))$ via linear transformations. Hence we obtain a canonical representation $k : G \to \boldsymbol{GL}(7, \boldsymbol{C})$ and the associated projective representation $k' : G \to \boldsymbol{PGL}(6, \boldsymbol{C})$.

Lemma 1. *k' is a faithful representation.*

Proof. Let $g \in G$ be an element of $\ker k'$. Since $g^2 = \mathrm{id}$, $k(g) = \pm \mathrm{id}$. Hence $p_g(\bar{S}/\langle g \rangle) = 7$ or 0. But $p_g(\bar{S}/\langle g \rangle) = 3$, if g is of order 2. Hence $g = \mathrm{id}$.

Let V denote the image of $\bar{S}$ by the canonical map $\Phi_{K_{\bar{S}}}$ associated with the canonical system $|K_{\bar{S}}|$. Then $k'(g)$ $(g \in G)$ induces an automorphism of V. Thus we obtain a natural homomorphism $a : G \to \mathrm{Aut}(V)$, where $\mathrm{Aut}(V)$ denotes the automorphism group of V.

Lemma 2. *a is injective.*

Proof. A trivial consequence of Lemma 1.

Lemma 3. *The canonical system $K_{\bar{S}}$ of $\bar{S}$ is not composed of a pencil.*

Proof. Assume that V is a curve. Since $q(\bar{S}) = 0$, V must be a (possibly singular) rational curve. An automorphism of V induces a unique automorphism of the non-singular model $\boldsymbol{P}^1$ of V. Hence, by virtue of the above lemma, we infer that there exists a faithful representation $a' : G \to \boldsymbol{PGL}(1, \boldsymbol{C})$. On the other hand, it is obvious that $\boldsymbol{PGL}(1, \boldsymbol{C})$ does not contain a subgroup isomorphic to $(\boldsymbol{Z}/(2))^3$. This is a contradiction.

Since G is a commutative group, we may assume that $k(G)$ is contained in the diagonal subgroup of $\boldsymbol{GL}(7, \boldsymbol{C})$. Let $w_1, \cdots, w_7$ be a basis of $H^0(\bar{S}, \mathcal{O}(K_{\bar{S}}))$ such that $g^*(w_j) = \pm w_j$ for any $g \in G$.

Lemma 4. *The linear subspace W of $H^0(\bar{S}, \mathcal{O}(2K_{\bar{S}}))$ spanned by $w_1^2, w_2^2, \cdots, w_7^2$ is 3-dimensional.*

Proof. Lemma 3 implies that the transcendence degree over $\boldsymbol{C}$ of the field $\boldsymbol{C}(w_2/w_1, \cdots, w_7/w_1)$ is 2. Hence the transcendence degree of $\boldsymbol{C}(w_2^2/w_1^2, \cdots, w_7^2/w_1^2)$ is also 2. This yields the inequality

$$\dim W \geqq 3.$$

On the other hand, since w_j^2 is G-invariant, W can be regarded as a subspace of $H^0(S, \mathcal{O}(2K_S))$. But the Riemann-Roch theorem gives the equality $\dim H^0(S, \mathcal{O}(2K_S)) = 3$. This completes the proof.

Lemma 5. *Let $\boldsymbol{K}$ be an extension of the rational function field $\boldsymbol{C}(x_1, \cdots, x_n)$ defined by*

$$\boldsymbol{K}_r = \boldsymbol{C}(x_1, \cdots, x_n, \sqrt{Q_1}, \cdots, \sqrt{Q_r}),$$

where Q_j is a quadric polynomial in x_i. Assume that $[\boldsymbol{K}_r : \boldsymbol{C}(x_1, \cdots, x_n)] = 2^r$. Then the integral closure of $\boldsymbol{C}[x_1, \cdots, x_n]$ in $\boldsymbol{K}_r$ is $R_r = \boldsymbol{C}[x_1, \cdots, x_n, \sqrt{Q_1}, \cdots, \sqrt{Q_r}]$.

Proof. Trivial.

Corollary. *Let $\boldsymbol{K}_r$ be as above. Let Q_{r+1} be another quadric polynomial in x_i. Assume that $\boldsymbol{K}_{r+1} = \boldsymbol{K}_r$. Then $\sqrt{Q_{r+1}}$ is a linear combination of x_i's and $\sqrt{Q_j}$'s.*

Let w_1^2, w_2^2, w_3^2 be a basis of W. From Lemma 4, we infer that there are quadric relations

$$w_j^2 = a_j w_1^2 + b_j w_2^2 + c_j w_3^2, \qquad j = 4, 5, 6, 7.$$

The above corollary asserts that, if the complete intersection defined by the above quadrics is reducible, then any irreducible component of it is contained in a hyperplane in $\boldsymbol{P}^6$. Since the image V of $\bar{S}$ is contained in the complete intersection V' defined by the above 4 equations and V is not contained in any hyperplane, $V' = V$ is a irreducible surface. Thus we obtain the following

Corollary. *V is a complete intersection of type* (2, 2, 2, 2) *in $\boldsymbol{P}^6$.*

As an immediate consequence of this corollary, we have

Theorem 8. *The canonical homomorphism*

$$\otimes^m H^0(\bar{S}, \mathcal{O}(K_{\bar{S}})) \to H^0(\bar{S}, \mathcal{O}(mK_{\bar{S}}))$$

is surjective.

Proof. Let $\mathcal{O}_V(m)$ denote the sheaf associated to the hypersurface section of degree m. Since V is a complete intersection of type (2, 2, 2, 2), we have

$$\dim H^0(V, \mathcal{O}_V(m)) \geq 8 + 8m(m-1) = \dim H^0(\bar{S}, \mathcal{O}_{\bar{S}}(mK_{\bar{S}}))$$

Moreover $H^0(V, \mathcal{O}_V(1))$ generates $H^0(V, \mathcal{O}_V(m))$. This proves the theorem.

Now we obtain the following refined version of Theorem B:

Theorem 9. *The canonical model* $\bar{X}$ *of* $\bar{S}$ *is isomorphic to a complete intersection of type* (2, 2, 2, 2) *in* $\boldsymbol{P}^6$. *The canonical model* X *of* S *is the quotient of* $\bar{X}$ *by the action of following subgroup* G *of* $\boldsymbol{PGL}(6, \boldsymbol{C})$:

$$G=\left\langle \begin{bmatrix} 1 & & & & & & \\ & 1 & & & & & 0 \\ & & 1 & & & & \\ & & & -1 & & & \\ 0 & & & & -1 & & \\ & & & & & -1 & \\ & & & & & & -1 \end{bmatrix}, \begin{bmatrix} 1 & & & & & & \\ & -1 & & & & & 0 \\ & & -1 & & & & \\ & & & 1 & & & \\ 0 & & & & 1 & & \\ & & & & & -1 & \\ & & & & & & -1 \end{bmatrix}, \begin{bmatrix} -1 & & & & & & \\ & -1 & & & & & \\ & & 1 & & & & 0 \\ & & & 1 & & & \\ 0 & & & & -1 & & \\ & & & & & 1 & \\ & & & & & & -1 \end{bmatrix} \right\rangle$$

The following theorem is a corollary of Theorem 9 and the form of the defining equations.

Theorem 10. *The moduli space of Campedelli surfaces is a unirational variety of dimension* 6.

References

[1] Bombieri, E. : Canonical models of surfaces of general type, Publ. Math. I. H. E. S., **42** (1973), 447–495.
[2] Campedelli, L. : Sui piani doppi con curva di diramazione del decimo ordine, Atti Accad. Naz. Lincei, **15** (1932), 358–362.
[3] Iitaka, S. : Deformations of compact complex surfaces II, J. Math. Soc. Japan, **22** (1970), 247–261.
[4] Miyaoka, Y. : Tricanonical maps of numerical Godeaux surfaces, to appear.
[5] Mumford, D. : The canonical ring of an algebraic surface, Ann. of Math., **76** (1962), 612–615.
[6] Ramanujam, C. P. : Remarks on the Kodaira vanishing theorem, J. Indian Math. Soc. (N. S.) **36** (1972), 41–51.

Department of Mathematics
Tokyo Metropolitan University

(Received January 7, 1976)

On Singular K3 Surfaces

T. Shioda and H. Inose

By a *singular $K3$ surface* we mean an algebraic $K3$ surface (defined over the field of complex numbers) whose Picard number equals the maximum possible number 20. The main purpose of this paper is to prove the following results:

Theorem. *There is a natural one-to-one correspondence from the set of singular $K3$ surfaces to the set of equivalence classes of positive-definite even integral binary quadratic forms with respect to $SL_2(\boldsymbol{Z})$* (Th. 4 in § 4).

Theorem. *Every singular $K3$ surface has an infinite group of automorphisms* (Th. 5 in § 5).

Theorem. *Every singular $K3$ surface has a model defined over an algebraic number field K, and its Hasse-Weil zeta function is given, up to finitely many Euler factors, by*

$$\zeta_K(s)\zeta_K(s-1)^{20}\zeta_K(s-2)L(s-1,\chi^2)L(s-1,\bar{\chi}^2),$$

where $\zeta_K(s)$ is the Dedekind zeta function of K and $L(s,\chi^2)$ or $L(s,\bar{\chi}^2)$ is the Hecke L-function with a suitable Grössencharacter χ^2 or $\bar{\chi}^2$ (Th. 6 in § 6).

The correspondence in the first theorem associates with a singular $K3$ surface X the binary quadratic form given by the intersection product on the oriented lattice T_X of transcendental cycles on X, and its injectivity was proved by Pjateckii-Šapiro and Šafarevič as a consequence of the Torelli theorem for $K3$ surfaces (cf. [4] § 8). We shall prove the surjectivity of this correspondence by constructing a singular $K3$ surface X with a prescribed T_X. In fact we shall construct a wider class of $K3$ surfaces containing all singular ones, starting from Kummer surfaces associated with abelian surfaces of product type (§§ 2, 3). We use the corresponding result for singular abelian surfaces, which was proved in [8]. For the proof of the second theorem, we find on a given singular $K3$ surface X a suitable elliptic pencil $\Psi : X \to \boldsymbol{P}^1$ which has infinitely many sections.

The present work is based on the general theory of elliptic surfaces, due to Kodaira [3]. It is a great pleasure for the authors to dedicate this paper to Professor Kodaira.

1. Elliptic Pencils on a K3 Surface

We recall in this section a few facts on elliptic pencils (i.e. structures of elliptic surface) on a $K3$ surface, which will be used in the sequel.

Lemma 1. 1. *Assume that an effective divisor D on a $K3$ surface X has the same type as a simple singular fibre of an elliptic surface in the sense of Kodaira* [3] § 6. *Then there is a unique elliptic pencil $\Phi : X \to \boldsymbol{P}^1$, of which D is a singular fibre. Moreover any irreducible curve C on X with $(CD)=1$ defines a (holomorphic) section of Φ.*

This follows immediately from Theorem 1 of [4] § 3.

Lemma 1. 2. *If a divisor D on an elliptic surface has its support contained in a (simple) singular fibre, then the self-intersection number D^2 is non-positive, and $D^2=0$ holds if and only if D is a multiple of the singular fibre.*

For the proof, we refer to [6] Ch. VII § 2 or [7] Lemma 1. 3.

Lemma 1. 3. *Let $\Phi : X \to \boldsymbol{P}^1$ denote an elliptic $K3$ surface, and $D_\nu = \Phi^{-1}(t_\nu)$ $(1 \leq \nu \leq k)$ be all the singular fibres of Φ. We denote by ε_ν, m_ν or $m_\nu^{(1)}$ respectively the Euler number of D_ν, the number of irreducible components of D_ν, or the number of simple components of D_ν. Then*

(i) *we have*

$$\sum_{\nu=1}^{k} \varepsilon_\nu = 24 \ (= \text{the Euler number of } X). \tag{1. 1}$$

(ii) *Assuming moreover that Φ has a section, let $r(\Phi)$ be the rank of the group of sections of Φ. Then the Picard number $\rho(X)$ of X is given by the formula*:

$$\rho(X) = r(\Phi) + 2 + \sum_{\nu=1}^{k} (m_\nu - 1). \tag{1. 2}$$

(iii) *When $r(\Phi)=0$, let $n(\Phi)$ denote the order of the group of sections of Φ. Then we have*

$$|\det T_X| = |\det S_X| = \prod_{\nu=1}^{k} m_\nu^{(1)} / n(\Phi)^2. \tag{1. 3}$$

Here S_X is the sublattice of $H_2(X, \boldsymbol{Z})$ consisting of algebraic cycles (the Néron-Severi group of X), and $T_X = S_X^\perp$ is the orthogonal complement of S_X in $H_2(X, \boldsymbol{Z})$ (the lattice of transcendental cycles).

In fact these three statements hold for an arbitrary elliptic surface $\Phi : X \to \Delta$ with at least one singular fibre. The assertion (i) is implicit in [3] Theorem 12. 2. As for (ii) and the second equality of (iii), we refer to [7] Corollaries 1. 5 and 1. 7. The first equality of (1. 3) follows from the elementary remark that, for a primitive sublattice S of a unimodular lattice H, $|\det S|$ is equal to $|\det S^\perp|$, $S^\perp$ being the orthogonal complement of S in H.

Finally we note the following table for singular fibres from Kodaira [3] pp. 565–566, p. 14 :

type	$\mathrm{I}_b (b\geq 1)$	II	III	IV	$\mathrm{I}_b^* (b\geq 0)$	II*	III*	IV*
m_ν	b	1	2	3	$b+5$	9	8	7
$m_\nu^{(1)}$	b	1	2	3	4	1	2	3
ε_ν	b	2	3	4	$b+6$	10	9	8

2. Construction of a Certain Elliptic Pencil on a Kummer Surface

Theorem 1. *Let $X=\mathrm{Km}(A)$ denote the Kummer surface associated with an arbitrary product abelian surface $A=C_1\times C_2$ (C_1, C_2 elliptic curves). Then there exists an elliptic pencil $\Phi : X\to \boldsymbol{P}^1$, which has a section and (at least) three singular fibres of types*
(i) II*, $\mathrm{I}_{b_1}^*$ *and* $\mathrm{I}_{b_2}^*$ $(b_1+b_2\leq 2)$, *or* (ii) II*, IV* and I_0^*.

Recall that the Kummer surface $X=\mathrm{Km}(A)$ is the minimal non-singular model of the quotient surface A/ι_A of A by the inversion automorphism ι_A ($\iota_A(z)=-z$), which has the 16 singular points corresponding to the points of order 2 of A. Letting u_i (or u_i') $(1\leq i\leq 4)$ be the 4 points of order 2 on the elliptic curve C_1 (or C_2), we denote by $E_{ij}(1\leq i, j\leq 4)$ the non-singular rational curve on X corresponding to the point (u_i, u_j') of A. Also we denote by F_i (or G_j) the non-singular rational curve on X, which is the image of $u_i\times C_2$ (or $C_1\times u_j'$) under the natural rational map $\alpha : A\to X$.

The Kummer surface X has the two obvious elliptic pencils $\Phi_\nu : X\to \boldsymbol{P}^1$, which are induced by the projections $A\to C_\nu$ $(\nu=1, 2)$. Each Φ_ν has 4 singular fibres, all of type I_0^* :

$$(2.1)\qquad \begin{aligned} 2F_i+\sum_{j=1}^{4}E_{ij} &\sim F \qquad (1\leq i\leq 4),\\ 2G_j+\sum_{i=1}^{4}E_{ij} &\sim G \qquad (1\leq j\leq 4),\end{aligned}$$

in which F (or G) is a general fibre of Φ_1 (or Φ_2). The symbol $\sim$ indicates the linear equivalence, which coincides with the algebraic or homological equivalence on a $K3$ surface. The intersection number of these curves are given as follows :

$$(2.2)\qquad \begin{aligned} &FG=2, \quad FE_{ij}=GE_{ij}=F_iG_j=0, \quad FG_j=GF_i=1,\\ &E_{ij}^2=F_i^2=G_i^2=-2, \quad F_kE_{ij}=\delta_{ki}, \quad G_kE_{ij}=\delta_{kj}.\end{aligned}$$

The configuration formed by the rational curves E_{ij}, F_i, G_j is called the *double Kummer pencil* on $X=\mathrm{Km}(C_1\times C_2)$ (see Fig. 1).

Now we shall construct a new elliptic pencil on X. For that purpose, let us consider the following divisor D :

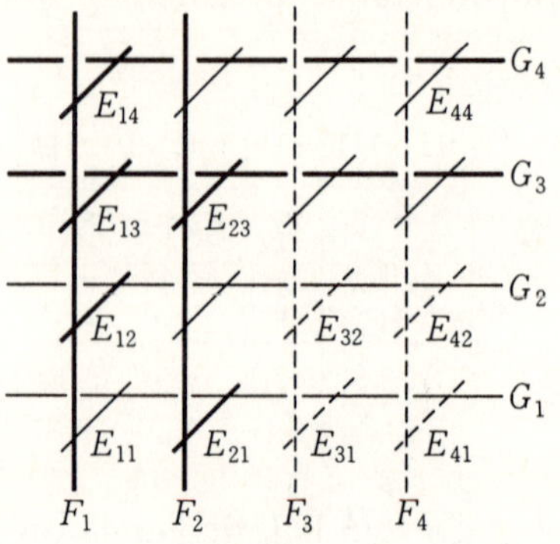

Fig. 1

(2. 3) $$D = E_{21}+2F_2+3E_{23}+4G_3+5E_{13}+6F_1+3E_{12}+4E_{14}+2G_4$$

(see the bold lines in Fig. 1). It has type II* in Kodaira's list [3] § 6. Applying Lemma 1. 1, we obtain an elliptic pencil $\Phi : X \to \boldsymbol{P}^1$ having D as a singular fibre. Note that the curve G_1 is a section of Φ, since $G_1 D=1$. To find other singular fibres of Φ, we consider the divisors (see the dotted lines in Fig. 1) :

(2. 4) $$B_1 = F_3+E_{31}+E_{32}, \qquad B_2 = F_4+E_{41}+E_{42}.$$

These divisors do not meet the fibre D and their supports are connected. Hence the image $\Phi(B_\lambda)$ $(\lambda=1, 2)$ is a point t_λ of $\boldsymbol{P}^1$, and B_λ is contained in the singular fibre $\Phi^{-1}(t_\lambda)$. We have $t_1 \neq t_2$, because both B_1 and B_2 meet the section G_1, and they cannot belong to one and the same fibre. Therefore the proof of Theorem 1 is reduced to the assertion that *the types of singular fibres* $\Phi^{-1}(t_\lambda)$ $(\lambda=1, 2)$ *are either* (i) $\mathrm{I}^*_{b_1}$ *and* $\mathrm{I}^*_{b_2}$ $(b_1+b_2 \leq 2)$ *or* (ii) IV* *and* I^*_0.

Lemma 2. 1. *There exist* 4 *effective divisors* A_ν $(1 \leq \nu \leq 4)$ *on* X *satisfying the following conditions* :

(1) A_ν *is linearly equivalent to a linear combination of* F, G *and* E_{ij} $(1 \leq i, j \leq 4)$.

(2) $A_\nu C=0$ *for every irreducible component* C *of the singular fibre* D (2. 3) *and for* $C= E_{31}, E_{32}, E_{41}, E_{42}$ *and* G_1.

(3) $A_\nu^2=-2$, $A_\nu A_\mu=0$ $(\nu \neq \mu)$.

(4) A_ν *is contained in a fibre of* Φ, *and its support is connected.*

(5) $A_1F_3=A_2F_3=1$, $A_3F_4=A_4F_4=1$.

Proof. By an easy computation using (2. 2), we see that the space of divisors satisfying the conditions (1) and (2) is generated (up to linear equivalence) by the 4 divisors A'_ν $(1 \leq \nu \leq 4)$:

(2. 5) $$A'_1 = E_{11}+E_{24}+E_{43}-F-G, \qquad A'_2 = E_{22}-E_{24}+E_{34},$$
$$A'_3 = E_{33}-E_{43}, \qquad A'_4 = E_{44}-E_{34}.$$

In diagonalizing the symmetric matrix $(A'_\nu A'_\mu)$, we find the following 4 divisors A_ν $(1 \leq \nu \leq 4)$ satisfying (1), (2) and (3) :

$$(2.6)\quad \begin{cases} A_1 \sim 2F+2G-2E_{11}-E_{22}-E_{24}-E_{33}-E_{43}-E_{44} \\ A_2 \sim F+G-E_{11}-E_{24}-E_{43} \\ A_3 \sim F+G-E_{11}-E_{24}-E_{33} \\ A_4 \sim 2F+2G-2E_{11}-E_{22}-E_{24}-E_{33}-E_{34}-E_{43}. \end{cases}$$

Now each A_ν is linearly equivalent to an effective divisor. In fact, by the Riemann-Roch inequality on a $K3$ surface, we have

$$l(A_\nu)+l(-A_\nu) \geq \frac{1}{2}A_\nu^2+2 \geq 1.$$

Hence either A_ν or $-A_\nu$ is equivalent to an effective divisor A_ν^*. Since $A_\nu^*D=0$ by (2), each component of A_ν^* lies in some fibre of Φ, and we have $A_\nu^*E\geq 0$ for any irreducible curve E not lying in a fibre. On the other hand, we have $A_1E_{33}=A_2E_{33}=2$ and $A_3E_{43}=A_4E_{43}=2$. Therefore $A_\nu \sim A_\nu^*$, and we can assume that A_ν itself is effective for all ν.

Let us then decompose A_ν into its connected components $A_{\nu i}$ $(i=1, 2, \cdots)$; each $A_{\nu i}$ is contained in a fibre. By Lemma 1. 2, we have $A_{\nu i}^2 \leq 0$. Since $A_{\nu i}^2$ is even, we can assume $A_{\nu 1}^2=-2$, $A_{\nu i}^2=0$ for $i\geq 2$. Again, by the same lemma, $A_{\nu i}$ is a multiple of a singular fibre for $i\geq 2$. Therefore $A_\nu \sim A_{\nu 1}+m\Phi^{-1}(t)$. Computing the intersection numbers of both sides with the section G_1 of Φ, we see $m=0$. Thus the conditions (1), $\cdots$, (4) are satisfied when A_ν is replaced by $A_{\nu 1}$. Finally (5) follows immediately from (2. 2). This completes the proof of Lemma 2. 1. q. e. d.

Lemma 2. 2. *The singular fibres $\Phi^{-1}(t_\lambda)$ $(\lambda=1, 2)$ are given as follows:*

$$(2.7)\quad \Phi^{-1}(t_1) = 2F_3+E_{31}+E_{32}+A_1+A_2, \qquad \Phi^{-1}(t_2) = 2F_4+E_{41}+E_{42}+A_3+A_4.$$

In particular, each $\Phi^{-1}(t_\lambda)$ has at least 5 irreducible components and is not of type $\mathrm{I}_b(b\geq 1)$.

Proof. It follows from the conditions (4), (5) of Lemma 2. 1 that the supports of A_1 and A_2 lie in the singular fibre $\Phi^{-1}(t_1)$. Moreover, if we put $H=2F_3+E_{31}+E_{32}+A_1+A_2$, it is easy to check $H^2=0$. By Lemma 1. 2, we conclude that H is a multiple of $\Phi^{-1}(t_1)$. This multiplicity must be 1, because $HG_1=E_{31}G_1=1$ and $\Phi^{-1}(t_1)G_1=1$ for the section G_1. This proves the first equality, and the second one can be similarly proven. The last assertion follows from the fact that every irreducible component of a singular fibre of type I_b $(b\geq 1)$ appears with multiplicity 1. q. e. d.

Now we continue the proof of Theorem 1. From Kodaira's list of singular fibres, [3] § 6, the possibilities for the singular fibre $\Phi^{-1}(t_\lambda)$ $(\lambda=1, 2)$ are the following types: I_b^* $(b\geq 0)$, II^*, III^* or IV^*. Let m_λ be the number of irreducible components of $\Phi^{-1}(t_\lambda)$. Applying the formula (1. 2) to our elliptic pencil Φ, we have

$$(2.8)\qquad \rho(X) \geq 2+(9-1)+(m_1-1)+(m_2-1).$$

Since $\rho(X)\leq 20$ for a complex $K3$ surface X, this implies that $m_1+m_2\leq 12$, and hence $5\leq m_1$, $m_2\leq 7$. This excludes the possibility for $\Phi^{-1}(t_\lambda)$ to be of type II^* or III^* (cf. the table in § 1). If $\Phi^{-1}(t_1)$ is of type IV^* $(m_1=7)$, then $\Phi^{-1}(t_2)$ has $m_2=5$

components, hence it must be of type I_0^*. This corresponds to the case (ii) of Theorem 1. Otherwise $\Phi^{-1}(t_1)$ and $\Phi^{-1}(t_2)$ are of types $I_{b_1}^*$ and $I_{b_2}^*$. Setting $m_\lambda = b_\lambda + 5$ ($\lambda = 1, 2$) in (2. 8), we have $b_1 + b_2 \leq 2$ in this case. This completes the proof of Theorem 1.

q. e. d.

Remark. (1) Actually the converse of Theorem 1 is also true. Namely any $K3$ surface having an elliptic pencil such as that in Theorem 1 is isomorphic to $\mathrm{Km}(A)$ for some $A = C_1 \times C_2$. We omit the proof, since it is rather complicated and also will not be used in the following.

(2) We shall prove later that $b_1, b_2 \leq 1$ in the case (i) (cf. Lemma 3. 1).

(3) The case (ii) of Theorem 1 occurs if and only if the abelian surface A is isomorphic to the product $C_\omega \times C_\omega$, C_ω being the elliptic curve with the fundamental periods 1 and $\omega = e^{2\pi i/3}$.

3. Certain Double Coverings of Kummer Surfaces

Throughout this section, we assume that X is a Kummer surface given with an elliptic pencil $\Phi : X \to \boldsymbol{P}^1$, which has a section and at least three singular fibres $\Phi^{-1}(t_0)$, $\Phi^{-1}(t_1)$ and $\Phi^{-1}(t_2)$ of types (i) II^*, $I_{b_1}^*$ and $I_{b_2}^*$ ($b_1 + b_2 \leq 2$), or (ii) II^*, IV^*, I_0^*. We shall construct a certain $K3$ surface Y and a rational map $\pi : Y \to X$ of degree 2, and examine the relation between T_X and T_Y, the lattices of transcendental cycles on X and on Y.

We consider a ramified double covering $f : \Delta \to \boldsymbol{P}^1$, ramified only at the 2 points t_1, t_2 of $\boldsymbol{P}^1$, and put

$$f^{-1}(t_0) = \{s_{01}, s_{02}\}, \quad f^{-1}(t_1) = s_1, \quad f^{-1}(t_2) = s_2.$$

We denote by $\Psi : Y \to \Delta$ the elliptic pencil induced by $\Phi : X \to \boldsymbol{P}^1$. Note that Y is a non-singular surface, birationally equivalent to the fibre product of X and Δ over $\boldsymbol{P}^1$, and that no fibre of Ψ contains an exceptional curve of the first kind. Obviously Ψ has two singular fibres of type II^* over s_{01} and s_{02}. In case (i), the fibre $\Psi^{-1}(s_\lambda)$ over s_λ ($\lambda = 1, 2$) is of type I_{2b_λ}, since it is the pull-back of the singular fibre $\Phi^{-1}(t_\lambda)$ of type $I_{b_\lambda}^*$ at the ramification point t_λ of the double covering $f : \Delta \to \boldsymbol{P}^1$. In fact this is clear from the behavior of the homological invariant (or local monodromy) of Ψ around s_λ : $\begin{bmatrix} -1 & -b_\lambda \\ 0 & -1 \end{bmatrix}^2 = \begin{bmatrix} 1 & 2b_\lambda \\ 0 & 1 \end{bmatrix}$ (cf. [3] p. 604). In case (ii), $\Psi^{-1}(s_1)$ is a singular fibre of type IV and $\Psi^{-1}(s_2)$ is a regular fibre, as is shown by the same argument as above.

Definition. We shall call the surface Y, together with the rational map of degree 2 $\pi : Y \to X$, *the double covering of X with respect to the elliptic pencil $\Phi : X \to \boldsymbol{P}^1$.*

Lemma 3. 1. *Y is a $K3$ surface. Moreover the collection of singular fibres of $\Psi : Y \to \Delta$ falls into one of the following* : 1) II^*, II^*, I_2, I_2, 2) II^*, II^*, I_2, I_1, I_1, 3) II^*, II^*, I_1, I_1, I_1, I_1, 4) II^*, II^*, II, II *and* 5) II^*, II^*, IV.

Proof. We denote by $\Phi^{-1}(t_\nu)$ ($\nu=0, 1, 2, \cdots, k$) all the singular fibres of $\Phi : X \to \boldsymbol{P}^1$, and by ε_ν the Euler number of $\Phi^{-1}(t_\nu)$. In case (i) (resp. (ii)), since $\varepsilon_0=10$ and $\varepsilon_\lambda=b_\lambda+6$ ($\lambda=1, 2$) (resp. $\varepsilon_1=8$, $\varepsilon_2=6$) (cf. the table in § 1), we deduce from the formula (1. 1) that

$$b_1+b_2+(\varepsilon_3+\cdots+\varepsilon_k) = 2 \qquad (\text{resp. } \varepsilon_3+\cdots+\varepsilon_k = 0). \tag{3. 1}$$

If we put $f^{-1}(t_\nu)=\{s_{\nu 1}, s_{\nu 2}\}$ ($\nu\geq 3$), the singular fibre $\Psi^{-1}(s_{\nu i})$ ($i=1, 2$) is of the same type as $\Phi^{-1}(t_\nu)$. Now the Euler number $\varepsilon(Y)$ of the surface Y is equal to the sum of those of singular fibres of $\Psi : Y \to \Delta$ (cf. (1. 1)) :

$$\varepsilon(Y) = 2\varepsilon_0+(2b_1)+(2b_2)+2(\varepsilon_3+\cdots+\varepsilon_k) \qquad (\text{resp. } \varepsilon(Y) = 2\varepsilon_0+4),$$

where $2b_\lambda$ ($\lambda=1, 2$) is the Euler number of the singular fibre $\Psi^{-1}(s_\lambda)$ of type I_{2b_λ}. It follows from (3. 1) that $\varepsilon(Y)=24$.

Next we note that the elliptic pencil Ψ has no multiple fibre because it has a section induced by one of Φ. By Kodaira [3] Theorem 12. 1, the canonical bundle K_Y of Y is given by $K_Y=\Psi^*(\mathfrak{k}-\mathfrak{f})$, in which $\mathfrak{k}$ is the canonical bundle of the base curve Δ and $\mathfrak{f}$ is a line bundle on Δ of degree

$$-p_g(Y)+q(Y)-1 = -(K_Y^2+\varepsilon(Y))/12 = -2.$$

Since $\Delta \simeq \boldsymbol{P}^1$, this implies that K_Y is trivial and $q(Y)=0$. Therefore Y is a $K3$ surface.

In case (ii), Ψ has no more singular fibres ($k=2$) and we have case 5). In order to determine the singular fibres of Ψ in case (i), we observe that $b_1\leq 1$, $b_2\leq 1$. In fact, if b_1 or $b_2=2$, then Ψ would have (at least) three singular fibres of types II*, II* and I_4, and then the formula (1. 2) would imply $\rho(Y)\geq 21$, a contradiction. Then (3. 1) has only 4 solutions : 1′) $b_1=b_2=1$, $k=2$; 2′) $\{b_1, b_2\}=\{0, 1\}$, $k=3$, $\varepsilon_3=1$; 3′) $b_1=b_2=0$, $k=4$, $\varepsilon_3=\varepsilon_4=1$; 4′) $b_1=b_2=0$, $k=3$, $\varepsilon_3=2$.
Noting that a singular fibre has Euler number 1 (resp. 2) if and only if it is of type I_1 (resp. I_2 or II) (cf. the table in § 1), we see that the cases 1′), 2′), 3′) correspond respectively to the cases 1), 2), 3) stated in Lemma 3. 1, and case 4′) corresponds to case 1) or 4). This proves Lemma 3. 1. q. e. d.

We remark that the surface X can be recovered from Y as follows. The non-trivial covering transformation of the double covering $f: \Delta \to \boldsymbol{P}^1$ induces an involutive birational transformation of Y, hence an automorphism ι of Y (by the minimality of a $K3$ surface). The involution ι has the 8 fixed points, 4 on each of the two fibres $\Psi^{-1}(s_\lambda)$ ($\lambda=1, 2$), and X is the minimal non-singular model of the quotient Y/ι (cf. [3] § 8, pp. 585–586, 591–592, 600–602). In particular, the rational map $\pi : Y \to X$ has the 8 fundamental points $p_\nu (1\leq\nu\leq 8)$ corresponding to the 8 curves $\Theta_\nu (1\leq\nu\leq 8)$ in X, which appear with odd multiplicities in the two singular fibres $\Phi^{-1}(t_\lambda)$ ($\lambda=1, 2$).

In order to formulate the following theorem, we introduce some notation. The rational map $\pi : Y \to X$ induces a homomorphism :

$$\pi_* : H_2(Y, \boldsymbol{Z}) \simeq H_2(Y-\{p_\nu\}, \boldsymbol{Z}) \to H_2(X, \boldsymbol{Z}). \tag{3. 2}$$

We denote by T_X the sublattice of transcendental cycles in $H_2(X, \boldsymbol{Z})$, and by p_X *the period on* T_X, i.e. the linear functional on T_X defined (up to constants) by

$$p_X(t) = \int_t \omega_X \qquad (t \in T_X), \tag{3. 3}$$

ω_X being a non-vanishing holomorphic 2-form on X. For $x_1, x_2 \in H_2(X, \boldsymbol{Z})$, $(x_1 \cdot x_2)$ denotes the intersection number.

Theorem 2. *Let $\pi : Y \to X$ be the double covering defined as above. Then π_* induces a bijection of T_Y onto T_X such that*

(a) $(\pi_* y_1 \cdot \pi_* y_2) = 2(y_1 \cdot y_2), \quad y_1, y_2 \in T_Y,$

(b) $p_X \circ \pi_* = \text{const. } p_Y.$

The rest of this section is devoted to the proof of this theorem. First we observe that the pull-back $\pi^*(\omega_X)$ of a holomorphic 2-form ω_X on X is holomorphic on $Y - \{p_\nu\}$, hence holomorphic everywhere on Y. By (3. 3), this shows the assertion (b). Next, keeping the same notation as before, we define some sublattices of $H_X = H_2(X, \boldsymbol{Z})$ and $H_Y = H_2(Y, \boldsymbol{Z})$. Let Γ be the sublattice of H_X generated by the 8 irreducible components with multiplicity ≥ 2 of the singular fibre $\Phi^{-1}(t_0)$ of type II*; Γ is a negative-definite even unimodular lattice of rank 8 (cf. [7] Lemma 1. 3). Also let Γ_n $(n=1, 2)$ be the sublattice of H_Y obtained from the singular fibre $\Psi^{-1}(s_{0n})$ of type II* in the same way. Since $\Gamma_1 \oplus \Gamma_2$ is a unimodular sublattice of H_Y, we have the orthogonal decomposition:

$$H_Y = \Gamma_1 \oplus \Gamma_2 \oplus L, \tag{3. 4}$$

L being unimodular and of rank 6. Obviously we have

$$L \supset T_Y, \qquad \pi_*(\Gamma_1) = \pi_*(\Gamma_2) = \Gamma. \tag{3. 5}$$

Furthermore let M denote the sublattice of H_X generated by the 8 curves $\Theta_\nu (1 \leq \nu \leq 8)$. By the definition of π_* (3. 2), π_* maps H_Y into the orthogonal complement $M^\perp$ of M in H_X. On the other hand, there is a natural map $\pi^* : M^\perp \to H_Y$ such that

$$(\pi^* x_1 \cdot \pi^* x_2) = 2(x_1 \cdot x_2) \qquad (x_1, x_2 \in M^\perp), \tag{3. 6}$$

$$\pi^* \pi_* y = y + \iota_*(y) \qquad (y \in H_Y), \tag{3. 7}$$

where ι_* is the induced map of the involution ι of Y (cf. [2] § 1). Moreover, since ι interchanges the singular fibres $\Psi^{-1}(s_{01})$ and $\Psi^{-1}(s_{02})$, we have

$$\iota_*(\Gamma_1) = \Gamma_2, \qquad \iota_*(\Gamma_2) = \Gamma_1. \tag{3. 8}$$

Finally it follows from (b) and (3. 7) that both π_* and π^* preserve the algebraic cycles:

$$\pi_*(S_Y) \subset S_X, \qquad \pi^*(S_X \cap M^\perp) \subset S_Y. \tag{3. 9}$$

Lemma 3. 2. *The action of ι_* on L is trivial. The map π_* induces a bijection of L onto its image in H_X, and*

$$(\pi_* y_1 \cdot \pi_* y_2) = 2(y_1 \cdot y_2) \quad \textit{for} \quad y_1, y_2 \in L.$$

Proof. The involution ι of Y has the 8 fixed points $\{p_\nu\}$. By the Lefschetz fixed point theorem, we have

$$1+0+\mathrm{tr}(\iota_*|H_2(Y, \boldsymbol{Z}))+0+1 = 8.$$

In view of (3. 4) and (3. 8), this implies that $\mathrm{tr}(\iota_*|L)=6$. Since the eigenvalues of ι_* are 1 or -1, we conclude that $\iota_*|L$ is the identity. It follows from (3. 7) that $\pi^*\pi_*(y)=2y$ for $y \in L$. Hence $\pi_*|L$ is injective, and moreover we have from (3. 6)

$$2(\pi_* y_1 \cdot \pi_* y_2) = (\pi^*\pi_* y_1 \cdot \pi^*\pi_* y_2) = (2y_1 \cdot 2y_2) = 4(y_1 \cdot y_2). \qquad \text{q. e. d.}$$

Lemma 3. 3. $\pi_*(H_Y) = \pi_*(L) \oplus \Gamma$.

Proof. It suffices to show that $\pi_*(L)$ and Γ are orthogonal to each other. Take $x_1=\pi_*(y_1)$ $(y_1 \in L)$ and $x_2=\pi_*(y_2)$ $(y_2 \in \Gamma_1)$. From (3. 7) and (3. 8), we see that $\pi^*(x_1)=2y_1 \in L$ and $\pi^*(x_2)=y_2+\iota_*(y_2) \in \Gamma_1 \oplus \Gamma_2$. Hence we have

$$2(x_1 \cdot x_2) = (\pi^* x_1 \cdot \pi^* x_2) = (2y_1 \cdot (y_2+\iota_*(y_2))) = 0. \qquad \text{q. e. d.}$$

Lemma 3. 4. *Let M' be the minimal primitive sublattice of H_X containing M. Then* $\det M'=|\det M^\perp|=2^6$.

Proof. Obviously $(M')^\perp=M^\perp$, and $|\det M'|=|\det (M')^\perp|$, because M' is primitive in the unimodular lattice H_X. Since $\det M=\det(\Theta_\nu\Theta_\mu)=2^8$, we have only to prove that the index $[M' : M]=2$. Let θ_ν be the homology class of the curve Θ_ν. Take an element $x=\sum_{\nu=1}^{8} a_\nu\theta_\nu (a_\nu \in \boldsymbol{Q})$ of M', $x \notin M$. Since $x \cdot \theta_\nu \in \boldsymbol{Z}$, a_ν must be half-integers. We may assume $a_\nu=0$ or $1/2$. Then the number m of a_ν with $a_\nu=1/2$ is either 4 or 8, because $x^2=-m/2$ is an even integer. Assume $m=4$ and $x=\sum_{\nu=1}^{4}\theta_\nu/2$. If we take a divisor D in the homology class x $(D^2=x^2=-2)$, the Riemann-Roch theorem implies that D or $-D$ is linearly equivalent to an effective divisor D'. But $D \sim D'$, since $2D \sim \sum_{\nu=1}^{4}\Theta_\nu$. As $D'\Theta_\nu=x\theta_\nu=-1$ for $1 \leq \nu \leq 4$, D' contains $\Theta_\nu (1 \leq \nu \leq 4)$ as components: $D'=\sum_{\nu=1}^{4}\Theta_\nu+D''$, $D''>0$. It follows that $2D'' \sim -\sum_{\nu=1}^{4}\Theta_\nu$, which is a contradiction. Therefore $m=8$ and $x=\sum_{\nu=1}^{8}\theta_\nu/2$. This shows $[M' : M] \leq 2$. Conversely, the element $\sum_{\nu=1}^{8}\theta_\nu/2$ belongs to M', because we have

$$\sum_{\nu=1}^{8}\Theta_\nu \sim 2(\Phi^{-1}(t)-D_1-D_2),$$

in which D_λ $(\lambda=1, 2)$ is the sum of irreducible components with multiplicity ≥ 2 in the singular fibre $\Phi^{-1}(t_\lambda)$ of type $\mathrm{I}^*_{b_\lambda}$ or IV^*. q. e. d.

Lemma 3. 5. $\pi_*(H_Y) = M^\perp, \quad \pi_*(T_Y) = T_X$.

Proof. For the first equality, we know that the inclusion $\pi_*(H_Y) \subset M^\perp$ holds. By Lemmas 3. 3 and 3. 2, we have

$$|\det \pi_*(H_Y)| = |\det \pi_*(L)| = 2^6 |\det L| = 2^6,$$

since Γ and L are unimodular. Then $\pi_*(H_Y)=M^\perp$ by Lemma 3. 4. Next we shall show that $\pi_*(T_Y)\subset T_X$, i.e. if $y\in T_Y$, then $(\pi_*(y)\cdot x)=0$ for all $x\in S_X=T_*^\perp$. Indeed this is obvious if $x\in M$. If $x\in M^\perp\cap S_X$, then $\pi^*(x)$ is defined and belongs to S_Y. Therefore, by (3. 6), we have

$$2(\pi_*(y)\cdot x) = (\pi^*\pi_*(y)\cdot\pi^*(x)) = (2y\cdot\pi^*(x)) = 0.$$

This proves $\pi_*(T_Y)\subset T_X$. Similarly we have $\pi^*(T_X)\subset T_Y$. Now, by Lemma 3. 3, we have (noting $M\subset S_X$)

$$T_X = S_X^\perp \subset M^\perp = \pi_*(H_Y) = \pi_*(L)\oplus\Gamma.$$

But T_X is orthogonal to $\Gamma\subset S_X$, and hence $T_X\subset\pi_*(L)$. For any $x\in T_X$, we can find a unique $y\in L$ with $x=\pi_*(y)$. Since $\pi^*(x)=2y$ belongs to T_Y and T_Y is primitive in H_Y, $y\in T_Y$. This proves $T_X\subset\pi_*(T_Y)$. q. e. d.

Proof of Theorem 2. Since $T_Y\subset L$, Lemmas 3. 2 and 3. 5 imply that $\pi_* : T_Y\to T_X$ is bijective and satisfies the condition (a). This proves Theorem 2. q. e. d.

Combining Theorems 1 and 2, we have the following

Theorem 3. *Let $A=C_1\times C_2$ be a product abelian surface. Then there exists a K3 surface Y such that*

(a) $\varphi : T_Y\simeq T_A$ (*isomorphism of Euclidean lattices*),

(b) $p_A\circ\varphi=$const. p_Y.

Proof. We consider the Kummer surface $X=\mathrm{Km}(A)$, and the rational map $\alpha : A\to X$ of degree 2. It is well-known (cf. [4] § 5) that α_* gives a bijection $T_A\to T_X$ such that (a′) $(\alpha_*t_1\cdot\alpha_*t_2)=2(t_1\cdot t_2)$ for $t_1, t_2\in T_A$; (b′) $p_X\circ\alpha_*=$const. p_A. On the other hand, X has the elliptic pencil $\Phi : X\to \boldsymbol{P}^1$, constructed in Theorem 1. Let $\pi : Y\to X$ be the double covering of X with respect to the elliptic pencil Φ, as defined at the beginning of this section. In view of Theorem 2, the map $\varphi=(\alpha_*)^{-1}\circ\pi_*$ satisfies the required properties (a) and (b). q. e. d.

4. Classification of Singular K3 Surfaces

For a singular $K3$ (or abelian) surface Y, the lattice of transcendental cycles T_Y has a natural orientation. Namely we call a basis $\{y_1, y_2\}$ of T_Y *oriented* if the imaginary part of $p_Y(y_1)/p_Y(y_2)$ is positive. Using an oriented basis $\{y_1, y_2\}$ of T_Y, we put

$$Q_Y = \begin{bmatrix} y_1^2 & y_1y_2 \\ y_1y_2 & y_2^2 \end{bmatrix}. \tag{4. 1}$$

Let us denote by $\mathcal{Q}$ the set of 2×2 positive-definite even integral matrices

$$Q=\begin{bmatrix} 2a & b \\ b & 2c \end{bmatrix} \qquad (a, b, c\in\boldsymbol{Z}, \quad a, c>0, \quad b^2-4ac<0). \tag{4. 2}$$

We define $Q_1\sim Q_2$ if and only if $Q_1={}^t\gamma Q_2\gamma$ for some $\gamma\in SL_2(\boldsymbol{Z})$. We denote by $\{Q\}$

the equivalence class of Q, and by $\mathcal{Q}/SL_2(\boldsymbol{Z})$ the set of equivalence classes. Note that $\{Q_Y\}$ is uniquely determined by Y.

Theorem 4. *The map $Y \mapsto \{Q_Y\}$ establishes a bijective correspondence from the set of singular $K3$ surfaces onto $\mathcal{Q}/SL_2(\boldsymbol{Z})$.*

Proof. The injectivity of this correspondence was proved essentially by Pjateckii-Šapiro and Šafarevič. (See [4] § 8. Compare also [8] § 5. We pointed out in [8] p. 278 that there was a certain gap in the proof of the Torelli theorem in [4], but it has since been filled in by several people, notably by M. Rapoport.) The surjectivity follows from the corresponding result on singular abelian surfaces [8] and Theorem 3. Namely we take Q as in (4. 2), and consider the singular abelian surface A:

(4. 3) $$A = C_1 \times C_2, \quad C_\nu = \boldsymbol{C}/\boldsymbol{Z}+\boldsymbol{Z}\tau_\nu \quad (\nu=1, 2),$$

where

(4. 4) $$\tau_1 = (-b+\sqrt{\Delta})/2a, \quad \tau_2 = (b+\sqrt{\Delta})/2 \quad (\Delta = b^2-4ac).$$

By [8] § 3, we have $Q_A = Q$. Let Y be the double covering of the Kummer surface $X = \mathrm{Km}(A)$ with respect to the elliptic pencil Φ, constructed in Theorem 1. Then, using Theorem 2, we have

$$Q_Y = \frac{1}{2} Q_X = Q_A = Q.$$

This proves the surjectivity of the correspondence. q. e. d.

For a singular $K3$ (or abelian) surface X, we put

(4. 5) $$C_X = H^2(X, \mathcal{O}_X)/\text{image } H^2(X, \boldsymbol{Z}),$$

which is an elliptic curve with complex multiplications (cf. [8] § 1). We mention the following result, suggested by Šafarevič [5] p. 416.

Corollary. *For two singular $K3$ surfaces X and Y, the following statements are equivalent to each other:*

(1) *X and Y are related by an algebraic correspondence.*

(2) *$\boldsymbol{Q}(\sqrt{\Delta_X}) = \boldsymbol{Q}(\sqrt{\Delta_Y})$, where $\Delta_X = -\det Q_X$ and $\Delta_Y = -\det Q_Y$.*

(3) *The elliptic curves C_X and C_Y are isogenous.*

Proof. In view of the construction of a singular $K3$ surface X with a prescribed $\{Q_X\}$ (Theorem 4), the assertion is reduced to the case of singular abelian surfaces. In that case, it follows from [8] § 3. q. e. d.

Remark. Actually (1) can be replaced by a stronger condition: (1′) There is a rational map of finite degree of X to Y. This condition (1′) turns out to be *symmetric* in X, Y, like the notion of isogeny for abelian varieties. The proof is based on another construction of a singular $K3$ surface, not as a double covering, but as a quotient of a suitable Kummer surface.

5. Automorphisms

Theorem 5. *Every singular K3 surface has an infinite group of automorphisms.*

Proof. Let Y be a singular $K3$ surface with $Q_Y=Q$. We distinguish the three cases:

$$(\alpha)\quad Q\not\sim\begin{bmatrix}2&1\\1&2\end{bmatrix},\ \begin{bmatrix}2&0\\0&2\end{bmatrix},\qquad (\beta)\quad Q\sim\begin{bmatrix}2&1\\1&2\end{bmatrix},\qquad (\gamma)\quad Q\sim\begin{bmatrix}2&0\\0&2\end{bmatrix}.$$

In each case, we shall exhibit *an elliptic pencil on Y having an infinite group of sections.* Since any section of an elliptic pencil defines an automorphism of Y, this will prove Theorem 5.

Recall that the $K3$ surface Y is obtained as the double covering of a Kummer surface $X=\mathrm{Km}(A)$ with $Q_A=Q$. As we have seen in § 3, there is an elliptic pencil $\Psi: Y\to\Delta$ with a section, whose singular fibres are described in Lemma 3. 1. Using the formula (1. 2), we can determine the rank $r(\Psi)$ of the group of sections of Ψ. The result is

$$r(\Psi)=0,\ 1,\ 2,\ 2 \text{ and } 0, \tag{5. 1}$$

according to the cases 1), 2), 3), 4) and 5) of Lemma 3. 1. In case 1), the formula (1. 3) implies that

$$\det Q_Y=\det T_Y=2\cdot 2/n(\Psi)^2.$$

Since Q_Y is even, we have $n(\Psi)=1$ and $\det Q_Y=4$. Hence $Q_Y\sim\begin{bmatrix}2&0\\0&2\end{bmatrix}$, which corresponds to the case (γ). A similar argument shows that, in case 5), we have $Q_Y\sim\begin{bmatrix}2&1\\1&2\end{bmatrix}$, which corresponds to the case (β). Thus the assertion for the case (α) is a consequence of (5. 1).

For the cases (β) and (γ), we need the following lemmas:

Lemma 5. 1. *Let $A=C_\omega\times C_\omega$, C_ω being the elliptic curve with the fundamental periods 1 and $\omega=e^{2\pi i/3}$. Let σ be the automorphism of A defined by $\sigma(z_1, z_2)=(\omega z_1, \omega^2 z_2)$. Then the minimal non-singular model Y of the quotient surface $A/\langle\sigma\rangle$ is a singular $K3$ surface such that $Q_Y\sim\begin{bmatrix}2&1\\1&2\end{bmatrix}$.*

Proof. The automorphism σ has the 9 fixed points (v_i, v_j) $(1\le i, j\le 3)$, where $\{v_i\}$ are the fixed points of the automorphism σ_1 of C_ω, defined by $\sigma_1(z)=\omega z$. The quotient $A/\langle\sigma\rangle$ has the 9 singular points p_{ij}, each of which is locally isomorphic to the singularity N_{-3} in the notation of Kodaira [3] p. 583. The minimal non-singular model Y of $A/\langle\sigma\rangle$ is obtained by a "canonical reduction", in which each p_{ij} is replaced by 2 non-singular rational curves E_{ij} and E'_{ij} with $E_{ij}E'_{ij}=1$. Moreover, Y contains 6 non-singular rational curves, i. e. the image F_i (or G_j) of $v_i\times C_\omega$ (or $C_\omega\times v_j$) in Y. These 24 curves on Y form the configuration of Fig. 2.

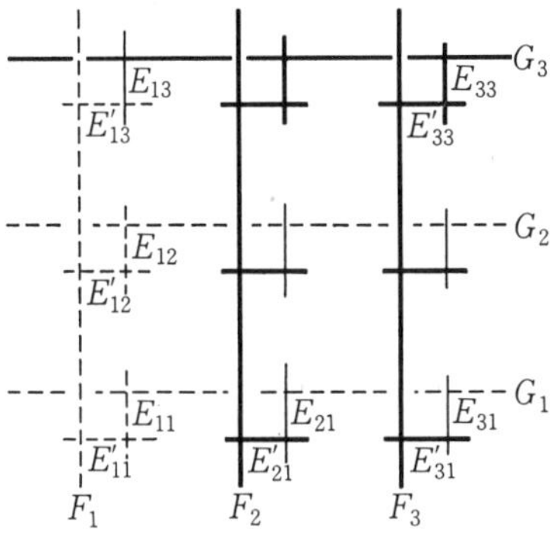

Fig. 2

Now it is easily verified that Y is a $K3$ surface. Furthermore the elliptic pencil $\Psi : Y \to \boldsymbol{P}^1$, induced by the projection $A \to C_\omega$ to the first factor, has the 3 singular fibres of type IV* :

$$\sum_{j=1}^{3} E_{ij} + 2\sum_{j=1}^{3} E'_{ij} + 3F_i \qquad (i=1, 2, 3),$$

and the 3 sections G_1, G_2 and G_3. Applying (1. 2), we see that $\rho(Y)=20$ and $r(\Psi)=0$. Then, by (1. 3), we have det $T_Y=3^3/n^2$ with $n=n(\Psi)\geq 3$. Hence det T_Y $=3$, and $Q_Y \sim \begin{bmatrix} 2 & 1 \\ 1 & 2 \end{bmatrix}$. q. e. d.

Lemma 5. 2. *Let $A=C_i \times C_i$, C_i being the elliptic curve with the fundamental periods 1 and $i=\sqrt{-1}$. Let σ be the automorphism of A defined by $\sigma(z_1, z_2)=(iz_1, -iz_2)$. Then the minimal non-singular model Y of the quotient surface $A/\langle\sigma\rangle$ is a singular $K3$ surface such that $Q_Y \sim \begin{bmatrix} 2 & 0 \\ 0 & 2 \end{bmatrix}$.*

Proof. Since σ^2 is the inversion ι_A of A, it has 16 fixed points. Among these, the 4 points (v_i, v_j) $(1 \leq i, j \leq 2)$ are the fixed points of σ, where $\{v_1, v_2\}$ are the fixed points of the automorphism σ_1 on C_i defined by $\sigma_1(z)=iz$. It follows that the quotient surface $A/\langle\sigma\rangle$ has the 4 singular points p_{ij} of type N_{-4} corresponding to

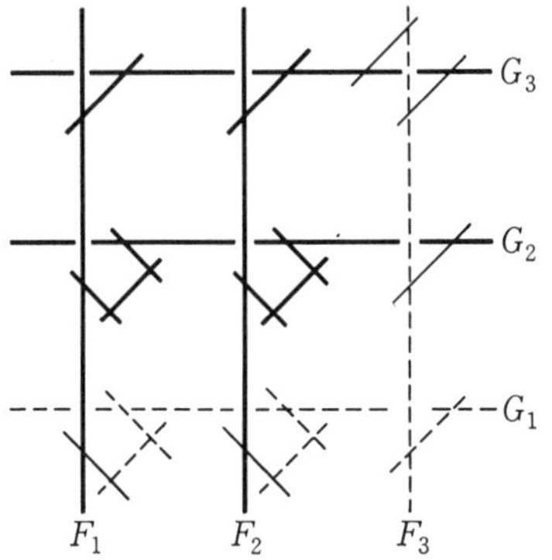

Fig. 3

(v_i, v_j) and the 6 other singular points of type N_{-2} (cf. [3] p. 583). The minimal non-singular model of $A/\langle\sigma\rangle$ is obtained by a "canonical reduction", in which each p_{ij} is replaced by 3 curves and each of the 6 other singular points is replaced by a curve. These curves, together with the 6 image curves of $v\times C_i$ and $C_i\times v$ (v=points of order 2 of C_i), form the configuration of non-singular rational curves on Y, which is described in Fig. 3.

Now the projection $A\to C_i$ (to the first factor) induces an elliptic pencil $\Phi: Y\to \boldsymbol{P}^1$, which has the 3 singular fibres of types III*, III* and I_0^* (delete the 3 horizontal curves G_1, G_2, G_3 from Fig. 3). Moreover G_1 and G_2 are sections of Φ. On the other hand, Y is easily seen to be a $K3$ surface. Hence, applying (1. 2), we have $\rho(Y)=20$ and $r(\Phi)=0$. Then (1. 3) implies that $\det T_Y=2\cdot 2\cdot 4/n(\Phi)^2$. Since $n(\Phi)\geq 2$, it follows that $\det T_Y=4$ and $Q_Y\sim\begin{bmatrix}2 & 0\\ 0 & 2\end{bmatrix}$. q. e. d.

Proof of Theorem 5 (*Continued*). *Case* (β). We shall find a new elliptic pencil Ψ on the $K3$ surface Y of Lemma 5. 1 such that $r(\Psi)>0$. With the notation in the proof of that lemma, we consider the two divisors (cf. the dotted or bold lines in Fig. 2):

$$D_1 = G_1+2E_{11}+3E'_{11}+4F_1+3E'_{12}+2E_{12}+G_2+2E'_{13},$$
$$D_2 = E'_{21}+E'_{22}+E'_{31}+E'_{32}+2(F_2+E'_{23}+E_{23}+G_3+E_{33}+E'_{33}+F_3),$$

which are respectively of types III* and I_6^* and disjoint from each other. By Lemma 1. 1, there is an elliptic pencil $\Psi: Y\to\boldsymbol{P}^1$, of which D_1 and D_2 are singular fibres and E_{21} is a section. Assume that the group of sections of Ψ is finite, i. e. $r(\Psi)=0$. Let us denote by $m_1, \cdots, m_k$ ($m_1\geq\cdots\geq m_k$) the number of irreducible components of other singular fibres of Ψ. Applying (1. 2), we have $\sum_{\nu=1}^{k}(m_\nu-1)=1$. Hence $m_1=2$ and $m_\nu=1$ for $\nu\geq 2$. Then (1. 3) implies the equality

$$3 = \det T_Y = 2\cdot 4\cdot 2/n(\Psi)^2,$$

which is a contradiction. Thus we have $r(\Psi)>0$.

Case (γ) Similarly we can find an elliptic pencil Ψ on the $K3$ surface Y of Lemma 5. 2 such that $r(\Psi)>0$. Namely there is an elliptic pencil $\Psi: Y\to\boldsymbol{P}^1$ having two singular fibres of types I_{12} and IV* (cf. the bold or dotted lines in Fig. 3) and having at least one section. Using the same argument as in (β), we can show that $r(\Psi)>0$. This completes the proof of Theorem 5. q. e. d.

Remark. As to the automorphisms of a $K3$ surface, the following result might be worth mentioning:

(*) *There is a* K3 *surface with the Picard number* 18 *whose automorphism group is finite.*

In fact, take two non-isogenous elliptic curves C_1, C_2 and put $A=C_1\times C_2$. Then T_A is a unimodular lattice of rank 4 with signature (2, 2). By Theorem 3, there exists a $K3$ surface Y (i.e. a double covering of $\mathrm{Km}(A)$) such that $T_Y\simeq T_A$. Therefore $S_Y=T_Y^{\perp}$ is a unimodular lattice of rank 18 with signature (1, 17). Then, by a result of Vinberg [10], the automorphism group of the Euclidean lattice S_Y has a subgroup of finite index generated by reflections. Applying Theorem 1 of [4] § 7, we conclude that the automorphism group of the $K3$ surface Y is finite.

6. An Arithmetic Application

The elliptic curve C_X (4.5) associated with a singular $K3$ surface X has complex multiplications. Hence it has a model defined over some algebraic number field K_0. We assume that K_0 contains $\mathrm{End}(C_X) \otimes \mathbf{Q}$. By Deuring [1], for any finite extension K of K_0, the Hasse-Weil zeta function $\zeta(C_X/K, s)$ of C_X over K is given by

(6.1) $\quad \zeta(C_X/K, s) \doteq \zeta_K(s)\zeta_K(s-1)L(s-1/2, \chi)^{-1}L(s-1/2, \bar{\chi})^{-1}$,

where $\zeta_K(s)$ is the Dedekind zeta function of K and $L(s, \chi)$ is the Hecke L-function with a suitable Grössencharacter χ. (The symbol $\doteq$ indicates equality up to finitely many Euler factors.)

Theorem 6. *Every singular $K3$ surface X has a model defined over some algebraic number field $K \supset K_0$, and its Hasse-Weil zeta function $\zeta(X/K, s)$ is given by*

(6.2) $\quad \zeta(X/K, s) \doteq \zeta_K(s)\zeta_K(s-1)^{20}\zeta_K(s-2)L(s-1, \chi^2)L(s-1, \bar{\chi}^2)$.

As a consequence $\zeta(X/K, s)$ is meromorphic on the whole s-plane, and satisfies a functional equation. This answers a question raised by Šafarevič [5] p. 416. Also we can verify the conjecture of Tate ([9] § 4, Conj. 2) for singular $K3$ surfaces:

Corollary. *The order of the pole of $\zeta(X/K, s)$ at $s=2$ is equal to 20, the Picard number of X.*

In fact, this follows from the fact that $L(1, \chi^2) L(1, \bar{\chi}^2) \neq 0$, which is a special case of [13] p. 288, Theorem 11.

Proof of Theorem 6. (I) First we consider the case of Kummer surfaces. Assume that $X = \mathrm{Km}(A)$ is the Kummer surface for the singular abelian surface $A = C_1 \times C_2$ (4.3). We note that $C_X \simeq C_A = C_1$ (cf. [8] (3.13)) and C_2 is isogenous to C_1. Then we can find models for C_X and A defined over some algebraic number field $K \supset K_0$ such that (a) all points of order 2 of A are K-rational, and (b) an isogeny of A to C_X^2 is defined over K. In this situation it is obvious that $X = \mathrm{Km}(A)$ is defined over K. Moreover, by blowing up the 16 points of order 2 of A, we obtain a surface $\tilde{A}$ fitting in the following diagram, also defined over K:

(6.3)
$$\begin{array}{ccc} A & \xleftarrow{\beta} & \tilde{A} \\ \downarrow & \searrow{\alpha} & \downarrow{\tilde{\alpha}} \\ A/\iota_A & \longleftarrow & X \end{array}$$

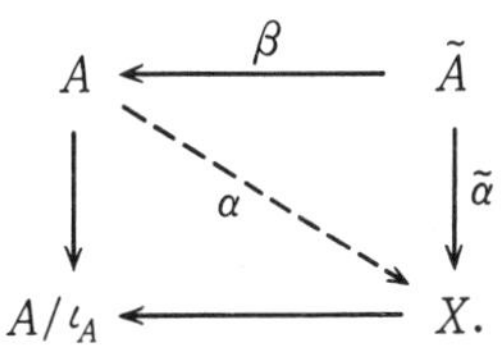

The inversion ι_A of A induces an involution $\tilde{\iota}_A$ of $\tilde{A}$ such that X is the quotient of $\tilde{A}$ by $\tilde{\iota}_A$. Fix a prime number l. For almost all prime ideals $\mathfrak{p}$ of K $(2l \nmid N\mathfrak{p})$, we can consider the reduction mod $\mathfrak{p}$ of the diagram (6. 3). For a moment, using the same notation $A, X, \cdots$ for the reduced varieties, we consider the diagram (6. 3) defined over the finite field $\boldsymbol{F}_q$ of $q = N\mathfrak{p}$ elements. For a surface Z over $\boldsymbol{F}_q$, we denote by H_Z the l-adic cohomology group $H^2(\bar{Z}, \boldsymbol{Q}_l)$ $(\bar{Z} = Z \otimes \bar{\boldsymbol{F}}_q$, $\bar{\boldsymbol{F}}_q$=algebraic closure of $\boldsymbol{F}_q)$, and by S_Z the subspace of H_Z generated over $\boldsymbol{Q}_l$ by algebraic cycles (defined over $\bar{\boldsymbol{F}}_q$). Then we have the induced isomorphisms :

$$(6.4) \qquad \tilde{\alpha}^* : S_X \simeq S_{\tilde{A}}.$$

$$(6.5) \qquad H_A/S_A \underset{\beta^*}{\simeq} H_{\tilde{A}}/S_{\tilde{A}} \underset{\tilde{\alpha}^*}{\simeq} H_X/S_X.$$

In fact, this follows immediately from the fact that $\tilde{\iota}_A^*$ acts trivially on $H_{\tilde{A}}$ (A : abelian surface). As usual we put

$$(6.6) \qquad P_2(Z, T) = \det(1 - F^* T | H_Z) \qquad (T = q^{-s}),$$

where F^* is the endomorphism of H_Z induced by the Frobenius morphism F of Z over $\boldsymbol{F}_q$. By (6. 4), (6. 5), we have

$$(6.7) \qquad \begin{aligned} P_2(X, T) &= \det(1 - F^* T | S_X) \det(1 - F^* T | (H_X/S_X)) \\ &= P_2(\tilde{A}, T) \\ &= (1 - qT)^{16} P_2(A, T). \end{aligned}$$

On the other hand, if we set

$$(6.8) \quad \zeta(C_X, T) = (1 - \pi T)(1 - \bar{\pi} T)/(1 - T)(1 - qT) \qquad (\pi\bar{\pi} = q, \ |\pi| = q^{1/2}),$$

we have

$$(6.9) \qquad P_2(A, T) = (1 - qT)^4 (1 - \pi^2 T)(1 - \bar{\pi}^2 T),$$

because A is isogenous to $C_X \times C_X$. Hence we have

$$(6.10) \qquad P_2(X, T) = (1 - qT)^{20} (1 - \pi^2 T)(1 - \bar{\pi}^2 T).$$

Going back to the global situation, we first note that $\chi(\mathfrak{p}) = \pi/|\pi|$ (π as in (6. 8)) gives the Grössencharacter χ in (6. 1). Therefore, using (6. 10), we obtain

$$(6.11) \quad \begin{aligned} \zeta(X/K, s) &\doteqdot \prod_{\mathfrak{p}} [(1 - N\mathfrak{p}^{-s}) P_2(X \bmod \mathfrak{p}, N\mathfrak{p}^{-s})(1 - N\mathfrak{p}^{2-s})]^{-1} \\ &\doteqdot \zeta_K(s) \zeta_K(s-1)^{20} \zeta_K(s-2) L(s-1, \chi^2) L(s-1, \bar{\chi}^2), \end{aligned}$$

as asserted in (6. 2).

(II) In the general case, a singular $K3$ surface Y is obtained as a double covering of a Kummer surface $X = \mathrm{Km}(A)$. The elliptic curve C_Y is isomorphic to C_X, since $2Q_Y \sim Q_X$. Using the same notation as in (I), we note that all irreducible curves in the double Kummer pencil on X (Fig. 1) are K-rational ; hence the divisor D in (2. 3) is also K-rational. Therefore the elliptic pencil Φ considered in Theorem 1 is defined over K, since it is the rational map (in fact morphism) associated with the complete linear system $|D|$. Also the double covering $f : \Delta \to \boldsymbol{P}^1$, ramified at $t_1 = \Phi(B_1)$ and $t_2 = \Phi(B_2)$ (cf. (2. 4)), is defined over K. Hence the double covering Y of X with respect to Φ (§ 3), together with the rational map $\pi : Y \to X$, is defined over K. By blowing up Y at the 8 fundamental points of π, we obtain the following diagram, similar to (6. 3) :

$$(6.12)\qquad \begin{array}{ccc} Y & \xleftarrow{\ \beta'\ } & \tilde{Y} \\ \downarrow & \overset{\pi}{\searrow} & \downarrow \tilde{\pi} \\ Y/\iota & \longleftarrow & X = \tilde{Y}/\tilde{\iota}\ . \end{array}$$

Replacing K by a finite extension, we can assume that this diagram is defined over K. For almost all prime ideals $\mathfrak{p}$ of K, we have a reduced diagram mod $\mathfrak{p}$, defined over $\boldsymbol{F}_q$ ($q=N\mathfrak{p}$). In this situation over $\boldsymbol{F}_q$, we have

$$(6.13)\qquad H_{\tilde{Y}} = H_Y \oplus \langle 8 \text{ blown up curves} \rangle$$

$$(6.14)\qquad H_Y = \Gamma_1 \oplus \Gamma_2 \oplus L,$$

where Γ_n ($n=1, 2$) is the subspace of H_Y generated over $\boldsymbol{Q}_l$ by the 8 irreducible components with multiplicity ≥ 2 in the singular fibre $\Psi^{-1}(s_{0n})$ of type II* (cf. (3. 4)). Since the involution ι of Y acts trivially on L (cf. Lemma 3. 2), we deduce the following isomorphism :

$$(6.15)\qquad H_Y/S_Y \underset{\beta'^*}{\rightleftarrows} H_{\tilde{Y}}/S_{\tilde{Y}} \underset{\tilde{\pi}'^*}{\rightleftarrows} H_X/S_X.$$

As in (I), this implies

$$(6.16)\qquad P_2(Y \bmod \mathfrak{p},\ T) = P_2(X \bmod \mathfrak{p},\ T)$$

for almost all $\mathfrak{p}$, and hence we have

$$(6.17)\qquad \zeta(Y/K, s) \doteqdot \zeta(X/K, s).$$

This completes the proof of Theorem 6. q. e. d.

Remark. Theorem 6 and its Corollary were known for some special singular $K3$ surfaces, e.g. the Fermat quartic surface defined over $\boldsymbol{Q}(e^{2\pi i/8})$ (Weil [11], [12], cf. [9] p. 105) and the elliptic modular surface of level 4 defined over $\boldsymbol{Q}(\sqrt{-1})$ (Shioda [7] p. 57). The former corresponds to the matrix $\begin{bmatrix} 8 & 0 \\ 0 & 8 \end{bmatrix}$ and the latter to $\begin{bmatrix} 4 & 0 \\ 0 & 4 \end{bmatrix}$.

References

[1] Deuring, M. : Die Zetafunktion einer algebraischen Kurve vom Geschlechte Eins, Nachr. Akad. Wiss. Göttingen (1953), 85–94.

[2] Inose, H. : On certain Kummer surfaces which can be realized as non-singular quartic surfaces in $\boldsymbol{P}^3$, J. Fac. Sci. Univ. Tokyo, Sec. IA, **23** (1976), 545–560.

[3] Kodaira, K. : On compact analytic surfaces II–III, Ann. of Math. **77** (1963), 563–626 ; **78** (1963), 1–40.

[4] Pjateckii-Šapiro, I. I. and Šafarevič, I. R. : A Torelli theorem for algebraic surfaces of type $K3$, Izv. Akad. Nauk SSSR, **35** (1971), 530–572.

[5] Šafarevič, I. R. : Le théorème de Torelli pour les surfaces algébriques de type $K3$, Actes,

Congrès intern. math. (1970) Tome 1, 413–417.
[6] Šafarevič, I. R. et al. : Algebraic surfaces, Proc. Steklov Inst. Math., **75** (1965) ; AMS translation (1967).
[7] Shioda, T. : On elliptic modular surfaces, J. Math. Soc. Japan, **24** (1972), 20–59.
[8] Shioda, T. and Mitani, N. : Singular abelian surfaces and binary quadratic forms, in "Classification of algebraic varieties and compact complex manifolds", Springer Lecture Notes, No. 412 (1974), 259–287.
[9] Tate, J. : Algebraic cycles and poles of zeta functions, in "Arithmetical Algebraic Geometry", Harper and Row (1965), 93–110.
[10] Vinberg, È. B. : Some arithmetical discrete groups in Lobačevskii spaces, in "Discrete subgroups of Lie groups and applications to Moduli", Tata-Oxford (1975), 323–348.
[11] Weil, A. : Numbers of solutions of equations in finite fields, Bull. Amer. Math. Soc., **55** (1949), 497–508.
[12] Weil, A. : Jacobi sums as "Grössencharaktere", Trans. Amer. Math. Soc., **73** (1952), 487–495.
[13] Weil, A. : Basic number theory, Springer (1967).

Department of Mathematics
University of Tokyo

(Received November 7, 1975)

On the Minimality of Certain Hilbert Modular Surfaces

G. van der Geer and A. Van de Ven[1)]

1. Introduction

For some time now Hirzebruch and others have studied certain fields of Hilbert modular functions from a geometric point of view (see [2], [3], [5]). This leads to the introduction of a non-singular algebraic surface $Y^0(p)$ for all square-free positive integers p. In [5] the question is settled how the surfaces $Y^0(p)$ fit into the rough classification of algebraic surfaces, at least for those values of p which are prime and congruent 1 mod 4.[2)] It turns out that for $p=5$, 13 and 17 the surface $Y^0(p)$ is rational, that for $p=29$, 37 and 41 this surface is an elliptic $K3$-surface, that for $p=53$, 61 and 73 it is a minimal honestly elliptic surface, and that for $p\geq 89$ the surface $Y^0(p)$ is of general type. As already follows from this description, the surfaces $Y^0(p)$ are minimal (i.e. without exceptional curves of the first kind) for $29\leq p\leq 73$. Now it is stated as a conjecture in [5] (p. 21) that this remains true for all $p\geq 89$. Of course, it would be very interesting if this conjecture could be proved. In fact, if you know that a certain (simply connected) surface is of general type, and even if you know in addition its arithmetical genus and its Euler characteristic, but you don't know whether it is minimal, your knowledge does not amount to very much. For example, you don't know the (higher) plurigenera of your surface, and in the case we consider here, these are of particular importance in number theory, for they give the dimensions of certain spaces of weighted holomorphic cusp forms ([3], p. 255).

Given this situation, and given the fact that even Hirzebruch believes that the conjecture in question poses a very difficult problem ([3], loc. cit.), we think it of some use to present here a few principles which lead to a proof of the conjecture for many primes, in particular for all primes p (congruent 1 mod 4) with $89\leq p\leq 317$. We emphasize that it does not seem possible to obtain a general proof along these lines ; our proofs are slightly different from prime to prime.

The principles, mentioned in this preceding paragraph, are explained in chapter 2, where also some facts about the surfaces $Y^0(p)$ are recalled. In chapter 3 we prove the conjecture for $p=89$, 193, 229 and 293, leaving the other cases between

1) This research has been supported by the Netherlands organization for the advancement of pure research Z. W. O.

2) The cases $p\not\equiv 1$ mod 4 are treated in [6].

$p=89$ and $p=317$ to the reader. Finally, in chapter 4 a few comments are given.

We wish to thank Mr. R. Brand from our institute, who wrote the computer programs.

2. Preliminaries

We shall use the notations of [5]. In particular, we refer to [5] for the definition of the surfaces $Y(p)$, the surfaces $Y^0(p)$, and the involution ι on $Y^0(p)$.

On $Y^0(p)$ we have the curves S_i and $S'_i=S_{-i}$, arising from the resolution of the cusps, the curves F_N (in as far as they are not blown down), the curves D_i, arising from quotient singularities of order 2 which lie on F_p, and not on F_1, the curves C_i and C'_i, coming from quotient singularities of order 3 on F_p. Those curves, coming from the resolution of quotient singularities of order 2, which are interchanged pairwise by ι will be denoted by O_i, $i=1, \cdots, \frac{1}{2}h(-p)-1-\delta$, with $\iota(O_i)=O_{i+1}$, i odd; and those which come from order 3 singularities and are interchanged pairwise by ι will be denoted by T_j, $j=1, \cdots, \frac{1}{2}h(-3p)-1-\varepsilon$, with $\iota(T_j)=T_{j+1}$, j odd. Here δ and ε are defined as in [5], p. 17.

For the next three propositions, see [5], p. 6 and 7.

Proposition 2. 1. *If on a non-singular algebraic surface X with $q(X)=0$ there exists an irreducible curve C with $KC<0$ and $C^2\geq 0$, then X is a rational surface.*

Proposition 2. 2. *If on a non-singular algebraic surface X with $q(X)=0$ there exists a curve C with $KC<0$ and which has at least one singular point or which is not rational, then X is rational.*

Also, if on such a surface X there exists an irreducible curve C with $KC\leq -2$, then X is again a rational surface.

Proposition 2. 3. *If on a non-singular algebraic surface X with $q(X)=0$ there exist two intersecting exceptional curves (of the first kind), then X is rational.*

An important role in the study of the surfaces $Y^0(p)$ is played by (-2)-configurations on these surfaces, i.e. the configurations consisting of finitely many non-singular rational curves with self-intersection -2, the union of which is connected. It is known which (-2)-configurations can exist on a non-singular surface of general type. In particular one has

Proposition 2. 4. *Let $\{C_1, \cdots, C_k\}$ be a (-2)-configuration on a surface of general type. Then the intersection matrix $(C_iC_j)_{1\leq i,j\leq k}$ is negative definite.*

Proposition 2. 5. *Let X be an algebraic surface of general type, and C an irreducible*

curve on X. *Then* $KC>0$, *except if either* C *is an exceptional curve of the first kind* (*then* $KC=-1$), *or if* C *is a non-singular rational curve with* $C^2=-2$ (*then* $KC=0$).

Proof. The proposition is well-known for the case that X is a minimal surface ([1], p. 174). As to the general case, it is sufficient to prove: if the statement holds for the surface X, then it also holds for the surface Y, obtained from X by blowing up a point $p\in X$. This, however, is an easy consequence of Proposition I. 1 in [5].

The next proposition is a special case of the lemma on p. 266 in [3].

Proposition 2. 6. *Let the surface* X *be non-ruled, and* $\iota: X\to X$ *a holomorphic involution on* X, *and* $\alpha: X\to Y$ *the canonical map from* X *onto the quotient* $Y=X/\iota$. *Suppose* Y *to be non-singular. Then if* E *is any exceptional curve of the first kind on* X *not contained in the ramification curve, then* E *does not meet the ramification curve on* X. *Consequently,* $E\cap\iota(E)=\phi$, *and there exists an exceptional curve* E' *on* Y, *such that* $\alpha^{-1}(E')=E\cup\iota(E)$. *Furthermore, if* X_0 *is obtained from* X *by blowing down* E *and* $\iota(E)$, *and if* Y_0 *is obtained from* Y *by blowing down* E', *then* ι *induces an involution* κ *on* X_0, *with* $X_0/\kappa=Y_0$.

Let $A\subset Y(p)$ be the union of all the curves, appearing in the basic configuration ([5], p. 18), with the exception of F_p, and let B be the image of A on $Y^0(p)$. Then we have:

Proposition 2. 7. *If* E *is any exceptional curve of the first kind on* $Y^0(p)$, *then* $E\cap B$ *consists of at least three different points.*

Proof. There exists on

$$X=\mathcal{H}\times\mathcal{H}/\mathrm{SL}(2,\mathfrak{o}_K)-(\text{union of all quotient singularities})$$

a curve, isomorphic to $E-E\cap B$ (here $K=\boldsymbol{Q}(\sqrt{p})$). If $E\cap B$ would consist of not more than two points, then there would exist on X a curve, isomorphic to either $\boldsymbol{P}_1$, $\boldsymbol{C}$ or $\boldsymbol{C}^*$. An unramified covering of this curve would be contained in $\mathcal{H}\times\mathcal{H}$. But every map from $\boldsymbol{P}_1$, $\boldsymbol{C}$ or $\boldsymbol{C}^*$ into $\mathcal{H}$ is constant (for $\boldsymbol{C}$ and hence for $\boldsymbol{C}^*$ this follows for example from Picard's theorem).

Apart from the condition, embodied in Proposition 2. 7, there are other conditions for exceptional curves on $Y^0(p)$, which follow easily from Propositions 2. 1 to 2. 6. So for $p\geq 89$ one has the following conditions for exceptional curves E on $Y^0(p)$:

(a) $E\cap\iota(E)=\phi$;

(b) if C is a non-singular rational curve with $C^2=-2$, -3 or -4, then $CE\leq 1$;

(c) E intersects at most two non-singular rational curves with self-intersection -3, and if E meets two of them, then E does not meet any such curve with self-intersection -2;

(d) if a (-2)-curve C and $\iota(C)$ belong to the same (-2)-configuration, then no exceptional curve meets C.

In some cases, like $p=89$, the minimality of $Y^0(p)$ follows already from these simple rules. However, in other cases this is not sufficient. Sometimes (like for $p=193$) a very special argument gives the desired result, but most effective seems to be a reasoning used many times in the cases $p=229$ and $p=293$, which in its simplest form can be described in the following way.

Suppose $Y^0(p)$ is not minimal. Then, by (a), there exist at least two disjoint exceptional curves on $Y^0(p)$. Let Y be the surface obtained from $Y^0(p)$ by blowing down these two curves. We have $p_g(Y)=p_g(Y^0(p))$. On the (hypothetical) surface Y we construct a particular positive canonical divisor

$$K = \Sigma a_i A_i, \qquad a_i \in \mathbf{Z},\ a_i \geq 1,$$

the A_i's being irreducible curves on Y (the divisor K is chosen differently in different situations). We then single out one of the curves A_i, say A_0, to be the "final curve". By intersecting K with the other curves A_i we obtain inequalities of the following type

$$a_i \geq \alpha_i a_0 + \beta_i, \qquad \alpha_i, \beta_i \in \mathbf{Q},\ \alpha_i \geq 0.$$

Finally, we consider the intersection number

$$(\Sigma a_i A_i) A_0 = a_0 A_0^2 + \Sigma a_i A_0 A_i$$

and find it to be more than KA_0, a contradiction.

To illustrate this, we describe a simple example.

Let $\mathcal{L}$ be a configuration formed by three curves: A_0, A_1 and A_2. The curve A_0 is a rational curve with one ordinary double point p. The curves A_1 and A_2 are non-singular rational curves. Both A_1 and A_2 have only the point p in common with A_0. $A_0^2=-3$ (hence $KA_0=3$), $A_1^2=A_2^2=-2$, $A_0A_1=A_0A_2=2$, $A_1A_2=1$. We claim: *if on a complex surface X a configuration $\mathcal{L}$ exists, then $p_g(X)\leq 1$.*

Indeed, suppose $p_g(X)\geq 2$. Then there exists a positive canonical divisor K on X, passing through p. Such a divisor K contains A_0, A_1 and A_2, and hence it can be written as

$$K = a_0A_0 + a_1A_1 + a_2A_2 + R,$$

with $a_0, a_1, a_2 \geq 1$, $RA_0 \geq 0$, $RA_1 \geq 0$, $RA_2 \geq 0$. Intersection with A_1 gives

$$\begin{aligned} 2a_0 - 2a_1 + a_2 + RA_1 &= 0 \\ 2a_0 - 2a_1 + a_2 &\leq 0, \end{aligned}$$

and similarly intersecting with A_2 gives

$$2a_0 + a_1 - 2a_2 \leq 0.$$

Finally, intersection with A_0 gives

$$\begin{aligned} -3a_0 + 2(a_1 + a_2) &\leq 3 \\ 5a_0 &\leq 3, \end{aligned}$$

a contradiction.

In order to apply this principle, it is obviously very convenient to know that if a canonical curve passes through a certain point, it automatically passes through one or more other points. In this direction we have

Proposition 2. 8. *Let X be a compact complex surface, and let* $\iota : X \to X$ *be a holomorphic involution on X, such that the quotient* X/ι *is a non-singular rational surface Y. If* K_X *is the canonical divisor class on X, then* ι *induces the identity on* $|K_X|$.

Proof. Let R be the ramification curve on X. By a well-known result of Hurwitz we have

$$K_X = \alpha^*(K_Y) + R,$$

where $\alpha : X \to Y$ is the canonical projection. From this formula it follows in our case that there cannot be a divisor $D \in |K_X|$, containing R, for $\alpha^*(D-R)$ would be a divisor in $|2K_Y|$, which is impossible, because Y is rational. Now let s be any section in the canonical bundle of X, different from the zero section. By the preceding remark, the restriction s_R of s to R does not vanish identically. The holomorphic function $\iota^*(s_R)s_R^{-1}$ is equal to a constant c, $c \neq 0$. The section $\iota^*(s) - cs$ of the canonical bundle vanishes on R and hence vanishes identically on X. So we find $\iota^*(s) = cs$, and $\iota^*(s)$ and s have the same zero divisor. This proves the proposition.

The following remark also will turn out to be quite useful for the cases $p=229$ and $p=293$.

By a *chain* of curves on a complex surface we mean a set of irreducible curves $C_1, \cdots, C_n$, with the following intersection behaviour :

$$C_j C_k = \begin{cases} 1 & \text{if } |j-k| = 1 \\ 0 & \text{otherwise.} \end{cases}$$

Proposition 2. 9. *Let* $B_1, \cdots, B_n$ *be a chain of* (-2)*-curves on the surface X, and let* $A_1, \cdots, A_m$ *be irreducible curves on X, all of them different from* $B_1, \cdots, B_n$. *Suppose* $A_i B_1 = A_i B_n = \alpha_i \neq 0$ *for* $i=1, \cdots, m$. *If* $\sum_{i=1}^{m} a_i A_i + \sum_{j=1}^{n} b_j B_j + R$ *with* $A_i R \geq 0$, $B_j R \geq 0$ *is a canonical divisor on X, then*

$$b_j \geq \sum_{i=1}^{m} \alpha_i a_i$$

for all $j=1, \cdots, n$.

Proof. If for this canonical divisor all a_i vanish, then the statement is trivial. Otherwise, we have $b_j \geq 1$ for all $j=1, \cdots, n$. Now suppose $b_j \geq k$ for all $j=1, \cdots, n$ and $1 \leq k < \sum_{i=1}^{m} \alpha_i a_i$. We claim : from this assumption it follows that $b_j \geq k+1$ for all $j=1, \cdots, n$. Indeed, from

$$\left(\sum_{i=1}^{m} a_i A_i + \sum_{j=1}^{n} b_j B_j + R\right) B_1 = 0$$

we derive

$$\sum_{i=1}^{m} \alpha_i a_i - 2b_1 + b_2 \leq 0, \quad \text{i.e.} \quad b_1 \geq k+1.$$

But this and

$$\left(\sum_{i=1}^{m} a_i A_i + \sum_{j=1}^{m} b_j B_j + R\right) B_2 = 0$$

implies $b_2 \geq k+1$, etc. Thus the proposition follows by induction with respect to k, $1 \leq k \leq \left(\sum_{i=1}^{m} \alpha_i a_i\right) - 1$.

Finally, a word or two about the diagrams. These diagrams don't show always the whole picture, that is, curves may have more points in common than shown in the diagram. For example, in the diagram for $p=229$, the curve F_p passes through the intersection point of T_3 and T_4 (which also lies on S_0).

The intersection properties (and other information) used in the diagrams are generally to be found in [3] and [5]. However, § 3 of [4] is needed to see that on $Y^0(293)$ the curves F_{43}, F_{31} and F_{37} are intersected by *different* curves T_i, and that *different* curves O_i meet F_{17} and F_{37}. From [5], p. 20 it follows that F_p (the ramification curve on $Y^0(p)$) is not a rational curve, and hence not an exceptional curve if $p \geq 89$.

3. The Main Result

Theorem 3. 1. *The surface* $Y^0(p)$ *is minimal for* $89 \leq p \leq 317$.

We shall give the proof for four typical cases, namely $p=89$, 193, 229 and 293. The remaining cases can be dealt with in a similar way.

The case $p=89$.

In this case, no exceptional curve E can meet S_i (or S_i') for $i=1$, 2, 3, 5, 7, 8, 9 and 10, because of principle (d). Also, no E can meet S_4 because of (b) and—after the blowing down of E and $\iota(E)$—principle (d) again. It follows from these remarks and from Proposition 2. 7 that any E must have at least three points in common with

$$S_0 \cup S_6 \cup S_6' \cup O,$$

where $O = O_1 \cup O_2 \cup O_3 \cup O_4$. Now $ES_0 \leq 1$, for $ES_0 = \iota(E)S_0$, and if ES_0 were ≥ 2, then after blowing down E and $\iota(E)$ we would arrive at a singular curve S_0 with $KS_0 \leq 0$, which is impossible by Proposition 2. 5. Similarly, $ES_6 \leq 1$ (because of (b)) and $EO \leq 1$ (also because of (b)). Furthermore, $ES_6 = ES_6' = 1$ is impossible, for these relations would imply $\iota(E)S_6 = \iota(E)S_6' = 1$, and after blowing down E and $\iota(E)$ we would obtain a (-2)-configuration, that cannot exist on a surface of general type. There remains only one possibility to exclude: $ES_0 = ES_6 = EO = 1$. But in this case, if say $EO_1 = 1$, and hence $\iota(E)O_2 = 1$, after blowing down E, $\iota(E)$, O_1 and O_2, we would again obtain a singular curve S_0 with $KS_0 \leq 0$, a situation which was already excluded before.

The case $p=193$.

In this case a very special argument can be used. In fact, we know that $Y=Y_0(193)/\iota$ is a blown-up $K3$-surface with $c_1^2(Y)=-18$ ([3]). But on this surface we can blow down eighteen curves, namely D_1^*, D_2^*, C_1^*, $\cdots$, C_4^*, S_{23}^*, $\cdots$, S_{18}^*, F_6^*, F_7^*, F_9^*, B_2^*, F_8^* and S_1^* (as is customary, we set $\alpha_*(A)=A^*$), thus obtaining a minimal surface. Therefore, the image on Y of an exceptional curve on $Y^0(193)$, which is again an exceptional curve on Y by principle (a), has to intersect the union of the eighteen curves mentioned before. Using the fact that Y is not rational, this is easily seen to be impossible.[1)]

The case $p=229$.

It is sufficient to prove the following five statements:

I Any exceptional curve E on $Y^0(p)$ meets at most one of the curves O_1, O_2, and also at most one of the curves T_1, T_2. If E meets any of these curves, then the intersection number is 1.

II The case $EO_1=ET_1=1$ cannot occur.

III The case $EO_1=1$, $ET_1=ET_2=0$ cannot occur.

IV The case $ET_1=1$, $EO_1=EO_2=0$ cannot occur.

V The case $ET_1=ET_2=EO_1=EO_2=0$ cannot occur.

After some preliminary remarks, we shall prove these five statements one by one.

Because of the principles (a)–(d) there is no exceptional curve E intersecting T_3, T_4 or S_i for $i=1, \cdots, 7, 11$. But also S_{10}, S_{12}, S_{13} and S_{14} cannot meet a curve E. For if E would meet S_{10}, then $ES_{10}=1$ (principle (b)) and after blowing down the curves S_{10}, S'_{10}, S_{11}, S'_{11} we would obtain a non-rational curve, namely F_{15}, with $KF_{15}\leq 0$, which is impossible. And if E intersects S_{12} (without meeting S_{10}, we may assume), then after blowing down S_{12}, S_{11}, S_{10} we would have $S_{13}^2\geq 0$ for the non-singular rational curve S_{13}, which is impossible since $Y^0(p)$ is not rational. Also $ES_{14}=1$, hence $\iota(E)S'_{14}=1$ would give $KF_{11}\leq 0$ for the non-rational curve F_{11}. Finally, $ES_{13}=1$ would after blowing down lead to a non-negative definite (-2)-configuration, namely

$$\{F_9, S_{13}, S'_{13}, S_{14}, S'_{14}, S_{12}\}.$$

The (-2)-curves O_3 and O_4, intersected by the (-2)-curve F_5 cannot meet an exceptional curve since otherwise intersecting exceptional curves would be obtained by blowing down.

In this way we find that an exceptional curve E has at least three points in common with $S_0 \cup S_8 \cup S'_8 \cup S_9 \cup S'_9 \cup O_1 \cup O_2 \cup T_1 \cup T_2$. For such a curve E we have in view of principle (b) that $CE\leq 1$ if $C=S_9$, S'_9, O_1, O_2, T_1, T_2, and $CE\leq 2$ if $C=S_8$, S'_8. If E meets O_1 or O_2 then $ES_8\leq 1$, and if E meets both a curve O_i and a curve T_i, then E does not meet S_8, S'_8, S_9, S'_9 by principle (b).

1) Hirzebruch has informed us that he is able to find minimal models of $Y^0(p)/\iota$ for all primes p, for which $Y^0(p)/\iota$ is neither rational, nor of general type. Therefore, for all these primes, we can prove the minimality of $Y^0(p)$ by the same argument.

We now come to the proof of the statements I–V.

Proof of I. This statement is an immediate consequence of our principles (a)–(d).

Proof of II. Suppose $EO_1=ET_1=1$. By the preliminary remarks we have $ES_8=ES'_8=ES_9=ES'_9=0$, hence $ES_0\geq 1$. After blowing down we find $KS_0\leq 2$ and $\pi(S_0)\geq 3$, hence $S_0^2\geq 2$. Now since $p_g=9$, we can find a canonical divisor of the form aS_0+R, $a\geq 1$, $S_0R\geq 1$, hence $KS_0=(aS_0+R)S_0\geq 3$, a contradiction.

Proof of III. We know from the remark above that E intersects $S_0\cup S_8\cup S'_8\cup S_9\cup S'_9$ in at least two points. We claim furthermore : if E meets any of these five curves, then E intersects that curve transversally in one point. For S_8, S'_8, S_9, S'_9 this follows already from the preliminary remarks ; as for S_0, this can be seen by the same argument as was used in case II. Now let E be an exceptional curve with $EO_1=ES_0=ES_8=1$. By blowing down four times we obtain $KS_0=6$, $KS_8=1$, $S_0S_8\geq 2$. Since $p_g=9$, there is a positive canonical divisor

$$aS_0+bS_8+bS'_8+R,$$

with $a, b\geq 1$, $RS_0\geq 0$, $RS_8\geq 0$, $RS'_8\geq 0$, which contains at least once the curves S_1, S'_1, S_9, S'_9, S_{14}, S'_{14}, F_9 and B_2. Intersection with S_8 gives $b\geq 4a$; then intersection with S_0 and application of Proposition 2. 9 gives $7a\leq 6$, a contradiction.

To finish case III, we also have to deal with the following four situations :

$$EO_1 = ES_0 = ES_9 = 1$$
$$EO_1 = ES_8 = ES'_8 = 1$$
$$EO_1 = ES_9 = ES'_9 = 1$$
$$EO_1 = ES_8 = ES_9 = 1.$$

Now the first case being very similar to the one just treated, and the third one being very easy, we shall restrict ourselves to the cases $EO_1=ES_8=ES'_8=1$ and $EO_1=ES_8=ES_9=1$. In the first of these cases we have also $\iota(E)S_8=\iota(E)S'_8=1$, hence after blowing down E, O_1, $\iota(E)$ and O_2 we would obtain intersecting exceptional curves, which is impossible. In the case $EO_1=ES_8=ES_9=1$, after blowing down four times, we can find as usual a canonical divisor

$$aS_8+bS_9+R$$

with $a, b\geq 1$, $RS_8\geq 1$, $RS_9\geq 1$. Intersection with S_8 and S'_8 yields

$$-3a+\beta b+RS_8 = 1$$
$$\beta a-2b+RS_9 = 0 \quad \text{with } \beta\geq 3$$
$$(\beta-3)a+(\beta-2)b+RS_8+RS_9 = 1.$$

This gives a contradiction, for $RS_8\geq 1$ and $RS_9\geq 1$.

Proof of IV. Because of the preliminary remarks it will be sufficient to exclude the following possibilities :

1) E meets S_0 (at least) twice
2) E meets S_8 twice
3) E meets S_0 and S_8 (both once)
4) E meets S_0 and S_9 (both once)

5) E meets S_8 and S_9
6) E meets S_8 and S'_9
7) E meets S_8 and S'_8
8) E meets S_9 and S'_9.

In case 1), after blowing down E and $\iota(E)$, we get $KS_0 \le 6$ and $S_0^2 \ge -2$. We consider the system of canonical divisors passing through three singular points and one simple point of S_0. This system has dimension at least 4, and all its divisors contain S_0, S_1, S'_1, T_1, T_2, T_3 and T_4. Since

$$(K - S_0 - S_1 - S'_1 - T_1 - T_2 - T_3 - T_4)S_0 \le 2,$$

there are canonical divisors containing S_0 at least twice. Thus a contradiction arises from

$$(aS_0 + bS_1 + bS'_1 + c_1T_1 + c_2T_2 + c_3T_3 + c_4T_4 + R)S_0 \ge 6a = 6.$$

In case 2) we can do a similar thing, but now using S_8.

The case 3) is slightly more difficult. After blowing down E and $\iota(E)$, we find $KS_0 = 8$, $S_0^2 = -8$. We can find a canonical divisor containing general points of T_1, T_3, S_1, B_2, S_{14}, F_9, F_{15} and F_{25}. This divisor can be written as

$$aS_0 + bS_8 + bS'_8 + cS_{13} + cS'_{13} + dF_9 + R$$

with $a, b, c, d \ge 1$, $RS_0 \ge 0, \cdots, RF_9 \ge 0$, and R containing all other curves S_i, S'_i, B_2, $T_1, \cdots, T_4$, F_{15} and F_{25}. The curves S_{10}, S_{11} and S_{12} are contained in R at least twice. From $KF_9 \le 0$ we derive $2a - 2d + 2 \le 0$, i.e. $d \ge a + 1$. Then, intersection of D with S_{13} gives us $3c \ge d + 2$, hence $c \ge \frac{1}{3}a + 1$. Intersecting D again with F_9 we find $-2d + 2a + \frac{2}{3}a + 2 \le 0$, that is, $d \ge \frac{4}{3}a + 1$. Next, $KS_8 = 2$ yields us $b \ge \frac{3}{2}a + \frac{1}{2}$. All this information, together with Proposition 2. 9 and the fact that F_{15} is contained in R, gives a contradiction if we intersect D with S_0.

For case 4) the argument is very similar.

In case 5) we take after blowing down, a canonical divisor

$$aS_8 + bS_9 + R,$$

$a \ge 1$, $b \ge 1$, $RS_8 \ge 0$, $RS_9 \ge 0$, and find by successive intersection with S_8 and S_9 that $a \ge b$ and that $-3b + 2b + \frac{1}{2}(a+b) + \frac{1}{2}b + 1 \le 1$, i.e. $a \le 0$, a contradiction.

As to case 6), the same can be done as in the preceding case, but now starting from a canonical divisor

$$aS_8 + bS'_9 + R.$$

In case 7), also $\iota(E)$ meets S_8 and S'_8. After blowing down both curves, we consider a canonical divisor $aS_8 + R$, and obtain immediately a contradiction by intersecting with S_8.

In the final case 8), after blowing down both E and $\iota(E)$, we would obtain two (-2)-curves intersecting in two points, which is impossible on a surface of general type.

Proof of V. We begin by observing that an exceptional curve E can meet S_0 at most once. Otherwise, after blowing down we would obtain $KS_0 \leq 6$, $\pi(S_0) \geq 3$, hence $S_0^2 \geq -2$. Now there is a positive canonical divisor $D = aS_0 + R$, containing S_0 at least twice and F_{15} at least once. D contains S_1 and S_1' both at least $a+1$ times, and T_3, T_4 and F_9 at least a times each. Thus we find

$$(aS_0+R)S_0 \leq 6$$
$$-2a+2a+2+2a+a \leq 6$$
$$3a+2 \leq 6,$$

contradicting $a \geq 2$.

A similar but simpler argument shows that a curve E intersects also S_8 at most once. So we are left with the following possibilities:

1) $ES_0 = ES_8 = ES_9 = 1$
2) $ES_0 = ES_8 = ES_9' = 1$
3) $ES_0 = ES_8 = ES_8' = 1$
4) $ES_0 = ES_9 = ES_9' = 1$
5) $ES_8 = ES_8' = ES_9 = 1$
6) $ES_8 = ES_9 = ES_9' = 1$.

In the cases 1), 5) and 6) we take a canonical divisor

$$a(S_8+S_8')+b(S_9+S_9')+R,$$

with $a, b \geq 1$, $RS_8, \cdots, RS_9' \geq 0$, which contains S_{13}, F_{11}, F_{25} and B_2. Intersection with S_9 gives

$$2a-3b+\frac{3b+1}{2}+1 \leq 1, \qquad b \geq a.$$

Then intersection with S_8 yields

$$-4a+a+a+2a+\frac{a+2}{2}+1 \leq 2, \text{ i.e. } a \leq 0,$$

a contradiction.

For case 2) we can use the same type of canonical divisor, but this time we intersect it with S_9' and S_8 to obtain a contradiction.

Finally, in the cases 3) and 4) we use divisors

$$aS_0+b(S_8+S_8')+R \quad \text{and} \quad aS_0+b(S_9+S_9')+R$$

respectively, to obtain our last contradiction.

The case $p=293$.

In view of the principles (a), $\cdots$, (d) it is sufficient to exclude the following cases:

1) $ES_0 \geq 3$
2) $ES_0 = 2$, $EO = 1$, where O is one of the curves $O_1, \cdots, O_8$
3) $ES_0 = 2$, $ET = 1$, where T is one of the curves $T_1, \cdots, T_8$
4) $ES_0 = 1$, $EO = ET = 1$, T and O as in 2) or 3)

5) $ES_0 = 1$, $ET = ET' = 1$, where T and T' are among T_1, T_3, T_5 and T_7, $T \neq T'$.

For the cases 1) and 2) it is sufficient to intersect (after blowing down E and $\iota(E)$) the curve S_0 with a canonical divisor of the type aS_0+R, $a\geq 1$, $RS_0\geq 0$, to obtain a contradiction.

To exclude case 3), we blow down E and $\iota(E)$, obtaining $KS_0\leq 8$, $S_0^2\geq -4$, and consider a canonical divisor of the type $aS_0+bF_{17}+cF_{37}+R$, $a, b, c\geq 1$, $RS_0\geq 0$, $RF_{17}\geq 0$, $RF_{37}\geq 0$. This divisor contains T, $\iota(T)$, T_9 and T_{10} all at least a times, and S_1 and S_1' at least $a+b+c$ times (Proposition 2. 9).

Hence we would find :

$$-4a+2a+4a+2b+2a+2b+2c \leq 8,$$

again a contradiction.

In case 4), after blowing down six times, a canonical divisor of the type $aS_0+bF_{37}+R$ leads to the usual contradiction.

The last case, case 5), requires a little bit more care. After blowing down E and $\iota(E)$ we have $KS_0=-S_0^2=10$. The curve S_0 intersects three (-2)-crosses, namely the pair T, T' ; the pair $\iota(T)$, $\iota(T')$ and the pair T_9, T_{10}. We claim that it is possible to find a canonical divisor D of the form

$$aS_0+bF_{17}+cF_{31}+dF_{37}+e(S_1+S_1')+f(S_2+S_2')+g(T+\iota(T))+h(T'+\iota(T'))+R,$$

with $a, \cdots, h\geq 1$, and $RS_0, \cdots, R\iota(T')\geq 0$. In fact, the divisors in the at least 4-dimensional linear system of canonical curves, which pass through the intersection point of T and T', through four general points of S_0 and through two general points of F_{17}, all contain S_0 and F_{17} at least once and S_1, S_1' at least twice (Proposition 2. 9). The linear subsystem, passing through two general points of F_{37} contains F_{37}, and hence S_2 and S_2' at least three times. So there is at least one canonical divisor in this subsystem passing through two general points of F_{31}, hence containing F_{31}, and this is a divisor of the type mentioned above. Application of Proposition 2. 9 gives $f\geq 4$, and subsequent intersection with S_1, S_2, F_{17}, S_1 and S_2 gives $e\geq 4$, $f\geq 5$, $b\geq 2$, $e\geq 5$, $f\geq 6$. Since F_{31}, F_{37} and F_{43} intersect different T_i's, we may assume that T and $\iota(T)$ meet one of the curves F_{31}, F_{37} or F_{43}. If T and $\iota(T)$ meet F_{31} or F_{37}, then $g\geq 2$, $h\geq 2$, and intersection with S_0 yields $a\geq 2$. If T and $\iota(T)$ meet F_{43}, then intersection with F_{43} shows that D contains F_{43}, and as before we conclude that $a\geq 2$. Proposition 2. 9 implies $f\geq a+b+d+1$; intersection with S_1 and S_2 now gives $e\geq a+b+d+1$.

Intersection of D with F_{37} and F_{17} leads to

$$-6d+d+2(a+b+d+1)+2a \leq 8,$$
$$d \geq \frac{4}{3}a+\frac{2}{3}b-2$$

and

$$-4b+b+2a+2b+4 \leq 4$$
$$b \geq 2a.$$

Finally, intersection with S_0 yields

$$-10a+\frac{8}{3}a+\frac{8}{3}a-4+6a+4+6a \leq 10,$$
$$a \leq 1,$$

the ultimate contradiction.

4. Some Comments

Once it is known that $Y^0(p)$ is a minimal surface, it is of interest to study the pluricanonical maps of the surface, in particular the canonical and bicanonical maps. As an example we sketch the situation for $p=89$. By a theorem of Bombieri ([1], Theorem 2) the bicanonical system on $Y^0(p)$ has no base points. Therefore, $|2K|$ maps $Y^0(p)$ either birationally onto a surface of degree 8 in P_5, or generically 2 to 1 onto a surface of degree 4 in P_5 (of course not on a surface of degree 1 or 2, for these surfaces are always contained in a proper linear subspace of P_5). Now $p_g(Y^0(89))=3$, so in the first case there would be a linear system of dimension 2 of curves of degree 4 on X. But such curves are always rational or elliptic, and $Y^0(p)$ would not be of general type. Hence $Y^0(p)$ is mapped by $|2K|$ generically 2 to 1 onto a surface of degree 4 in P_5. This surface contains a linear system of dimension 2 of conics, and it is not contained in a hyperplane of P_5. But then, by a very classical theorem, it has to be the Veronese surface, which is biregularly equivalent to P_2. Therefore, $|K|$ gives an everywhere defined map from $Y^0(p)$ onto P_2, generically 2 to 1, and with "exceptional fibres", which are trees of (-2)-curves.

All this also follows from the results of Horikawa ([7]) on minimal surfaces of general type for which $p_g=\frac{1}{2}K^2+2$ (K^2 even) or $p_g=\frac{1}{2}K^2+\frac{3}{2}$ (K^2 odd). These surfaces are in a sense extreme cases, for one has for minimal surfaces of general type always the *in*equalities

$$p_g \leq \frac{1}{2}K^2+2 \qquad (K^2 \text{ even})$$
$$p_g \leq \frac{1}{2}K^2+\frac{3}{2} \qquad (K^2 \text{ odd}).$$

Also for higher values of p it is possible to obtain much information on the canonical and bicanonical maps, and in particular on their relations with the involution ι on $Y^0(p)$. The first of the present authors intends to treat these questions in a forthcoming publication.

As was observed already in chapter 2, what we really do in this paper is to prove theorems of the following type : *If on a compact complex surface X there exists a certain configuration $\mathcal{L}$ of curves, then $P_i(X) \leq C_i(\mathcal{L})$.* The methods used here can also be used to obtain many more examples, some of them quite pretty, but the authors have to admit that they don't even have an idea which (if any) theorem of this type would imply the minimality of $Y^0(p)$ in the general case.

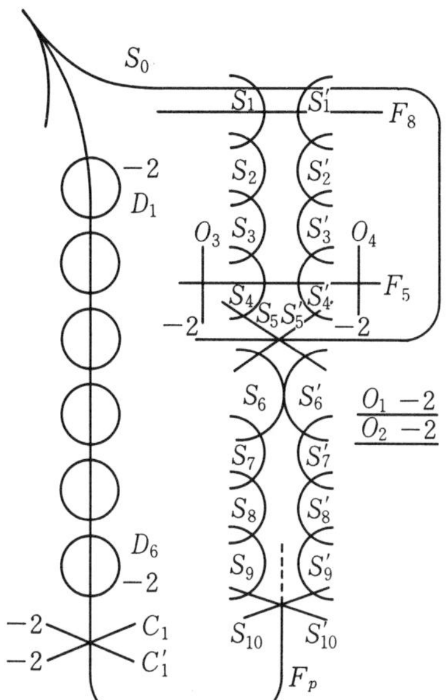

The configuration on $Y^0(89)$. $K^2=2$; $p_g=3$; $KS_0=4$, $\pi(S_0)=1$; $S_i^2=-2$, $i=1,2,3,5,7,8,9,10$. $S_4^2=-3$, $S_6^2=-4$; $KF_5=KF_8=\pi(F_5)=\pi(F_8)=0$

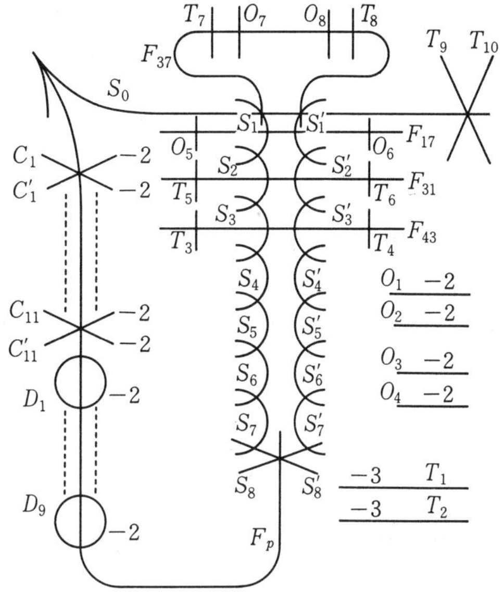

The configuration on $Y^0(293)$. $K^2=40$; $p_g=12$; $KS_0=12$, $\pi(S_0)=1$; $S_i^2=-2$, $i=1,\cdots,8$. $KF_{17}\leq 4$, $\pi(F_{17})=1$; $KF_{31}\leq 8$, $\pi(F_{31})=2$; $KF_{37}\leq 8$, $\pi(F_{37})=2$; $KF_{43}\leq 12$, $\pi(F_{43})=3$.

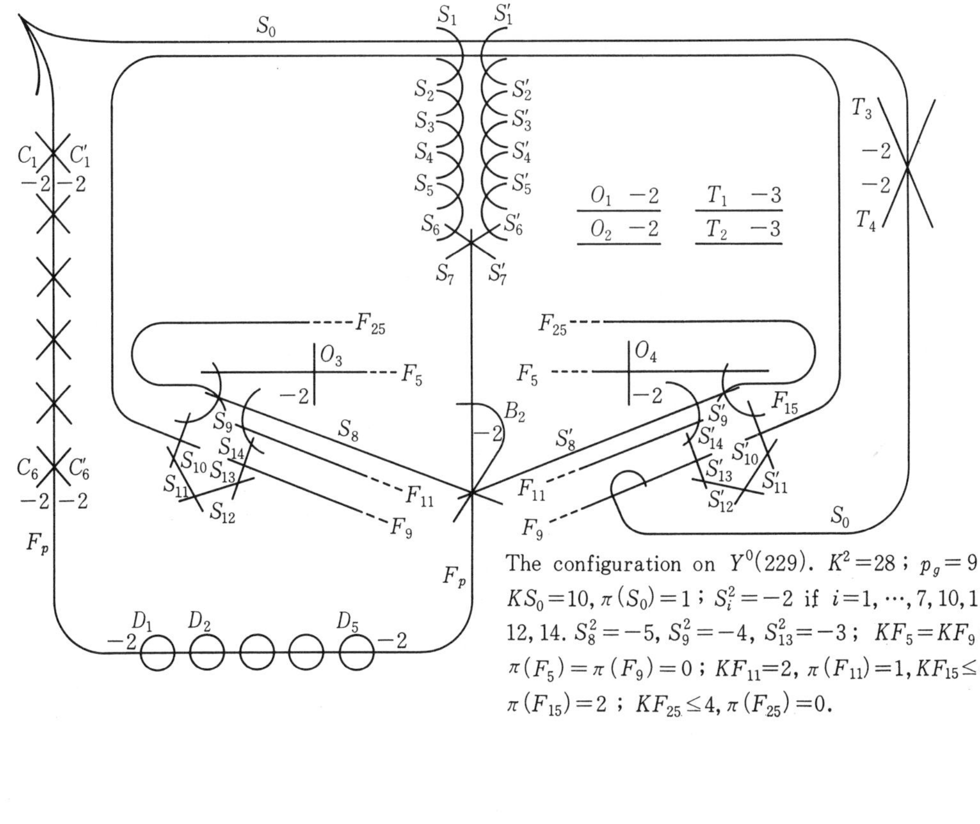

The configuration on $Y^0(229)$. $K^2=28$; $p_g=9$; $KS_0=10$, $\pi(S_0)=1$; $S_i^2=-2$ if $i=1,\cdots,7,10,11,12,14$. $S_8^2=-5$, $S_9^2=-4$, $S_{13}^2=-3$; $KF_5=KF_9=\pi(F_5)=\pi(F_9)=0$; $KF_{11}=2$, $\pi(F_{11})=1$, $KF_{15}\leq 4$, $\pi(F_{15})=2$; $KF_{25}\leq 4$, $\pi(F_{25})=0$.

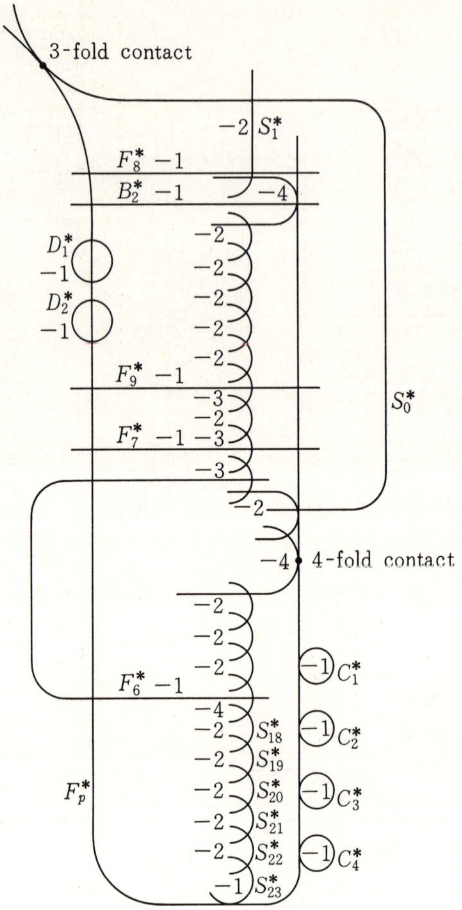

The configuration on $Y^0(193)/\iota$ $K^2{}_{Y^0(193)/\iota} = -18$

References

[1] Bombieri, E.: Canonical models of surfaces of general type, Publ. Math. IHES, **42** (1973), 171–219.

[2] Hirzebruch, F.: The Hilbert modular group and some algebraic surfaces. Intern. Symp. Number Theory, Moscow 1971.

[3] Hirzebruch, F.: Hilbert modular surfaces. Enseignement Math. **19** (1973), 183–281.

[4] Hirzebruch, F.: Kurven auf Hilbertschen Modulflächen und Klassenzahlrelationen. In: Classification of algebraic Varieties and Compact Complex Manifolds. Lecture Notes 412, Springer Berlin-Heidelberg- New York (1974).

[5] Hirzebruch, F. and Van de Ven, A.: Hilbert modular surfaces and the classification of algebraic surfaces. Inventiones Math. **23** (1974), 1–29.

[6] Hirzebruch, F. and Zagier, D.: Classification of Hilbert modular surfaces. In this Volume.

[7] Horikawa, E.: Algebraic surfaces of general type with small c_1^2 I and II. Annals of Math. **104**(1976), 357–387, and Inventiones Math. **37**(1976), 121–155.

University of Leiden

(Received December 15, 1975)

Part II

Complex Structures on $S^{2p+1} \times S^{2q+1}$ with Algebraic Codimension 1

K. Akao[1)]

0. Introduction

In the present paper we study certain complex structures of a compact complex manifold X of dimension n which is homeomorphic to the product of two odd-dimensional spheres $\boldsymbol{S}^{2p+1} \times \boldsymbol{S}^{2q+1}$ with $p+q>0$. Since the second Betti number of X vanishes, the transcendence degree over $\boldsymbol{C}$ of the field of all meromorphic functions on X does not exceed $n-1$. In the following we restrict ourselves to the case where X has exactly $(n-1)$ algebraically independent meromorphic functions. A so-called Hopf manifold is an example of such a manifold with $p=0$. E. Brieskorn and A. Van de Ven [2] have constructed a somewhat different kind of complex structure on $\boldsymbol{S}^1 \times \boldsymbol{S}^{2p+1}$ which also has p algebraically independent meromorphic functions. A complex structure on $\boldsymbol{S}^{2p+1} \times \boldsymbol{S}^{2q+1}$ with $p \geq 1$ and $q \geq 1$ was first constructed by E. Calabi and B. Eckmann [3]. It also satisfies the above condition. (See § 2 below.) Recently Ma. Kato [8] [9] has studied complex structures on $\boldsymbol{S}^1 \times \boldsymbol{S}^5$ with algebraic dimension 2 which satisfy some additional conditions. Our results are generalizations of his to higher dimensional cases. Now we summarize our main results. First in § 1 we study the structure of a compact complex manifold X of dimension n such that $\pi_1(X)$[2)] $\simeq \{1\}$ or $\boldsymbol{Z}$, $b_2(X)=0$ and such that $\mathrm{a}(X)=n-1$. For such an X, we have the following

Theorem 1. *There exists a finite unramified covering $\tilde{X}$ of X such that $\tilde{X}$ is subject to a holomorphic semi-free action of a 1-dimensional complex torus E and such that the meromorphic function field of the quotient space $\tilde{X}/E$ is naturally isomorphic to that of $\tilde{X}$.*

This implies in particular that the Euler number of X vanishes if X admits such a complex structure. Now in § 3 we assume that X is homeomorphic to $\boldsymbol{S}^{2p+1} \times \boldsymbol{S}^{2q+1}$ $(p+q>0)$, and that X has a flat fibration of elliptic curves over a non-singular projective variety W. Then we have

Theorems 2 and 3. *There exists a finite holomorphic map φ from X onto a submanifold*

---

1) Partially supported by the Fujukai Foundation.

2) For notation, see below.

of a Calabi-Eckmann manifold V (resp. a Hopf manifold if $p=0$ or $q=0$) such that the torus action on X constructed in Theorem 1 above is induced from the standard one on V. X is an abelian branched covering of $\varphi(X)$.

As corollaries, we compute the values of some numerical invariants of X and W. Some of the results of this paper were announced in [1].

The author would like to express his hearty thanks to Prof. K. Kodaira, Prof. S. Iitaka and Dr. K. Ueno for their valuable advice.

Notation. Now we fix some notation used throughout this paper.
For a topological space M:
$\pi_1(M)$ = the fundamental group of M,
$b_i(M)$ = the i-th Betti number of M,
$\chi(M) = \Sigma(-1)^i b_i(M)$ = the Euler number of M.
For a complex manifold X:
$\mathcal{M}(X)$ = the field of all meromorphic functions on X,
$\mathrm{a}(X) = \text{trans. deg.}_{C}\mathcal{M}(X)$ = the algebraic dimension of X,
$\mathcal{O}_X$ = the structure sheaf of X,
$\mathcal{O}_X(L)$ = the sheaf of germs of holomorphic cross-sections of a line bundle L on X,
$c(L)$ = the first Chern class of a line bundle L on X,
$h^{0,p}(X) = \dim H^p(X, \mathcal{O}_X)$,
$q(X) = h^{0,1}(X) = \dim H^1(X, \mathcal{O}_X)$,
$\mathrm{Aut}(X)$ = the group of all biholomorphic automorphisms of X,
$\mathrm{Aut}^0(X)$ = the component of the identity of Aut (X),
$\boldsymbol{Q}_X$ = the constant sheaf on X with $\boldsymbol{Q}$ as its stalk.
For a subvariety Y of X:
$\mathrm{codim}_X(Y)$ = the codimension of Y in X.

1. Existence Theorem for a Holomorphic Torus Action

Let X be a compact complex manifold of dimension n. In this section we assume that X satisfies the following conditions:

(1) $\mathrm{a}(X) = n-1$,

(2) $\pi_1(X)$ is a finitely generated abelian group whose free rank does not exceed one,

(3) $b_2(\tilde{X}) = 0$ for any finite unramified covering $\tilde{X}$ of X.

Note that under the condition (3), the following holds:

(4) For any irreducible effective divisor D and any compact analytic curve C in X, $D \cap C \neq \phi$ implies that $D \supset C$.

In fact, let $\tilde{C} \xrightarrow{\pi'} C$ be the normalization of C, and $\pi: \tilde{C} \to X$ the composite of π' and the inclusion $C \hookrightarrow X$. Assume that $D \not\supset C$. Then, for the line bundle $[D]$ on X corresponding to D, we have $c_1(\pi^*([D]))(\tilde{C}) = \pi^*(c_1([D]))(\tilde{C}) > 0$. But, since $H^2(X, \boldsymbol{Z})$

is a torsion group, $\pi^* c_1([D])=0$, which is a contradiction, q. e. d.

Under these assumptions, we have the following theorem.

Theorem 1. *Assume that X satisfies (1), (2) and (3) above. Then there exist a finite unramified covering $\tilde{X}$ of X and a holomorphic action of a 1-dimensional complex torus E on $\tilde{X}$, satisfying the following conditions:*

(i)[1] *the stabilizer subgroup of E at each point of $\tilde{X}$ is finite,*

(ii) *the quotient map $\pi : \tilde{X}\to\tilde{X}/E = W$ induces an isomorphism $\pi^* : \mathcal{M}(\tilde{X})\simeq\mathcal{M}(W)$, and W is a Moišezon space, and*

(iii) *W has only quotient singularities arising from action by finite abelian groups. Moreover such a torus action is unique.*

Proof. First we quote a lemma due to S. Iitaka.

Let V and S be compact complex manifolds of dimension n and $(n-1)$ respectively, and $\pi : V\to S$ be a surjective holomorphic map whose general fibre is isomorphic to a fixed elliptic curve E. E can be represented in the form $E=\boldsymbol{C}/L$ where L is a lattice in $\boldsymbol{C}$ generated by 1 and ω with $\mathrm{Im}\,\omega>0$. We denote by ζ the standard coordinate on $\boldsymbol{C}$, and by $[\zeta]$ the one on E induced by it. Let D be an irreducible subvariety of S with $\mathrm{codim}_S D=1$, such that for a general point p of D, $\pi^{-1}(p)$ is a (possibly multiple) non-singular elliptic curve.

Lemma 1. *Under these assumptions, there exist a positive integer m, an open polydisc neighbourhood U of p in S, an m-fold branched covering $\varphi : \tilde{U}\to U$, and local coordinate systems $(z)=(z_1, \cdots, z_{n-1})$ on U and $(w)=(w_1, \cdots, w_{n-1})$ on $\tilde{U}$ respectively such that*

a) $D\cap U= \{(z)\in U|z_1 = 0\}$,

b) $\varphi(w_1, w_2, \cdots, w_{n-1}) = (w_1^m, w_2, \cdots, w_{n-1})$, *and*

c) *the fibre space $\pi^{-1}(U)\xrightarrow{\pi}U$ is isomorphic to $(\tilde{U}\times E)/G\xrightarrow{\tilde{\varphi}}U$, where G is the cyclic group generated by $\tilde{g} : (w_1, \cdots, w_{n-1}, [\zeta])\to(\varepsilon w_1, \cdots, w_{n-1}, [\zeta+1/m])$, and $\tilde{\varphi}(w_1, \cdots, w_{n-1}, [\zeta]) = \varphi(w_1, \cdots, w_{n-1})$ with ε a primitive m-th root of unity.*

For a proof of this lemma, see S. Iitaka [7] or Ma. Kato [9]; there the lemma is proved for $n=3$, but the generalization to any n is straightforward.

Now we return to the proof of Theorem 1. Obviously we can assume without loss of generality that $\pi_1(X)$ is a free abelian group, that is, $\pi_1(X)\simeq \{1\}$ or $\boldsymbol{Z}$. Let V be a non-singular projective model of $\mathcal{M}(X)$, and $\psi : X\to V$ the corresponding meromorphic map from X to V. By the assumption (1), $\dim V=n-1$. By Hironaka, there exist a compact complex manifold X^* of dimension n and surjective holomorphic maps $\varpi : X^*\to X$ and $f: X^*\to V$ such that ϖ is an isomorphism outside of an analytic subset S of X with $\mathrm{codim}_X S\geq 2$, and such that the following diagram commutes:

1) We call the action of E semi-free if it satisfies the condition (i).

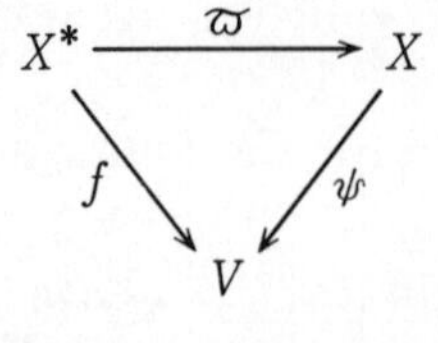

Note that, via ϖ and f, we get the isomorphisms of fields : $\mathcal{M}(X)\overset{\varpi^*}{\simeq}\mathcal{M}(X^*)\overset{f^*}{\rightleftarrows}\mathcal{M}(V)$. By Kawai [10] and Hironaka [4], the general fibres of f are elliptic curves, namely, X^* is an elliptic fibre space over V. By changing the models V and X^* if necessary, we may further assume that for each degenerate divisor Y of X^*, $\operatorname{codim}_X\varpi(Y)$ is greater than one. Here, by a degenerate divisor Y, we mean a subvariety Y of X^* with $\operatorname{codim}_{X^*}Y=1$ such that $\operatorname{codim}_V f(Y)$ is greater than one. Let $D\subset V$ be a subvariety of pure codimension 1 such that, for a general point p of D, $f^{-1}(p)$ is not a regular fibre. Assume that $f^{-1}(p)$ is not a multiple elliptic fibre (which means that $f^{-1}(p)$ is not of type ${}_mI_0$ in the notation of K. Kodaira [11]). Then the singular loci of $f^{-1}(D)$ have an irreducible component Σ such that $f|_\Sigma : \Sigma\to D$ is a surjective, generically finite map. Note also that $a(\Sigma)=n-2=\dim \Sigma$. But then, since V is supposed to be projective algebraic, $f^{-1}(H)\cap\Sigma\neq\phi$ for any generic hyperplane section of V. Hence applying the above property (4) to each compact analytic curve contained in Σ, we have $\varpi(f^{-1}(H))\supset\varpi(\Sigma)$ which again shows that, for general $p\in D$, $\varpi(f^{-1}(p))\subset\varpi(f^{-1}(H))$. Therefore $\operatorname{codim}_X \varpi(f^{-1}(D))\geq 2$. Let D_i $(1\leq i\leq r)$ be all such subvarieties of V, $Y_j(1\leq j\leq s)$ all degenerate divisors on X^*, and $Z_k(1\leq k\leq t)$ the irreducible components of the singular loci of the image by f of singular fibres. Put $\varpi(f^{-1}(D_i))=\Delta_i$, $\varpi(Y_j)=\Gamma_j$ and $\varpi(f^{-1}(Z_k))=\Theta_k$. Let $X'=X-(\bigcup_i\Delta_i\cup\bigcup_j\Gamma_j\cup\bigcup_k\Theta_k)$, $V'=V-(\bigcup_i D_i\cup\bigcup_j f(Y_j)\cup\bigcup_k Z_k)$, and $\pi'=f\circ\varpi^{-1}|_{X'} : X'\to V'$. Then π' is a surjective holomorphic map, each fibre of which is a (possibly multiple) smooth elliptic curve. Note that $\operatorname{codim}_X (X-X')\geq 2$, hence also $\pi_1(X')\cong\pi_1(X)$. Let $j: V'\to \boldsymbol{C}$ be the functional invariant of this elliptic fibre space (see [10] and [11]). Since $\pi'\circ j : X'\to\boldsymbol{C}$ is extendable to the whole of X, j is a constant map, which implies that every regular fibre of π' is isomorphic to a fixed elliptic curve E. Then by Lemma 1, $\pi' : X'\to V'$ is a Seifert fibre space in the sense of Holmann [6]. Since j is a constant map, each local monodromy around D_i is finite. Therefore, by taking a suitable finite unramified covering of X, we may assume that every local monodromy is trivial. But then we can extend the flat sheaf $\boldsymbol{R}^1\pi'_*\boldsymbol{Q}_{X'}$ on V' to the flat sheaf $\mathcal{F}$ on the whole V. Since V is projective and since $\pi_1(X)\to\pi_1(V)\to\{1\}$ is exact, $\pi_1(V)$ is at most a finite group. Again, by taking a suitable finite unramified covering if necessary, we can assume that $\mathcal{F}$ and $\boldsymbol{R}^1\pi'_*\boldsymbol{Q}_{X'}$ are constant sheaves. Then the Seifert fibre space $\pi' : X'\to V'$ is reduced to a principal Seifert fibre space with $\operatorname{Aut}^0(E)\cong E$ as its structure group. Hence X' is subject to a holomorphic action of E such that the stabilizer subgroup at each point of X' is finite, and that $X'/E\simeq V'$. Since $\operatorname{codim}_X (X-X')$ is greater than one, this action can be extended to the whole of X. This clearly satisfies (i) and (ii), because $W=X/E$ and V are birationally equivalent. In fact, since $\operatorname{codim}_X (X-X')\geq 2$, every meromorphic function on V' is extended to an E-invariant meromorphic function on the whole of X. By [5], the action of E has a holomorphic slice at each point of X, which proves (iii). The uniqueness of this torus action is trivial, q. e. d.

Corollary 1. *Under the same assumptions as in Theorem 1, we obtain* $\chi(X)=0$.
Proof. Obvious.

This shows, for example, that S^6 admits no complex structure with algebraic dimension 2.

Corollary 2. *Under the same assumptions as in Theorem 1, we have the following values of numerical invariants of* $\tilde{X}$, X *and* W: $q(X)=q(\tilde{X})=1$, $b_1(W)=q(W)=0$, *and* $b_2(W)=2-\text{rank}\,\pi_1(X)$.

Proof. Let $\tilde{\pi}$ be the natural projection from $\tilde{X}$ to $W=\tilde{X}/E$. Then, from the above construction, we have

$$(*)\qquad \boldsymbol{R}^q\tilde{\pi}_*\mathcal{O}_{\tilde{X}}\cong\begin{cases}\mathcal{O}_W & \text{if } q=0 \text{ and } q=1,\\ 0 & \text{otherwise,}\end{cases}$$

and

$$(**)\qquad \boldsymbol{R}^q\tilde{\pi}_*\boldsymbol{Q}_{\tilde{X}}\cong\begin{cases}\boldsymbol{Q}_W & \text{if } q=0 \text{ and } q=2,\\ \boldsymbol{Q}_W\oplus\boldsymbol{Q}_W & \text{if } q=1,\\ 0 & \text{otherwise.}\end{cases}$$

Since W has only quotient singularities, and since every quotient singularity is rational, the non-singular model $\tilde{W}$ of W has the same fundamental group and irregularity as those of W. But $\pi_1(W)$ is a quotient group of $\pi_1(X)$, hence $b_1(W)=b_1(\tilde{W})\leq 1$. Since $\tilde{W}$ is a Moišezon manifold, $b_1(\tilde{W})$ is even, so $b_1(W)=b_1(\tilde{W})=0$, which also proves that $q(W)=q(\tilde{W})=0$. Consider the Leray spectral sequence for $\mathcal{O}_{\tilde{X}}$: $E_2^{p,q}=H^p(W,\boldsymbol{R}^q\tilde{\pi}_*\mathcal{O}_{\tilde{X}})\Rightarrow H^{p+q}(\tilde{X},\mathcal{O}_{\tilde{X}})$. $q(W)=0$ implies $E_2^{1,0}=0$. Since $\dim_C E_2^{0,1}=1$, $q(\tilde{X})\leq 1$. If $q(\tilde{X})=0$, then from the exact sequence

$$0\to\boldsymbol{Z}\to\mathcal{O}\to\mathcal{O}^*\to 0,$$

we get the inclusion $H^1(\tilde{X},\mathcal{O}^*)\hookrightarrow H^2(\tilde{X},\boldsymbol{Z})$. Hence, by the above assumption (3), every line bundle on $\tilde{X}$ is of finite order, which contradicts the assumption (1). Therefore $q(\tilde{X})=1$. On the other hand, we have $\dim_C H^1(X,\mathcal{O}_X)=\dim(H^1(\tilde{X},\mathcal{O}_{\tilde{X}})^G)$, where G is the covering transformation group of $\tilde{X}$ over X. Hence $q(X)\leq 1$. But, by the same argument as above, we get $q(X)>0$, so $q(X)=q(\tilde{X})=1$. For proving the last equality, we use the Leray spectral sequence for $\boldsymbol{Q}_{\tilde{X}}$: $'E_2^{p,q}=H^p(W,\boldsymbol{R}^q\tilde{\pi}_*\boldsymbol{Q}_{\tilde{X}})\Rightarrow H^{p+q}(\tilde{X},\boldsymbol{Q})$. From $(**)$, $\dim_{\boldsymbol{Q}}{}'E_2^{0,1}=2$, and $\dim_{\boldsymbol{Q}}{}'E_2^{2,0}=b_2(W)$. Let d_2' : $'E_2^{0,1}\to{}'E_2^{2,0}$ be the corresponding differential map. Note that $\dim_{\boldsymbol{Q}}{}'E_3^{0,1}=\dim_{\boldsymbol{Q}}{}'E_\infty^{0,1}=\text{rank}\,\pi_1(X)$, and that $'E_3^{2,0}={}'E_\infty^{2,0}=0$. This means that $\dim_{\boldsymbol{Q}}(\ker d_2')=\text{rank}\ \pi_1(X)$. Then $\dim_{\boldsymbol{Q}}{}'E_3^{2,0}=b_2(W)-(2-\text{rank}\ \pi_1\ (X))=0$, thus $b_2(W)=2-\text{rank}\ \pi_1(X)$, q. e. d.

Corollary 3. *Suppose that* $\dim X=3$, *and that* X *satisfies the conditions* (2) *and* (3) *above. If X admits a holomorphic action of a 1-dimensional complex torus E satisfying the above* (i), *then* $\mathrm{a}(X)=2$.

Proof. Let S be the quotient surface X/E, and $\varpi : \tilde{S}\to S$ the resolution of singularities. By the same argument as in Corollary 2, we obtain $q(S)=q(\tilde{S})=0$, which implies $b_1(S)=b_1(\tilde{S})=0$ by [12]. Hence again by the same spectral sequence argument, we have $b_2(S)\leq 2$. Let $p_i(1\leq i\leq k)$ be the singular points of S, and let $C_{i\lambda}$ be the irreducible components of $\varpi^{-1}(p_i)$. Via ϖ, $H^2(\tilde{S}, \boldsymbol{Q})\simeq\varpi^*H^2(S, \boldsymbol{Q})\oplus L$, where L is the vector space over $\boldsymbol{Q}$ generated by all $c([C_{i\lambda}])$.[1]. As is well-known, the quadratic form A on $H^2(\tilde{S}, \boldsymbol{Q})$ defined by the cup product is negative definite on L. We denote by $b^+(\tilde{S})$ the number of positive eigenvalues of the symmetric matrix corresponding to A. From the above decomposition of $H^2(\tilde{S}, \boldsymbol{Q})$, $b_2(S)\leq 2$ implies that $b^+(\tilde{S})\leq 2$. Suppose that $h^{0,2}(S)=h^{0,2}(\tilde{S})>0$. By a theorem of Y. Miyaoka [13], $b_1(\tilde{S})=q(\tilde{S})=0$ implies that $\tilde{S}$ is either a Kähler surface or a $K3$ surface. In either case we get $b^+(\tilde{S})\geq 1+2h^{0,2}(\tilde{S})\geq 3$, which is a contradiction. So $h^{0,2}(\tilde{S})=0$, and $\tilde{S}$ is a Kähler surface ([13]). Hence, $\tilde{S}$ is actually a Hodge manifold. Therefore $\mathrm{a}(X)\geq\mathrm{a}(\tilde{S})=2$. Since X is not a Moišezon space, $\mathrm{a}(X)\leq 2$, so we have proved the assertion.

Corollary 4. *Assume further that $\tilde{X}$ is homeomorphic to $\boldsymbol{S}^{2p+1}\times\boldsymbol{S}^{2q+1}$ where $p+q>0$. Then we get*

$$H^*(W, \boldsymbol{Q}) \cong H^*(\boldsymbol{P}^p\times\boldsymbol{P}^q, \boldsymbol{Q}).$$

Proof. Put $n=\dim_C X=p+q+1$. Note that $b_{2r}(\tilde{X})=0$ holds except for $r=0$ and $r=n$. First we shall show that $b_{2r+1}(W)=0$ for all $r\in\boldsymbol{Z}$. Consider the Leray spectral sequence for $\boldsymbol{Q}_{\tilde{X}}$; $E_2^{p,q}=H^p(W, \boldsymbol{R}^q\tilde{\pi}_*\boldsymbol{Q}_{\tilde{X}})\Rightarrow H^{p+q}(\tilde{X}, \boldsymbol{Q})$ where $\tilde{\pi}$ is the quotient map from $\tilde{X}$ to $W=\tilde{X}/E$. We get then

$$(*)\qquad \boldsymbol{R}^q\tilde{\pi}_*\boldsymbol{Q}_{\tilde{X}}\cong\begin{cases}\boldsymbol{Q}_W & \text{if } q=0 \text{ and } q=2,\\ \boldsymbol{Q}_W\oplus\boldsymbol{Q}_W & \text{if } q=1,\\ 0 & \text{otherwise.}\end{cases}$$

For an odd s, we get the following semi-exact sequence

$$0\longrightarrow E_2^{s-2,2}\overset{\alpha}{\longrightarrow}E_2^{s,1}\overset{\beta}{\longrightarrow}E_2^{s+2,0}\longrightarrow 0,$$

where α and β are differentials of this spectral sequence. By $(*)$, $E_3^{s,1}=E_\infty^{s,1}$, which is a zero module in view of the vanishing of $b_{\text{even}}(\tilde{X})$. Hence this sequence is exact at $E_2^{s,1}$, which means that $\dim_{\boldsymbol{Q}}E_2^{s,1}\leq\dim_{\boldsymbol{Q}}E_2^{s-2,2}+\dim_{\boldsymbol{Q}}E_2^{s+2,0}$, namely, $2b_s(W)\leq b_{s+2}(W)+b_{s-2}(W)$. Clearly $b_s(W)=0$ holds if $s\geq 2n-1$. From this, we get $b_s(W)\leq b_{s-2}(W)$ for s odd, using repeatedly the above inequality. By Corollary 2, $b_1(W)=0$, so $b_s(W)=0$ for all odd s. Next, consider the sequence

$$0\longrightarrow E_2^{2r-2,2}\overset{\alpha'}{\longrightarrow}E_2^{2r,1}\overset{\beta'}{\longrightarrow}E_2^{2r+2,0}\longrightarrow 0,$$

where α' and β' are differentials. Since $E_2^{2r+1,p}=0$ for all integers r and p, we get $E_3^{p,q}=E_\infty^{p,q}$ for all p and q. Hence α' is injective except for $r=n$ and β' is surjective

1) By $[C_{i\lambda}]$ we denote the line bundle on S corresponding to the divisor $C_{i\lambda}$.

because of the vanishing of $b_{\text{even}}(\tilde{S})$. Therefore we obtain the equalities

$$2b_{2r}(W) = b_{2r+2}(W)+b_{2r-2}(W)+b_{2r+1}(\tilde{X}) \quad \text{for} \quad r \neq n,$$

and

$$b_0(W) = b_{2n-2}(W) = 1.$$

From this equality we easily get the assertion. q. e. d.

2. The Construction of Calabi-Eckmann Manifolds

In this section we state a construction of Calabi-Eckmann manifolds following J. Tits [14]. (See also [2].) Let $z=(z_1, \cdots, z_p)$ and $w=(w_1, \cdots, w_q)$ be standard coordinates on $\boldsymbol{C}^p$ and $\boldsymbol{C}^q$ respectively. For each $t \in \boldsymbol{C}$, let g_t^λ be the biholomorphic automorphism of $(\boldsymbol{C}^p-(0))\times(\boldsymbol{C}^q-(0))$ defined by

$$g_t^\lambda(z_1, \cdots, z_p, w_1, \cdots, w_q) = (e^t z_1, \cdots, e^t z_p, e^{\lambda t}w_1, \cdots, e^{\lambda t}w_q),$$

where λ is a fixed complex number with Im $\lambda \neq 0$. Let G_λ be the one-parameter complex Lie group consisting of all g_t^λ for $t \in \boldsymbol{C}$. G_λ operates freely and properly on $(\boldsymbol{C}^p-(0))\times(\boldsymbol{C}^q-(0))$. Hence, by Holmann [5], we can construct the quotient manifold $M=(\boldsymbol{C}^p-(0))\times(\boldsymbol{C}^q-(0))/G_\lambda$, which is a compact complex manifold of dimension $(p+q-1)$. M is homeomorphic to $\boldsymbol{S}^{2p-1}\times\boldsymbol{S}^{2q-1}$, and its complex structure coincides with the one constructed by E. Calabi and B. Eckmann in [3]. We call M a Calabi-Eckmann manifold. Note that M is a complex analytic principal fibre bundle over $\boldsymbol{P}^{p-1}\times\boldsymbol{P}^{q-1}$, the fibre of which is a non-singular elliptic curve E with 1 and λ as periods. Hence a(M)=dim $M-1$. Moreover M admits a unique holomorphic free action of E.

Remark. The Calabi-Eckmann manifolds are the simplest example of so-called non-Kähler C-manifolds, namely, simply connected compact complex homogeneous manifolds ([14]).

3. Complex Structures on $S^{2p+1}\times S^{2q+1}$

In this section we shall study a complex manifold X of dimension n which is homeomorphic to $\boldsymbol{S}^{2p+1}\times\boldsymbol{S}^{2q+1}$, which satisfies the following condition:

There exist an $(n-1)$-dimensional smooth projective algebraic variety W, and a surjective flat holomorphic map π from X to W with connected fibres.

If X satisfies this assumption, the conditions (1), (2) and (3) in § 1 are also satisfied. Hence there exists a finite unramified covering $\tilde{X}$ of X such that $\tilde{X}$ admits a holomorphic semi-free action of a 1-dimensional complex torus E. In particular every fibre of π is a non-singular elliptic curve, and a general fibre is isomorphic to a fixed elliptic curve isogenous to E. Note also that if $p\geq 1$ and $q\geq 1$, $\tilde{X}$ coincides with X. Then we have the following theorems.

Theorem 2. *Suppose that X satisfies the above conditions, and that $p \geq 1$ and $q \geq 1$. Then there exists a finite holomorphic map from X onto a submanifold X' of a Calabi-Eckmann manifold V such that the following diagram commutes:*

$$\begin{array}{ccccc} X & \longrightarrow & X' & \longrightarrow & V \\ {\scriptstyle \pi}\downarrow & & \downarrow{\scriptstyle \varphi|X'} & & \downarrow{\scriptstyle \varphi} \\ W & \cong & W & \hookrightarrow & \boldsymbol{P}^{N_1} \times \boldsymbol{P}^{N_2}. \end{array}$$

Here, φ is the natural projection mentioned in § 2. Moreover W has the same rational homology group as that of $\boldsymbol{P}^p \times \boldsymbol{P}^q$, and the image by π of the set of singular fibres of the elliptic fibre space $X \to W$ has only simple normal crossings for singularities. Furthermore X is an abelian branched covering of X', and admits a complex torus action compatible with that of V.

Theorem 3. *Suppose $p=0$. Then, under the same assumptions as in Theorem 2, there exists a finite holomorphic map from X onto a submanifold X' of a Hopf manifold V such that the following diagram commutes:*

$$\begin{array}{ccccc} X & \longrightarrow & X' & \hookrightarrow & V \\ {\scriptstyle \pi}\downarrow & & \downarrow{\scriptstyle \varphi|X'} & & \downarrow{\scriptstyle \varphi} \\ W & \cong & W & \hookrightarrow & \boldsymbol{P}^{N}. \end{array}$$

In this case, W has the same rational homology group as that of $\boldsymbol{P}^q$, and π has the same property as in Theorem 2.

Before proving Theorem 2, we need several lemmata. In the first place, by Corollary 2 in § 1, we have $q(X)=1$, $q(W)=b_1(W)=0$ and $b_2(W)=2$ because $\pi_1(X) = \{1\}$.

Lemma 2. $H^1(X, \mathcal{O}) \simeq H^1(X, \mathcal{O}^*) \simeq \boldsymbol{C}$.

Proof. Since $p \geq 1$ and $q \geq 1$, $H^1(X, \boldsymbol{Z}) = H^2(X, \boldsymbol{Z}) = 0$. Therefore, the conclusion follows from the exact sequence:

$$\cdots \to H^1(X, \boldsymbol{Z}) \to H^1(X, \mathcal{O}) \to H^1(X, \mathcal{O}^*) \to H^2(X, \boldsymbol{Z}) \to \cdots$$

q. e. d.

Lemma 3. *There exist two very ample line bundles L_1 and L_2 on W such that $c(L_1)$ and $c(L_2)$ are linearly independent in $H^2(X, \boldsymbol{Z})$, and such that $\pi^*(L_1)$ and $\pi^*(L_2)$ generate a lattice in $H^1(X, \mathcal{O}^*) \cong \boldsymbol{C}$.*

Proof. Since $b_2(W)=2$, and W is non-singular projective, $h^{0,2}(W)=0$. Then the vanishing of $q(W)$ implies that $H^1(W, \mathcal{O}^*) \simeq H^2(W, \boldsymbol{Z})$. Therefore, using again the fact that W is projective, we can find two very ample line bundles L_1 and L_2 such that $c(L_1)$ and $c(L_2)$ generate $H^2(W, \boldsymbol{Q}) \cong H^2(W, \boldsymbol{Z}) \otimes_{\boldsymbol{Z}} \boldsymbol{Q}$. It is clear that $\pi^* L_1$ and $\pi^* L_2$ are linearly independent over $\boldsymbol{Q}$. We must show that they are independent over $\boldsymbol{R}$. Let p be a generic point of W and $j : \pi^{-1}(p) \hookrightarrow X$ be the natural inclusion. $\pi^{-1}(p)$ is a smooth elliptic curve. Then we have the following commutative diagram:

$$\begin{array}{ccccccccc} & & 0 \to & H^1(X, \mathcal{O}_X) & \overset{\alpha}{\simeq} & H^1(X, \mathcal{O}_X^*) & \to 0 & & \\ & & & \wr\!\downarrow j_1^* & & \downarrow j_2^* & & & \\ 0 \to H^1(\pi^{-1}(p), \boldsymbol{Z}) & \overset{i_*}{\to} & & H^1(\pi^{-1}(p), \mathcal{O}_{\pi^{-1}(p)}) & \to & H^1(\pi^{-1}(p), \mathcal{O}^*_{\pi^{-1}(p)}) & \to & H^2(\pi^{-1}(p), \boldsymbol{Z}), & \end{array}$$

where both rows are exact. Note that $j_1^*: H^1(X, \mathcal{O}_X)\to H^1(\pi^{-1}(p), \mathcal{O}_{\pi^{-1}(p)})$ is an isomorphism, as can be proved by using the Leray spectral sequence in § 1. Since clearly $j_2^*\pi^*L_1$ and $j_2^*\pi^*L_2$ are trivial line bundles, $j_1^*\alpha^{-1}\pi^*L_1$ and $j_1^*\alpha^{-1}\pi^*L_2$ lie in the image of i^*. Since they are linearly independent over $\boldsymbol{Z}$, they generate a sub-lattice of $i^*H^1(\pi^{-1}(p), \boldsymbol{Z})$. But, since $\pi^{-1}(p)$ is a non-singular elliptic curve, $i^*H^1(\pi^{-1}(p), \boldsymbol{Z})$ is a lattice in $H^1(\pi^{-1}(p), \mathcal{O}_{\pi^{-1}(p)})\cong H^1(X, \mathcal{O}_X)\cong H^1(X, \mathcal{O}_X^*)$, hence π^*L_1 and π^*L_2 generate a lattice in $H^1(X, \mathcal{O}_X^*)$. q. e. d.

Proof of Theorem 2. First we choose coordinate coverings $\{U_i^\mu\}$ of X and $\{W_i\}$ of W such that $\pi(U_i^\mu)=W_i$. Let $l_{\alpha,ij}$ be the transition functions of the line bundle L_α $(\alpha=1, 2)$ with respect to $\{W_i\}$. Then, by Lemma 2 and Lemma 3, we can find non-vanishing holomorphic functions p_i^μ and q_i^μ on each U_i^μ, a holomorphic function $\lambda_{ij}^{\mu\nu}$ on each $U_i^\mu\cap U_j^\nu$ $(\neq\phi)$, and a complex number c with $\operatorname{Im} c\neq 0$ such that the following equations are satisfied:

$$\pi^*l_{1,ij} = p_i^\mu \exp(\lambda_{ij}^{\mu\nu})(p_j^\nu)^{-1},$$
$$\pi^*l_{2,ij} = q_i^\mu \exp(c\lambda_{ij}^{\mu\nu})(q_j^\nu)^{-1},$$

where we put $l_{\alpha,ij}=1$ for $i=j$. Since L_α is very ample, we can find a basis $\{\zeta_{\alpha,i}^{(k)}\}$ $(k=0, 1, \cdots, N)$ of $H^0(W, \mathcal{O}(L_\alpha))$ which gives an imbedding j_α of W into $\boldsymbol{P}^{N_\alpha}$, where $N_\alpha=\dim_{\boldsymbol{C}}H^0(W, \mathcal{O}(L_\alpha))-1$ $(\alpha=1, 2)$. Then by the above equations, we can construct a holomorphic map ψ from X into $V=(\boldsymbol{C}^{N_1+1}-(0))\times(\boldsymbol{C}^{N_2+1}-(0))/G$, which maps $x\in U_i^\mu$ to $[(p_i^\mu\pi^*\zeta_{1,i}^{(0)}, \cdots, p_i^\mu\pi^*\zeta_{1,i}^{(N_1)}, q_i^\mu\pi^*\zeta_{2,i}^{(0)}, \cdots, q_i^\mu\pi^*\zeta_{2,i}^{(N_2)})]$. Here $G=\{g_t^c\}$ is the one parameter group defined in § 2, and for $z\in\boldsymbol{C}^{N_1+1}-(0)\times\boldsymbol{C}^{N_2+1}-(0)$, $[z]$ denotes the image of z in V. The manifold V is nothing but a Calabi-Eckmann manifold and we get the following commutative diagram:

$$\begin{array}{ccc} X & \overset{\psi}{\longrightarrow} & V \\ {\scriptstyle\pi}\downarrow & & \downarrow{\scriptstyle\varphi} \\ W & \underset{j=(j_1,j_2)}{\hookrightarrow} & \boldsymbol{P}^{N_1}\times\boldsymbol{P}^{N_2}. \end{array}$$

On the other hand, every positive-dimensional closed subvariety of V is of the form $\varphi^{-1}(Y)$ for a closed subvariety Y of $\boldsymbol{P}^{N_1}\times\boldsymbol{P}^{N_2}$ ([3]). Therefore $\psi: X\to\psi(X)$ is a finite map. For a general point p of W, the elliptic curve $\pi^{-1}(p)\cong E$ is a finite unramified covering of $\varphi^{-1}(j(p))$. Hence there exists a finite abelian subgroup Γ of E, such that $\pi^{-1}(p)/\Gamma\simeq\varphi^{-1}(j(p))$. Then it is obvious that $\psi(X)=X/\Gamma$ because Γ operates on X by Theorem 1. The statement $H^*(W, \boldsymbol{Q})\cong H^*(\boldsymbol{P}^p\times\boldsymbol{P}^q, \boldsymbol{Q})$ was already proved in Corollary 4 in § 1. Now we have only to show that the image Δ by π of the set of the singular fibres of $\pi: X\to W$ has only simple normal crossings for singularities. By Holmann [5], the action of E admits a holomorphic slice at every point of X. Let $p \in W$ be a singular point of Δ, and Γ_p the stabilizer subgroup of E at a point of $\pi^{-1}(p)$. Then there exist an open polydisc neighbourhood U of p in W, and a holomorphic slice $\tilde{U}$ at a point of $\pi^{-1}(p)$, which is stable under Γ_p and also isomorphic to an $(n-1)$-dimensional polydisc. Moreover $\tilde{U}/\Gamma_p \simeq U$ via π, and the image by π of the set of fixed points of Γ_p in $\tilde{U}$ coincides with $\Delta \cap U$. Since $\Gamma_p \subset E$, Γ_p is a finite abelian group. Therefore we choose a coordinate system $z=(z_1, \cdots, z_{n-1})$ on $\tilde{U}$ such that every element $\gamma \in \Gamma_p$ is expressed by a diagonal matrix with respect to these coordinates. Let U' be another $(n-1)$-dimensional polydisc and $w=(w_1, \cdots, w_{n-1})$ be a coordinate system on U'. Put $g=\sharp|\Gamma_p|$, and consider the map $\varpi : \tilde{U} \to U'$ defined by $\varpi(z_1, \cdots, z_{n-1})=(z_1^g, \cdots, z_{n-1}^g)$. Then ϖ factors through $\pi|_{\tilde{U}}$:

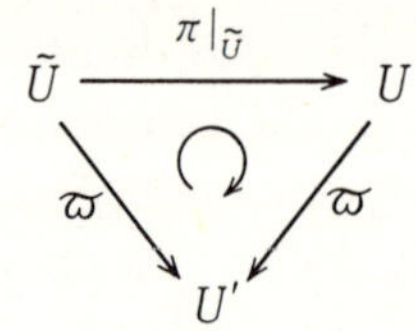

and $\varpi'(\Delta) \subset \{(w) \in U' | w_1 w_2 \cdots w_{n-1}=0\}$. Hence Δ has only a normal crossing at p. The smoothness of irreducible components of Δ can also be proved easily. q. e. d.

Proof of Theorem 3. In this case $\pi_1(X) \cong \boldsymbol{Z}$, so $q(X)=1$, and $b_1(W)=q(W)=0$. Since W is projective, $\pi_1(W)$ is a finite cyclic group. Let $\tilde{W}$ be the universal covering of W. Then by Theorem 1, $\tilde{X}=X \times_W \tilde{W}$ admits a holomorphic torus action. Hence $H^*(\tilde{W}) \cong H^*(\boldsymbol{P}^q)$ by Corollary 4 in § 1, and $b_\nu(W) \leq b_\nu(\tilde{W})=b_\nu(\boldsymbol{P}^q)$. But the projectivity of W implies that $b_{2r}(W) \geq 1$ for $0 \leq r \leq \dim_{\boldsymbol{C}} W$. Hence we get $H^*(W) \cong H^*(\boldsymbol{P}^q)$. In particular $b_2(W)=1$. Then, by an argument similar to that in the proof of Theorem 2, we easily deduce that every line bundle on X is flat. Let L be a very ample line bundle on W. Let $\{U_i^\mu\}$ and $\{W_i\}$ be as in the proof of Theorem 2, and l_{ij} the transition functions of L with respect to $\{W_i\}$. Then, on account of the flatness of π^*L, we can find non-vanishing holomorphic functions p_i^μ on each U_i^μ, an integer $m_{ij}^{\mu\nu}$ for each pair $\{(i, \mu), (j, \nu)\}$ with $U_i^\mu \cap U_j^\nu \neq \phi$, and a non-zero complex number α with $|\alpha| \neq 1$, such that $\alpha^* l_{ij}=p_i^\mu \alpha^{m_{ij}^{\mu\nu}} (p_j^\nu)^{-1}$ on $U_i^\mu \cap U_j^\nu$, where we put $l_{ii}=1$ for each i. Let $\{\xi_i^{(k)}\}$ $(k=0, 1, \cdots, N)$ be a basis of $H^0(W, \mathcal{O}(L))$, where $N+1=\dim H^0(W, \mathcal{O}(L))$. This gives an imbedding of W into $\boldsymbol{P}^N$. Then, as in Theorem 2, we can construct a holomorphic map from X to $V=\boldsymbol{C}^{N+1}-(0)/\langle \alpha^n \rangle_{n \in \boldsymbol{Z}}$ which maps $x \in U_i^\mu$ to $[(p_i^\mu \pi^* \xi_i^{(0)}, \cdots, p_i^\mu \pi^* \xi_i^{(N)})]$. V is a (homogeneous) Hopf manifold, and the rest of the proof is the same as that of Theorem 2. q. e. d.

Remark. If X is homeomorphic to $\boldsymbol{S}^1 \times \boldsymbol{S}^5$, then we can get rid of the assumption of the flatness of $\pi : X \to W$.

Proof. Since $\pi_1(W)$ is finite, we may assume obviously that $\pi_1(W)=\{1\}$. Then, by Theorem 1, X admits a holomorphic action of 1-dimensional complex torus E.

Let $\varphi: X\to V=X/E$ be the quotient map. Then π factors through φ:

$$\begin{array}{ccc} X & \xrightarrow{\quad\pi\quad} & W \\ & \searrow^{\varphi} \quad \circlearrowleft \quad \nearrow_{\varpi} & \\ & V & \end{array}$$

where ϖ is a birational morphism. Note that we have $b_2(V)=1$. Let $\psi: \tilde{V}\to V$ be the resolution of singularities of V. Then, since $\dim_C W=\dim_C V=2$, $\psi\circ\varpi$ is a composite of quadratic transformations. Now consider the Leray spectral sequence for $V\overset{\varpi}{\to}W$ and $\boldsymbol{Q}_V$; $E_2^{p,q}=H^p(W, \boldsymbol{R}^q\varpi_*\boldsymbol{Q}_V)\Rightarrow H^{p+q}(V, \boldsymbol{Q})$. We shall show that $E_2^{0,2}=0$, which means that ϖ is a finite map. But then since W is non-singular and ϖ is birational, ϖ is an isomorphism, and π is flat. Since ϖ is birational, the dimension of the support of $\boldsymbol{R}^1\varpi_*\boldsymbol{Q}_V$ is zero, so $E_2^{2,1}=0$. Since W is non-singular, $b_3(W)=b_1(W)=0$ by the Poincaré duality, hence $E_2^{3,0}=0$. This implies that $E_2^{0,2}=E_\infty^{0,2}$. But since W is projective, $b_2(W)\geq 1$, and from the inclusion $H^2(W, \boldsymbol{Q})\overset{(\psi\circ\varpi)^*}{\hookrightarrow} H^2(\tilde{V}, \boldsymbol{Q})$, we have $E_2^{2,0}=E_\infty^{2,0}\simeq H^2(W, \boldsymbol{Q})$. Since $b_2(V)=1$, we get $E_2^{0,2}=0$. q. e. d.

References

[1] Akao, K.: On certain complex structures on the product of two odd dimensional spheres. Proc. Japan Acad., **50** (1974), 802–805.

[2] Brieskorn, E. and Van de Ven, A.: Some complex structures on products of homotopy spheres. Topology **7** (1967), 389–393.

[3] Calabi, E. and Eckmann, B.: A class of compact complex manifolds which are not algebraic. Ann. of Math., **58** (1953), 494–500.

[4] Hironaka, H.: Review of S. Kawai's paper, Math. Rev., **32**, #466 (1966), 87–88.

[5] Holmann, H.: Quotientenräume komplexer Mannigfaltigkeiten nach komplexen Lieschen Automorphismengruppen. Math. Ann., **139** (1960), 383–402.

[6] Holmann, H.: Seifertsche Faserräume. Math. Ann., **157** (1964), 138–166.

[7] Iitaka, S.: On algebraic varieties whose universal covering manifolds are complex affine 3-spaces, 1, in Number Theory, Algebraic Geometry and Commutative Algebra, in honor of Yasuo Akizuki, Kinokuniya, Tokyo (1973), 147–167.

[8] Kato, Ma.: Complex structures on $\boldsymbol{S}^1\times\boldsymbol{S}^5$. Proc. Japan Acad. **49** (1973), 575–577.

[9] ——: Complex structures on $\boldsymbol{S}^1\times\boldsymbol{S}^5$, J. Math. Soc. Japan **28** (1976), 550–576.

[10] Kawai, S.: On compact complex analytic manifolds of dimension 3. J. Math. Soc. Japan, **17** (1965), 438–442.

[11] Kodaira, K.: On compact analytic surfaces II. Ann. of Math., **77** (1963), 563–626.

[12] ——: On the structure of compact complex analytic surfaces I. Amer. J. Math., **86** (1964), 751–798.

[13] Miyaoka, Y.: Kähler Metrics on Elliptic Surfaces. Proc. Japan Acad., **50** (1974), 533–536.

[14] Tits, J.: Espaces homogènes complexes compacts. Comment. Math. Helv., **37** (1962–63), 111–120.

Department of Mathematics
Gakushuin University

(Received December 15, 1975)

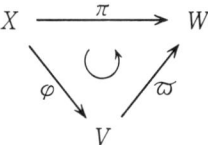

Defining Equations for Certain Types of Polarized Varieties

T. Fujita

Introduction

In this paper we improve a result of Mumford [5]. To be explicit, we fix our notation and terminology[1]. Every variety is assumed to be defined over an algebraically closed field K. For line bundles L, M on a variety V we denote by $R(L, M)$ the kernel of the natural multiplication homomorphism $\Gamma(L)\otimes\Gamma(M)\to\Gamma(L+M)$. A line bundle L on V is said to be *simply generated* if $\Gamma(tL)\otimes\Gamma(L)\to\Gamma((t+1)L)$ is surjective for every $t\geqq 1$. L is said to be *quadratically presented* if it is simply generated and if the natural homomorphism $R(sL, tL)\otimes\Gamma(L)\to R(sL, (t+1)L)$ is surjective for all $s, t\geqq 1$[2]. Now we state the following

Theorem (Mumford). *Let L be a line bundle on a smooth curve C of genus g. Then L is simply generated if* $\deg L\geqq 2g+1$, *and L is quadratically presented if* $\deg L\geqq 3g+1$.

We improve the above result in the following three ways. First, we can weaken the assumption that C is smooth. Second, we show that L is quadratically presented if $\deg L\geqq 2g+2$. Third, in the complex case, we give a higher dimensional version of these results, an embedding theorem and a structure theorem for certain types of polarized varieties, which will play an important role in our study of polarized varieties (see [1a]). As an example of applications, we give in § 5 a criterion characterizing smooth hypercubics.

1. One-dimensional Case

Throughout this section, let C be an irreducible reduced curve with $h^1(C, \mathcal{O})=g$. We remark that C is locally Macaulay, i.e., $\mathcal{O}_x$ is a Macaulay local ring for any $x\in C$. Hence there is a canonical dualizing sheaf ω on C (see [2]).

1) Basically we employ the same notation as in [1].

2) L is simply generated if and only if the graded K-algebra $G(V, L)=\bigoplus_{t=0}^{\infty}\Gamma(V, tL)$ is generated by $\Gamma(V, L)$. Then L is quadratically presented if and only if all the relations among the elements of $\Gamma(V, L)$ in $G(V, L)$ are derived from quadratic ones.

Definition 1. 1. A coherent sheaf $\mathcal{F}$ on C is said to be *quasi-invertible* if it is torsion-free and of rank one.

Remark 1. 2. ω is quasi-invertible. Any non-trivial subsheaf of a quasi-invertible sheaf is also quasi-invertible. A sheaf is quasi-invertible if and only if it is locally isomorphic to a non-zero ideal of $\mathcal{O}_C$.

Definition 1. 3. For a quasi-invertible sheaf $\mathcal{F}$, we define $\deg \mathcal{F}$ and $\Delta(\mathcal{F})$ as follows:

$$\deg \mathcal{F} = \chi(C, \mathcal{F})-1+g,$$
$$\Delta(\mathcal{F}) = 1+\deg \mathcal{F}-h^0(C, \mathcal{F}) = g-h^1(C, \mathcal{F}).$$

Lemma 1. 4. *Let $\mathcal{F}$, $\mathcal{G}$ be quasi-invertible sheaves and let $\varphi: \mathcal{F}\to\mathcal{G}$ be a non-trivial homomorphism. Then φ is injective,* $\deg \mathcal{F}\leqq\deg \mathcal{G}$ *and $h^1(\mathcal{F})\geqq h^1(\mathcal{G})$. Moreover, φ is an isomorphism if* $\deg \mathcal{F}=\deg \mathcal{G}$.

Proof. $\varphi(\mathcal{F})$ is quasi-invertible since it is a non-trivial subsheaf of $\mathcal{G}$. Hence φ is bijective at a general point of C. So $\dim \operatorname{Supp}(\operatorname{Ker} \varphi)\leqq 0$ and consequently $\operatorname{Ker} \varphi=0$. The remaining assertion follows from $\dim \operatorname{Supp}(\operatorname{Coker} \varphi)\leqq 0$. q. e. d.

Proposition 1. 5. *Let $\mathcal{F}$ be a quasi-invertible sheaf such that* $\deg \mathcal{F}\geqq 2\Delta(\mathcal{F})+1\geqq -1$. *Then $\Delta(\mathcal{F})=g$.*

Proof. Assume $g-\Delta=h^1(\mathcal{F})>0$. Then there is an effective divisor D of degree $\Delta+1$ such that $h^0(\omega[-D])=g-\Delta-1$. On the other hand, $\operatorname{Hom}(\mathcal{O}[D], \mathcal{F})\neq 0$ since $h^0(\mathcal{F}[-D])\geqq h^0(\mathcal{F})-\deg D=\deg \mathcal{F}-2\Delta\geqq 1$. We apply Lemma 1. 4 to infer that $h^1(\mathcal{F})\leqq h^1(D)=h^0(\omega[-D])=g-\Delta-1$. This contradicts $h^1(\mathcal{F})=g-\Delta$. q. e. d.

Proposition 1. 6. *Let $\mathcal{F}$ be a quasi-invertible sheaf such that* $\deg \mathcal{F}\geqq 2\Delta(\mathcal{F})\geqq 0$. *Then $\mathcal{F}$ is generated by its global sections.*

Proof. Let $\mathcal{G}$ be the subsheaf of $\mathcal{F}$ generated by global sections. Suppose $\mathcal{G}\neq\mathcal{F}$. Then $\deg \mathcal{G}<\deg \mathcal{F}$ and $\Delta(\mathcal{G})<\Delta(\mathcal{F})$, since $h^0(\mathcal{G})=h^0(\mathcal{F})$ implies $\deg \mathcal{F}-\Delta(\mathcal{F})=\deg \mathcal{G}-\Delta(\mathcal{G})$. Hence $\deg \mathcal{G}>2\Delta(\mathcal{G})$ and consequently $g=\Delta(\mathcal{G})<\Delta(\mathcal{F})$. This contradicts $\Delta(\mathcal{F})\leqq g$. q. e. d.

Lemma 1. 7. *Let L be a general*[3] *line bundle of degree l. Then*

a) $h^1(L) = 0$ *if* $l\geqq g-1$,

b) $Bs\,|L|$[4] $= \phi$ *if* $l\geqq g+1$.

Proof. We may assume $l=g-1$ in *a*) and $l=g+1$ in *b*). Let F be a line bundle of degree $2g-1$ and let D be a general effective divisor of degree g. Then $h^0(F-D)=h^0(F)-\deg D=0$ and consequently $h^1(F-D)=0$. This proves *a*). Take $g+1$ simple points $p_1, \cdots, p_{g+1}$ of C in general position and put $E=\sum_{j=1}^{g+1} p_j$ and $E_\alpha=E-p_\alpha$ $(1\leqq\alpha\leqq$

3) By a *general* line bundle we mean one belonging to a suitable open dense subset U of the space of line bundles of degree l.

4) By $Bs\Lambda$ we denote the intersection of all the members of a linear system Λ on a variety.

$g+1$). Then each of these divisors is general as a line bundle since $\deg E_\alpha = g$. Thus $h^1(E_\alpha)=0$ by a). This implies that $p_\alpha \notin Bs\,|E|$. Therefore $Bs\,|E|=\phi$. This proves b). q. e. d.

Lemma 1. 8. *Let L, M be line bundles on C and let D be an effective divisor such that $H^1(L-D)=0$. Then*

a) $Bs\,|L| = \phi$ if $Bs\,|L-D| = \phi$,

b) $\Gamma(M)\otimes\Gamma(L)\to\Gamma(L+M)$ is surjective if $\Gamma(M-D)\otimes\Gamma(L)\to\Gamma(M+L-D)$ is surjective and if $H^1(M-D)=0$,

c) $R(M, L)\otimes\Gamma(D)\to R(M, L+D)$ is surjective if $Bs\,|D|=\phi$ and if $\Gamma(M)\otimes\Gamma(L-D)\to\Gamma(M+L-D)$ is surjective.

Proof. a). $H^1(L-D)=0$ implies that the restriction $\Gamma(C, L)\to\Gamma(D, L_D)$ is surjective. Hence $D\cap Bs|L|=\phi$ and $Bs\,|L|\subset Bs\,|L-D|$. This proves a).

b). We have the following commutative diagram :

$$\begin{array}{ccccccccc}
0\to & \Gamma(M-D)\otimes\Gamma(L) & \to & \Gamma(M)\otimes\Gamma(L) & \to & \Gamma(D, M_D)\otimes\Gamma(L) & \to 0 \\
 & \downarrow & & \downarrow & & \downarrow & \\
0\to & \Gamma(M+L-D) & \to & \Gamma(M+L) & \to & \Gamma(D, [M+L]_D) & \\
 & \downarrow & & \downarrow & & \downarrow & \\
 & 0 & & 0 & & 0 &
\end{array}$$

It is easy to see that the rows and the first and the third columns are exact. Hence the second column is also exact, proving b).

c). By $R(M, L_D)$ we denote Ker $(\Gamma(D, L_D)\otimes\Gamma(C, M)\to\Gamma(D, [M+L]_D))$. Then we have the following commutative diagram :

$$\begin{array}{ccccccc}
 & 0 & & 0 & & 0 & \\
 & \downarrow & & \downarrow & & \downarrow & \\
 & R(M, L-D) & \to & R(M, L) & \overset{\varphi}{\to} & R(M, L_D) & \\
 & \downarrow & & \downarrow & & \downarrow & \\
 & \Gamma(M)\otimes\Gamma(L-D) & \to & \Gamma(M)\otimes\Gamma(L) & \to & \Gamma(M)\otimes\Gamma(D, L_D) & \to 0 \\
 & \downarrow & & \downarrow & & \downarrow & \\
0\to & \Gamma(M+L-D) & \to & \Gamma(M+L) & \to & \Gamma(D, [M+L]_D). & \\
 & \downarrow & & & & & \\
 & 0 & & & & &
\end{array}$$

Every sequence above is exact. It follows that φ is surjective. Next we consider the following diagram :

$$\begin{array}{ccccccc}
 & & R(M, L)\otimes\Gamma(D) & \to & R(M, L_D)\otimes\Gamma(D) & \to 0 \\
 & \nearrow & \downarrow & & \downarrow \psi & \\
R(M, L) & \longrightarrow & R(M, L+D) & \longrightarrow & R(M, [L+D]_D). &
\end{array}$$

The dotted arrow is defined by $\alpha\mapsto\alpha\otimes\delta$, where $\delta\in\Gamma(D)$ is the section defining D. Then the above diagram is commutative and the rows are exact. Moreover, ψ is surjective since $Bs\,|D|=\phi$. Tracing this we prove c). q. e. d.

Lemma 1. 9. *Let L, M be line bundles of degree l, m respectively. Suppose that $m\geqq 2g$, $l\geqq g+1$ and L is general. Then $\mu:\Gamma(L)\otimes\Gamma(M)\to\Gamma(L+M)$ is surjective.*

Proof. In view of Lemma 1. 8. b), it suffices to consider the case $l=g+1$, by replacing L by $L-D$, D being a general effective divisor of degree $l-g-1$. In this case, we have $h^1(L)=0$, $h^0(L)=2$ and $Bs\,|L|=\phi$. Let λ_1, λ_2 be a basis for $H^0(L)$. Then $\{\lambda_1=0\}\cap\{\lambda_2=0\}=\phi$. Therefore, $\lambda_1\varphi_2=\lambda_2\varphi_1\in\Gamma(L+M)$ for $\varphi_1, \varphi_2\in\Gamma(M)$ if and only if there is a section $\varphi\in\Gamma(M-L)$ such that $\varphi_1=\varphi\lambda_1$ and $\varphi_2=\varphi\lambda_2$. Consequently $\dim(\mathrm{Ker}\,\mu)=h^0(M-L)=\chi(M-L)=m-2g$, since $M-L$ is general. Hence $\dim(\mathrm{Im}\,\mu)=2h^0(M)-(m-2g)=m+2$. On the other hand, $h^0(L+M)=\chi(L+M)=m+2$. Hence μ is surjective. q. e. d.

Proposition 1. 10. *Let L, M be line bundles of degree l, m respectively. Then $\Gamma(L)\otimes\Gamma(M)\to\Gamma(L+M)$ is surjective if $l\geqq 2g+1$, $m\geqq 2g$.*

Proof. Let D be a general effective divisor of degree g. Then the above lemma implies that $\Gamma(L-D)\otimes\Gamma(M)\to\Gamma(L+M-D)$ is surjective. So we can apply Lemma 1. 8 *b*) since $h^1(L-D)=h^1(M-D)=0$. q. e. d.

Corollary 1. 11. *A line bundle of degree $\geqq 2g+1$ is simply generated.*

Lemma 1. 12. *Let L, M be line bundles of degree l, m respectively and let D be a general effective divisor of degree $g+1$. Then $R(M, L)\otimes\Gamma(D)\to R(M, L+D)$ is surjective if $l\geqq 2g+2$, $m\geqq 2g$.*

Proof. It is easy to see that Lemma 1. 9 enables us to apply Lemma 1. 8 *c*). q. e. d.

Proposition 1. 13. *Let L, M and N be line bundles of degree l, m and n, respectively. Suppose that $l\geqq 2g+2$, $m\geqq 2g$ and $n\geqq 2g+2$. Then $R(M, L)\otimes\Gamma(N)\to R(M, L+N)$ is surjective.*

Proof. We prove this by induction on n. First, suppose that $n\leqq 3g+2$. Let D be a general effective divisor of degree $g+1$. Lemma 1. 12 says that $R(M, L)\otimes\Gamma(D)\to R(M, L+D)$ is surjective. So it suffices to show that $R(M, L+D)\otimes\Gamma(N-D)\to R(M, L+N)$ is surjective. For this purpose we want to apply Lemma 1. 8 *c*). We should show that $h^1(L-N+2D)=0$, $Bs\,|N-D|=\phi$ and that $\Gamma(M)\otimes\Gamma(L-N+2D)\to\Gamma(M+L-N+2D)$ is surjective. This follows easily from Lemma 1. 7 and Lemma 1. 9. Second, suppose that $n\geqq 3g+3$. Let D be a general effective divisor of degree $g+1$. Again Lemma 1. 12 says that $R(M, L)\otimes\Gamma(D)\to R(M, L+D)$ is surjective. Moreover by the induction hypothesis, $R(M, L+D)\otimes\Gamma(N-D)\to R(M, L+N)$ is surjective. This completes our proof. q. e. d.

Corollary 1. 14. *A line bundle of degree $\geqq 2g+2$ is quadratically presented.*

2. Climbing a Ladder

Definition 2. 1. Let (V, L) be a prepolarized variety[5]. A member D of $|L|$ is called a *rung* of (V, L) if it is irreducible and reduced[6]. A rung D is said to be *regular*

5) A pair consisting of a variety V and a line bundle L on it. It is called a *polarized variety* if L is ample.

if the restriction $\Gamma(V, L) \to \Gamma(D, L_D)$ is surjective. A sequence $V = D_n \supset D_{n-1} \supset \cdots \supset D_1$ of subvarieties of V with $\dim D_j = j$ is called a *ladder* of (V, L), if each D_j is a rung of (D_{j+1}, L). A ladder is said to be *regular* (resp. *smooth*) if each rung of it is so.

Proposition 2. 2. *Let D be a rung of a prepolarized variety (V, L) defined by $\delta \in \Gamma(V, L)$. Let $\xi_\alpha (\alpha = 1, \cdots, k)$ be homogeneous elements of the graded algebra $G(V, L) = \bigoplus_{t=0}^{\infty} \Gamma(V, tL)$ with $\deg \xi_\alpha = d_\alpha$, and let η_α be the restriction of ξ_α to $G(D, L) = \bigoplus_{t=0}^{\infty} \Gamma(D, tL)$*[7]. *Suppose that $\{\eta_\alpha\}_{\alpha=1,\cdots,k}$ generate the algebra $G(D, L)$. Then $G(V, L)$ is generated by δ and $\{\xi_\alpha\}_{\alpha=1,\cdots,k}$.*

Proof. Let A be the graded subalgebra of $G(V, L)$ generated by $\{\delta, \xi_1, \cdots, \xi_k\}$ and let $A_t = A \cap \Gamma(V, tL)$. Let ρ_t be the restriction $\Gamma(V, tL) \to \Gamma(D, tL)$. Then we have $\operatorname{Ker} \rho_t = \delta\Gamma(V, (t-1)L)$. Moreover, $\Gamma(V, tL) \subset A_t + \operatorname{Ker} \rho_t$ since $G(D, L)$ is generated by $\{\eta_\alpha\}$. Combining them, we obtain $\Gamma(V, tL) = A_t$ by induction on t.

Corollary 2. 3. *Let D be a regular rung of a prepolarized variety (V, L). If L_D is simply generated, then so is L.*

Proposition 2. 4. *Let (V, L), D, δ, $\{\xi_\alpha\}$ and $\{\eta_\alpha\}$ be as in Proposition 2. 2. Let g_j $(1 \leqq j \leqq l)$ be homogeneous polynomials in k variables $Y_1, \cdots, Y_k$ with $\deg Y_\alpha = d_\alpha$ for $1 \leqq \alpha \leqq k$. Suppose that all the relations among $\{\eta_\alpha\}$ in $G(D, L)$ are derived from $g_1(\eta_1, \cdots, \eta_k) = \cdots = g_l(\eta_1, \cdots, \eta_k) = 0$. Then there exist l homogeneous polynomials $f_1, \cdots, f_l$ in $(k+1)$ variables $X_0, \cdots, X_k$ with $\deg X_0 = 1$, $\deg X_\alpha = d_\alpha$ for $1 \leqq \alpha \leqq k$ such that $f_j(0, Y_1, \cdots, Y_k) = g_j(Y_1, \cdots, Y_k)$ for $1 \leqq j \leqq l$ and that all the relations among $\delta, \xi_1, \cdots, \xi_k$ in $G(V, L)$ are derived from $f_1(\delta, \xi_1, \cdots, \xi_k) = \cdots = f_l(\delta, \xi_1, \cdots, \xi_k) = 0$.*

Proof[8]. Let $Q = K[X_0, X_1, \cdots, X_k]$ and $R = K[Y_1, \cdots, Y_k]$. Let $q : Q \to G(V, L)$, $r : R \to G(D, L)$ and $\pi : Q \to R$ be the K-algebra homomorphisms defined by $q(X_0) = \delta$, $q(X_j) = \xi_j$, $r(Y_j) = \eta_j$, $\pi(X_0) = 0$ and $\pi(X_j) = Y_j$ for $1 \leqq j \leqq k$. Letting $\mu : Q \to Q$ be the multiplication by X_0, we have the following commutative diagram :

$$\begin{array}{ccccccccc}
 & & 0 & & 0 & & 0 & & \\
 & & \downarrow & & \downarrow & & \downarrow & & \\
0 & \to & \operatorname{Ker} q & \to & \operatorname{Ker} q & \to & \operatorname{Ker} r & \to & 0 \\
 & & \downarrow & \mu & \downarrow & \pi & \downarrow & & \\
0 & \to & Q & \to & Q & \to & R & \to & 0 \\
 & & \downarrow & & \downarrow & & \downarrow & & \\
0 & \to & G(V, L) & \to & G(V, L) & \to & G(D, L) & \to & 0 \\
 & & \downarrow & & \downarrow & & \downarrow & & \\
 & & 0 & & 0 & & 0 & &
\end{array}$$

All the columns and the middle and the bottom rows of this diagram are exact. Hence so is the top row. Therefore, there is $f_\alpha \in \operatorname{Ker} q$ such that $\pi(f_\alpha) = g_\alpha$ for each $1 \leqq \alpha \leqq l$. Clearly we can take f_α to be homogeneous. Now it suffices to show that

6) This definition means that a rung has no embedded component.

7) We denote here by the same letter L the line bundle L_D on D induced by L.

8) The author would like to express his thanks to Mr. Mori to whom he owes this proof.

the ideal J of Q generated by $\{f_\alpha\}$ is the same as Ker q. By assumption, we have $\pi(J)=\text{Ker } r$. So Ker $q\subset J+\mu(\text{Ker } q)$ and hence Ker $q/J=\mu(\text{Ker } q/J)$. Recall the following fact: *Let I be an ideal of a ring A and let M be a finitely generated A-module. Suppose that $(1+a)M\neq 0$ for any $a\in I$. Then $IM\neq M$.* The proof is easy and quite similar to that of Nakayama's Lemma. Returning to the proof of the proposition, we assume $J\subsetneq \text{Ker } q$. Then $(1+\varphi X_0)$ (Ker $q/J)\neq 0$ for any $\varphi\in Q$. This contradicts the above fact since Ker $q/J=\mu(\text{Ker } q/J)$. q. e. d.

Corollary 2. 5. *Let D be a regular rung of a prepolarized variety (V, L). If L_D is quadratically presented, then so is L.*

3. Existence of a Ladder

Now we want to establish a sufficient condition for the existence of a ladder of a given prepolarized variety. Unfortunately, we have to assume $K=\boldsymbol{C}$ from now on[9], since we need a theorem of the Bertini-type. So every variety is an analytic variety. First we recall the following result of Hironaka [3].

Theorem 3. 1. *Let Λ be a linear system of Cartier divisors on a variety V. Then there exists a triple of a smooth variety V' together with a birational morphism $\pi: V'\to V$, a linear system Λ' on V' with $Bs\ \Lambda'=\phi$ and an effective divisor E on V' such that $\pi^*\Lambda=E+\Lambda'$. Moreover, $W=\rho_{\Lambda'}(V')$[10] is determined by (V, Λ) and is independent of the choice of (V', π, Λ', E).*

Remark 3. 2. A general member of Λ is irreducible if dim $W\geqq 2$ and if codim $Bs\ \Lambda\geqq 2$ (see, for example, [1], p. 114).

Definition 3. 3. Λ is said to be *non-degenerate* if dim $W=$dim V.

Proposition 3. 4. *Let (V, L) be a prepolarized variety and suppose that V is locally Macaulay,* dim $Bs\ |L|\leqq 0$ *and that $|L|$ is non-degenerate. Then (V, L) has a ladder.*

Proof. Remark 3. 2 says that a general member D of $|L|$ is irreducible. Moreover, D is locally Macaulay and generically smooth, hence D is reduced. So D is a rung of (V, L). Repeating this process we obtain a ladder. q. e. d.

Proposition 3. 5. *Let (M, L) be a prepolarized manifold such that* dim $Bs\ |L|\leqq 0$, $d(M, L)\geqq 2\Delta(M, L)-1$[11]. *If $|L|$ is degenerate, then a general member of $|L|$ is smooth.*

Proof. Since every base point of $|L|$ is isolated, it suffices to show the existence of a member D of $|L|$ which is smooth at any given point $p\in Bs\ |L|$. Assume that

9) Recently the author proved similar results also for the cases in which char $K>0$. (1976, Oct.)

10) $\rho_{\Lambda'}$ denotes the rational mapping associated with Λ'.

11) For a prepolarized variety (V, L) with dim $V=n$ we define (cf. [1]) $d(V, L)=L^n$ and $\Delta(V, L)=n+d(V, L)-h^0(V, L)$.

such a member does not exist. Let M_1 be the monoidal transform of M with center p, and E_p the inverse image of p. Then mE_p is a fixed part of $|L|_{M_1}$ for some integer $m \geqq 2$. Applying Theorem 3. 1 to the pair (M_1, Λ_1) where $\Lambda_1 = |L|_{M_1} - mE_p$, we have a manifold M' together with a birational morphism $\pi : M' \to M_1$, a linear system Λ' with $Bs\ \Lambda' = \phi$ and an effective divisor E such that $\pi^* \Lambda_1 = E + \Lambda'$. Let $W = \rho_{\Lambda'}(M')$ and let H be the natural hyperplane section on W. Then $H_{M'} = [\Lambda']$ and consequently $LH^{n-1}\{M'\} = L(L - mE_p - E)^{n-1} = L^n > 0$, where $n = \dim M$. Thus $\dim W = n-1$, since $|L|$ is degenerate. Put $w = \deg W$ and let X be a general fiber of $\rho_{\Lambda'}$. Then $E_p\{X\} > 0$ since $L^n - m^n = [\Lambda_1]^n \geqq [\Lambda_1]H^{n-1} = LH^{n-1} - mE_pH^{n-1} = L^n - mwE_pX$. Therefore, $d = LH^{n-1} = (H + E + mE_p)H^{n-1} \geqq mE_pH^{n-1} = mwE_pX \geqq mw \geqq 2w$. On the other hand, $0 \leqq \Delta(W, H) \leqq n - 1 + w - h^0(M, L) = w - d + \Delta - 1$. Putting these together, we obtain $d \geqq 2w \geqq 2(d - \Delta + 1)$. This contradicts the assumption $d \geqq 2\Delta - 1$.

q. e. d.

Corollary 3. 6. *Let (M, L) be a prepolarized manifold such that* $\dim Bs|L| \leqq 0$, *$d(M, L) \geqq 2\Delta(M, L) - 1$ and $d(M, L) > 0$. Then (M, L) has a ladder.*

Proof. We may assume that $|L|$ is degenerate, since otherwise we can apply Proposition 3. 4. So a general member D of $|L|$ is smooth. Since $\dim Bs\ |L| \leqq 0$, we have $Bs\ |lL| = \phi$ for $l \gg 0$. $|lL|$ is non-degenerate since $d(M, L) > 0$. Hence $H^1(M, -L) = 0$ (see Mumford [6]). Therefore D is connected, and is a rung of (M, L). Now we have $d(D, L) = d(M, L)$ and $\Delta(D, L) \leqq \Delta(M, L)$. Repeating this process, we obtain a ladder.

q. e. d.

4. Main Theorem in Higher Dimensional Cases

Theorem4. 1. *Let (M, L) be a prepolarized manifold such that $d = d(M, L) > 0$, $\Delta = \Delta(M, L) \leqq g(M, L)$*[12] *$= g$ and that* $\dim Bs\ |L| \leqq 0$. *Then*

a) (M, L) has a regular ladder if $d \geqq 2\Delta - 1$,

b) $Bs\ |L| = \phi$ if $d \geqq 2\Delta$,

c) $g(M, L) = \Delta(M, L)$ and L is simdly generated if $d \geqq 2\Delta + 1$,

d) L is quadratically presented if $d \geqq 2\Delta + 2$.

Proof. a) Corollary 3. 6 implies the existence of a ladder $\{D_j\}$ with $\dim D_j = j$. We have $d(D_1, L) = d$ and $g(D_1, L) = g$. If this ladder were not regular, then $\Delta(D_1, L) < \Delta$ and consequently $d \geqq 2\Delta(D_1, L) + 1$. This would contradict Proposition 1. 5. As for *b*), we have a regular ladder $\{D_j\}$ of (M, L). Proposition 1. 6 says $Bs\ |L_{D_1}| = \phi$. This implies $Bs\ |L| = \phi$ since the ladder is regular. *c*) follows similarly from Proposition 1. 5, Corollary 1. 11, Corollary 2. 3, and *d*) from Corollary 1. 14 and Corollary 2. 5.

q. e. d.

---

12) A precise definition of $g(V, L)$ for a given prepolarized variety (V, L) is found in [1]. Here we remark only that

a) $g(V, L) = h^1(V, \mathcal{O})$ if $\dim V = 1$,

b) $g(D, L_D) = g(V, L)$ for a rung D of (V, L),

c) $2g(V, L) - 2 = (K_V + (n-1)L)L^{n-1}$ for a smooth V, where K_V is the canonical bundle of V.

Corollary 4. 2. *Let (M, L) be as in Theorem 4. 1, a). Then (M, L) has a smooth regular ladder.*

Proof. We may assume $Bs\,|L| \neq \phi$. Take a regular ladder $\{D_j\}$ of (M, L). Let $\mathcal{A}_j$ be the subsheaf of $\mathcal{L}_j = \mathcal{O}_{D_j}[L]$ generated by global sections, and let $\mathcal{B}_j = \mathcal{L}_j/\mathcal{A}_j$. Then Supp $\mathcal{B}_j = Bs\,|L|$ and $h^0(\mathcal{B}_1) = h^0(\mathcal{B}_n)$, since the ladder $\{D_j\}$ is regular. $\mathcal{A}_1$ is quasi-invertible on D_1 and deg $\mathcal{A}_1 = d - h^0(\mathcal{B}_1)$. So $\Delta(\mathcal{A}_1) = \Delta - h^0(\mathcal{B}_1) < g$. Hence, in view of Proposition 1. 5, we infer that deg $\mathcal{A}_1 \leqq 2\Delta(\mathcal{A}_1)$. Therefore $d \leqq 2\Delta - h^0(\mathcal{B}_1)$ and consequently $h^0(\mathcal{B}_n) = 1$. This implies that $Bs\,|L|$ consists of only one point p and that $\mathcal{A}_n$ is, locally, the maximal ideal of $\mathcal{O}_M$ at p. Therefore a general member of $|L|$ is smooth. Repeating this argument, we obtain a smooth ladder. Moreover, this is easily shown to be regular. q. e. d.

5. Applications

Lemma 5. 1. *Let (M, L) be a polarized manifold with $g(M, L) \leqq 0$ and suppose that (M, L) has a ladder. Then $\Delta(M, L) = 0$.*

Proof. Let $\{D_j\}$ be a ladder of (M, L). We claim that $h^p(D_j, -tL) = 0$ for $0 \leqq p \leqq j$, $1 \leqq t \leqq j-1$. We prove this by descending induction on j. When $j = n$, our claim for $p < n$ follows from the vanishing theorem of Kodaira. For $p = n$ and $t \leqq n-1$, we have $h^n(M, -tL) = h^0(M, K_M + tL)$ and $(K_M + tL)L^{n-1} \leqq 2g(M, L) - 2 < 0$. This settles the claim for $j = n$. When $j < n$, we have the following exact sequence: $H^p(D_{j+1}, -tL) \to H^p(D_j, -tL) \to H^{p+1}(D_{j+1}, -(t+1)L)$. So the claim follows easily from the induction hypothesis. Now we have proved the above claim, which means $h^1(D_j, -L) = h^2(D_j, -L) = 0$ for $2 \leqq j \leqq n$. So $h^1(D_n) = h^1(D_{n-1}) = \cdots = h^1(D_1) = g(D_1, L) = g(M, L) \leqq 0$. This implies that $\{D_j\}$ is regular. Hence $\Delta(M, L) = \Delta(D_1, L) = 0$, since $\Delta(D_1, L) \leqq h^1(D_1) = 0$. q. e. d.

Theorem 5. 2. *Let (M, L) be a polarized manifold such that* dim $M = n$, $L^n = 3$ *and* $h^0(M, L) = n+2$. *Then M is a hypercubic in $\boldsymbol{P}^{n+1}$ and L is the line bundle associated to a hyperplane section on it.*

Proof. We have $\Delta(M, L) = 1$. This implies dim $Bs\,|L| \leqq 0$ (see [1]). Therefore (M, L) has a ladder (Corollary 3. 6). So $g(M, L) \geqq 1$ (Lemma 5. 1). Hence L is simply generated and very ample (Theorem 4. 1, c)). Our theorem follows from this. q. e. d.

Corollary 5. 3. *Let (M, L) be a polarized manifold such that* $n = $ dim M, $L^n = 3$ and $K_M + (n-1)L = 0$. *Then M is a hypercubic and L is the line bundle associated to a hyperplane section on it.*

Outline of proof. We can calculate $\chi(M, tL)$ using the vanishing theorem of Kodaira. Then it follows that $h^0(M, L) = \chi(M, L) = n+2$.

We note that $K_M + (n-1)L = 0$ holds for a polarized manifold (M, L) if and only if $\Delta(M, L) = g(M, L) = 1$. So we have similarly the following

Theorem 5. 4. *Let* (M, L) *be a polarized manifold such that* $\dim M=n$ *and* $L^n=4$. *Then the following conditions are equivalent to each other* :

a) *M is isomorphic to a complete intersection of type* $(2, 2)$ *and L is the line bundle associated to a hyperplane section on it.*

b) $h^0(M, L)=n+3$.

c) $K_M+(n-1)L=0$.

References

[1] Fujita, T. : On the structure of polarized varieties with Δ-genera zero, J. of Fac. Sci. Univ. of Tokyo, Sec. IA, **22** (1975), 103–115.

[1a] Fujita, T. : On the structure of certain types of polarized varieties, Proc. Japan Acad. **49** (1973), 800--802 & **50** (1974), 411–412.

[2] Grothendieck, A. : Local cohomology, Lecture Notes in Math. 41, Springer, 1966.

[3] Hironaka, H. : Resolution of singularities of an algebraic variety over a field of characteristic zero I–II, Ann. of Math., **79** (1964), 109–326.

[4] Mori, S. : On a generalization of complete intersections, J. of Math. Kyoto Univ., **15** (1975), 619–646.

[5] Mumford, D. : Varieties defined by quadratic equations, C. I. M. E. (1969) –III, 29–100.

[6] Mumford, D. : Pathology III, Amer. J. Math., **89** (1967), 94–103.

College of General Education
University of Tokyo

(Received December 12, 1975)

On Logarithmic Kodaira Dimension of Algebraic Varieties

S. Iitaka

1. Introduction and Notation

Let k be an algebraically closed field of characteristic zero. We shall work in the category of schemes over k.

Let V be an n-dimensional algebraic variety. By Nagata, we have a complete algebraic variety $\bar{V}$ which contains V as a Zariski open subset. We call $\bar{V}-V$ an algebraic boundary of V. By Hironaka's Main Theorems I and II, we have after a finite succession of monoidal transformations, a non-singular algebraic variety $\bar{V}^*$ and a proper birational morphism $\bar{\mu}: \bar{V}^* \to \bar{V}$ such that if $V^*=\mu^{-1}(V)$, then $\bar{D}=\bar{V}^*-V^*=\bar{\mu}^{-1}(\bar{V}-V)$ is a divisor of simple normal crossing type. We call $\bar{V}^*$ a compactification of V^* with smooth boundary $\bar{D}$.

The sheaf of germs of logarithmic q-forms of $\bar{V}$ along $\bar{D}$ is the subsheaf of the sheaf of germs of rational q-forms $\Omega^q(*)$ of $\bar{V}^*$ which consists of those germs which for any point $p \in \bar{V}^*$, are written as

$$\sum_{\substack{r+s=q \\ I=(i(1),\cdots,i(r)) \\ J=(j(1),\cdots,j(s))}} a_{I.J}(z, w)\frac{dz_{i(1)}}{z_{i(1)}} \wedge\cdots\wedge \frac{dz_{i(r)}}{z_{i(r)}} \wedge dw_{j(1)} \wedge\cdots\wedge dw_{j(s)},$$

where $(z_1, \cdots, z_m, w_1, \cdots, w_{n-m})$ is a local system of regular parameters at p such that $z_1\cdots z_m=0$ defines $\bar{D}$ around p and $a_{I,J}(z, w) \in \mathcal{O}_{\bar{V}^*,p}$. By $\Omega^q(\log \bar{D})$ we denote the sheaf of germs of logarithmic q-forms along $\bar{D}$.

For any integers $m_1 \geqq 0, \cdots, m_n \geqq 0$, put

$$T_{m_1,\cdots,m_n}(V) = H^0(\bar{V}^*, (\Omega^1 \log \bar{D})^{\otimes m_1} \otimes \cdots \otimes (\Omega^n \log \bar{D})^{\otimes m_n}).$$

We shall prove that

$$\dim T_{m_1,\cdots,m_n}(V)$$

is independent of the choice of compactification of V with smooth boundary and of non-singular model of V. In particular, we define

$$\begin{aligned}
\textit{the logarithmic irregularity } \bar{q}V &= \dim H^0(\bar{V}^*, \Omega^1 \log \bar{D}),\\
\textit{the logarithmic geometric genus } \bar{p}_g V &= \dim H^0(\bar{V}^*, \Omega^n \log \bar{D})\\
&= \dim H^0(\bar{V}^*, \Omega^n(\bar{D})) = \dim H^0(\bar{K}^*+\bar{D}),\\
\textit{the logarithmic m-genus } \bar{P}_m(V) &= \dim H^0(\bar{V}^*, (\Omega^n(\bar{D}))^{\otimes m})\\
&= \dim H^0(m(\bar{K}^*+\bar{D})),
\end{aligned}$$

and finally

the logarithmic Kodaira dimension $\bar{\kappa} V = \kappa(\bar{K}+\bar{D}, \bar{V}^*)$.

Here, $\bar{K}^*$ denotes a canonical divisor of $\bar{V}^*$ and in general for a divisor D, we abbreviate $H^0(\bar{V}^*, \mathcal{O}(D))$ by H^0D. We write

$$\bar{P}_{m_1,\cdots,m_n}(V) = \dim T_{m_1,\cdots,m_n}(V).$$

2. Examples

Example 1. Let V be a non-singular curve. Then
$\bar{\kappa} V = -\infty$ if and only if $V = \boldsymbol{P}^1$ or an affine line $\boldsymbol{A}_k^1 = G_a$,
$\bar{\kappa} V = 0$ if and only if V is a complete elliptic curve or a 1-dimensional algebraic torus $G_m = \boldsymbol{A}_k^1 - \{p\}$,
$\bar{\kappa} V = 1$ if and only if V is one of the others.

Example 2. Let $\bar{V}$ be a complete non-singular curve of genus g and $p_0, \cdots, p_t$, $t+1$ points on $\bar{V}$. Putting

$$V = \bar{V} - \{p_0, \cdots, p_t\},$$

we have

$$\bar{q} V = \bar{p}_g V = g + t.$$

Example 3. Let $D_0, \cdots, D_t$ be $1+t$ lines in $\boldsymbol{P}^2$ and $V = \boldsymbol{P}^2 - D_0 \cup \cdots \cup D_t$. Then

$\bigcup_{j=0}^{t} D_j$ in $\boldsymbol{P}^2$	$\bar{\kappa}(V)$	$\bar{q}(V)$	V
	$-\infty$	t	$\boldsymbol{A}_k^1 \times (\boldsymbol{P}^1 - \{a_0, \cdots, a_t\})$
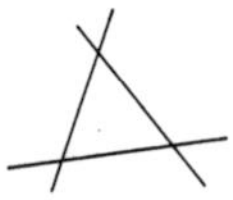	0	2	$G_m \times G_m$
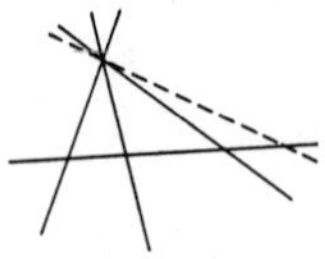	1	t ($t \geqq 3$)	$G_m \times (\boldsymbol{A}_k^1 - \{a_2, \cdots, a_t\})$
otherwise	2	t	

Write $\kappa V = \kappa \bar{V}^* = \kappa(\bar{K}^*, \bar{V}^*)$, which is the original Kodaira dimension of V. F. Sakai [7] defines another kind of Kodaira dimension of V, which we indicate by $\underline{\kappa} V$. Letting ρ_m be the rational map associated with $m\bar{K} + (m-1)\bar{D}$, he defines

$$\underline{\kappa} V = \max_{m>1} \dim \rho_m(\bar{V}^*).$$

In general, $\kappa V \leqq \underline{\kappa} V \leqq \bar{\kappa} V$. If $\underline{\kappa} V \geqq 0$, then $\underline{\kappa} V = \bar{\kappa} V$ and it is easily seen that $\underline{\kappa} V = n$ if and only if $\bar{\kappa}V=n$.

Example 4. Let $\bar{V}$ be a minimal complete non-singular algebraic surface with $\kappa\bar{V}\geqq 0$. Let $D=D_0\cup\cdots\cup D_t$ be a union of $1+t$ irreducible curves. Put $V=\bar{V}-D$. Then $\bar{\kappa}V$ is computed as follows:

(i) if $\bar{V}$ is an abelian variety or a hyperelliptic surface, then $\bar{\kappa}V\geqq 1$ and $\bar{\kappa}V=1$ implies that $D_0, \cdots, D_t$ are fibers of an elliptic surface $\varphi: \bar{V}\rightarrow\Delta$,

(ii) if $\bar{V}$ is a $K3$ surface or an Enriques surface, then $\bar{\kappa}V\geqq 0$ and $\bar{\kappa}V=0$ implies that each D_j is $\boldsymbol{P}^1$ with self-intersection number -2. Moreover, $\bar{\kappa}V=1$ implies that D is contained in a finite union of fibers of an elliptic surface $\varphi: \bar{V}\rightarrow\Delta$

(iii) if $\kappa\bar{V}=1$, then V has the structure of an elliptic surface $\varphi: \bar{V}\rightarrow\Delta$. $\bar{\kappa}V=1$ implies that D is contained in a finite union of fibers of φ.

In order to prove these we make the following observation: Assuming V to be non-singular, we let $\bar{V}$ be a non-singular complete variety containing V as a Zariski open subset. If codim $(\bar{V}-V)\geqq 2$, then $\bar{P}_{m_1,\cdots,m_n}V=P_{m_1,\cdots,m_n}V$, in particular, $\bar{\kappa}V=\kappa V$. Assume $D=\bar{V}-V$ is a divisor. Then

$$\bar{P}_m V\leqq \dim H^0 m(\bar{K}+D), \quad \text{and if } \kappa V\geqq 0, \quad \text{then}$$
$$\bar{\kappa}V=\kappa(\bar{K}+D, \bar{V}) \qquad \text{(see [6] Lemma 5).}$$

Now, we let $\bar{V}$ be a $K3$ surface. $\bar{\kappa}V=0$ implies that for any $\nu_0\geqq 0, \cdots, \nu_t\geqq 0$,

$$\dim H^0(\bar{V}, \mathcal{O}(\textstyle\sum\nu_j D_j))=1.$$

Hence by Riemann-Roch Theorem, $D_j^2=-2$ and the matrix $((D_iD_j))$ is negative definite. Hence the connected components of $D=D_0\cup\cdots\cup D_t$ may be indicated by Dynkin diagrams: A_n, D_n, E_6, E_7, E_8. $\bar{\kappa}V=1$ implies that the moving irreducible component C of the general member of $|mD|$, $m\gg 0$, satisfies $C^2=0$ and so the virtual genus $\pi(C)=C^2/2+1=1$. Hence we get an elliptic surface a finite union of whose fibers contains D. Second, we assume $\bar{V}$ to be an Enriques surface, then there is a 2-sheeted unramified covering map $\psi: \bar{V}'\rightarrow\bar{V}$ where $\bar{V}'$ is a $K3$ surface. Since $\bar{\kappa}V=\kappa(D, \bar{V})=\kappa(\psi^*D, \bar{V}')$, we can infer a similar result in this case.

The proofs of the other cases are easy.

As in [3], we call V a variety of elliptic type if $\bar{\kappa}V=-\infty$, of parabolic type if $\bar{\kappa}V=0$ and of hyperbolic type if $\bar{\kappa}V=n$. A variety of hyperbolic type might be called a variety of general type.

Our purpose here is to prove the theorem to the effect that the group of strictly birational maps of V onto itself is a finite group if V is of hyperbolic type. Furthermore, we study the structure of algebraic varieties which have many automorphisms in the case where they are not of elliptic type. In a forthcoming paper: Logarithmic forms of algebraic varieties, we shall study the quasi-Albanese maps with some applications (see [11]).

3

Let V_1 and V_2 be n-dimensional non-singular algebraic varieties and f a dominant morphism $V_1 \to V_2$. Choose two compactifications with smooth boundaries $\bar{D}_1$ and $\bar{D}_2$ of V_1 and V_2, respectively. Then f is a representative of a rational map $\bar{f}$ from $\bar{V}_1$ into $\bar{V}_2$. By applying Hironaka's Main Theorem II, we may assume that $\bar{f}$ is a morphism. Under this assumption, we prove

Proposition 1. *If a rational q-form ω on $\bar{V}_2$ is logarithmic along $\bar{D}_2$, then $f^*\omega$ is a logarithmic q-form on $\bar{V}_1$ along $\bar{D}_1$. Moreover, if f is proper and birational, then*

$$f^* : H^0\Omega^q \log \bar{D}_2 \to H^0\Omega^q \log \bar{D}_1$$

is an isomorphism.

Proof. By definition, $f^*\omega$ is a rational q-form. For any point $p_1 \in \bar{V}_1$, choose a local system of regular parameters $(w_1, \cdots, w_n)$ and for $p_2 = \bar{f}(p_1)$ choose a local system of regular parameters $(z_1, \cdots, z_n)$ such that if $p_2 \in \bar{D}_2$, $z_1 \cdots z_r = 0$ defines $\bar{D}_2$ around p_2 and if $p_1 \in \bar{D}_1$, $w_1 \cdots w_s = 0$ defines $\bar{D}_1$ around p_1. By definition of f, $\bar{f}^{-1}(\bar{D}_1) \subset \bar{D}_2$ and so

(L) $\quad z_i = \prod w_j^{n_{ij}} \varepsilon_i, \quad n_{ij} \geqq 0$, where ε_i is non-vanishing around p_1.

Thanks to

$$\left(\frac{\mathrm{dL}}{\mathrm{L}}\right) \qquad \frac{dz_i}{z_i} = \sum n_{ij} \frac{dw_j}{w_j} + \frac{d\varepsilon_i}{\varepsilon_i},$$

we see that $f^*\omega$ is logarithmic. Now, assume that f is proper. Then fV_1 is closed in V_2, hence $fV_1 = \bar{f}V_1 = V_2$. Letting $j : V_2 \to \bar{f}^{-1}V_1$, we get $f = (\bar{f}|\bar{f}^{-1}V_1) \cdot j$, which is proper. Hence j is also proper. Thus $V_2 = \bar{f}^{-1}V_1$. In other words, $\bar{f}^{-1}\bar{D}_1 = \bar{D}_2$. Furthermore assume f to be birational. Take a logarithmic q-form ω_1 on $\bar{V}_1$ along $\bar{D}_1$. Then we have a rational q-form ω_2 on $\bar{V}_2$ such that $\omega_1 = f^*\omega_2$. Since $\bar{f}$ is birational and $\bar{V}_1$ is complete, there exists a Zariski open subset $\bar{V}_2^0$ with codim $(\bar{V}_2 - \bar{V}_2^0) \geqq 2$ such that the inverse $g : \bar{V}_2^0 \to \bar{V}_1$ of $\bar{f}$ exists, that is, $\bar{f} \cdot g$ is the open immersion $\bar{V}_2^0 \to \bar{V}_2$. Since ω_1 is a logarithmic q-form along $\bar{D}_1 = \bar{f}^{-1}\bar{D}_2$, $g^*\omega_1 = \omega_2 | \bar{V}_2^0$ is also logarithmic along $g^{-1}\bar{D}_1 = \bar{D}_2 | \bar{V}_2^0$. Hence ω_2 is a logarithmic q-form along $\bar{D}_2$.

q. e. d.

This proposition holds for forms in $T_{m_1, \cdots, m_n}(V)$. Thus we have established the *birational invariance* of dim $T_{m_1, \cdots, m_n}(V)$. Precisely speaking, a rational map $f : V_1 \to V_2$ is called a *strictly rational map* if there is a *proper* birational morphism $\mu : V_3 \to V_1$, V_3 being an algebraic variety, such that $f \cdot \mu$ is a morphism. For example, a dominant rational map from a complete V_1 into a non-complete V_2 is not strictly rational. A rational map from V_1 into a complete V_2 is always strictly rational. Actually, consider a compactification $\bar{V}_1$ of V_1 and regard f as a representative of a rational map $\bar{f} : \bar{V}_1 \to V_2$. Then there is a birational morphism $\bar{\mu}$ from a complete variety $\bar{V}_3$ onto $\bar{V}_1$ such that $\bar{f} \cdot \bar{\mu}$ is a morphism. Put $V_3 = \bar{\mu}^{-1}V_1$. Then $\mu = \bar{\mu}|V_3$ is proper and $f \cdot \mu$ is a morphism.

Consider a strictly rational map from an algebraic variety V_1 into an algebraic variety V_2. As in § 1, taking compactifications $\bar{V}_1^*$ and $\bar{V}_2^*$ of non-singular models $V_1^* \to V_1$, $V_2^* \to V_2$ of V_1 and V_2, with smooth boundaries $\bar{D}_1$ and $\bar{D}_2$, respectively, we obtain a linear map :

$$f^* : H^0\Omega^q \log \bar{D}_2 \to H^0\Omega^q(*)$$
$$\cup$$
$$H^0\Omega^q \log \bar{D}_1.$$

By definition, there is a birational proper morphism $\mu: V_3 \to V_1$ such that $g=f\cdot\mu$ is a morphism. Choose a suitable compactification $\bar{V}_3^*$ with smooth boundary $\bar{D}_3$ such that $\bar{\mu}$ is a morphism and $\bar{g}=\bar{f}\cdot\bar{\mu}$ is a morphism. By Proposition 1,

$$g^* : H^0\Omega^q \log \bar{D}_2 \to H^0\Omega^q \log \bar{D}_3,$$

and

$$\mu^* : H^0\Omega^q \log \bar{D}_1 \simeq H^0\Omega^q \log \bar{D}_3.$$

Hence $f^*=\mu^{*-1}\cdot g^* : H^0\Omega^q \log \bar{D}_2 \to H^0\Omega^q \log \bar{D}_1$.
Note that in case f is dominant, f^* is injective. Thus, in general, if $\dim V_1=\dim V_2 =n$, and f is dominant, then

$$\bar{P}_{m_1,\cdots,m_n} V_1 \geqq \bar{P}_{m_1,\cdots,m_n} V_2.$$

In particular, if V_1 is a non empty Zariski open subset of V_2, then

$$\bar{P}_m V_1 \geqq \bar{P}_m V_2, \quad \bar{\kappa} V_1 \geqq \bar{\kappa} V_2 \quad \text{and} \quad \bar{q} V_1 \geqq \bar{q} V_2.$$

We say that a birational map $f: V_1 \to V_2$ is proper if f and f^{-1} are both strictly rational. For example, a proper birational morphism is a proper birational map. Note that

$$\bar{P}_{m_1,\cdots,m_n} V_1 = \bar{P}_{m_1,\cdots,m_n} V_2,$$

if there is a proper birational map $V_1 \to V_2$.

Proposition 2. *Let $\bar{V}$ be a non-singular complete algebraic variety and $\bar{D}$ a divisor of normal crossing type. Assume that $\bar{K}+\bar{D}$ is ample. Put $V=\bar{V}-\bar{D}$. Then any strictly rational dominant map $f: V \to V$ is a restriction of an automorphism of $\bar{V}$.*

Proof. From the argument above, we infer that

$$f^* : H^0 m(\bar{K}+\bar{D}) \to H^0 m(\bar{K}+\bar{D})$$

is an isomorphism. Since $m(\bar{K}+\bar{D})$ is a hyperplane section for $m \gg 0$, we get the conclusion. q. e. d.

Example 5. Let D be a divisor of normal crossing type in $\boldsymbol{P}^n$. Assume that $\deg D \geqq n+2$. Then

$$\mathrm{Aut}(\boldsymbol{P}^n - D) \subset PGL(n, k).$$

For a variety V in $\boldsymbol{P}^n$, we write

Lin $V = \{\alpha \in PGL(n, k); \alpha V = V\}$,
S Bir $V =$ the group generated by strictly birational maps : $V \to V$.

Then under the conditions of Proposition 2,

S Bir $V =$ Lin V, which is a finite group (see Theorem 6).

4

Let V_1 and V_2 be non-singular algebraic varieties and f a dominant morphism $V_1 \to V_2$. Choose compactifications $\bar{V}_1$ and $\bar{V}_2$ with smooth boundaries $\bar{D}_1$ and $\bar{D}_2$ of V_1 and V_2, respectively such that the rational map $\bar{f} : \bar{V}_1 \to \bar{V}_2$, $\bar{f}|V_1 = f$, is a morphism. Then there is an effective divisor $\bar{R}_f$ which satisfies

(R) $\quad \bar{K}_1 + \bar{D}_1 \sim \bar{f}^*(\bar{K}_2 + \bar{D}_2) + \bar{R}_f$

$\bar{R}_f$ is called the *logarithmic ramification divisor* for f.

We prove (R) by using the notation in the proof of Proposition 1. A rational logarithmic q-form ω on $\bar{V}_2$ along $\bar{D}_2$ is a rational section of the line bundle $[\bar{K}_2 + \bar{D}_2]$. Hence for any $p \in \bar{V}_2$,

$$\omega = \varphi(z) \frac{dz_1}{z_1} \wedge \cdots \wedge \frac{dz_r}{z_r} \wedge dz_{r+1} \wedge \cdots \wedge dz_n,$$

where $(z_1, \cdots, z_n)$ is a local system of regular parameter at p such that $z_1 \cdots z_r = 0$ defines $\bar{D}_2$ around p, and φ is a rational function. By (L) in § 3, we get the rational logarithmic n-form

$$f^*\omega = (\varphi \cdot f)\psi(w) \frac{dw_1}{w_1} \wedge \cdots \wedge \frac{dw_s}{w_s} \wedge dw_{s+1} \wedge \cdots \wedge dw_n,$$

$\psi(w) \in \mathcal{O}_{\bar{V}_1, p_1}$. Hence a collection $\{\psi(w)\}$ defines the effective divisor $\bar{R}_f$ and a collection $\{\varphi\}$ is a rational section of $[\bar{K}_2 + \bar{D}_2]$, which defines a divisor $\sim \bar{K}_2 + \bar{D}_2$. A collection $\{\varphi \cdot f\}$ defines a divisor $\sim \bar{f}^*(\bar{K}_2 + \bar{D}_2)$. Hence we obtain the formula (R).

Proposition 3. *Furthermore, assume that* $\bar{\kappa} V_2 \geqq 0$ *and* $\bar{P}_m V_1 = \bar{P}_m V_2$ *for sufficiently large* m. *Then the rational map* Φ_{m_1, V_1} *associated to* $|m(\bar{K}_1 + \bar{D}_1)|$ *is factored as follows:*

$$\Phi_{m_1, V_1} = \Phi_{m_2, V_2} \cdot f.$$

Proof. As in the proof of Theorem 3 in [3], choose m_i so large that $\Phi_{m_i, V_i} : \bar{V}_i \to \Phi_{m_i, V_i}(\bar{V}_i)$ is birationally equivalent to the $\bar{K}_i + \bar{D}_i$-canonical fibered manifold of $\bar{V}_i$ for $i = 1, 2$ (see [4]). Choose m = a multiple of L. C. M. (m_1, m_2) so that $\bar{P}_m V_1 = \bar{P}_m V_2$. Then the natural inclusions:

$$H^0 m(\bar{K}_2 + \bar{D}_2) \subset H^0 m\bar{f}^*(\bar{K}_2 + \bar{D}_2) \subset H^0 m(\bar{f}^*(\bar{K}_2 + \bar{D}_2) + \bar{R}_f) = H^0(m(\bar{K}_1 + \bar{D}_1))$$

are isomorphisms. From this follows the conclusion. q. e. d.

Using this we can prove easily

Theorem 1. *If* $\bar{\kappa} V_1 = n$, $\bar{P}_m V_1 = \bar{P}_m V_2$ *for* $m \gg 0$, *then any dominant strictly rational map from* V_1 *into* V_2 *is birational.*

As a generalization of a theorem of Peters [10], we prove

Theorem 2. *Let* V *be a non-singular algebraic variety with* $\bar{\kappa} V \geqq 0$. *Then any dominant morphism from* V *into itself is an étale covering map from* V *onto* V.

Proof. Let $\bar{V}$ be a compactification with smooth boundary $\bar{D}$. f is regarded as a representative of a rational map $\bar{f} : \bar{V} \to \bar{V}$. Performing a finite succession of monoidal transformations with non-singular centers $\subset \bar{D}$ on $\bar{V}$, we get a birational morphism $\bar{\mu} : \bar{V}^* \to \bar{V}$ such that $\bar{f} \cdot \bar{\mu} = \bar{g}$ is a morphism $\bar{V}^* \to \bar{V}$. Then by (R)

$$m(\bar{K}^* + \bar{D}^*) \sim \bar{g}^* m(\bar{K} + \bar{D}) + m\bar{R}_g,$$

where $\bar{K}^*$ denotes a canonical divisor on $\bar{V}^*$ and $\bar{D}^* = \bar{\mu}^{-1} D$; and

$$m(\bar{K}^* + \bar{D}^*) \sim \bar{\mu}^* m(\bar{K} + \bar{D}) + m\bar{R}_\mu.$$

Hence,

$$m(\bar{K} + \bar{D}) \sim \bar{\mu}_*(m(\bar{K}^* + \bar{D}^*)) \sim \bar{\mu}_* \bar{g}^* m(\bar{K} + \bar{D}) + \bar{\mu}_* m\bar{R}_g.$$

With the decomposition $\bar{f} = \bar{\mu} \cdot \bar{g}$ in mind, we write

$$\bar{f}^*(\bar{K} + \bar{D}) = \bar{\mu}_* \bar{g}^*(\bar{K} + \bar{D}),$$
$$\bar{R}_f = \bar{\mu}_*(\bar{R}_g).$$

Then

$$m(\bar{K} + \bar{D}) \sim m\bar{f}^*(\bar{K} + \bar{D}) + m\bar{R}_f.$$

By assumption $\bar{\kappa} V \geqq 0$, there is an $m > 0$ such that $|m(\bar{K} + \bar{D})| \ni C(m)$. Write $\bar{g}^* C(m)$ as a sum of effective divisors D^* and $\mathcal{E}$ where $\mathcal{E}$ consists of the components that are mapped to 0 by $\bar{\mu}_*$. Similarly write $\bar{\mu}^* \bar{\mu}_* D^* = D^* + \mathcal{E}'$. Note that $\mathcal{E}$, $\mathcal{E}' > 0$. Then, since $\mathcal{E}$ is exceptional for μ, we have

$$\dim |D^*| \leqq \dim |D^* + \mathcal{E}' + \mathcal{E}| = \dim |\bar{\mu}^* \bar{\mu}_* D^* + \mathcal{E}|$$
$$= \dim |\bar{\mu}^* \bar{\mu}_* D^*| = \dim |\bar{\mu}_* D^*|.$$

Hence $\dim |m(\bar{K} + \bar{D})| = \dim |m\bar{g}^*(\bar{K} + \bar{D})| \leqq \dim |m\bar{f}^*(\bar{K} + \bar{D})| = \dim |m\bar{f}^*(\bar{K} + \bar{D}) + m\bar{R}_f| = \dim |m(\bar{K} + \bar{D})|$.

Let $V_1 = \bar{g}^{-1}(V)$ and $g_1 = g | V_1$, which is proper. $\bar{V}^*$ is regarded as a compactification of V with smooth boundary $\bar{D}_1^*$ as well as that of V. Hence g_1 defines the rational map $\bar{g}_1 : \bar{V}^* \to \bar{V}^*$. Let $\bar{\mu}' : \bar{V}_2 \to \bar{V}^*$ be a birational morphism such that $\bar{h} = \bar{g}_1 \cdot \bar{\mu}'$ is a morphism. We assume that $\bar{V}_2$ is a compactification of $V_2 = \bar{\mu}'^{-1}(V_1)$ with smooth boundary $\bar{D}_2$. Hence by $\bar{K}_2$ denoting a canonical divisor on $\bar{V}_2$,

$$m(\bar{K}_2 + \bar{D}_2) = \bar{h}^* m(\bar{K}^* + \bar{D}_1^*) + m\bar{R}_h \quad \text{and}$$
$$m(\bar{K}^* + \bar{D}_1^*) = \bar{g}_1^* m(\bar{K}^* + \bar{D}_1^*) + m\bar{R}_{g_1}.$$

Therefore by $|D|_{\text{fix}}$ denoting the fixed part of $|D|$ in general, we obtain

$$|m(\bar{K}^* + \bar{D}_1^*)|_{\text{fix}} \geqq m\bar{R}_{g_1} + m\bar{g}_1^* \bar{R}_{g_1} + m\bar{g}_1^* \bar{g}_1^* \bar{R}_{g_1} + \cdots.$$

Hence $\bar{g}_1^* \bar{R}_{g_1} = 0$. On the other hand, since $g_1 V_1 = V$ and $\bar{g}_1^* \bar{R}_{g_1} | V_1 = g_1^*(\bar{R}_{g_1} \cap V) = 0$ we see that $\bar{R}_f = R_{g_1} | V = 0$. Similarly, $R_{g_1} = \bar{R}_{g_1} \cap V_1 = 0$. This implies that g_1 is étale. On the other hand, V_1 has only to satisfy the condition that $V_1 \supset V$ and V_1 be non-singular. We have proved that any $g_1 : V_1 \to V$ extending f is étale. Hence $V_1 = V$ and $g = f$. Therefore f is the *étale covering map* from V onto V. q. e. d.

The following proposition is remarkable.

Proposition 4 (Y. Kawamata). *Let V be an algebraic scheme and F a closed subscheme of dimension d. If there is an isomorphism $j : V \to V - F$, then $F = \phi$.*

Proof. We prove this by making use of the homology groups $H_m(V)$ of Hartshorne [2]. It is easy to see that if $m>2\dim V=2n$, then $H_m(V)=0$ and $H_{2n}(V)\neq 0$. Therefore, making use of Mayer-Vietoris sequences, we can prove that $\dim H_{2n}(V)\geqq s$ if there are at least s irreducible components whose dimensions are n. Let $d=\dim F$. Then

$$H_{2d+1}(F)\rightarrow H_{2d+1}(V)\rightarrow H_{2d+1}(V-F)\rightarrow H_{2d}(F)\rightarrow H_{2d}(V)\quad\text{(exact)}.$$

Since j is isomorphic, $\dim H_{2d+1}(V)=\dim H_{2d+1}(V-F)$. Hence

$$0\rightarrow H_{2d+1}(V)\rightarrow H_{2d+1}(V-F)$$

implies

$$0\rightarrow H_{2d}(F)\rightarrow H_{2d}(V).$$

Let F' be the closure of jF in V and $F_1=F'\cup F$. Since $F'=F_1-F$, we know that $\dim(F_1-F)=d$, hence $\sharp$ {d-dimensional irreducible components of F_1} $>\sharp$ {d-dimensional irreducible components of F}. Composing isomorphisms :

$$V\rightarrow V-F\rightarrow V-F_1\rightarrow V-F_2\rightarrow\cdots\rightarrow V-F_l,$$

we get

$$\sharp\ \{d\text{-dimensional irreducible components of } F_l\}\geqq l+1.$$

Hence $\dim H_{2d}(F_l)\rightarrow\infty$ as $l\rightarrow\infty$. This contradicts $\dim H_{2d}(F_l)\leqq\dim H_{2d}(V)<\infty$.

q. e. d.

Corollary. *Moreover, if f is birational, then f turns out to be an isomorphism. Let V be a non-singular algebraic variety of hyperbolic type. Then any dominant morphism $f: V\rightarrow V$ turns out to be an isomorphism. Furthermore, if V is affine, then*

$$\text{S Bir } V=\text{Aut } V.$$

Note that any affine variety is strongly minimal in our birational geometry.

Example 6. Let $R=k[x, y, 1/(xy-1)]$. Then any endomorphism f of R over k satisfies

$$\frac{\partial(f(x), f(y))}{\partial(x, y)}=0 \text{ or } c(xy-1)^m \text{ for some } c\in k^*,\ m\in\boldsymbol{Z}.$$

Actually, what we do is to check that $\bar{\kappa}$ Spec $R=0$ and use the Corollary. Then we get the assertion.

Similarly, let $R=k[x, y, z, 1/(x^3+y^3+z^3-1)]$ and let f be an endomorphism of R over k. Then

$$\frac{\partial(f(x), f(y), f(z))}{\partial(x, y, z)}=0\quad\text{or}\quad c(x^3+y^3+z^3-1)^m.$$

5

Next, we consider unramified covering maps between distinct varieties.

Theorem 3. *Let V_1 and V_2 be algebraic varieties and f an étale covering morphism from*

V_1 onto V_2. Then

$$\bar{\kappa} V_1 = \bar{\kappa} V_2.$$

Proof. Let $\mu : V_2^* \to V_2$ be a non-singular model of V_2, that is, V_2^* is non-singular and μ is a proper birational morphism. Denote by V_1^* a fibre product of V_1 and V_2^* over V_2. Then we have the étale covering morphism $V_1^* \to V_2^*$ and a proper birational morphism $V_1^* \to V_1$. Hence, V_1^* is a non-singular algebraic variety. Since by definition $\bar{\kappa} V_1 = \bar{\kappa} V_1^*$ and $\bar{\kappa} V_2 = \bar{\kappa} V_2^*$, we can assume that V_1 and V_2 are both non-singular. As in § 1, we choose compactifications $\bar{V}_1$ and $\bar{V}_2$ with smooth boundaries $\bar{D}_1$ and $\bar{D}_2$ such that $\bar{f} : \bar{V}_1 \to \bar{V}_2$ is a morphism, $\bar{f} | V_1 = f$. Since f is proper, $\bar{f}^{-1} \bar{D}_2 = \bar{D}_1$. Delete the condition that f is unramified in the hypothesis and add the condition to the effect that f is proper and dominant with $\dim V_1 = \dim V_2$. Under this condition, we shall prove that

$$\bar{f}_* \bar{R}_f \cap \bar{D}_1 = 0.$$

By C we denote an irreducible component of $\bar{D}_1$ such that $\bar{f} C = \Gamma$ is a divisor. Take general points p_1 of C and p_2 of Γ such that $\bar{f} p_1 = p_2$. Then we have a local system of regular parameters $(z_1, \cdots, z_n)$ at p_2 such that $z_1 = 0$ defines Γ around p_2. Let ζ_1 be a minimal equation of C around p_1. Then $(\zeta_1, z_2, \cdots, z_n)$ could be regarded as a local system of regular parameters at p_1. Hence $z_1 = \zeta_1^{\nu} \varepsilon$, ε being a unit in $\mathcal{O}_{V_1, p_1}$. Hence

$$\frac{dz_1}{z_1} \wedge dz_2 \wedge \cdots \wedge dz_n = \left(\nu + \frac{\zeta_1}{\varepsilon} \frac{\partial \varepsilon}{\partial \zeta_1} \right) \frac{d\zeta_1}{\zeta_1} \wedge dz_2 \wedge \cdots \wedge dz_n.$$

Thus $\bar{R}_f = 0$ around p_1.

We shall proceed with the proof of Theorem 3. Using the notation of the proof of Proposition 1, we get, in view of (R)

$$\bar{K}_2 + \bar{D}_2 = \bar{f}^* (\bar{K}_1 + \bar{D}_1) + \bar{R}_f.$$

Since f is unramified, $\bar{f}_* \bar{R}_f = 0$.
Hence using the following lemma, we obtain the result.

Lemma 1. *Let $\bar{f} : \bar{V}_1 \to \bar{V}_2$ be a surjective morphism between complete algebraic varieties. Let $\bar{D}_2$ be a Cartier divisor on $\bar{V}_2$ and E an effective divisor on $\bar{V}_1$ such that* $\operatorname{codim} \bar{f}(E) \geqq 2$. *Then*

$$\kappa(\bar{f}^* \bar{D}_2 + E, \bar{V}_1) = \kappa(\bar{D}_2, \bar{V}_2).$$

Proof. By definition, we can assume that $\bar{V}_1$ and $\bar{V}_2$ are normal. First, consider the case when $k(\bar{V}_2)$ is algebraically closed in $k(\bar{V}_1)$. Then it is easy to see that $l(m(\bar{f}^* \bar{D}_2 + E)) = l(m\bar{D}_2)$. From this we get the result in this case. Next, let $\bar{f} = \bar{\mu} \cdot \bar{g}$, where $\bar{g} : \bar{V}_1 \to \bar{V}_3$, $\bar{\mu} : \bar{V}_3 \to \bar{V}_2$, be the Stein factorization of $\bar{f}$. Then since $\bar{\mu}$ is finite, $\operatorname{codim} \bar{g}(E) \geqq 2$. Hence, by the former considerations, we have

$$\kappa(\bar{f}^* \bar{D}_2 + E, \bar{V}_1) = \kappa(\bar{\mu}^* \bar{D}_2, \bar{V}_3).$$

Furthermore, from the Theorem 4 in [3], we get

$$\kappa(\bar{\mu}^* \bar{D}_2, \bar{V}_3) = \kappa(\bar{D}_2, \bar{V}_2). \qquad \text{q. e. d.}$$

Example 7. Let $\bar{S}$ be a projective algebraic surface which is a covering surface of $\boldsymbol{P}^2$. By Σ we denote the ramification locus of $\pi: \bar{S} \to \boldsymbol{P}^2$, and let $B = \pi(\Sigma)$. If $\bar{S}$ is of hyperbolic type, then

$$2 = \kappa\bar{S} \leqq \bar{\kappa}(\bar{S} - \pi^{-1}B) = \bar{\kappa}(\boldsymbol{P}^2 - B). \text{ Hence } \bar{\kappa}(\boldsymbol{P}^2 - B) = 2.$$

6

As special results of the D-dimension theory, we have

Theorem 4. *Let $f: V \to W$ be a strictly rational dominant map. By V_w we denote a general irreducible component of a general fiber of f over a general point $w \in W$. Then*

$$\bar{\kappa}V \leqq \bar{\kappa}V_w + \dim W.$$

Theorem 5. *Let V be an algebraic variety with $\bar{\kappa} = \bar{\kappa}V \geqq 0$. Then there is a proper birational morphism μ from a non-singular algebraic variety V^* onto V such that there is a surjective morphism $f: V^* \to W$, in which W is an irreducible constructible set of dimension $\bar{\kappa}$ and a general fiber V^*_w is an algebraic variety whose logarithmic Kodaira dimension vanishes. Such a fibered variety is unique up to proper birational equivalence.*

Proof. We assume V to be non-singular. Hence, consider a compactification of V with smooth boundary $\bar{D}$. Then for $m \gg 0$, $\Phi_{m,V}: \bar{V} \to \bar{W}$ is the $\bar{D} + \bar{K}$-canonical fibered variety ([4]). By performing a finite succession of monoidal transformations, we can assume $\bar{f} = \Phi_{m,V}$ to be a morphism. $\bar{f}V = W$ is an irreducible constructible set of dimension $\bar{\kappa}$ and $\kappa((\bar{D} + \bar{K})|\bar{V}_w, \bar{V}_w) = 0$ for a general point $w \in W$. Since $\bar{D}|\bar{V}_w$ is a divisor of normal crossing type in $\bar{V}_w$, we see from the definition that

$$\bar{\kappa}V_w = \kappa((\bar{D} + \bar{K})|\bar{V}_w, \bar{V}_w) = 0.$$

Proof of the uniqueness is easy and so is omitted. q. e. d.

An irreducible constructible set is not so easy to handle. We will find a variety in W. Put $\bar{F} = \bar{f}\bar{D}$ and $W^0 = \bar{W} - \bar{F}$ which is a Zariski open subset of $\bar{W}$. Then $\bar{D} \subset \bar{f}^{-1}\bar{F}$ and hence $\bar{V} - \bar{f}^{-1}\bar{F} = \bar{f}^{-1}(\bar{V} - \bar{F}) = \bar{f}^{-1}W^0 \subset V = \bar{V} - \bar{D}$. Note that $f|\bar{f}^{-1}W^0$ is proper. Let σ be a strictly birational map: $V \to V$. By definition, there is a proper birational morphism $\mu: V^* \to V$ such that $\sigma^! = \sigma \cdot \mu$ is a morphism and that a compactification with smooth boundary $\bar{D}^*$ has the following property: (i) $\bar{\sigma}^! = \bar{\sigma} \cdot \bar{\mu}$ is a morphism and (ii) $\bar{f}^! = \Phi_{m,V*}$ is a morphism. Then $\bar{f}^! = \bar{f} \cdot \bar{\mu}$ and $\bar{\sigma}^! V^* \subset V$, and so $\bar{\sigma}^{!-1}\bar{D} \subset \bar{D}^*$. σ induces the isomorphism $\bar{\sigma}_1: \bar{W} \to \bar{W}$. Using $\bar{\mu}^{-1}\bar{D} = \bar{D}^*$ and $\bar{f} \cdot \bar{\sigma}^! = \bar{\sigma}_1 \cdot \bar{f}^! = \bar{\sigma}_1 \cdot \bar{f} \cdot \bar{\mu}$, we get

$$\bar{\sigma}_1\bar{f}(\bar{D}) = \bar{\sigma}_1\bar{f}^!(\bar{D}^*) = \bar{f}\sigma^!(\bar{D}^*) = \bar{f}\bar{D}.$$

$$\bar{\sigma}^\# \begin{pmatrix} \bar{V} & \xrightarrow{\bar{f}} & \bar{W} \\ \cup & & \cup \\ V & \xrightarrow{f} & W \\ \sigma^\# \uparrow & & \uparrow \sigma_1 \\ V^* & \xrightarrow{f^\#} & W \\ \cap & & \cap \\ \overline{V^*} & \xrightarrow{\bar{f}^\#} & \overline{W} \end{pmatrix} \bar{\sigma}_1$$

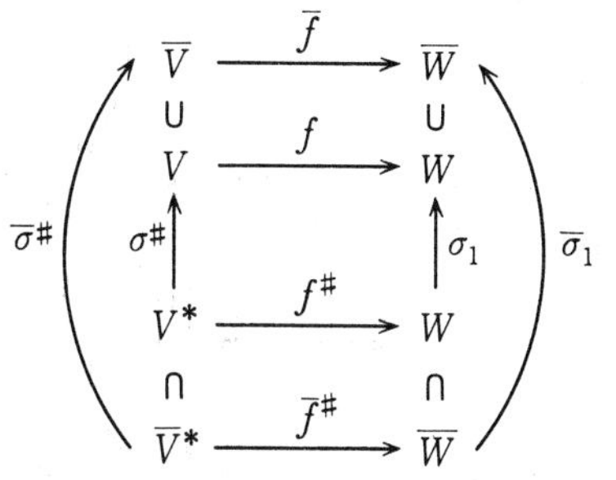

Since $\bar{\sigma}_1$ is bijective, it follows that $\sigma_1(W^0) = \bar{\sigma}_1(\bar{W} - \bar{F}) = \bar{W} - \bar{F} = W^0$. Thus, we obtain a group homomorphism

$$\beta_V : \text{S Bir } V \to \text{Lin}(W^0) \subset \text{Aut}(W^0)$$

which is defined by $\beta_V(\sigma) = \sigma_1$.

The following exact sequence of groups is important:

$$1 \to \text{Ker } \beta_V \to \text{S Bir } V \to \text{Im } \beta \to 1 \quad \text{(exact)}.$$

One may expect that $\text{Im } \beta_V$ is finite and that $(\text{Ker } \beta_V)^0$ is a quasi-abelian variety (see § 7). A related problem will be partly solved in a forthcoming paper.

Theorem 6. *Let V be an n-dimensional algebraic variety of hyperbolic type. Then S Bir V is finite.*

Proof. By assumption, $\text{Ker } \beta_V = 1$. $(\text{Lin } W^0)^0$ is a connected affine algebraic group G. If G is not trivial, then $G \supset G_a$ or $\supset G_m$. We use

Lemma 2. *Let V be an algebraic variety on which a connected algebraic group G acts. Then*

$$\bar{\kappa} V \leq \bar{\kappa} Gp + \dim V - \dim Gp,$$

where p is a general point of V and Gp is the G-orbit of p.

Proof. There exists an admissible dense open subset V^0 of V whose quotient variety by G exists. Then we get a fibered variety: $V^0 \to V^0/G$, whose general fiber is Gp. Hence by Theorem 4,

$$\bar{\kappa} V \leq \bar{\kappa} V^0 \leq \bar{\kappa} Gp + \dim (V^0/G). \qquad \text{q. e. d.}$$

Therefore if $G \supset G_a$, then $\bar{\kappa} W^0 = -\infty$ and if $G \supset G_m$, then $\bar{\kappa} W^0 \leq n-1$. On the other hand, $\bar{\kappa} W^0 = \bar{\kappa} \bar{f}^{-1} W^0$ for $\bar{f} | \bar{f}^{-1} W^0$ is proper and birational, and we get

$$n = \bar{\kappa} V \leq \bar{\kappa} \bar{f}^{-1} W^0 = \bar{\kappa} W^0 \leq n-1.$$

This is a contradiction. Thus we finish the proof of Theorem 6.

Remark. For a compact complex variety M and a closed analytic subset Σ in M, we can define $\bar{\kappa}(M - \Sigma)$ similarly by making use of Hironaka's Main Theorems. Let M_1 and M_2 be compact complex varieties and $\Sigma_1 \subset M_1$, $\Sigma_2 \subset M_2$ closed analytic subsets. If dim M_1=dim $M_2=\bar{\kappa}(M_2-\Sigma_2)=n$, then a meromorphic dominant (i.e., non-degenerate) map $f: M_1-\Sigma_1\to M_2-\Sigma_2$ is the restriction of a meromorphic map $\bar{f}: M_1\to M_2$(see Sakai [6]). In this case, both M_1 and M_2 are Moishezon varieties and hence we can assume that these are algebraic. Thus

$$\text{S Bim}(M-\Sigma) \quad \text{is finite if} \quad \bar{\kappa}(M-\Sigma)=n.$$

7

If G is an algebraic group, we have the Chevalley decomposition :

$$1\to\mathcal{G}\to G\to\mathcal{A}\to 1 \qquad \text{(exact)},$$

in which $\mathcal{G}$ is the connected maximal affine algebraic subgroup and $\mathcal{A}$ is an abelian variety.

We assume $\bar{\kappa}V\geqq 0$. Then by Lemma 1 Aut$(V)^0$ cannot contain G_a.

Lemma 3. *If a connected affine algebraic group $\mathcal{G}$ does not contain G_a, then $\mathcal{G}$ is an algebraic torus G_m^t.*

Proof. Since the radical is an algebraic torus, $\mathcal{G}$ is reductive. Hence $\mathcal{G}$ is a semi product of an algebraic semi-simple group and a torus. However, any non trivial semi-simple group $\supset G_a$. Hence $\mathcal{G}$ is a torus. q. e. d.

Hence, if $\bar{\kappa}V\geqq 0$, we have for any connected algebraic subgroup G of Aut$(V)^0$,

$$1\to G_m^t\to G\to\mathcal{A}\to 1 \qquad \text{(exact)}.$$

A connected algebraic group $\tilde{\mathcal{A}}$ is called a *quasi-abelian variety,* if $\tilde{\mathcal{A}}$ is an extension of G_m^t by an abelian variety $\mathcal{A}$ as an algebraic group.

Lemma 4. *A quasi-abelian variety $\tilde{\mathcal{A}}$ is commutative.*

Proof. Let $\tau\in\tilde{\mathcal{A}}$ and consider the group homomorphism :

$$\Psi_\tau(\sigma)=\tau\sigma\tau^{-1}: G_m^t\to\tilde{\mathcal{A}}.$$

Since G_m^t is rational, $\Psi_\tau: G_m^t\to G_m^t$. Hence

$$G_m^t\ni\tau\mapsto\Psi_\tau\in\text{Hom}(G_m^t, G_m^t),$$

which is discrete. Thus $\Psi_1=\Psi_\tau$, hence G_m^t is contained in the center of $\tilde{\mathcal{A}}$. Moreover, if σ, $\tau\in\tilde{\mathcal{A}}$, we get

$$[\tau,\sigma]=\tau\sigma\tau^{-1}\sigma^{-1}\in G_m^t,$$

since $\mathcal{A}$ is commutative. We have the right coset decomposition :

$$\tilde{\mathcal{A}}=\coprod\rho G_m^t.$$

Then $[\tau,\sigma]=[\rho\rho^{-1}\tau,\sigma]=[\rho,\sigma]$, because $\rho^{-1}\tau\in G_m^t$ for a certain ρ. Hence the morphism

$$\tilde{\mathcal{A}}\ni\tau\mapsto[\tau,\sigma]\in G_m^t$$

factors through $\mathcal{A}\to G_m^t$, which is trivial. Hence $[\tau,\sigma]=1$. This implies $\tilde{\mathcal{A}}$ is commutative. q. e. d.

By this lemma, we can verify that $\mathrm{Hom}(\tilde{\mathscr{A}}, \tilde{\mathscr{A}})$ and the set {algebraic subgroups of $\tilde{\mathscr{A}}$} are countable sets. Assume that $\bar{\kappa}V \geqq 0$. Then the quasi-abelian variety $\tilde{\mathscr{A}} \subseteqq \mathrm{Aut}(V)^0$ operates on V. By the following lemma, we see that $\mathrm{Aut}(V)^0$ turns out to be an algebraic group.

Lemma 5. *Let G be an algebraic group acting effectively on an algebraic variety V. Assume that the set of isotropy groups G_p for $p \in V$ is countable. Then there is a non-empty Zariski open subset $V^{\star} \subset V$ such that $G_p = 1$* for $p \in V^{\star}$.

Proof. Let $F = \{(g, p) \in G \times V\,;\, g \cdot p = p\}$ and let π be the projection $F \subset G \times V \to V$. Then $\pi^{-1}(p) = G_p \times p$. Hence, there exists a non-empty Zariski open subset V^0 such that

$$\dim G_p < \dim G \quad \text{for} \quad p \in V^0.$$

Let $G^* = \bigcup G_p$ where $p \in V^0$. By hypothesis, we can write

$$G^* = \bigcup_{j=1}^{\infty} G_{p(j)}.$$

$\hat{\pi}F^0 = G^*$ and $\hat{\pi}F^0$ is a constructible subset of G, where $\hat{\pi}$ is another projection: $F^0 = G \times V^0 \cap F \subset G \times V^0 \to G$.

Therefore it is easy to see that there is a finite set of subgroups $G_1, \cdots, G_m$ of G such that

$$\hat{\pi}F^0 = \bigcup_{j=1}^{m} G_j.$$

Hence $G_j = G_{p(j)}$ for some $p(j) \in V^0$, and so $\dim G_j < \dim G$. By using induction on $\dim G$, we have a finite set of Zariski open subsets $\tilde{V}_j$ such that G_j has no fixed points on $\tilde{V}_j$. Put $V^{\star} = \bigcap \tilde{V}_j$. This satisfies the desired condition. In fact, if $g \in G_p \neq 1$ for some $p \in V^{\star}$, then $G_p \in G^*$. Hence, $g \in G_{p(j)}$ for some j. Therefore, G_p has no fixed points on $V^{\star}$. q. e. d.

Consider the canonical fibered variety:

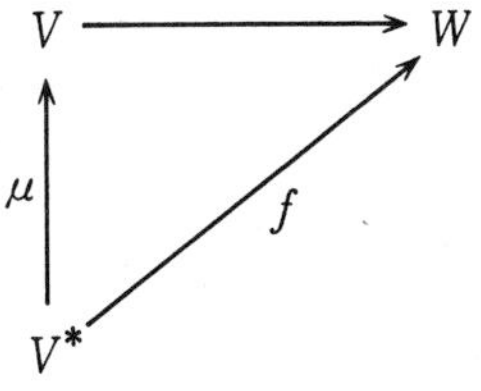

and consider $V_w^* = f^{-1}(w)$, $V_w = \mu(V_w^*)$ for a general $w \in W$. By the result in § 3, $\bar{\kappa}V_w \leqq \bar{\kappa}V_w^* = 0$ and so $\bar{\kappa}V_w = 0$, because $\bar{\kappa}V \geqq 0$. Hence $\tilde{\mathscr{A}}(p) =$ the $\tilde{\mathscr{A}}$-orbit of $p \subset V_w$ for a certain pair (p, w) of general points $p \in V$ and $w \in W$. Thus $n - \bar{\kappa}V \geqq \dim \tilde{\mathscr{A}} = \dim \mathrm{Aut}(V)^0$. Consequently, we obtain

Theorem 7. *If* $\bar{\kappa}V \geqq 0$, *the* $\mathrm{Aut}(V)^0$ *is a quasi-abelian variety and* $\dim \mathrm{Aut}(V)^0 \leqq n - \bar{\kappa}V$.

Proposition 5. *If V is affine and* $\bar{\kappa}V \geqq 0$, *then* $\mathrm{Aut}(V)^0$ *is an algebraic torus*.
The proof is easy.

Example 8. Let $V = \boldsymbol{P}^n - F$, F being a Zariski closed set, and assume $\bar{\kappa}V \geqq 0$ and $\dim \mathrm{Aut}(V)^0 = n$. Then $V \supset G_m^n$. In the next section we shall prove $V = G_m^n$.

Remark. It may be true that the dimension of the abelian part of $\mathrm{Aut}(V)^0$ is not greater than $q(\bar{V})$ = the irregularity of $\bar{V}$.

8

We shall prove

Theorem 8. *If* $\bar{\kappa}V \geqq 0$ *and* $\dim \mathrm{Aut}(V)^0 = \dim V = n$, *then V is a quasi-abelian variety*.

Proof. By assumption, the quasi-abelian variety $\tilde{\mathscr{A}} = \mathrm{Aut}(V)^0$ operates on V and $\tilde{\mathscr{A}} = \tilde{\mathscr{A}}(p)$ is a Zariski open subset for a general $p \in V$. Note that $\tilde{\mathscr{A}} \subset V$ is an equivariant imbedding of $\tilde{\mathscr{A}}$.

Let $\tilde{V} \to V$ be a normalization of V. Then since $\mathrm{Aut}(V) \subset \mathrm{Aut}(\tilde{V})$, we have the $\tilde{\mathscr{A}}$-equivariant imbedding : $\tilde{\mathscr{A}} \subset \tilde{V}$. Hence we assume V to be normal. By Reg V we denote the set of non-singular points of V. Then since Aut $V \subset$ Aut Reg V, we have the $\tilde{\mathscr{A}}$-equivariant imbedding : $\tilde{\mathscr{A}} \subset$ Reg V. First consider the case in which $\tilde{\mathscr{A}}$ is an algebraic torus G_m^n. By Sumihiro's Theorem 5 [see [5] p. 20], we have a finite covering of G_m^n-admissible affine open subsets $U_1, \cdots, U_m$, such that Reg $V = \bigcup_{j=1}^{m} U_j$. If $G_m^n \subsetneqq U_j$, then $U_j \simeq G_a^r \times G_m^{n-r}$, r being positive, by Theorem 4 in [5] p. 14. Hence $\bar{\kappa}U_j = -\infty$, which contradicts $\bar{\kappa}U_j \geqq \bar{\kappa}$ Reg $V \geqq \bar{\kappa}V \geqq 0$. Thus Reg $V = \tilde{\mathscr{A}}$ and hence $\mathrm{codim}(V - \tilde{\mathscr{A}}) \geqq 2$. By the following lemma, we conclude that $V = \tilde{\mathscr{A}}$. q. e. d.

Lemma 6. *Let V be an affine variety and suppose that an algebraic variety V' contains V as a Zariski open subset. Then* $\mathrm{codim}_{V'}(V' - V) = 1$ or $V = V'$.

Proof. Let $p \in V' - V$ and consider an affine open subset U of V'. Then $U \cap V$ is also affine. Hence we assume V' to be affine. Moreover, we can assume that V' is normal. Then if $\mathrm{codim}(V' - V) \geqq 2$, we get $\Gamma(V', \mathcal{O}_{V'}) = \Gamma(V, \mathcal{O}_V)$ by the extension theorem. Hence $V' = \mathrm{Spec}\ \Gamma(V', \mathcal{O}_{V'}) = \mathrm{Spec}\ \Gamma(V, \mathcal{O}_V) = V$, a contradiction. Thus we conclude that $V' - V$ is purely 1-codimensional. (The author owes this proof to Nagata.)

In general, $\tilde{\mathscr{A}}$ has the following decomposition :

$$0 \to T = G_m^{n'} \xrightarrow{\pi} \tilde{\mathscr{A}} \to \mathscr{A} \to 0,$$

$\mathscr{A}$ being an abelian variety. Let u be a general point of $\mathscr{A}$ and let $T_u = \pi^{-1}(u)$. Consider the closure of T_u in V, which is written $\hat{T}_u$. Then $\bar{\kappa}\hat{T}_u \geqq 0$, because $\bar{\kappa}V \geqq 0$.

Since $T \simeq T_u \subset \hat{T}_u$, we have a torus imbedding. In fact, if $\alpha \in T \subset \tilde{\mathscr{A}} \subset \mathrm{Aut}(V)$, α operates on V. Hence $\alpha \hat{T}_u = \hat{T}_u$. By the previous argument, we conclude that $\hat{T}_u = T_u$. Let v be an arbitrary point of $\mathscr{A}$. Then we have some $\tau \in \mathscr{A}$ such that $\tau v = u$. Let $\tilde{\tau} \in \pi^{-1}(\tau)$. Then $\tilde{\tau} \in \mathrm{Aut}(V)$ and $\tilde{\tau} \hat{T}_v = \hat{T}_u$. Hence $\hat{T}_v = T_v$. This implies $\tilde{\mathscr{A}} = V$. q. e. d.

Theorem 9. *Let V be an algebraic variety of dimension n. If* $\dim \mathrm{Aut}(V)^0 = n - \bar{\kappa} V$, *then V contains a Zariski open subset V^0 which is a principal $\tilde{\mathscr{A}}$-bundle, where* $\tilde{\mathscr{A}} = \mathrm{Aut}(V)^0$ *is a quasi-abelian variety. $V^0 \subset V$ is an $\tilde{\mathscr{A}}$-equivariant imbedding and $\bar{\kappa} V^0 = \bar{\kappa} V$.*

Remark. It seems interesting to know whether $V = V^0$ or not.

References

[1] Deligne, P.: Théoric dc Hodge II. Publ. Math. IHES **40** (1973), 5–58.
[2] Hartshorne, R.: Algebraic de Rham cohomology, Manuscripta Math., **7** (1972), 125–140.
[3] Iitaka, S.: On D-dimensions of algebraic varieties. J. Math. Soc. Japan **23** (1971), 356–373.
[4] Iitaka, S.: On algebraic varieties whose universal covering manifolds are complex affine 3-spaces I. Number Theory, Algebraic Geometry, and Commutative Algebra, in honor of Y. Akizuki, Kinokuniya, Tokyo (1973), 147–167.
[5] Mumford, D. and others: Toroidal embeddings I. Lecture Notes in Mathematics, **339** (1973), Springer.
[6] Sakai, F.: Degeneracy of holomorphic maps with ramification. Inv. Math., **26** (1974), 213–229.
[7] Sakai, F.: Kodaira dimensions of complements of divisors. In this volume.
[8] Ueno, K.: Classification theory of algebraic varieties and compact complex spaces. Lecture Note in Mathematics, **439** (1974), Springer.
[9] Wakabayashi, I.: On automorphism groups of $\mathbf{P}^2 - \{\text{curves}\}$, in preparation.
[10] Peters, K.: Über holomorphe und meromorphe Abbildungen gewisser kompacter komplexer Mannigfaltigkeiten, Arch. Math., **15** (1964), 222–231.
[11] Iitaka, S.: Logarithmic form of algebraic varieties, J. Fac. Sci. Univ. Tokyo, **23** (1976), 525–544.

Department of Mathematics
University of Tokyo

(Received November 22, 1975)

On a Characterization of Submanifolds of Hopf Manifolds

Ma. Kato

0. Introduction

The Hopf manifolds afford a quite elementary, though quite typical example of *non-Kähler* compact complex manifolds. Those were defined by H. Hopf in 1948, and investigated completely by K. Kodaira [3], [4] in the case of dimension 2. In this paper, we intend to study higher dimensional Hopf manifolds and their subvarieties.

A *Hopf manifold* of dimension $n \geq 2$ is defined to be a compact complex manifold of which the universal covering manifold is biholomorphic to the domain $\boldsymbol{C}^n - \{0\}$ ([3]). Any Hopf manifold contains nowhere discrete subvarieties. These subvarieties have rather special properties (§ 3, see also [2]). It is our aim to give a complete characterization of submanifolds of Hopf manifolds. In the case of dimension 2, the following result is known :

Theorem 1 ([1], [2]). *A compact complex manifold S of dimension 2 is biholomorphic to a submanifold of a Hopf manifold if and only if S is of class VI_0, VII_0-elliptic or a Hopf surface.*

The proof of the theorem depends on Kodaira's classification theory of surfaces.

The purpose of this paper is to give a sufficient condition for a compact complex manifold of dimension ≥ 4 to dominate bimeromorphically a subvariety of a Hopf manifold (Main Theorem § 1).

In § 1, we give basic definitions and the statement of our main theorem.

In § 2, we recall some recent results due to Y-T. Siu [8], [9] and H-S. Ling [5], which play an essential role in the proof of our main theorem.

In § 3, we recall some properties of subvarieties of Hopf manifolds and show that the conditions in our main theorem are satisfied if the compact complex manifold is actually a submanifold of a Hopf manifold.

In § 4, we give the proof of our main theorem and a remark.

1. Definitions and Statement of Main Theorem

Definition 1. By a *Hopf manifold* of dimension $n \geq 2$, we shall mean a compact complex manifold of which the universal covering manifold is biholomorphic to the domain $\boldsymbol{C}^n - \{0\}$, where 0 denotes the origin of $\boldsymbol{C}^n$ ([3]). By a *primary* Hopf manifold, we shall mean a Hopf manifold whose fundamental group is infinite cyclic.

Definition 2. Let X be a complex space[1]. Let g be a holomorphic map of X into itself with unique fixed point $0 \in X$, $g(0)=0$. Then g is called a *contraction* of X if g satisfies the following two conditions:

(i) $\lim_{\nu \to +\infty} g^\nu(x) = 0$ for all $x \in X$,

(ii) for any sufficiently small neighborhood U of 0 in X, there exists an integer ν_0 such that $g^\nu(U) \subset U$ for all $\nu \geq \nu_0$.

Any Hopf manifold is a submanifold of a higher dimensional *primary* Hopf manifold ([2]). By Kodaira [3] p. 694, any n-dimensional primary Hopf manifold is biholomorphic to the quotient space $\boldsymbol{C}^n - \{0\} / \langle G \rangle$, where G is a contracting holomorphic automorphism of $\boldsymbol{C}^n$ which fixes the origin. Let Y be a *connected* analytic subset of a Hopf manifold. Then we can assume that Y is isomorphic to an analytic subset of a primary Hopf manifold. Let $j: Y \to \boldsymbol{C}^n - \{0\} / \langle G \rangle$ be a (closed holomorphic) embedding. It is easy to see that j induces a non-trivial homomorphism j_*: $\pi_1(Y) \to \pi_1(\boldsymbol{C}^n - \{0\} / \langle G \rangle) \simeq \boldsymbol{Z}$ of the fundamental groups. Denote by γ_G the element of $\pi_1(\boldsymbol{C}^n - \{0\} / \langle G \rangle)$ corresponding to the contracting automorphism G.

Definition 3. An element $\gamma \in \pi_1(Y)$ is said to be *j-contractive* if $j_* \gamma = \gamma_G^m$ for some *positive* integer m.

If a complex space is irreducible, we sometimes call it a *variety*.

A line bundle L on a compact normal variety X can be regarded as an element of $H^1(X, \mathcal{O}^*)$. If L is in the image of $H^1(X, \boldsymbol{C}^*) \to H^1(X, \mathcal{O}^*)$, then L is said to be *flat*, where $\boldsymbol{C}^*$ denotes the constant sheaf of the multiplicative group $\boldsymbol{C}^* = \boldsymbol{C} - 0$. Note that a flat line bundle L canonically determines a group representation ρ_L: $\pi_1(X) \to \boldsymbol{C}^*$.

Now we shall state our main theorem.

Main Theorem. *Let X be a compact complex manifold of dimension $n \geq 4$. Assume that there exists an effective divisor D on X which satisfies the following four conditions:*

(i) *The support Y of D is connected and biholomorphic to an analytic subset of a Hopf manifold,*

---

1) By a *complex space,* we shall mean a reduced Hausdorff complex space.

(ii) *The line bundle* $L=[D]$ *is flat.*

(iii) Im $\rho_L \simeq \boldsymbol{Z}$.

Let i_* *be the homomorphism* $i_*: \pi_1(Y) \to \pi_1(X)$ *induced by the inclusion* $i: Y \to X$, *and* $\rho_{L,Y} = \rho_L \circ i_*$.

(iv) *There exists a (closed holomorphic) embedding* $j: Y \to \boldsymbol{C}^m - \{0\}/\langle G \rangle$ *such that*

$$|\rho_{L,Y}(\gamma)| < 1$$

for any j-contractive element $\gamma \in \pi_1(Y)$.

Then there exists a bimeromorphic holomorphic map of X onto a subvariety of a Hopf manifold.

2. Stein Completions and Extensions of Holomorphic Maps

A real-valued function φ on a complex space Z is said to be C^∞ (resp. *strongly plurisubharmonic (s-psh)*) if, for any point $z \in Z$, there exist an open neighborhood U of z, a biholomorphic map Φ of U onto an analytic subset of an open set $\tilde{U}$ of some $\boldsymbol{C}^k$, and a C^∞ (resp. *s-psh*) function ψ on $\tilde{U}$ such that $\varphi|U = \psi \circ \Phi$.

Definition 4. Let Z and W be complex spaces. Let $\pi: Z \to W$ be a holomorphic map.

(i) π is called 1-*convex* if there exist a C^∞-function $\varphi: Z \to \boldsymbol{R}$ and a constant $c^\dagger \in \boldsymbol{R}$ such that

(a) $\pi|\{z \in Z; \varphi(z) \leq c\}$ is proper for any $c \in \boldsymbol{R}$,

(b) φ is s-psh on $\{z \in Z; \varphi(z) > c^\dagger\}$.

(ii) π is called (1, 1)-*convex-concave* if there exist a C^∞-function $\varphi: Z \to \boldsymbol{R}$ and constants c_*, $c_\ddagger$ and $c^\dagger$ in $\boldsymbol{R} \cup \{-\infty\}$ with $c_* < c_\ddagger < c^\dagger$ such that

(a) $\pi|\{z \in Z; c_1 \leq \varphi(z) \leq c_2\}$ is proper for any $c_* < c_1 < c_2 < +\infty$,

(b) φ is s-psh on $\{z \in Z; \varphi(z) < c_\ddagger\} \cup \{z \in Z; \varphi(z) > c^\dagger\}$.

In both cases the function φ is called an *exhaustion function.*

Definition 5. Let $\pi: Z \to W$ be a 1-convex holomorphic map with exhaustion function $\varphi: Z \to \boldsymbol{R}$. A subset U of Z is called a *neighborhood of infinity* if there exists a constant $c \in \boldsymbol{R}$ such that $U \supset \{z \in Z; \varphi(z) > c\}$.

We recall here some results obtained by Siu [8], [9] and Ling [5].

Theorem 2 [9]. *Let* $\pi: Z \to W$ *be a* 1-*convex holomorphic map. Then, for any point* $w \in W$ *and any sufficiently small polyhedral domain U in W with center w,* $\pi^{-1}(U)$ *is a holomorphically convex space.*

Theorem 3 [5]. *Suppose that Z is an* $(n+1)$-*normal*[1] *complex space and that W is a*

1) See [5] for the definition. Note that if X is an irreducible normal complex space of dimension n, then X is p-normal for any $0 \leq p \leq n-2$.

Stein space of dimension n. Suppose that $\pi : Z \to W$ is a $(1, 1)$-convex-concave holomorphic surjection with exhaustion function $\varphi : Z \to \boldsymbol{R}$ which is s-psh on the whole Z. Then there exists a unique $(n+1)$-normal Stein space $\hat{Z}$ with the following properties :

(i) *Z is an open subset of $\hat{Z}$,*

(ii) *Each irreducible component of $\hat{Z}$ intersects Z,*

(iii) *There exists a holomorphic map $\hat{\pi} : \hat{Z} \to W$ such that $\hat{\pi}|Z = \pi$,*

(iv) *$\hat{\pi}|(\hat{Z} - \{z \in Z\,;\, \varphi(z) > c\})$ is proper for any $c \in \boldsymbol{R}$.*

Note that, by the construction of $\hat{Z}$, the set $(Z - \{z \in Z\,;\, \varphi(z) < c\}) \cap (\hat{Z} - Z)$ is empty for any $c \in \boldsymbol{R}$. The space $\hat{Z}$ is called the *Stein completion* of Z. If Z is irreducible and normal, then $\hat{Z}$ is also irreducible and normal.

Proposition 1 [5]. *In Theorem 3, the restriction map $r : \Gamma(\hat{Z}, \mathcal{O}) \to \Gamma(Z, \mathcal{O})$ is bijective.*

Let $\pi_i : Z_i \to U$ be 1-convex holomorphic surjections with exhaustion functions $\varphi_i : Z_i \to \boldsymbol{R} (i=1, 2)$. Suppose that the following five conditions are satisfied :

(α) U is a Stein open subset of $\boldsymbol{C}^k$,

(β) Z_1 is an irreducible normal holomorphically convex space,

(γ) Z_2 is an irreducible normal Stein space which is biholomorphic to an analytic subset of an affine space $\boldsymbol{C}^l$ for some l,

(δ) $\dim Z_1 = \dim Z_2 \geq k+3$,

(ε) there exist neighborhoods N_i of infinities of Z_i and a finite ramified covering $h : N_1 \to N_2$ which makes the following diagram commutative :

$$\begin{array}{ccc} N_1 & \xrightarrow{\quad h \quad} & N_2 \\ & {\scriptstyle \pi_1|N_1} \searrow \qquad \swarrow {\scriptstyle \pi_2|N_2} & \\ & U. & \end{array}$$

Under these conditions we shall prove the following two lemmas.

Lemma 1. *There exists a unique proper holomorphic surjection $\hat{h} : Z_1 \to Z_2$ such that $\hat{h}$ agrees with h on a neighborhood of infinity of Z_1 and that the diagram*

$$\begin{array}{ccc} Z_1 & \xrightarrow{\quad \hat{h} \quad} & Z_2 \\ & {\scriptstyle \pi_1} \searrow \qquad \swarrow {\scriptstyle \pi_2} & \\ & U & \end{array}$$

is commutative.

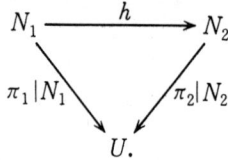

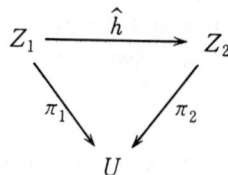

Proof. By (β), we can consider the Remmert quotient $\eta : Z_1 \to Q$, where Q is a normal Stein space. The surjection η is proper and every fibre of η is connected. Since U is Stein by (α), there exists a unique holomorphic map $\sigma: Q \to U$ such that $\sigma \circ \eta = \pi_1$. Since π_1 is 1-convex, η is a biholomorphic map on $M_c = \{z \in Z_1 ; \varphi_1(z) > c\}$ for some $c \in \boldsymbol{R}$. We can assume that $M_c \subset N_1$. Put $Q_c = \eta(M_c)$ and $\lambda = (\eta | M_c)^{-1}$. Define $\tau' : Q_c \to Z_2$ by $\tau' = h \circ \lambda$. We shall prove that τ' extends uniquely to a proper holomorphic map τ of Q onto Z_2. By (γ), there is a closed embedding $i : Z_2 \to \boldsymbol{C}^l$. We identify Z_2 with $i(Z_2) \subset \boldsymbol{C}^l$. Then $\tau' : Q_c \to Z_2$ is a (vector-valued) holomorphic function on Q_c. It is clear that $\sigma | Q_c : Q_c \to U$ is a (1, 1)-convex-concave holomorphic map with exhaustion function $\lambda^* \varphi_1 : Q_c \to (c, +\infty)$ and that Q is the Stein completion of Q_c. By (δ) and Proposition 1, we can extend τ' to a (vector-valued) holomorphic function $\tau : Q \to \boldsymbol{C}^l$. By the uniqueness of the extension we also have $\sigma = \pi_2 \circ \tau$. Now we shall prove that τ is proper. Take any compact set K in $\boldsymbol{C}^l$. Since $K \cap Z_2$ is compact, $\sigma \circ \tau^{-1}(K) = \pi_2(K \cap Z_2)$ is a compact subset in U. Put $H = \pi_2(K \cap Z_2)$.Fix a constant d such that $d > c$. Then $\tau^{-1}(K) \subset \tau'^{-1}(\tau(\bar{Q}_d) \cap K) \cup (\sigma^{-1}(H) \cap (Q - Q_d))$. Since $\tau' : Q_c \to Z_2$ is an open embedding and $\tau(\bar{Q}_d) \cap K$ is a compact subset of $\tau'(Q_c)$, $\tau'^{-1}(\tau(\bar{Q}_d) \cap K)$ is a compact subset of Q_c. It is clear that $\sigma^{-1}(H) \cap (Q - Q_d)$ is compact. Hence $\tau^{-1}(K)$ is compact. This proves that τ is proper. Put $Z_3 = \tau(Q)$, which is a closed analytic subset in $\boldsymbol{C}^l$ by the proper mapping theorem. Furthermore $\tau(Q_c) \subset Z_2 \cap Z_3$. Hence by ($\varepsilon$) every irreducible component of Z_3 is contained in Z_2 in a neighborhood of infinity. Hence we infer that $Z_2 = Z_3$. Now define $\hat{h} = \tau \circ \eta$. Then, by the above argument, $\hat{h}$ is a unique proper surjection which agrees with h on a neighborhood of infinity in Z_1 such that $\pi_1 = \pi_2 \circ \hat{h}$.

q. e. d.

Lemma 2. *If we further assume that Z_1 is a Stein space and that h is biholomorphic, then $\hat{h}$ is also biholomorphic.*

Proof. In the above proof, $\eta : Z_1 \to Q$ becomes an isomorphism. Clearly τ' is an open embedding. Therefore $\tau : Q \to Z_2$ is biholomorphic outside an analytic subset S such that $\dim S \cap \sigma^{-1}(t) = 0$ for any $t \in U$. Since Z_2 is normal, we infer that τ is biholomorphic on the whole of Q by Zariski's Main Theorem. Therefore $\hat{h} = \tau \circ \eta$ is biholomorphic.

q. e. d.

3. Subvarieties of Hopf Manifolds

In this section we recall some results in [2] on subvarieties of Hopf manifolds and discuss further properties of the subvarieties (see also [1]).

Propositon 2 [2]. *Let $\hat{V}$ be a complex space and g a contracting holomorphic automorphism of $\hat{V}$ with a unique fixed point 0. Then the compact complex space $\hat{V} - \{0\} / \langle g \rangle$ is biholomorphic to an analytic subset of a Hopf manifold.*

Proposition 3 [2]. *Any n-dimensional subvariety of a Hopf manifold contains subvarieties of arbitrary dimensions less than n.*

Proposition 4 [2]. *Let V be a normal subvariety of a Hopf manifold and D an effective Weil divisor on V. Assume that* $\dim V \geq 2$. *Then there exists an effective Weil divisor E on V such that $D+E$ is a Cartier divisor and that the line bundle $[D+E]$ is flat and has the transition functions which are some powers of a certain constant $\alpha \in \boldsymbol{C}^*(|\alpha|<1)$.*

Proposition 4 is proved as follows. Let $j: V \to \boldsymbol{C}^m - \{0\}/\langle G\rangle$ be an embedding, where G is a contracting automorphism of $\boldsymbol{C}^m$ with $G(0)=0$. We identify V with $j(V)$. Let $\tilde{\pi}: \boldsymbol{C}^m - \{0\} \to \boldsymbol{C}^m - \{0\}/\langle G\rangle$ be the canonical projection. Put $\tilde{V} = \tilde{\pi}^{-1}(V)$. By a theorem of Remmert-Stein, $\tilde{V}$ extends to an analytic subset $\hat{V}$ in $\boldsymbol{C}^m$. Moreover $\hat{V}$ is G-invariant. Let $\hat{Y}_1 = \tilde{\pi}^{-1}(\operatorname{Supp} D) \cup \{0\}$, which is a G-invariant analytic subset in $\hat{V}$. Then we can show that there exists a holomorphic function f on $\hat{V}$ which does not vanish identically on any irreducible component of $\hat{V}$, and which satisfies $f|\hat{Y}_1 = 0$ and

$$G^*f = \alpha f \tag{1}$$

for some $\alpha \in \boldsymbol{C}^*(|\alpha|<1)$. Then some power f^k $(k>0)$ of f defines an effective Cartier divisor D' on V such that $E := D' - D$ is effective. Moreover the line bundle $[D']$ is flat and has the transition functions which are some powers of α.

Denote by $\hat{Y}$ the zero-locus of f. Let $\tilde{Z} := \hat{V} - \hat{Y}$, $Y := \operatorname{Supp} D' = \hat{Y} - \{0\}/\langle G\rangle$, $Z := V - Y$, and $\tilde{\pi}_0 = \tilde{\pi}|\tilde{Z}$.

Lemma 3. $f: \hat{V} \to \boldsymbol{C}$ *is a holomorphic surjection.*

Proof. Take a point x of $\hat{Y}$ and a small neighborhood N of x in $\hat{V}$. Let C be a 1-dimensional complex subvariety in N such that $C \cap (\hat{Y} \cap N) = \{x\}$. Since $f|C: C \to \boldsymbol{C}$ is not constant, $f|C$ is an open map and $f(C)$ contains $0 \in \boldsymbol{C}$. Therefore $f(C)$ contains a small disk U with center 0. Hence $f(\hat{V}) \supseteq U$. Hence, by (1),

$$\boldsymbol{C} \supseteq f(\hat{V}) = \bigcup_{\nu\geq 0} \alpha^{-\nu} f(G^\nu(\hat{V})) \supseteq \bigcup_{\nu \geq 0} \alpha^{-\nu} f(\hat{V}) \supseteq \bigcup_{\nu\geq 0} \alpha^{-\nu} U = \boldsymbol{C}.$$

q. e. d.

Lemma 4. *f induces a holomorphic surjection $f_*: Z \to \Delta := \boldsymbol{C}^*/\langle\alpha\rangle$.*

Proof. Clear by Lemma 3.

Lemma 5. *For any point $\tau \in \Delta$ and any sufficiently small open disk U with center τ, the holomorphic surjection $f_*|f_*^{-1}(U): f_*^{-1}(U) \to U$ is 1-convex with respect to some exhaustion function φ_U. Moreover $f_*^{-1}(U)$ is a Stein space.*

Proof. Let $p: \boldsymbol{C}^* \to \Delta$ be the canonical projection. Fix a point $t_0 \in p^{-1}(\tau) \subset \boldsymbol{C}^*$. Let ε be a sufficiently small positive number such that the set $\tilde{U} := \{t \in \boldsymbol{C};\ |t-t_0| < \varepsilon\}$ is contained in $\boldsymbol{C}^*$ and that $p|\tilde{U}: \tilde{U} \to \Delta$ is one-to-one. Put $U = p(\tilde{U})$. Then $\tilde{\pi}$ gives a biholomorphic map between $T := \{z \in \hat{V};\ |f(z)-t_0| < \varepsilon\}$ $(\subset \tilde{Z})$ and $f_*^{-1}(U)$. Hence $f_*^{-1}(U)$ is a Stein space. Now we define φ_U in the following manner. Let $\tilde{\phi}$ be an exhaustion function on the Stein space $\hat{V}$. Then $\tilde{\phi}' = \tilde{\phi} + |f|^{-2}$ is an exhaustion function on the Stein space $\tilde{Z}$. Let $\varphi_U := (\tilde{\pi}|T)^{-1*}\tilde{\phi}'$. It is easy to see that φ_U is an exhaustion function. q. e. d.

Proposition 5. *$Y=\mathrm{Supp}\ D'$ is connected.*

Proof. Assume that there exist non-empty analytic subsets Y_1, Y_2 such that $Y=Y_1\cup Y_2$ and $Y_1\cap Y_2=\phi$. Let N_i be tubular neighborhood of Y_i such that $N_1\cap N_2=\phi$. Let $N_i^0:=N_i-Y_i$. By Lemma 5, we can find a finite open covering $\bigcup_\alpha U_\alpha$ of Δ such that, for each α, $f_*|f_*^{-1}(U_\alpha):f_*^{-1}(U_\alpha)\to U_\alpha$ is 1-convex with respect to exhaustion function φ_α and, moreover, $f_*^{-1}(U_\alpha)$ is a Stein space. Let $N_\alpha=\{x\in f_*^{-1}(U_\alpha)\ ;\ \varphi_\alpha(x)>c_\alpha\}$ be a neighborhood of infinity in $f_*^{-1}(U_\alpha)$ with respect to φ_α. Taking c_α to be large enough, we can assume that $N_\alpha\subset N_1^0\cap N_2^0$. It is easy to see that $N_\alpha\cap N_i^0\neq\phi$ for $i=1, 2$. Hence N_α is not connected. Since $\dim f_*^{-1}(U_\alpha)\geq\dim\Delta+3$, by Proposition 1, the restriction map $\Gamma(f_*^{-1}(U_\alpha), \mathcal{O})\to\Gamma(N_\alpha, \mathcal{O})$ is bijective. This implies that $f_*^{-1}(U_\alpha)$ has at least two connected components. Let Z_α^i be the union of the connected components of $f_*^{-1}(U_\alpha)$ which intersect N_i^0. Put $Z^i=\bigcup_\alpha Z_\alpha^i$. Then Z^i $(i=1, 2)$ are non-empty, open and closed subsets of Z, and $Z^1\cap Z^2=\phi$. Hence Z is not connected. This is a contradiction. This implies that Y is connected.

q. e. d.

Corollary 1. *If X is a submanifold of a Hopf manifold, then there exists an effective divisor D on X satisfying the four conditions* (i)–(iv) *of Main Theorem.*

Proof. D' satisfies the conditions (i)–(iii) by Propositions 3, 4 and 5. In the proof of Proposition 4, we see that the condition (iv) is also satisfied.

q. e. d.

4. Proof of Main Theorem

Corresponding to the kernel of $\rho_L:\pi_1(X)\to \boldsymbol{C}^*$, we form the infinite cyclic unramified covering $\varpi:\tilde{X}\to X$. Then there exists a holomorphic automorphism g of $\tilde{X}$ such that $\varpi\circ g=\varpi$ and $\tilde{X}/\langle g\rangle=X$. Put $\tilde{Y}=\varpi^{-1}(Y)$. Let $\tilde{Y}_\nu(\nu=0, 1, \cdots)$ be connected components of $\tilde{Y}$. Let $\tilde{i}:\tilde{Y}\to\tilde{X}$ be the inclusion. Put $\varpi_\nu=\varpi|\tilde{Y}_\nu$ and $\tilde{i}_\nu=\tilde{i}|\tilde{Y}_\nu$. Consider the commutative diagram :

$$
(2)\qquad
\begin{array}{ccccccc}
0 \longrightarrow & \pi_1(\tilde{X}) & \xrightarrow{\varpi_*} & \pi_1(X) & \xrightarrow{\rho} & \mathrm{Im}\,\rho & \longrightarrow 1\,, \\
 & \uparrow{\scriptstyle \tilde{i}_{\nu*}} & & \uparrow{\scriptstyle i_*} & \nearrow{\scriptstyle \rho_Y} & & \\
0 \longrightarrow & \pi_1(\tilde{Y}_\nu) & \xrightarrow{\varpi_{\nu*}} & \pi_1(Y)\,, & & &
\end{array}
\qquad \rho:=\rho_L,\ \rho_Y:=\rho_{L,Y},
$$

where the horizontal arrows are exact. If $\tilde{Y}$ had infinitely many connected components, then $\varpi_{\nu*}$ would be an isomorphism. Then, by (2), $\rho_Y=(\rho\circ\varpi_*)\circ(\tilde{i}_{\nu*}\circ\varpi_{\nu*}^{-1})=1$. This contradicts condition (iv). Hence the number of connected components of $\tilde{Y}$ is finite. Let $\tilde{Y}=\bigcup_{\nu=0}^{h-1}\tilde{Y}_\nu$. Then g^h acts on each $\tilde{Y}_\nu$ and $\tilde{Y}_\nu/\langle g^h\rangle=Y$.

Lemma 6. $\mathrm{Ker}\,\rho_Y\subset\mathrm{Ker}\,j_*$.

Proof. Let $\gamma\in\pi_1(Y)$ be an element such that $j_*\gamma\neq 1$. Then either γ of γ^{-1} is j-

contractive. Hence by condition (iv) we have $|\rho_Y(\gamma)| \neq 1$. This implies $\gamma \notin \operatorname{Ker} \rho_Y$.
q. e. d.

Consider the following commutative diagram :

$$\begin{array}{ccccc}
\tilde{X} & \xleftarrow{\tilde{i}_0} & \tilde{Y}_0 & \overset{\tilde{j}_0}{\dashrightarrow} & \boldsymbol{C}^m-\{0\} \\
\Big\downarrow{\scriptstyle \varpi} & & \Big\downarrow{\scriptstyle \varpi_0} & & \Big\downarrow{\scriptstyle \tilde{\pi}} \\
X & \xleftarrow{i} & Y & \xrightarrow{j} & \boldsymbol{C}^m-\{0\}/\langle G\rangle,
\end{array}$$

where $\tilde{\pi}$ is the canonical projection.

Lemma 7. *There exists a lifting* $\tilde{j}_0 : \tilde{Y}_0 \to \boldsymbol{C}^m - \{0\}$ *of* j.

Proof. It is sufficient to show that $j_* \circ \varpi_{0*} : \pi_1(\tilde{Y}_0) \to \pi_1(\boldsymbol{C}^m - \{0\}/\langle G\rangle)$ is trivial. By (2), $\rho_Y \circ \varpi_{0*} = \rho \circ \varpi_* \circ \tilde{i}_{0*} = 1$. Hence $\varpi_{0*}(\pi_1(\tilde{Y}_0)) \subset \operatorname{Ker} \rho_Y$. Hence, by Lemma 6, $\varpi_{0*}(\pi_1(\tilde{Y}_0)) \subset \operatorname{Ker} j_*$.
q. e. d.

Note that $\tilde{j}_0$ is an embedding of $\tilde{Y}_0$ such that $\tilde{j}_0 \circ g^h = G^\varepsilon \circ \tilde{j}_0$, where ε is not equal to zero. Replacing g by g^{-1} if necessary, we can assume that

$$\tilde{j}_0 \circ g^h = G^\varepsilon \circ \tilde{j}_0 \qquad (\varepsilon > 0) \quad \text{on} \quad \tilde{Y}_0. \tag{3}$$

Lemma 8. *There exists a holomorphic function f on $\tilde{X}$ which has the following properties :*

(a) $\tilde{Y} = \{x \in \tilde{X} ; f(x) = 0\}$ *as a set,*

(b) $g^* f = \alpha f$ *for some constant* $\alpha \in \boldsymbol{C}^*$ *with* $|\alpha| < 1$,

(c) $f : \tilde{X} \to \boldsymbol{C}$ *is a surjection.*

Proof. (cf. [3] p. 701) Let $\mathfrak{U} = \{U_j\}$ be a covering of X by small open sets and represent the line bundle L as a cocycle $\{\alpha_0^{m_{jk}}\} \in Z^1(\mathfrak{U}, \boldsymbol{C}^*)$, $\{m_{jk}\} \in Z^1(\mathfrak{U}, \boldsymbol{Z})$, where α_0 is the generator of $\operatorname{Im} \rho$ such that $|\alpha_0| < 1$. Let $\{\varphi_j\} \in \Gamma(\mathfrak{U}, \mathcal{O}(L))$ be the section whose zero-locus coincides with Y. Then we have

$$\varphi_j = \alpha_0^{m_{jk}} \varphi_k \qquad \text{on } U_j \cap U_k.$$

Then

$$\eta = \frac{d\varphi_j}{\varphi_j} = \frac{d\varphi_k}{\varphi_k} = \cdots$$

is a meromorphic 1-form on X. Define a multiplicative multi-valued holomorphic function f_m on X by

$$f_m(x) = \exp \int_{x_0}^{x} \eta \qquad (x_0 \in X).$$

Then it is easy to see that $f = \varpi^* f_m$ is a (single-valued) holomorphic function on $\tilde{X}$ satisfying (a). It is clear that $g^* f = \alpha_0^l f$ for some non-zero integer l. Let $\tilde{\theta}$ be a path in $\tilde{Y}_0$ with the initial point $p \in \tilde{Y}_0$ and the terminal point $g^h(p) \in \tilde{Y}_0$. Clearly $\theta := \varpi(\tilde{\theta})$ is a closed path in Y. Then $g^{*h}f = \left(\exp \int_\theta \eta\right) f = \rho([\theta])\, f$, where $[\theta]$ denotes the element of $\pi_1(Y)$ represented by θ. Since $[\theta]$ is clearly j-contractive by (3), $|\rho([\theta])| < 1$ by condition (iv). Hence $|\alpha_0^{hl}| = |\rho([\theta])| < 1$. This implies $|\alpha_0^l| < 1$. Put $\alpha = \alpha_0^l$. Then we obtain (b). The proof of (c) is the same as that of Lemma 3.

q. e. d.

Let $(z_1, \cdots, z_m)$ be a standard system of coordinates on $\boldsymbol{C}^m$ with $0 = (0, \cdots, 0)$. Let $r(z) = \sum_{i=1}^{m} |z_i|^2$. On $\tilde{Y}_0$, we introduce an s-psh function r_0 by $r_0 = \tilde{j}_0^* r$. By Richberg [7], r_0 extends to an s-psh function $\tilde{\rho}$ on a certain neighborhood $\tilde{N}_0'$ of $\tilde{Y}_0$. We can assume that $\tilde{N}_0' \cap g^\nu(\tilde{N}_0') = \phi$ for all ν, $0 \leq \nu \leq h-1$. Fix $c > 0$. Put $\tilde{U}_c = \{x \in \tilde{N}_0' ; \tilde{\rho}(x) < c\}$, $\partial \tilde{U}_c = \{x \in \tilde{N}_0' ; \tilde{\rho}(x) = c\}$, $U_c = \tilde{U}_c \cap \tilde{Y}_0$, $\partial U_c = \partial \tilde{U}_c \cap \tilde{Y}_0$, $\underline{\tilde{U}}_c = \tilde{U}_c \cup \partial \tilde{U}_c$ and $\underline{U}_c = U_c \cup \partial U_c$. We can find a positive integer h_0 such that $g^{h_0 h}(\underline{U}_c) \subset U_c$. Put $l = h_0 h$. Note that $\underline{U}_c - g^l(U_c)$ is compact. We can choose positive constants a and b such that the set $K' = \{x \in \tilde{N}_0' ; a < \tilde{\rho}(x) < b\}$ contains $\underline{U}_c - g^l(U_c)$. Put $D_\varepsilon = \{t \in \boldsymbol{C} ; |t| < \varepsilon\}$ for $\varepsilon > 0$. Fix $\varepsilon > 0$ such that

(i) $f : K' \cap f^{-1}(D_\varepsilon) \to D_\varepsilon$ is (1, 1)-convex-concave,

(ii) $K' \cap f^{-1}(D_\varepsilon)$ is relatively compact in $\tilde{N}_0'$, and

(iii) $g^l(\partial \tilde{U}_c \cap f^{-1}(D_\varepsilon)) \subset \tilde{U}_c \cap f^{-1}(D_\varepsilon)$.

Put $\tilde{K} = K' \cap f^{-1}(D_\varepsilon)$. By Theorem 3, there exists uniquely an irreducible normal Stein completion $\hat{S}$ of $\tilde{K}$. Let $\hat{f} : \hat{S} \to D_\varepsilon$ be the extension of f. It is easy to see that the sets

$$E_1 = \{x \in \tilde{K} ; \tilde{\rho}(x) < c\} \cup (\hat{S} - \tilde{K})$$

and

$$E_2 = [\{x \in \tilde{K} ; \tilde{\rho}(g^{-l}(x)) < c\} \cup (\hat{S} - \tilde{K})] \cap \hat{f}^{-1}(D_{|\alpha|^l \varepsilon})$$

are Stein open subsets of $\hat{S}$. Clearly g^l maps a neighborhood T_1 of the boundary of E_1 in $\hat{f}^{-1}(D_\varepsilon)$ onto a neighborhood T_2 of the boundary of E_2 in $f^{-1}(D_{|\alpha|^l \varepsilon})$. We extend $(c - \tilde{\rho})^{-1}$ to a C^∞-function φ_1 on E_1. Put $\pi_1 = \hat{f} | E_1$. Then $\pi_1 : E^1 \to D_\varepsilon$ is a 1-convex holomorphic map with the exhaustion function φ_1. Similarly we extend $(c - (g^{-l})^* \tilde{\rho})^{-1}$ to a C^∞-function φ_2 on E_2. Put $\pi_2 = \alpha^{-l}(\hat{f} | E_2)$. Then $\pi_2 : E_2 \to D_\varepsilon$ is a 1-convex holomorphic map with exhaustion function φ_2. Furthermore we have the following commutative diagram by Lemma 8 (b) :

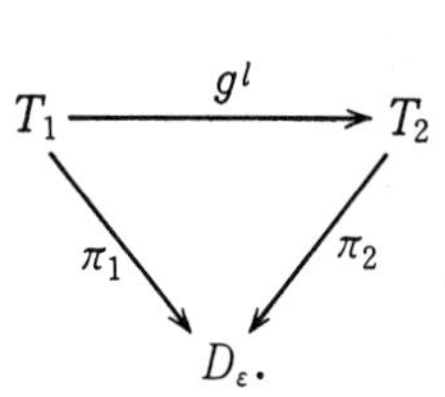

Note that T_i is a neigborhood of infinity in E_i with respect to φ_i, $j = 1, 2$. Since $g^l : T_1 \to T_2$ is an isomorphism, we obtain, by Lemma 2, a biholomorphic map

$\hat{G}: E_1 \to E_2$ such that $\hat{G}$ agrees with g^l on a neighborhood of infinity. Hence the following diagram is commutative :

$$\begin{array}{ccc} E_1 & \xrightarrow[\sim]{\hat{G}} & E_2 \\ {\scriptstyle \hat{f}}\downarrow & & \downarrow{\scriptstyle \hat{f}} \\ D_\varepsilon & \xrightarrow[\sim]{\alpha^l} & D_{|\alpha^l|\varepsilon}. \end{array} \tag{4}$$

Now we regard $\hat{G}$ as a biholomorphic map of E_1 into itself.

Lemma 9. *$\bigcap_{m\geq 0} \hat{G}^m(E_1) = \{\hat{0}\}$ for a certain point $\hat{0} \in \hat{f}^{-1}(0) \cap E_1$. Hence $\hat{G}$ is a contraction of E_1.*

Proof. Since E_1 is biholomorphic to the relatively compact open subset E_2 of the Stein space E_1, E_1 can be embedded in an affine space of a certain dimension as a closed analytic subset. Hence, for each $m \geq 0$, $\hat{G}^m$ can be viewed as a (vector-valued) holomorphic function on E_1. Since, for $m \geq 1$, $\hat{G}^m(E_1) \subset E_2$ is relatively compact in E_1, the sequence $\{\hat{G}^m\}_{m\geq 1}$ is uniformly bounded on E_1. Hence we can choose a subsequence $\{\hat{G}^{m_j}\}_{j\geq 1}$ of $\{\hat{G}^m\}_{m\geq 1}$ which converges uniformly to a holomorphic function u on E_2. Put $A = \bigcap_{m\geq 0} \hat{G}^m(\tilde{K} \cap \tilde{Y}_0)$ and denote by ∂A the boundary of A in $\hat{f}^{-1}(0) \cap E_1$. Then clearly $u(\hat{f}^{-1}(0) \cap E_2) = \partial A$. Since ∂A is compact, this implies that the absolute value of each component of $u|\hat{f}^{-1}(0) \cap E_2$ attains its maximum on the open set E_2. Hence $u|\hat{f}^{-1}(0) \cap E_2$ is constant. Since $u(E_2) \subset \hat{f}^{-1}(0) \cap E_2$, we infer that $u \circ u$ is a constant function. Let $\hat{0} = u(u(E_2))$. Then $\bigcap_{m\geq 0} \hat{G}^m(E_1) \subseteq \bigcap_{j\geq 1} \hat{G}^{2m_j}(E_1) = \{\hat{0}\}$. Since $\hat{G}^{m_1}(E_1) \supset \hat{G}^{m_2}(E_1) \neq \phi$ for $m_1 < m_2$, we have $\bigcap_{m\geq 0} \hat{G}^m(E_1) \neq \phi$. Hence it follows that $\bigcap_{m\geq 0} \hat{G}^m(E_1) = \{\hat{0}\}$, where $0 \in \hat{f}^{-1}(\hat{0}) \cap E_1$. q. e. d.

Lemma 10. *There exist for some m an open holomorphic embedding $I: E_1 \to \boldsymbol{C}^m (I(\hat{0}) = 0 = (0, \cdots, 0))$, a contracting holomorphic automorphism G of $\boldsymbol{C}^m$ with $G(0) = 0$, a G-invariant normal subvariety $\hat{V} \subset \boldsymbol{C}^m$, and a holomorphic function F on $\hat{V}$ such that*

(i) *$I(E_1)$ is an open subset of $\hat{V}$,*
(ii) *$I \circ \hat{G} = G \circ I$, and*
(iii) *$\hat{f} = F \circ I$ and $G^*F = \alpha^l F$.*

Proof. We can prove this by method similar to that in [1]. Here we explain the outline of the proof. Let U' be a sufficiently small neighborhood of 0 in E_1 such that there exists a neat embedding $j': U' \to \boldsymbol{C}^m (j'(0) = 0)$ (m = the dimension of the Zariski tangent space at 0), where $j'(U')$ is an analytic subvariety of an open subset of $\boldsymbol{C}^m$. Moreover, by Lemma 9, we can find an integer ν_0 such that $\hat{G}^\nu(U') \subset U'$ for all $\nu \geq \nu_0$. Put $U'' = \bigcap_{\nu=0}^{\nu_0 - 1} \hat{G}^\nu(U')$. Then $\hat{G}(U'') \subset U''$. Let $j'' = j'|U''$. Put $G' = j'' \circ \hat{G} \circ j''^{-1}$. Then, since j'' is a neat embedding, G' extends to a local biholomorphic map at $0 \in \boldsymbol{C}^m$. G' induces a linear automorphism of the tangent space at $0 \in \boldsymbol{C}^m$. We can show that the absolute values of all eigenvalues of the linear automorphism are less than 1. Hence, by Sternberg [10] and Reich [6], there exists a local coordinate transformation τ at 0 ($\tau(0)=0$) such that $G:=\tau\circ G'\circ\tau^{-1}$ automatically defines a contracting automorphism of $\boldsymbol{C}^m$ with $G(0)=0$. Put $j=\tau\circ j''$. Then, taking a small neighborhood $U(\subset U'')$ of $\hat{0}$, we find an embedding $j: U\to\boldsymbol{C}^m(j(\hat{0})=0)$ such that $j\circ\hat{G}=G\circ j$, $\hat{G}(U)\subset U$, and that $j(U)$ is an analytic subvariety of an open subset of $\boldsymbol{C}^m$. Now put

$$I_\nu = G^{-\nu}\circ j\circ\hat{G}^\nu \qquad (\nu\geq 0).$$

The holomorphic map I_ν is defined on $\hat{G}^{-\nu}(U)\cap E_1$ and $I_\nu=I_{\nu+1}$ on U. Since $E_1=\bigcup_{\nu\geq 0}\hat{G}^{-\nu}(U)\cap E_1$, we obtain by the analytic continuation of $\{I_\nu\}$ a holomorphic map $I: E_1\to\boldsymbol{C}^m$ which satisfies (ii). Put $\hat{V}=\bigcup_{\nu\geq 0}G^{-\nu}(j(U))$. Then it is easy to see that $\hat{V}$ is a closed G-invariant analytic subvariety of $\boldsymbol{C}^m$. Moreover, since E_1 is normal, $\hat{V}$ is also normal. It is clear that $I: E_1\to\hat{V}$ is locally biholomorphic. To prove (i), it suffices to show that I is one-to-one. Let $p_1, p_2\in E_1$ be points such that $I(p_1)=I(p_2)$. If $\nu(>0)$ is sufficiently large, $\hat{G}^\nu(p_1)$ and $\hat{G}_\nu(p_2)$ are contained in U. Then $j(\hat{G}^\nu(p_1))=G^\nu\circ I(p_1)=G^\nu\circ I(p_2)=j(\hat{G}^\nu(p_2))$. Since j is one-to-one on U, we have $\hat{G}^\nu(p_1)=\hat{G}^\nu(p_2)$. Therefore $p_1=p_2$. Thus I is one-to-one. Next we define the holomorphic functions F_ν by

$$F_\nu = \alpha^{-\nu l}\hat{f}\circ j^{-1}\circ G^\nu, \qquad (\nu\geq 0).$$

Each F_ν is defined on $G^{-\nu}(j(U))$, which is an open subset of $\hat{V}$. Since $\hat{V}=\bigcup_{\nu\geq 0}G^{-\nu}(j(U))$ and $F_\nu=F_{\nu+1}$ on $j(U)$, we obtain by the analytic continuation of $\{F_\nu\}$ a holomorphic function F on $\hat{V}$. By (4), it is easy to see that F satisfies (iii).

q. e. d.

Put $B=(\tilde{\underline{U}}_c-g^l(\tilde{U}_c))\cap f^{-1}(D_{|\alpha|^l\varepsilon})$, $B_0=$ the interior of B, $\tilde{N}'_0=\bigcup_{\nu\in\boldsymbol{Z}}g^{\nu l}(B)$, and $\tilde{N}_0=\bigcap_{\nu=0}^{h_0-1}g^{\nu h}(\tilde{N}'_0)$. Then $\tilde{N}_0$ is a neighborhood of $\tilde{Y}_0$ such that $g^h(\tilde{N}_0)=\tilde{N}_0$. Put $\tilde{V}=\hat{V}-\{0\}$, $\hat{W}=F^{-1}(0)$ and $\tilde{W}=\hat{W}-\{0\}$.

Lemma 11. *There exists an open holomorphic embedding $\tilde{J}: \tilde{N}_0\to\tilde{V}$ such that*
(i) $\tilde{J}(\tilde{Y}_0)=\tilde{W}$,
(ii) $\tilde{J}\circ g^l=G\circ\tilde{J}$, *and*
(iii) $f=F\circ\tilde{J}$.

Proof. Let $i:\tilde{K}\to E_1$ be the inclusion. Put $j=I\circ i$. Then $j:\tilde{K}\to\tilde{V}$ is an open holomorphic embedding. For each integer ν, define on $g^{-\nu l}(B)$ the mapping $\tilde{j}_\nu$ into $\tilde{V}$ by

$$\tilde{j}_\nu = G^{-\nu}\circ j\circ g^{\nu l} \qquad (\nu\in\boldsymbol{Z}).$$

Put $C_\nu=g^{-\nu l}(B)\cap g^{-(\nu+1)l}(B)$. By the definition of $\hat{G}$, $\hat{G}$ is automatically defined on a small neighborhood of C_0 and satisfies

(5) $$i\circ g^l=\hat{G}\circ i.$$

By Lemma 10, we have

(6) $$G\circ I=I\circ\hat{G}.$$

Combining (5) and (6), we have

$$G\circ j=j\circ g^{l} \tag{7}$$

on a neighborhood of C_0. Let x be a point near C_ν. Then

$$\begin{aligned}\tilde{j}_{\nu+1}(x) &= G^{-(\nu+1)}\circ j\circ g^{(\nu+1)l}(x) = G^{-(\nu+1)}\circ j\circ g^{l}(y) && (y=g^{\nu l}(x))\\ &= G^{-\nu}\circ j(y) && (\text{by } (7))\\ &= G^{-\nu}\circ j\circ g^{\nu l}(x) = \tilde{j}_\nu(x).\end{aligned}$$

Since $\{C_\nu\}_{\nu\in \boldsymbol{Z}}$ are disjoint from each other, by the analytic continuation of $\{\tilde{j}_\nu\}$, we can define a holomorphic map $\tilde{J}: \tilde{N}_0\to\tilde{V}$. It is clear that $\tilde{J}$ is locally biholomorphic and that (ii) holds. In order to show that $\tilde{J}$ is one-to-one, it suffices to show that $\tilde{J}(g^{\nu l}(B_0))\cap\tilde{J}(B_0)=\phi$ for all $\nu>0$. Clearly $\hat{G}^\nu(i(B_0))\subset\hat{G}(E_1)=E_2$. On the other hand $i(B_0)\cap E_2=\phi$. Hence $\hat{G}^\nu(i(B_0))\cap i(B_0)=\phi$. Since I is one-to-one by Lemma 10, we have $\phi=I\circ\hat{G}^\nu(i(B_0))\cap I(i(B_0))=G^\nu\circ\tilde{J}(B_0)\cap\tilde{J}(B_0)=\tilde{J}(g^{\nu l}(B_0))\cap\tilde{J}(B_0)$. It is easy to see that (i) and (iii) hold. q. e. d.

Put $\tilde{N}_\nu=g^\nu(\tilde{N}_0)$ $(0\le\nu\le h-1)$. Define the open holomorphic embedding $\tilde{J}_\nu: \tilde{N}_\nu\to\tilde{V}$ by $\tilde{J}_\nu=\tilde{J}\circ g^{-\nu}$. Note that $\tilde{J}_{\nu+1}\circ g=\tilde{J}_\nu$ for $0\le\nu\le h-2$, and $\tilde{J}_0=\tilde{J}$. Let $\varpi_1: \tilde{X}\to X_1 := \tilde{X}/\langle g^l\rangle$ and $\tilde{\pi}: \tilde{V}\to V := \tilde{V}/\langle G\rangle$ be the canonical projections. We fix the following notation: $\tilde{N}=\bigcup_{\nu=0}^{h-1}\tilde{N}_\nu$, $\tilde{M}=\tilde{J}_0(\tilde{N}_0)=\cdots=\tilde{J}_{h-1}(\tilde{N}_{h-1})\subset\tilde{V}$, $N=\varpi_1(\tilde{N})$, $N_\nu=\varpi_1(\tilde{N}_\nu)$, $M=\tilde{\pi}(\tilde{M})$, $Y_\nu=\varpi_1(\tilde{Y}_\nu)$, $W=\tilde{\pi}(\tilde{W})$, $\tilde{Z}_1=\tilde{X}-\tilde{Y}$, $\tilde{Z}_2=\tilde{V}-\tilde{W}=\tilde{V}-\hat{W}$, $Z_1=\varpi_1(\tilde{Z}_1)$, $Z_2=\tilde{\pi}(\tilde{Z}_2)$, $N_1^0=\varpi_1(\tilde{N}-\tilde{Y})$, and $N_2^0=\tilde{\pi}(\tilde{M}-\tilde{W})$. Note that each Y_ν is the h_0-fold cyclic covering manifold of Y. The set N_ν is a neighborhood of Y_ν such that $N_\nu\cap N_{\nu'}=\phi$ for $\nu\neq\nu'$. Denote by Δ the elliptic curve defined by $\boldsymbol{C}^*/\langle\alpha^l\rangle$ and by p the canonical projection $\boldsymbol{C}^*\to\Delta$. By Lemma 4, we have holomorphic surjections $f_*: Z_1\to\Delta$ and $F_*: Z_2\to\Delta$, where f_* and F_* are the maps induced by the holomorphic functions f and F defined in Lemma 8 and Lemma 10, respectively.

By Lemma 11, we obtain easily

Lemma 12. *$\{\tilde{J}_\nu\}$ induces an h-fold unramified covering $J: N\to M$ such that $J|Y_\nu: Y_\nu\to W$ $(0\le\nu\le h-1)$ is an isomorphism. Moreover, we have the following commutative diagram:*

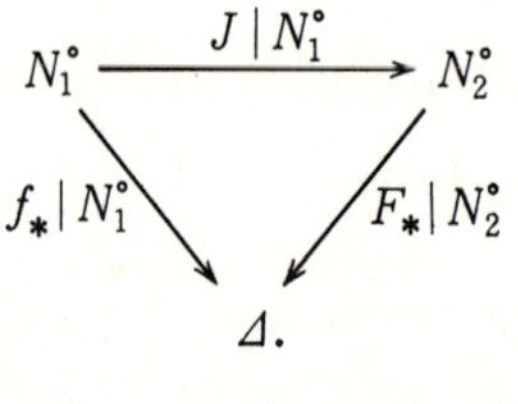

Now we shall prove the following

Lemma 13. *The covering map $J|N_1^0: N_1^0\to N_2^0$ extends uniquely to a proper holomorphic surjection $\hat{J}: Z_1\to Z_2$ which makes the following diagram commutative:*

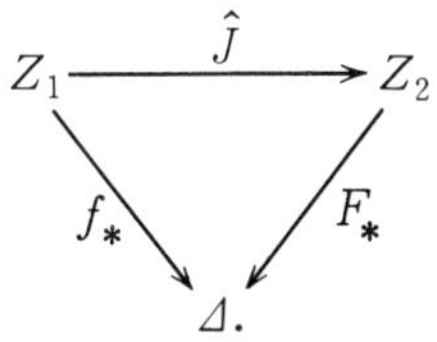

Proof. Let $\tau \in \Delta$ be any point. By Lemma 5, for any small disk U in Δ with center τ, $F_U := F_*|F_*^{-1}(U) : F_*^{-1}(U) \to U$ is a 1-convex holomorphic surjection with some exhaustion function $\tilde{\varphi}_2$. Moreover $F_*^{-1}(U)$ is a Stein space. Now we put $\tilde{\varphi}_1' = (J|N_1^0 \cap f_*^{-1}(U))^* \tilde{\varphi}_2$ and extend $\tilde{\varphi}_1'$ to a C^∞-function $\tilde{\varphi}_1$ on the whole of $f_*^{-1}(U)$. Put $f_U = f_*|f_*^{-1}(U)$. Then $f_U : f_*^{-1}(U) \to U$ is also a 1-convex holomorphic surjection with exhaustion function $\tilde{\varphi}_1$. Hence, by Theorem 2, shrinking U if necessary, we can assume that $f_*^{-1}(U)$ is holomorphically convex. Then, by Lemma 1[1], there exists a unique proper holomorphic surjection $\hat{J}_U : f_*^{-1}(U) \to F_*^{-1}(U)$ which agrees with $J|N_1^0$ on a neighborhood of infinity in $f_*^{-1}(U)$ and makes the following diagram commutative:

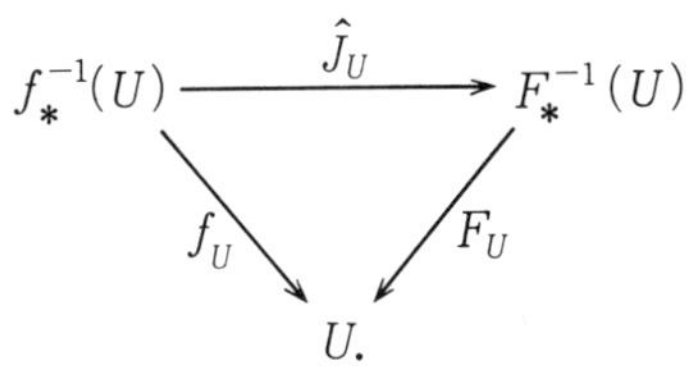

Patching together these pieces $\{\hat{J}_U\}_U$, we obtain the proper holomorphic surjection $\hat{J} : Z_1 \to Z_2$ of the lemma. q. e. d.

Lemma 14. *There exists a proper holomorphic surjection* $\Phi_1 : X_1 \to V$.

Proof. Define Φ_1 to be

$$\Phi_1(x) = \begin{cases} J(x) & \text{if } x \in N, \\ \hat{J}(x) & \text{if } x \in Z_1. \end{cases}$$

It is clear by definition that $J = \hat{J}$ on $N_1^0 = N \cap Z_1$. q. e. d.

Take the Stein factorization of Φ_1. Then we have the following commutative diagram:

(8)

$$\begin{array}{ccc} X_1 & \xrightarrow{\Phi} & V_0 \\ & \searrow{\Phi_1} \quad \swarrow{\Phi_0} & \\ & V, & \end{array}$$

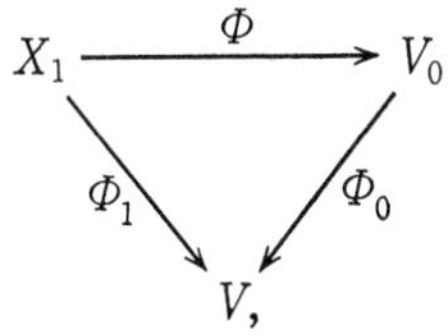

where

---
1) Apply Lemma 1 to each connected component of $f_*^{-1}(U)$.

(i) V, V_0 are compact normal varieties and Φ, Φ_0 are proper surjections,

(ii) each fibre of Φ is connected, and Φ is biholomorphic outside a proper analytic subset of X_1, and

(iii) each fibre of Φ_0 is discrete, i.e., Φ_0 is a finite ramified covering.

Now we shall prove the final lemma.

Lemma 15. *There exists a bimeromorphic holomorphic map of X onto a subvariety of a Hopf manifold.*

Proof. Lifting Φ_1 to $\tilde{\Phi}_1 : \tilde{X} \to \tilde{V}$, we have the following commutative diagram by (8) :

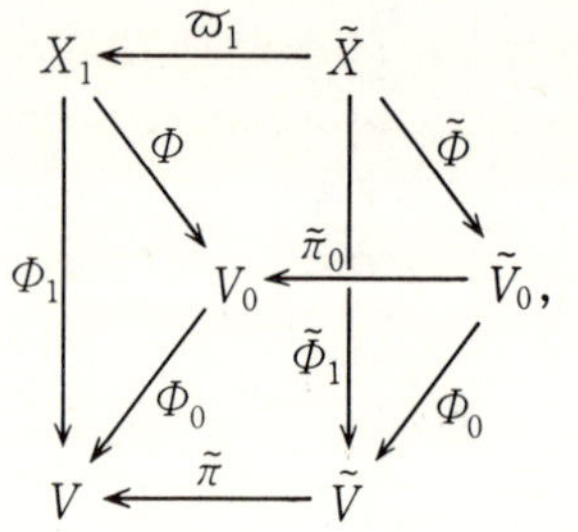

where $\tilde{V}_0$ is the Stein factorization of $\tilde{\Phi}_1 : \tilde{X} \to \tilde{V}$. Since every fibre of $\tilde{\Phi}$ is connected and $\tilde{V}_0$ contains no positive dimensional compact analytic subset, the automorphism g of $\tilde{X}$ induces an automorphism G_0 of $\tilde{V}_0$, which satisfies $G_0 \circ \tilde{\Phi} = \tilde{\Phi} \circ g$. Hence $\tilde{\Phi}$ induces $\Psi : X \to \tilde{V}_0 / \langle G_0 \rangle$. Note that Ψ is biholomorphic outside a proper analytic subset of X. Hence, to prove the lemma, it suffices to show that $\tilde{V}_0 / \langle G_0 \rangle$ is a subvariety of a Hopf manifold. Now $\tilde{\Phi}_0 : \tilde{V}_0 \to \tilde{V}$ is a finite ramified covering. The images of the ramification loci in $\tilde{V}$ extend to complex subvarieties in $\hat{V} = \tilde{V} \cup \{0\}$. Then, by virtue of the continuation theorem of Grauert-Remmert, we can obtain a normal variety $\hat{V}_0 = \tilde{V}_0 \cup \{\tilde{0}\}$ attaching *one point* $\tilde{0}$ to $\tilde{V}_0$ and a finite ramified covering $\hat{\Phi}_0 : \hat{V}_0 \to \hat{V}$ such that $\hat{\Phi}_0 | \tilde{V}_0 = \tilde{\Phi}_0$. Moreover the automorphism G_0 extends to an automorphism $\hat{G}_0$ such that $\hat{G}_0(\tilde{0}) = \tilde{0}$. Since $\hat{G}_0^l$ induces the contracting automorphism G of $\hat{V}$ such that $G(0) = 0$, and since the number of points of the fibres of $\tilde{\Phi}_0$ is bounded, $\hat{G}_0$ is a contraction of $\hat{V}_0$. Hence $\tilde{V}_0 / \langle G_0 \rangle$ is biholomorphic to a subvariety of a Hopf manifold by Proposition 2. q. e. d.

Remark. Let X be a compact complex manifold of dimension $n \geq 3$. Assume the following :

(i) $\pi_1(X) \simeq \boldsymbol{Z}$,

(ii) X contains a primary Hopf manifold $Y = \boldsymbol{C}^{n-1} - \{0\} / \langle G \rangle$ of dimension $n-1$, where G is a contracting holomorphic automorphism of $\boldsymbol{C}^{n-1}$ with $G(0) = 0$,

(iii) $[Y]$ is flat.

Let $\gamma_G \in \pi_1(Y)$ be the generator corresponding to G. In this case, we can easily see that the contractiveness of the elements of $\pi_1(Y)$ does not depend on the choice of embedding of Y. Put $\rho=\rho_{[Y]}$.

There are three cases to be considered :

(a) $|\rho(\gamma_G)| < 1$, (b) $|\rho(\gamma_G)| > 1$, (c) $|\rho(\gamma_G)| = 1$.

Case (a). If $n\geq 4$, then all conditions of Main Theorem are satisfied. Hence X bimeromorphically dominates a subvariety of a Hopf manifold. If $n=3$, we have no answer.

Case (b). We shall give an example of such manifold X.

Let $[z_0 : z_1 : \cdots : z_{n-1}]$ be homogeneous coordinates on the projective space $\boldsymbol{P}^{n-1}$. Put $0'=[1 : 0 : \cdots : 0]$ and $\boldsymbol{P}^{n-2}=\{z=[z_0 : z_1 : \cdots : z_{n-1}] \in \boldsymbol{P}^{n-1} ; z_0=0\}$. Let $\tilde{X}$ be the open subset of $\boldsymbol{P}^{n-1}\times \boldsymbol{C}$ defined by

$$\tilde{X} = (\boldsymbol{P}^{n-1}- \{0'\}) \times \boldsymbol{C}-\boldsymbol{P}^{n-2}\times \{0\}.$$

Define a holomorphic automorphism g of $\tilde{X}$ by

$$g: ([z_0: z_1: \cdots : z_{n-1}], t) \to ([z_0: 2^{-1}z_1: 2^{-1}z_2: \cdots : 2^{-1}z_{n-1}], 2t).$$

We shall show that the quotient space $X=\tilde{X}/\langle g\rangle$ is a compact complex manifold which satisfies (i)–(iii) and (b). It is easy to see that $\tilde{X}$ is simply connected. Consider the following function defined on X,

$$\sigma(z, t) = \frac{|z_1|^2+\cdots+|z_{n-1}|^2}{|z_0|^2+|t|^2(|z_1|^2+\cdots+|z_{n-1}|^2)}$$
$$(z=[z_0 : \cdots : z_{n-1}] \in \boldsymbol{P}^{n-1}- \{0'\}, t\in \boldsymbol{C}).$$

Then

$$\sigma(g(z, t)) = \frac{1}{4}\sigma(z, t). \tag{9}$$

Hence the action of g on $\tilde{X}$ is properly discontinuous. It is easy to see that the set

$$\{(z, t) \in (\boldsymbol{P}^{n-1}- \{0'\}) \times \boldsymbol{C}-(\boldsymbol{P}^{n-2}\times \{0\}) ; 1 \leq \sigma(z, t) \leq 4\} \tag{10}$$

is compact. Hence the quotient space X is a compact complex manifold with $\pi_1(X) \simeq \boldsymbol{Z}$. The equation $t=0$ defines the submanifold Y which is biholomorphic to $\boldsymbol{C}^{n-1}- \{0\} /\langle G\rangle$, where G is the scalar matrix $1/2\cdot I$. By the definition of g, $[Y]$ is flat and $\rho(\gamma_G)=2$.

Note that if we consider the above X in the case $n=2$, then X is a primary Hopf surface. Hence, if $n=2$, both cases (a) and (b) can occur. This is based on the fact that, if Y is an elliptic curve, then both γ_G and γ_G^{-1} can be contractive elements if we take different embeddings of Y into Hopf manifolds.

Case (c) *does not occur.* To prove this, we shall first show that, for the generator δ of $\pi_1(X)$, $|\rho(\delta)|\neq 1$ holds. Assume that $|\rho(\delta)|=1$. Take a non-zero section $f\in \Gamma(X, \mathcal{O}([Y]))$. By the assumption the absolute value $|f|$ of f is a single-valued continuous function with the mean value property. Hence f is reduced to a constant. This implies that $[Y]$ is trivial. This is a contradiction. Therefore $|\rho(\delta)|\neq 1$. Thus we infer that the map $i_* : \pi_1(Y)\to\pi_1(X)$ induced by the inclusion $i : Y\to X$ is trivial. Let $\varpi : \tilde{X}\to X$ be the universal covering of X. Let g be the holomorphic automorphism of $\tilde{X}$ such that $\tilde{X}/\langle g\rangle=X$. Let $f\in \Gamma(X, \mathcal{O}([Y]))$ be a non-zero section whose zero locus coincides with Y. Then $\tilde{f}=\varpi^*f$ is a single-valued holomorphic function defined on $\tilde{X}$. Taking g^{-1} instead of g if necessary, we can assume that $g^*\tilde{f}=\alpha\tilde{f}$ ($\alpha\in \boldsymbol{C}^*$, $|\alpha|<1$). By the same argument as in Lemma 3, we can show that $\tilde{f}: \tilde{X}\to\boldsymbol{C}$ is a surjection. Put $Z=X-Y$. Then $\tilde{f}$ induces a surjection $f_*: Z\to\varDelta:=\boldsymbol{C}^*/\langle\alpha\rangle$. Let N be an open tubular neighborhood of Y. Since i_* is trivial, we can find a single-valued branch f_1 of f on N. Put $D_r=\{t\in\boldsymbol{C};\ |t|\leq r\}$. We can find a small positive number ε such that, for any $t\in D_{|\alpha|^{-1}\varepsilon}$, $f_1^{-1}(t)$ is a compact analytic subset contained in N. Put $N'=f_1^{-1}(D_\varepsilon)$ and $Z'=X-N'$. Since the canonical projection $p: \boldsymbol{C}^*\to\varDelta$ maps the set $D_{|\alpha|^{-1}\varepsilon}-D_\varepsilon$ onto $\varDelta$, $p\circ f_1$ maps $N-N'$ onto $\varDelta$. Hence $f'_*: Z'\to\varDelta$ is a surjection, where $f'_*=f_*|Z'$. Let $\tau\in\varDelta$ be any point. Then $f_*^{-1}(\tau)$ and $f_*'^{-1}(\tau)$ are analytic subsets in Z and Z', respectively, such that

$$f_*'^{-1}(\tau)=f_*^{-1}(\tau)\cap Z'. \tag{11}$$

Let S be any connected component of $f_*^{-1}(\tau)$ such that $S\cap\partial N'\neq\phi$, where $\partial N'$ denotes the boundary of N'. Since $f_*(S)=\tau$, f_1 is constant on S. Hence we have $S\subset f_1^{-1}(f_1(p))$, where $p\in S\cap\partial N'$. Therefore in particular we have $S\subset\partial N'$. Hence, by (11), $f_*'^{-1}(\tau)$ coincides with the union of all connected components of $f_*^{-1}(\tau)$ contained in Z'. Since Z' is relatively compact in Z, this implies that $f_*'^{-1}(\tau)$ is compact. Thus any fibre of f'_* is compact. Hence, since $\varDelta$ is compact, Z' is also compact. But this is clearly absurd. This shows that case (c) does not occur.

References

[1] Kato, Ma.: Complex structures on $S^1\times S^5$. J. Math. Soc. Japan, **28** (1976), 550–576.

[2] Kato, Ma.: Some remarks on subvarieties of Hopf manifolds. A symposium of complex manifolds. Kokyu-roku. **240** (Kyoto Univ.) 1975.

[3] Kodaira, K.: On the structure of compact complex analytic surfaces, II. Amer. J. Math., **88** (1966), 682–721.

[4] Kodaira, K.: On the structure of compact complex analytic surfaces, III. Amer. J. Math., **90** (1968), 55–83.

[5] Ling, H-S.: Extending families of pseudoconcave complex spaces. Math. Ann., **204** (1973), 13–48.

[6] Reich, L.: Normalformen biholomorpher Abbildungen mit anziehendem Fixpunkt. Math. Ann., **180** (1969), 233–255.

[7] Richberg, R.: Stetige streng-pseudokonvexe Funktionen. Math. Ann., **175** (1968), 257–286.

[8] Siu, Y-T.: A pseudoconcave generalization of Grauert's direct image theorem, I, II. Ann. Scuola Norm. Sup. Pisa **24** (1970), 278–330.

[9] Siu, Y-T.: The 1-convex generalization of Grauert's direct image theorem. Math. Ann. **190** (1971), 203–214.

[10] Sternberg, S.: Local contractions and a theorem of Poincaré. Amer. J. Math. **79** (1957), 809–824.

Department of Mathematics
Rikkyo University

(Received December 16, 1975)

Relative Compactification of the Néron Model and its Application

I. Nakamura

Introduction

In this article we shall define an analytic Néron model, i.e., an analytic counterpart of Néron's minimal model [11], and prove the minimality of it among principal homogeneous spaces (§ 2, 3). This portion deals with a partial generalization of the notion of analytic fiber system of groups and related results in [4].

Secondly we shall relatively compactify an analytic Néron model by applying the theory of torus embeddings [3], [5]. One should recall that the usefulness of a relative compactification of Néron model has been conjectured in [11]. The original construction in [6] was given in a slightly different, elementary, however rather complicated form. To facilitate better understanding, we employ here the notations of torus embeddings.

Kodaira has made a deep investigation of elliptic fibrations of surfaces [4]. Iitaka [2] and Ogg [12] gave a numerical classification of singular fibers in a pencil of curves of genus two. Namikawa and Ueno [8], [9] gave their complete classification and made a systematic study of them, following in principle Kodaira.

The final objective of this article is to construct a family of reduced singular fibers of genus two in a systematic and geometric way by making use of a relative compactification of the Néron model (§ 5, 6, 7). In this respect, the present article should be viewed as a continuation of [9], which its authors had intended to write at that time. Namely it involves a part of the construction of a family of curves of genus two of parabolic type (cf. [9] for the terminology). Combining the results in [9] with those here, one could construct all singular fibers (§ 8).

The results in § 5, 6 and 7 have given the author a clue as to how to introduce a concept of stable quasi abelian variety (7. 4) (cf. [7] [10]). The latter was first obtained by Ueno in the 2-dimensional case [13].

Although the main part of this article was written in Nagoya in 1973, it has been completed during author's stay at the Mathematical Institute of Bonn University in 1975. The author would like to express his hearty thanks to Professor Hirzeburch and other people in Bonn University for their hospitality and kindness in inviting him there.

Notation

We denote by $\boldsymbol{Z}$, $\boldsymbol{R}$ and $\boldsymbol{C}$ the ring of integers, the field of real numbers and the field of complex numbers respectively. Also we write $\boldsymbol{C}^*=\boldsymbol{C}-\{0\}$, $D=\{s\in\boldsymbol{C};|s|<\varepsilon\}$, $D'=D-\{0\}$, $\boldsymbol{e}(x)=\exp(2\pi i\, x)$. The Siegel upper half plane $\mathfrak{S}_g$ is, by definition, $\mathfrak{S}_g=\{Z; g\times g$ symmetric matrix, $\mathrm{Im}\, Z>0\}$. Then the Siegel space $\mathfrak{S}_g^*$ is defined as the quotient space $\mathfrak{S}_g/Sp(g, \boldsymbol{Z})$.

1. Preliminaries

(1. 1) First we give preliminaries from the theory of torus embeddings (cf. [3],. [5]; compare also [10]).

By an n-dimensional algebraic torus over $\boldsymbol{C}$ we mean the group scheme $T=$ Spec R, $R=\boldsymbol{C}[w_1, w_1^{-1}, \cdots, w_n, w_n^{-1}]$. Let M and N be the group of characters of T and the group of one parameter subgroups of T respectively. Both M and N are isomorphic to $\boldsymbol{Z}^n$. An element of M can be expressed as $w^r=\prod w_i^{r_i}$, $r=(r_i)\in\boldsymbol{Z}^n$, whereas an element of N can be expressed as $\lambda_a(t)=(t^{a_i})$, $a=(a_i)\in\boldsymbol{Z}^n$. M and N are canonically dual to each other with respect to the pairing $<, >$:

$$\begin{array}{ccc} M\times N & \longrightarrow & \boldsymbol{Z} \\ \Cup & & \Cup \\ (r, a) & \longrightarrow & <r, a> = \sum r_i a_i. \end{array}$$

Denote $M\otimes\boldsymbol{R}$ and $N\otimes\boldsymbol{R}$ by $M_{\boldsymbol{R}}$ and $N_{\boldsymbol{R}}$ respectively.

Definition (1. 2) (1) An *(affine) torus embedding* is an (affine) scheme X of finite type, endowed with an action of T and containing T as an open orbit.

(2) A *morphism* of two torus embeddings $f: X_1\to X_2$ is a morphism of schemes satisfying the conditions (i) and (ii) :

(i) f induces a group epimorphism $f': T_1\to T_2$ of tori ($T_i\subset X_i$),

(ii) the following diagram commutes :

$$\begin{array}{ccc} T_1\times X_1 & \longrightarrow & X_1 \\ {\scriptstyle f'\times f}\downarrow & & \downarrow {\scriptstyle f} \\ T_2\times X_2 & \longrightarrow & X_2 \end{array}$$

where the horizontal arrows denote the action of tori.

By a *convex rational polyhedral cone* σ (abbr. *c. r. p. cone*) in $M_{\boldsymbol{R}}$ (or $N_{\boldsymbol{R}}$) we mean a set of the form

$$\sigma = \{x\in M_{\boldsymbol{R}}(\text{or } N_{\boldsymbol{R}}) ; l_i(x) \geqq 0, i = 1, \cdots, r\},$$

l_i being linear functionals over $\boldsymbol{Q}$. In what follows, we deal with only c. r. p. cones σ containg no other linear subspaces than 0. Denote by $\check{\sigma}$ the dual cone with respect to the pairing $<, >$.

Definition (1. 3) (1) A *rational partial polyhedral decomposition* of $N_{\boldsymbol{R}}$(abbr. *r. p. p.*

decomposition) is a collection $\mathscr{C}=\{\sigma_\alpha\}$ of c. r. p. cones σ_α in $N_{\boldsymbol{R}}$ such that (i) if $\sigma \prec \sigma_\alpha$, then σ is in $\mathscr{C}$; and (ii) for all σ_α, and σ_β, $\sigma_\alpha \cap \sigma_\beta$ is in $\mathscr{C}$.

(2) A *morphism* of an r. p. p. decomposition $\{\sigma_\alpha\}$ of $N_{\boldsymbol{R}}$ to another $\{\sigma'_\beta\}$ of $N'_{\boldsymbol{R}}$ is a $\boldsymbol{Z}$-homomorphism $\sigma : N \to N'$ of finite cokernel which maps each σ_α into some σ'_β.

For a given c. r. p. cone σ in $N_{\boldsymbol{R}}$, we denote Spec $\boldsymbol{C}[\check{\sigma}\cap M]$ by X_σ, where the ring $\boldsymbol{C}[\check{\sigma}\cap M]$ is by definition the ring generated by monomials w^r, $r\in\check{\sigma}\cap M$, over $\boldsymbol{C}$. Then X_σ is a normal affine torus embedding. If $\sigma_1 \prec \sigma_2$, then X_{σ_1} can be canonically embedded into X_{σ_2}. Through this embedding we identify X_{σ_1} with an open subscheme of X_{σ_2}. Hence, if we are given an r. p. p. decomposition $\mathscr{C}=\{\sigma_\alpha\}$, we have a normal complex space $X_{\mathscr{C}}$ locally of finite type which we call *the torus embedding associated with* $\mathscr{C}$. A morphism between two torus embeddings (locally of finite type) is defined in the same way as in (1. 2). Moreover we have

Theorem (1. 4). *The correspondence* $\{\sigma_\alpha\}\to X_{\{\sigma_\alpha\}}$ *defines a faithful functor from the category of r. p. p. decompositions to that of normal torus embeddings locally of finite type.*

(1. 5) We consider an r. p. p. decomposition $\mathscr{C}=\{\sigma\}$ of $N_{\boldsymbol{R}}(\dim N_{\boldsymbol{R}}=g+1)$ satisfying the following conditions:

(1) there exists a g' (>0) not greater than g such that any σ is of the form

$$\sigma = \{(u, x, y)\in \boldsymbol{R}^{g+1}\,;\, x=0\in \boldsymbol{R}^{g-g'},\ (u,y)\in\sigma'\},\ \sigma'\in\mathscr{C}'$$

where $\mathscr{C}'=\{\sigma'\}$ is an r. p. p. decomposition of $N'_{\boldsymbol{R}}$, $\dim N'_{\boldsymbol{R}}=g'+1$, and $\max(\dim\sigma)=g'$.

(2) $\mathscr{C}$ is invariant by transformations T_k of $N_{\boldsymbol{R}}$ defined by

$$T_k(u)=u,\ T_k(x)=x,\ T_k(y)=y+ku.\quad (k\in \boldsymbol{Z}^{g'})$$

(3) any one-dimensional cone in $\mathscr{C}$ is one of τ_k,

$$\tau_k=\{(u,x,y)\,;\,x=0,\ y=ku,\ u\geqq 0\},\ k\in\boldsymbol{Z}^{g'}.$$

Denote by $\mathscr{C}_0$ the subdecomposition of $\mathscr{C}$ consisting of all one-dimensional cones in $\mathscr{C}$ and $\{0\}$. Let $X_{\mathscr{C}}$ (or $X_{\mathscr{C}_0}$) be normal torus embeddings associated with $\mathscr{C}$ (or $\mathscr{C}_0$) respectively. $X_{\mathscr{C}_0}$ is an open subscheme of $X_{\mathscr{C}}$. We denote by X_σ the normal affine torus embedding associated with σ. We express $X_{(0)}$ as Spec $\boldsymbol{C}[s, s^{-1}, z_i^{\pm 1}, w_j^{\pm 1}]$. Here s corresponds to a character u, and z_i (or w_j) corresponds to the coordinate x_i (or y_j) of x (or of y).

In view of (3), X_σ has a regular projection to $\boldsymbol{C}=$Spec $\boldsymbol{C}[s]$ if $\dim\sigma>0$. Hence we can define

$$P_\sigma = X_\sigma\times_{\boldsymbol{C}} D,\ Q_k=P_{\tau_k}\qquad (\dim\sigma>0)$$
$$P'(\text{or } P_{(0)}) = Q'(\text{or } Q_{(0)}) = X_{(0)}\times_{\boldsymbol{C}^*} D'$$
$$P_{\mathscr{C}} = X_{\mathscr{C}}\times_{\boldsymbol{C}} D = \cup_\sigma P_\sigma,\ Q=X_{\mathscr{C}_0}\times_{\boldsymbol{C}} D=\cup_k Q_k.$$

We denote the canonical projection from P_σ to D (or from $P_{(0)}$ to D') by π. We call

$\pi : P_{\mathcal{C}} \to D$ the *relative compactification* of $\pi : Q \to D$ associated with $\mathcal{C}$.

(1. 6) To close this section, we shall give an example $\mathcal{D}$ satisfying the conditions in (1. 5). Let σ_k be a c. r. p. cone defined by

$$\sigma_k = \{(u, y) \in \boldsymbol{R}^{g+1} \,;\, u - y'_1 \geqq 0, y'_i - y'_{i+1} \geqq 0 \quad (0<i<g), \quad y'_g \geqq 0\},$$

and

$$\sigma^e_k = \{(u, y) \in \boldsymbol{R}^{g+1} \,;\, u - y'_{e(1)} \geqq 0, y'_{e(i)} - y'_{e(i+1)} \geqq 0 \quad (0<i<g), \quad y'_{e(g)} \geqq 0\}$$

where $y'_i = y_i - k_i u$, e a permutation of g letters and $k = (k_i) \in \boldsymbol{Z}^g$.

Let $\mathcal{D}$ be the collection of σ^e_k and their faces. The set-theoretical union of all σ^e_k covers $\boldsymbol{R}^+_0 \times \boldsymbol{R}^g$, where $\boldsymbol{R}^+_0 = \{u \in \boldsymbol{R} \,;\, u \geqq 0\}$. It is easy to check $\mathcal{D}$ satisfies (1)–(3) for $g' = g$. The fiber $\pi^{-1}(0)$ of $P_{\mathcal{D}}$ consists of infinitely many rational varieties, each of which is isomorphic to one and the same Δ. $\Delta = \Delta_g$ is a compact normal torus embedding of finite type, associated with a finite r. p. p. decomposition $\{{}_j\sigma^e$ and their faces$\}$ where ${}_j\sigma^e = \sigma^e_{\underset{j}{1\cdots 1}, 0\cdots 0}$. For instance Δ_2 is isomorphic to a projective plane blown up at three distinct points.

2. Existence of a Quotient

(2. 1) Let H_ε be the universal covering of $D' = \{s \in \boldsymbol{C} \,;\, 0 < |s| < \varepsilon\}$, namely, if we set $s = \boldsymbol{e}(l)$ $(s \in D')$, we have $H_\varepsilon = \{l \in H \,;\, \operatorname{Im} l > -\log \varepsilon / 2\pi\}$. We consider a holomorphic mapping from H_ε to the Siegel upper half plane $\mathfrak{S}_g$ satisfying the unipotency condition : $\tau(l+1) = \tau(l) + B$, for an integral symmetric matrix B. In what follows, we write $\tau(s)$ instead of $\tau(l)$, and express the unipotency condition as

(2. 2) $$\tau(e^{2\pi i} s) = \tau(s) + B.$$

We call a period matrix satisfying (2. 2) a *stable matrix*.

Lemma (2. 3). *A stable matrix $\tau(s)$ can be expressed as*

$$\tau(s) = \tau_0(s) + B \log s / 2\pi i$$

where $\tau_0(s)$ is a holomorphic matrix over D and B is a symmetric positive semi-definite integral matrix.

Proof. Take an integral vector $m (\neq 0)$ and define $f(s) = m\tau(s)^t m$, $g(s) = \boldsymbol{e}(f(s))$. Since $g(s)$ is single-valued and bounded on D' by assumption, $g(s)$ extends to a holomorphic function on D in view of Riemann's theorem. Let a_m be the order of zero of $g(s)$ at $s=0$. Then we can write $g(s) = s^{a_m} g_0(s)$ with a non-vanishing holomorphic function g_0 on D. By taking branches of log suitably, we may assume $f(s) = a_m \log s/2\pi i + \log g_0(s)/2\pi i$. Hence $a_m = mB^t m$. Since a_m is non-negative for any m, B is positive semi-definite. Moreover any component of τ is a rational linear combination of such $f(s)$, hence a sum of an integral multiple of $\log s/2\pi i$ and a holomorphic function over D. Therefore we have $\tau(s) = \tau_0(s) + B_1 \log s/2\pi i$ for a holomorphic matrix τ_0. From this it follows that $B = B_1$. q. e. d.

From now on, we assume $B=\begin{bmatrix}0 & 0\\ 0 & B'\end{bmatrix}$, where B' is positive definite and of rank g'. We remark that any symmetric positive semi-definite integral matrix can be transformed into the above form by an element u of $GL(g, \boldsymbol{Z})$, $B \to uB{}^tu$.

We set

$$(2.4) \qquad \tau(s) = \tau_0(s) + \begin{bmatrix}0 & 0\\ 0 & B'\end{bmatrix} \log s/2\pi i, \quad \tau_0(s) = \begin{bmatrix}\tau_1(s) & \tau_2(s)\\ {}^t\tau_2(s) & \tau_3(s)\end{bmatrix}$$

where τ_1, τ_2 and τ_3 are holomorphic matrices of sizes $(g-g', g-g')$, $(g-g', g')$ and (g', g') respectively.

(2.5) We consider r. p. p. decompositions $\mathcal{C}$ and $\mathcal{C}_0$ of $N_{\boldsymbol{R}}$ ($\dim N_{\boldsymbol{R}}=g+1$) satisfying the conditions (1.5) (1)–(3), and the corresponding toroidal embeddings $P_{\mathcal{C}}$ and Q.

Then we shall define an action of a discrete group Γ isomorphic to $\boldsymbol{Z}^g$ with reference to a stable matrix τ (2.4). Let $\gamma=(n, m)$ ($n \in \boldsymbol{Z}^{g-g'}$, $m \in \boldsymbol{Z}^{g'}$) be an element of Γ.

Define a ring homomorphism S_γ^* from $\boldsymbol{C}[\check{\sigma} \cap M]$ to $\boldsymbol{C}[\check{\sigma}' \cap M]$, $\sigma' = T_{-mB'}(\sigma)$, by

$$S_\gamma^*(z^b) = \boldsymbol{e}(n\tau_1{}^tb + m{}^t\tau_2{}^tb)z^b,$$
$$S_\gamma^*(s^a w^c) = \boldsymbol{e}(n\tau_2{}^tc + m\tau_3{}^tc)s^{a+mB'{}^tc}w^c$$

where $\boldsymbol{e}(u)=\exp(2\pi iu)$ and $w^c=\prod_{i=1}^{g'} w^{c_i}$, c_i being the i-th coefficient of the vector c. The holomorphic mapping S_γ induced by S_γ^* from $P_{\mathcal{C}}$ to $P_{\mathcal{C}}$ is an automorphism over D preserving Q. Then we have

Theorem (2.6). *The action of Γ on $P_{\mathcal{C}}$ (or Q) is properly discontinuous and fixed point free. Hence the quotient space of $P_{\mathcal{C}}$ (or Q) by Γ exists.*

Proof. If $g'=0$, then the assertion is trivial. Assume $g'>0$. Let p be a point of $P_{\mathcal{C}}$ and choose a cone σ such that P_σ contains p. Let the generators of $\check{\sigma} \cap M$ be $\alpha_j=(0, e_j, 0)$ $(1 \leqq j \leqq g-g')$ and $\alpha_{k+g-g'}=(a_k, 0, c_k)$ $(1 \leqq k \leqq N)$, where e_j stands for j-th unit vector of $\boldsymbol{R}^{g-g'}$. We denote the functions corresponding to α_j and $\alpha_{k+g-g'}$ by z_j and ξ_k. We write $\xi_k=s^{a_k}w^{c_k}$. We notice that there exists at least one relation (it is not necessarily unique, but we choose and fix one) of the form

$$(2.7) \qquad \sum_{k=1}^{N} l_{jk}c_k = 0, \quad l_{jk} \geqq 0, \quad l_{kk} > 0, \quad l_{jk} \in \boldsymbol{Z}.$$

Let $C_1=\min(1, |z_j(p)|, |\xi_k(p)| \neq 0)$, $C_2=2\max(1, |\xi_k(p)| \neq 0)$ and $l=\max l_{jk}$. Define $U(p, \varepsilon)$, $\varepsilon=(\varepsilon_1, \varepsilon_2, \varepsilon_3)$, by

$$U(p, \varepsilon) = \{(x, z_j, \xi_k) \in P_\sigma\,;\, |s| < \varepsilon_1, |z_j - z_j(p)| < \varepsilon_2, |\xi_k - \xi_k(p)| < \varepsilon_3\}$$

where $\varepsilon_i \ll 1$.

We want to show $S_{-\gamma}(U(p, \varepsilon)) \cap U(p, \varepsilon) = \phi$ except for a finite number of γ if we choose ε_i sufficiently small. Assume the contrary. Then we may assume there exist infinite sequences s_ν, γ_ν and p_ν such that

(i) $\lim_{\nu \to \infty} s_\nu = 0$, $s_\nu = \pi(p_\nu) \neq 0$

(ii) $\gamma_\nu = (n_\nu, m_\nu) = (\alpha, \beta)\nu + 0(\nu^{1-\delta})$, $(\alpha, \beta) \neq (0, 0)$

(iii) $p_\nu \in U(p, \varepsilon)$, $S_{-\gamma_\nu}(p_\nu) \in U(p, \varepsilon)$

where $\lim_{\nu\to\infty} 0(\nu^{1-\delta})/\nu = 0$.

In view of (iii) we have

$$(2.8) \quad \begin{cases} |z_j(p_\nu) - z_j(p)| < \varepsilon_2, \quad |\xi_k(p_\nu) - \xi_k(p)| < \varepsilon_3, \\ |e(n_\nu \tau_1{}^t e_j + m_\nu{}^t \tau_2{}^t e_j) z_j(p_\nu) - z_j(p)| < \varepsilon_2, \\ |e(n_\nu \tau_2{}^t c_k + m_\nu \tau_3{}^t c_k) s_\nu{}^{mB'^t c_k} \xi_k(p_\nu) - \xi_k(p)| < \varepsilon_3. \end{cases}$$

Hence

$$(2.9) \quad \begin{cases} |e(n_\nu \tau_1^t e_j + m_\nu^t \tau_2{}^t e_j) - 1| < 4C_1^{-1}\varepsilon_2, \\ |e(n_\nu \tau_2^t c_k + m_\nu \tau_3{}^t c_k) s_\nu^{mB'^t c_k} - 1| < 4C_1^{-1}\varepsilon_3 \quad \text{if} \quad \xi_k(p) \neq 0 \\ |e(n_\nu \tau_2^t c_k + m_\nu \tau_3{}^t c_k) s_\nu{}^{d_k + m_\nu B'^t c_k}| < C_2^{lN}\varepsilon_3 \quad \text{if} \quad \xi_k(p) = 0, \end{cases}$$

where $d_k = (\Sigma l_{jk} a_j)/l_{kk} \geqq 0$. The last relation in (2. 9) is obtained by multiplying the last inequality in (2. 8) by the l_{jk}-th power of the inequality

$$|\xi_j(p_\nu)| < |\xi_j(p)| + \varepsilon_3 \ (< C_2)$$

for all j except k. If $\beta \neq 0$, there exists c_k such the $\beta B'^t c_k < 0$. Then we have

$$\lim_{\nu\to\infty} |e(n_\nu \tau_2{}^t c_k + m_\nu \tau_3{}^t c_k) s_\nu^{d_k + mB'^t c_k}| = \infty.$$

Hence we have $\beta = 0$. Since Im $\tau_1 > 0$, we have

$$\lim_{\nu\to\infty} |e(n_\nu \tau_1{}^t e_j + m_\nu{}^t \tau_2{}^t e_j)| = 0 \quad \text{or} \quad \infty,$$

according as α Im $\tau_1{}^t e_j > 0$ or < 0. Hence, if $\alpha \neq 0$, we are led to a contradiction of (2. 9). Consequently $\alpha = 0$, which contradicts (ii). It is obvious that the action of Γ is fixed point free. q. e. d.

(2. 10) Denote by $A_\mathscr{C}$ the quotient $P_\mathscr{C}/\Gamma$, and by $\mathscr{X}$ the quotient Q/Γ. $A_\mathscr{C}$ will be called the *relative compactification* of $\mathscr{X}$ associated with $\mathscr{C}$. $A'_\mathscr{C} = A_\mathscr{C} \times_D D'$ is *a family of principally polarized abelian varieties* whose general fiber has periods $(1, \tau(s))$. $\mathscr{X}' = \mathscr{X} \times_D D' = A'_\mathscr{C}$.

The geometric fiber $\mathscr{X}_0$ of $\mathscr{X}$ at $s=0$ consists of det B' irreducible components. Each component is a principal bundle with fiber $(\boldsymbol{C}^*)^{g-g'}$ over an abelian variety $A_{\tau_1(0)}$ whose periods are 1 and $\tau_1(0)$. Its bundle structure is given by one cocycle in Ext^1 $(A_{\tau_1(0)}, (\boldsymbol{C}^*)^{g-g'}) \cong A_{\tau_1(0)}^{g-g'}$, which is determined by the g'-tuple of the column vectors of $\tau_2(0)$. The projection π from $A_\mathscr{C}$ to D is proper if and only if $\mathscr{C}$ covers $\boldsymbol{R}_0^+ \times \boldsymbol{R}^{g'}$. The subset $A_\mathscr{C} - \mathscr{X}$ is of codimension two in $A_\mathscr{C}$.

3. The Néron Model and a Proof of Minimality

(3. 1) In this section we shall define a principal homogeneous space (abbr. p. h. s.) in the restricted sense in nearly the same way as in [11], and prove the minimality of $(\mathscr{X}, \pi)$ among p. h. s.'s satisfying the condition (3. 3) below.

Definition (3. 2). (1) A *principal homogeneous space* $(\mathcal{Y}, \varpi)$ *over* D is a complex manifold $\mathcal{Y}$ and a holomorphic mapping ϖ from $\mathcal{Y}$ to D satisfying the following :

(i) $\mathcal{Y}'=\mathcal{Y}\underset{D}{\times}D'$ is a family of principally polarized abelian varieties with a stable matrix $\tau(s)$ (2. 9).

(ii) There exist a section e and a holomorphic mapping φ (or ψ) from $\mathcal{Y}\underset{D}{\times}\mathcal{Y}$ (or $\mathcal{Y}$) to $\mathcal{Y}$ over D such that

(a) $\varphi\circ(1, \varphi)=\varphi\circ(\varphi, 1)$, $\quad\varphi\circ(1, e)=\varphi\circ(e, 1)=1$

(b) $\varphi\circ(1, \psi)=\varphi\circ(\psi, 1)=e$

(c) (commutativity) $\varphi\circ(\varphi, \varphi\circ(\psi, \psi))=e$

where 1 indicates the identity mapping of $\mathcal{Y}$.

(iii) $\mathcal{Y}_s=\varpi^{-1}(s)$ $(s\in D)$ is a complex Lie group with respect to the group law induced by e, 1, φ and ψ.

(2) A *morphism* of two p. h. s.'s $(\mathcal{Y}_1, \varpi_1)$ and $(\mathcal{Y}_2, \varpi_2)$ is a holomorphic mapping from $\mathcal{Y}_1$ to $\mathcal{Y}_2$ satisfying

(i) $\varpi_1=\varpi_2\circ f$

(ii) $\varphi_2\circ(f, f)=f\circ\varphi_1$.

We consider the following condition on a p. h. s. $(\mathcal{Y}, \varpi)$:

(3. 3) $\mathcal{Y}_0=\varpi^{-1}(s)$ consists of at most a finite number of connected components.

It is easy to check $(\mathcal{X}, \pi)$ is a p. h. s. satisfying (3. 3). It is conjectured that any p. h. s. satisfies (3. 3) automatically. Now we have

Theorem (3. 4). *Let* $(\mathcal{Y}, \varpi)$ *be a p. h. s. over* D *satisfying* (3. 3). *Any morphism from* $\mathcal{Y}\underset{D}{\times}D'$ *to* $\mathcal{X}'$ *can be uniquely extended to a morphism from* $\mathcal{Y}$ *to* $\mathcal{X}$.

Definition (3. 5). $(\mathcal{X}, \pi)$ is called (*analytic*) *Néron model* of $(\mathcal{X}', \pi')$ where π' is the restriction of π to $\mathcal{X}'=\mathcal{X}\underset{D}{\times}D'$.

Before entering into the proof of Theorem (3. 4), we prove a lemma.

Lemma (3. 6). *There exists a basis* $\omega_1, \cdots, \omega_{g_1}$ *of holomorphic relative* 1-*forms on* $\mathcal{Y}$, *where* $g_1=\dim\mathcal{Y}-1$.

Proof. The fiber $\mathcal{Y}_0$ at $s=0$ is an abelian complex Lie group equipped with the structure induced by e, φ and ψ, hence is isomorphic to a complex torus extended by an algebraic torus and an additive complex vector group modulo a discrete group. Let N be the group $\mathcal{Y}_0/\mathcal{Y}_0^0$ where $\mathcal{Y}_0^0$ indicates the identity component of $\mathcal{Y}_0$. By assumption N is of finite order. Any connected component of $\mathcal{Y}_0$ is isomorphic to $\mathcal{Y}_0^0$. So, if we choose a global coordinate z on $\mathcal{Y}_0^0$, then $z^{(a)}=(z, a)$ $(a\in N)$ is a global coordinate on $\mathcal{Y}_0^a$, the a-th connected component of $\mathcal{Y}_0$. The group law of $\mathcal{Y}_0$ will be written additively ; $(z, a)+(z', a')=(z+z', a+a')$.

On $\mathcal{Y}'$ there exists a global coordinate $\zeta=(\zeta_\nu)$. (z, s) forms a local coordinate system in a neighborhood U_0 of $\mathcal{Y}_0^0$, because ϖ is smooth. Hence $\zeta_\nu=\zeta_\nu(z, s)$ is holomorphic in z and s in U_0. Since $\zeta=z=0$ on the section e, ζ_ν is holomorphic at $z=s=0$, hence in U_0 because of homogenity of $\mathcal{Y}_0^0$. The translation by $(0, a)$ on

$\mathcal{Y}_0$ is a specialization of the translation by a section $e(s, a)$ near the origin of D. Hence, if we put $\zeta^{(a)}(z, s)=\zeta(z, s)+e(s, a)$, then $\zeta^{(a)}(z, 0)=z^{(a)}$. Thus $\zeta^{(a)}$ is holomorphic in $z^{(a)}$ and s in a neighborhood U_a of $\mathcal{Y}_0^a$, where U_a is a translate of U_0 by $e(s, a)$. Therefore we have a relative 1-form $d\zeta_\nu^{(a)}$ on $U_a \cup \mathcal{Y}'$, which is equal to $d\zeta_\nu$ on $\mathcal{Y}'$. Consequently $\omega_\nu = d\zeta_\nu^{(a)}$ on $U_a \cup \mathcal{Y}'$ defines a relative 1-form on $\mathcal{Y}$ since $\mathcal{Y}=(\bigcup_{a\in N} U_a)\cup \mathcal{Y}'$. q. e. d.

Proof of Theorem (3. 4). Let σ be a stable matrix associated with $\mathcal{Y}'$. Let $\sigma(e^{2\pi i}s)= \sigma(s)+B_1$, $B_1=\begin{bmatrix} 0 & 0 \\ 0 & B_1' \end{bmatrix}$, $B_1'>0$. Denote by $(\mathcal{X}_1, \pi_1)$ the p. h. s. associated with σ (2.1 0). Let f be a given morphism from $\mathcal{Y}'$ to $\mathcal{X}'$. Our proof of Theorem (3. 4) is divided into two steps.

(First step) We shall prove that the canonical isomorphism from $\mathcal{Y}'$ to $\mathcal{X}_1'$ can be extended to a unique morphism from $\mathcal{Y}$ to $\mathcal{X}_1$. In view of Lemma (3. 6), we can choose relative 1-forms $\omega_1, \cdots, \omega_{g_1}$ and relative one cycles $\gamma_1, \cdots, \gamma_{2g_1}$ of $\varpi^{-1}(V)$ for a contractible open set V in D' such that

$$\int_{\gamma_\mu} \omega_\nu = \delta_{\nu\mu}, \quad \int_{\gamma_{g_1+\mu}} \omega_\nu = \sigma_{\nu\mu} \qquad 1\leqq \nu,\ \mu \leqq g_1.$$

The canonical morphism from $\mathcal{Y}'$ to $\mathcal{X}_1'$ is given by the Albanese mapping. Let $f_\nu(z^{(a)}, s)=\boldsymbol{e}\left(\int_{e(s)}^{\zeta} \omega_\nu\right)$ for a point $\zeta=\zeta(z^{(a)}, s)$ in $U_a \cap \mathcal{Y}'$. There exists a local section $e(s, a)$ passing through $\mathcal{Y}_0^a$ such that $e(0, a)$ is the point $(z^{(a)}, s)=(0, 0)$. Then f_ν is non-vanishing and holomorphic on $U_a \cup \mathcal{Y}'$. Since the order of N is finite, there exists a positive integer n such that $e_1(s)=ne(s, a)$ is a section passing through $\mathcal{Y}_0^0$. However we notice that

$$n\int_{e(s)}^{e(s,a)} \omega_\nu = \int_{e(s)}^{e_1(s)} \omega_\nu + \int_\gamma \omega_\nu \tag{3. 7}$$

where γ is a relative 1-cycle and the second integration is performed along a path contained in U_0. Hence the integration (3. 7) is the sum of a holomorphic function and an integral multiple of $\log s/2\pi i$.

On the other hand, f_ν is single-valued with respect to s. In particular $f_\nu(1\leqq \nu \leqq g_1-g_1')$ is always a non-vanishing holomorphic function on U_a for any a, while $f_\nu(z^{(a)}, s)=s^{b_\nu}F_\nu(z^{(a)}, s)$ for a non-vanishing holomorphic function $F_\nu(z^{(a)}, s)$ on U_a $(\nu > g_1-g_1')$.

If we define

$$F^*z_\nu = \boldsymbol{e}\left(\int_{e(s)}^{\zeta} \omega_\nu\right) \ (\nu \leqq g_1-g_1'),$$
$$F^*w^{(b)}_{\nu-g'_1+g_1} = F_\nu(z^{(a)}, s) \ \ (\nu > g_1-g_1'),\ b=(b_\nu),$$

then F is an extension of f. That F is well-defined and the uniqueness of the extension are obvious.

(Second step) Now we shall prove that a morphism f from $\mathcal{X}_1'$ to $\mathcal{X}'$ can be uniquely extended to a morphism from $\mathcal{X}_1$ to $\mathcal{X}$. Choose global coordinates η (or ζ) of $\mathcal{X}_1$ (or $\mathcal{X}$) in such a way that the identity section e is given by $\eta=0$ (or $\zeta=0$)

respectively. The period matrices of $\mathcal{X}_1$ and $\mathcal{X}$ are given by

$$\sigma(s) = \sigma_0(s) + \begin{bmatrix} 0 & 0 \\ 0 & B_1' \end{bmatrix} \log s/2\pi i, \text{ and}$$

$$\tau(s) = \begin{bmatrix} \tau_1 & \tau_2 \\ \tau_3 & \tau_4 \end{bmatrix} + \begin{bmatrix} 0 & 0 \\ 0 & B' \end{bmatrix} \log s/2\pi i$$

where τ_1, τ_2, τ_3 and τ_4 stand for $(g-g')\times(g-g')$, $(g-g')\times g'$, $g'\times(g-g')$ and $g'\times g'$ matrices.

The morphism f from $\mathcal{X}_1'$ to $\mathcal{X}'$ can be expressed as $\zeta=\eta A(s)$ with a holomorphic $g_1\times g$ matrix $A(s)$ on D. From this it follows that $A(s)=N_1+N_2\tau(s)$, $\sigma(s)A(s)=N_3+N_4\tau(s)$, where N_i indicates a $g_1\times g$ integral matrix. Then we have

$$N_2B = 0, \quad N_4B = B_1A(s). \tag{3. 8}$$

Let $N_i = \begin{bmatrix} N_{i1} & N_{i2} \\ N_{i3} & N_{i4} \end{bmatrix}$, where N_{i1}, N_{i2}, N_{i3} and N_{i4} stand for $(g_1-g_1')\times(g-g')$, $(g_1-g_1')\times g'$, $g_1'\times(g-g')$ and $g_1'\times g'$ matrices. Since $B'>0$, $N_{22}=N_{24}=0$. By virtue of (3. 8), we have

$$\begin{bmatrix} 0 & N_{42}B' \\ 0 & N_{44}B' \end{bmatrix} = \begin{bmatrix} 0 & 0 \\ 0 & B_1' \end{bmatrix} \begin{bmatrix} N_{11}+N_{21}\tau_1, & N_{12}+N_{21}\tau_2 \\ N_{13}+N_{23}\tau_1, & N_{14}+N_{23}\tau_2 \end{bmatrix}$$
$$= \begin{bmatrix} 0 & 0 \\ B_1'(N_{13}+N_{23}\tau_1), & B_1'(N_{14}+N_{23}\tau_2) \end{bmatrix}.$$

It follows that $N_{13}=N_{23}=0$ since Im $\tau_1>0$. Consequently we have $A(s)=\begin{bmatrix} N_{11}+N_{21}\tau_1, & N_{12}+N_{21}\tau_2 \\ 0 & N_{14} \end{bmatrix}$. We define a mapping F from $\mathcal{X}_1$ to $\mathcal{X}$ as follows:

$$\begin{aligned} &Q_\mu \to Q_\nu \qquad \nu = \mu(0\ N_{14}), \quad \mu \in \boldsymbol{Z}^{g_1}, \ \nu \in \boldsymbol{Z}^{g'}, \\ &F^*(\zeta') = \eta'(N_{11}+N_{21}\tau_1) \\ &F^*(w^{(\nu)}) = \boldsymbol{e}(\eta'(N_{12}+N_{22}\tau_2)) * \boldsymbol{e}(\eta''^{(\mu)}N_{14}) \end{aligned}$$

where $\eta=(\eta', \eta'')$, $\zeta=(\zeta', \zeta'')$, η', η'', ζ' and ζ'' are g_1-g_1', g_1', $g-g'$, g' vectors, $w^{(\nu)}=s^{-\nu}*\boldsymbol{e}(\zeta'')$, $\eta''^{(\mu)}=\eta''-\mu\log s/2\pi i$. Here $x*x'$ denotes componentwise multiplication of vectors x and x', and $s^{-\nu}=(s^{-\nu_1}, \cdots, s^{-\nu_n})$, $\nu=(\nu_1, \cdots, \nu_n)$, $\boldsymbol{e}(\zeta'')=(\boldsymbol{e}(\zeta_j''))$. Then F is an extension of f. That F is well-defined and the uniqueness of the extension are obvious. q. e. d.

4. A Family of Curves of Genus Two

Definition (4. 1). A *family of curves of genus two* (or simply a family) (X, π) is a complex manifold X of dimension 2 free from exceptional curves of the first kind, given with a holomorphic mapping π from X onto D such that

(1) π is proper, and smooth over D',

(2) for any $s\in D'$, $X_s=\pi^{-1}(s)$ is a smooth curve of genus two.

Definition (4. 2). The geometric fiber X_0 at $s=0$ is called the *singular fiber* of the family (X, π).

Given a family (X, π) we have a holomorphic mapping from D' to the Siegel space $\mathfrak{S}_2^*$ by assigning to each $s \in D'$ the normalized period $\tau(s)$ of the jacobian variety of X_s. We denote also by $\tau(s)$ the holomorphic mapping from the universal covering H_ε of D' to $\mathfrak{S}_2$ (2. 1).

If X_0 is a reduced curve with at worst ordinary double singularities, then the period matrix $\tau(s)$ is stable [1]. $GL(2, \boldsymbol{Z})$ operates on the space of all positive semi-definite integral matrices by $B \to uB{}^tu$, $u \in GL(2, \boldsymbol{Z})$. It is a classical result that any symmetric positive semi-definite integral matrix B is $GL(2, \boldsymbol{Z})$-equivalent to one of the following :

$$(4.\ 3) \qquad \text{(I)} \begin{bmatrix} 0 & 0 \\ 0 & 0 \end{bmatrix} \qquad \text{(II)} \begin{bmatrix} 0 & 0 \\ 0 & p \end{bmatrix} \quad (p>0)$$
$$\text{(III)} \begin{bmatrix} p & 0 \\ 0 & q \end{bmatrix} \quad (p, q>0) \qquad \text{(IV)} \begin{bmatrix} p+r & r \\ r & q+r \end{bmatrix} \quad (p, q, r>0)$$

By replacing $\tau(s)$ by $u\tau(s){}^tu$, we consider only stable matrices $\tau(s)$ with B in the above list. In the rest of this paper, we devote ourselves to construction of a family of reduced curves of genus two with at worst ordinary double singularities. To this end, we prove two more lemmas.

A matrix $\tau \in \mathfrak{S}_2$ determines a principally polarized abelian variety J. Then a theta function of the abelian variety J is defined by

$$(4.\ 4) \qquad \Theta(\tau, z+a) = \Sigma_\gamma \boldsymbol{e}(\gamma\tau{}^t\gamma/2+\gamma{}^t(z+a))$$

where $\gamma \in \boldsymbol{Z}^2$, $z=(z_1, z_2)$, $a=(a_1, a_2) \in \boldsymbol{C}^2$.

Lemma (4. 5). *Assume J to be the jacobian variety of a curve C of genus two. Then the equation $\Theta(\tau, z+a)=0$ defines a curve on J isomorphic to C.*

Proof. See Weil [14] Satz 2.

Lemma (4. 6). *Let τ be a stable matrix* (2. 4), *and a be the diagonal vector of B, $a_i=b_{ii}$. Define $\mu(k)=\min(\gamma B{}^t\gamma+(a+2k){}^t\gamma)/2$ where γ runs over $\boldsymbol{Z}^g$, $k=(0, k')$, $k' \in \boldsymbol{Z}^{g'}$, $g'=$rank B. Then*

(1) $s^{-\mu(k)}\Theta(\tau(s), z+a \log s/4\pi i)$ *is absolutely and uniformly convergent in the wide sense on Q_k* (1. 5),

(2) $\Theta(\tau, z+a \log s/4\pi i)$ *can extended to a meromorphic function on P_σ* (1. 5).

Proof. We set $\Theta_1(s, z)=\Theta(\tau(s), z+a \log s/4\pi i)$. Θ_1 is single-valued on P'. The second assertion follows directly from (1). In fact, since P_σ is Cohen-Macaulay ([3] p. 52) and $P_\sigma - Q$ is of codimension two, any holomorphic function can be extended to P_σ. On the other hand, Θ_1 is holomorphic on P' and in view of (1), for any point p of P_σ, there exists an integer μ and a neighborhood U of p such that $s^\mu\Theta_1$ is holomorphic on $(P_\sigma - Q) \cap U$. Hence $s^\mu\Theta_1$ is holomorphic on U, and consequently Θ_1 is meromorphic on P.

We shall prove (1). By choosing D smaller if necessary, we take s_0 in such a way that

$$\operatorname{Im} \tau_0(s)+B \log s_0/2\pi i > 2\mu\cdot 1_g \qquad \text{on } D,$$

where μ is a positive constant. We also choose a positive constant δ such that $B > 2\pi\delta\cdot 1_{g'}$. Then we have

$$\begin{aligned}&\mathrm{Im}\,(\gamma\tau^t\gamma/2+\gamma^t(a+2k)\log s/4\pi i+\gamma^t z)\\ &\geqq \mu\gamma^t\gamma+\delta n^t n(-\log|s/s_0|)+\gamma^t(a+2k)\log|s|/4\pi+\gamma^t(\mathrm{Im}\ z)\\ &\geqq \mu_1 n^t n+\delta_1 m^t m(-\log|s/s_0|),\end{aligned}$$

except for at most finitely many $\gamma=(n, m)$, if $|z-z_0|<\varepsilon_1$ and μ_1 and δ_1 are chosen sufficiently small. By taking D smaller we may assume there exists a positive constant δ_2 such that $\delta_2<-\log|s/s_0|$ on D. Then we conclude

$$|s^{-\mu(k)}\Theta_1| \leqq |\text{ finite sum in } s^{-\mu(k)}\Theta_1|+\textstyle\sum_{n,m}\exp(-2\pi(\mu_1 n^t n+\delta_1\delta_2 m^t m)),$$

which proves (1). q. e. d.

5. The Case $B=\begin{bmatrix}0 & 0\\ 0 & p\end{bmatrix}$ $(p>0)$

Let $\tau(s)$ be a stable matrix (2. 3), in which $B=\begin{bmatrix}0 & 0\\ 0 & p\end{bmatrix}$, $\tau_0=\begin{bmatrix}\tau_1 & \tau_2\\ \tau_2 & \tau_3\end{bmatrix}$, $\tau_i=\tau_i(s)$. $\tau_i(s)$ is holomorphic in s on D. We distinguish two cases according as $\tau_2(0)\neq 0$ or $\tau_2(0)=0$. In both cases, we take an r. p. p. decomposition $\mathcal{C}=\{\sigma_k, \tau_k, \{0\}\}$ defined by

$$\begin{aligned}\sigma_k &= \{(u, 0, y)\in \boldsymbol{R}^3;\ (1+k)u-y\geqq 0,\quad -ku+y\geqq 0\},\\ \tau_k &= \{(u, 0, y)\in \boldsymbol{R}^3;\ u\geqq 0,\ y=ku\},\quad k\in \boldsymbol{Z}.\end{aligned}$$

We consider the relative compactification P $(=P_{\mathcal{C}})$ and A $(=A_{\mathcal{C}})$ with the period matrix τ, (1. 5) and (2. 10).

(1) *The case* $\tau_2(0)\neq 0$. For simplicity, we assume τ_1 and τ_2 are constant and $\tau_3\equiv 0$. We set

$$\begin{aligned}\Theta(s, z, w) &= \Theta(\tau(s), \tilde{z}+a\log s/4\pi i),\\ &= \sum_{n,m} s^{p(m^2-m)}\boldsymbol{e}(n^2\tau_1/2+nm\tau_2+nz)w^m\end{aligned}\tag{5. 1}$$

where $\tilde{z}=(z_1, z_2)$, $z=z_1$ and $w=\boldsymbol{e}(z_2)$, and $a=(0, -p)$.

Let Y' be the divisor of P' defined by $\Theta=0$, Y the closure of Y' in P, $Y_0=Y\cap(P)_0$ Our problem is to find explicit defining equations of Y and Y_0. First we shall examine Y_0 on Q. Q is covered by Q_k $(k\in\boldsymbol{Z})$, where $Q_k=\mathrm{Spec}\ \boldsymbol{C}[x, \boldsymbol{e}(\pm z), w_k^{\pm 1}]\times_{\boldsymbol{C}} D$, $w_k=s^{-k}w$. On Q_k we have an expression for $s^{-\mu(k)}\Theta$ as

$$s^{-\mu(k)}\Theta(s, z, w) = \sum s^{p(m^2-m)/2+km-\mu(k)}\boldsymbol{e}(n^2\tau_1/2+mn\tau_2)(w_k)^m.\tag{5. 2}$$

The convergence of the expansion (5. 2) is absolute and uniform in s and $s^{-k}w$ (in the wide sense) in view of Lemma (4. 6). As s approaches the origin, the limit of (5. 2) is given by the sum of termwise limits. Hence we have

$$\lim_{s\to 0} s^{-\mu(k)}\Theta(s, z, w) = \begin{cases}(\theta(\tau_1, z+\tau_2)+\theta(\tau_1, z+2\tau_2)w_{-1})w_{-1} & (p=1, k=-1)\\ \theta(\tau_1, z+\tau_2)w_{-1} & (p>1, k=-1)\\ \theta(\tau_1, z)+\theta(\tau_1, z+\tau_2)w_0 & (k=0)\\ \theta(\tau_1, z) & (0<k<p)\\ \theta(\tau_1, z)+\theta(\tau_1, z-\tau_2)w_p^{-1} & (k=p)\end{cases}\tag{5. 3}$$

where $\theta(\tau_1, z)=\sum_n e(n^2\tau_1/2+nz)$. Since w_{-1} does not vanish on Q_{-1}, $\theta(\tau_1, z+\tau_2)+\theta(\tau_1, z+2\tau_2)w_{-1}$ or $\theta(\tau_1, z+\tau_2)$ gives a defining equation of Y_0 on $(Q_{-1})_0$. $(P_{\sigma_k})_0=\operatorname{Spec} \boldsymbol{C}[e(\pm z), w_k, w_{k+1}^{-1}]/(w_k w_{k+1}^{-1}) \underset{C}{\times} D$ is covered by U_k and V_k defined by

$$U_k=\operatorname{Spec} \boldsymbol{C}[e(\pm z), w_k], \quad V_k=\operatorname{Spec} \boldsymbol{C}[e(\pm z), w_{k+1}^{-1}].$$

$(Q_k)_0$ (or $(Q_{k+1})_0$) is an open dense subset in U_k (or V_k) respectively. Hence Y_0 is given by the same equations on $(P_{\sigma_k})_0$ $(0\leqq k<p)$.

On the other hand, Y_0 on $(P_{\sigma_{-1}})_0$ is given by the equations:

$$(5.4)\qquad \begin{cases} \theta(\tau_1, z)w_0^{-1}+\theta(\tau_1, z+\tau_2)=0 & \text{on } V_{-1} \\ \theta(\tau_1, z+\tau_2)+\theta(\tau_1, z+2\tau_2)w_{-1}=0 & \text{on } U_{-1} \quad (p=1) \\ \theta(\tau_1, z+\tau_2)=0 & \text{on } U_{-1} \quad (p>1). \end{cases}$$

Since A is covered by $P_{\sigma_k}(-1\leqq k\leqq p-1)$, the description of Y_0 is now complete. Next we shall examine Y. Let Θ_k be the holomorphic function on P_{σ_k} defined by

$$\Theta_k(z, w_k, w_{k+1}^{-1})=\sum e(n^2\tau_1/2+nm\tau_2+nz)(w_k)^{A(m)}(w_{k+1}^{-1})^{B(m)}(0\leqq k\leqq p-1),$$
$$\Theta_{-1}(z, w_{-1}, w_0^{-1})=\sum e(n^2\tau_1/2+mn\tau_2+nz)(w_{-1})^{A(m)}(w_0^{-1})^{B(m)+1},$$

where $A(m)=p(m^2-m)/2+(k+1)m$, $B(m)=p(m^2-m)/2+km$. We notice $\Theta_k=\Theta(0\leqq k\leqq p-1)$, $\Theta_{-1}=(w_0)^{-1}\Theta$ on P'. Now Θ_k and Θ_{-1} give defining equations of Y on P_{σ_k} and $P_{\sigma_{-1}}$. In fact, $\lim_{s\to 0}\Theta_k$ coincides with the equations in (5.3) or (5.4) according as $0\leqq k\leqq p-1$ or $k=-1$. Outside Y_0, Θ_k gives also a defining equation of Y in view of Lemma (4.5).

The remaining problem is to check that Y is smooth. First we examine the problem on Q. On Q_k, Y is defined by

$$\Theta(s, z, w)=0 \qquad (0\leqq k\leqq p-1).$$

We have

$$\begin{aligned} &(\partial\Theta/\partial z)(0, z, w)=\theta'(\tau_1, z)+\theta'(\tau_1, z+\tau_2)w \qquad (k=0) \\ &(\partial\Theta/\partial w)(0, z, w)=\theta(\tau_1, z+\tau_2), \quad w=w_0 \\ &(\partial\Theta/\partial z)(0, z, w)=\theta'(\tau_1, z) \qquad (0<k\leqq p-1) \\ &(\partial\Theta/\partial w_k)(0, z, w)=0 \qquad \text{where } \theta'(\tau_1, z)=\partial\theta(\tau_1, z)/\partial z. \end{aligned}$$

As is well-known, $\theta(\tau_1, z)$ has only one simple zero at $z=(1+\tau_1)/2$ on the elliptic curve A_{τ_1}. At a point of $(Q_k)_0$, $\Theta(0, z, w)$, $(\partial\Theta/\partial z)(0, z, w)$ and $(\partial\Theta/\partial w)(0, z, w)$ do not vanish simultaneously, hence Y is smooth in Q. Let $\mathfrak{p}_k$ be the point of Y defined by $w_k=w_{k+1}^{-1}=0$. Then we have

$$\begin{aligned} &(\partial\Theta_k/\partial z)(z, 0, 0)=\theta'(\tau_1, z) \qquad (0\leqq k\leqq p-1) \\ &(\partial\Theta_{-1}/\partial z)(z, 0, 0)=\theta'(\tau_1, z+\tau_2) \qquad (k=-1), \end{aligned}$$

which proves the smoothness of Y at $\mathfrak{p}_k$.

Let C_k be the closure of $Y_0\cap Q_k$. C_k is a non-singular elliptic curve if $k\equiv 0 \bmod p$, or a non-singular rational curve if $k\not\equiv 0 \bmod p$. C_k and C_{k+1} meet at $\mathfrak{p}_k$ transversally. $\Gamma(=\boldsymbol{Z}^2)$ transforms C_k (or $\mathfrak{p}_k$) into $C_{k'}$ (or $\mathfrak{p}_{k'}$) if $k\equiv k' \bmod p$.

Let X be the image of Y by the canonical projection from P_σ to A_σ. X_0 consists of p irreducible components, one of which is a smooth elliptic curve and the others are smooth rational curves if $p>1$. If $p=1$, X_0 is an elliptic curve with one ordinary double point. In fact, the points $\mathfrak{p}_{p-1}$ and $\mathfrak{p}_{-1}$ are given by

$$\mathfrak{p}_{p-1} = ((1+\tau_1)/2, 0, 0) \quad \text{in } P_{\sigma_{p-1}}$$
$$\mathfrak{p}_{-1} = ((1+\tau_1)/2-\tau_2, 0, 0) \quad \text{in } P_{\sigma_{-1}},$$

and they are identified by the action of Γ. In particular, if $p=1$, X_0 is obtained by identifying two points $(1+\tau_1)/2$ and $(1+\tau_1)/2-\tau_2$ of A_{τ_1}. The self-intersection number C_k^2 is -2 if $p>1$.

Finally we shall give the configuration for X_0 in the relative compactification $A_{\mathfrak{S}}$ of the corresponding Néron model.

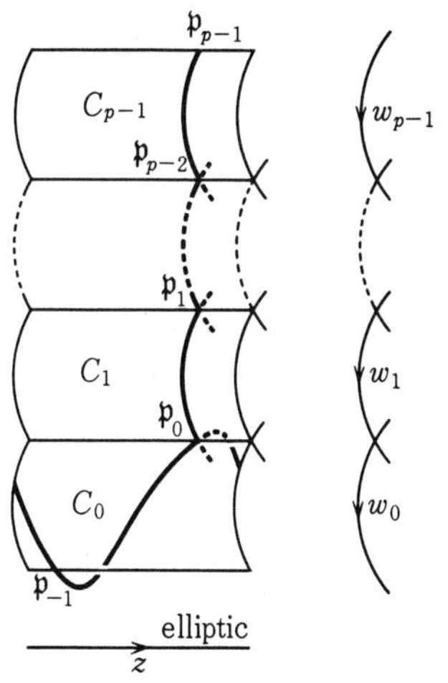

(2) *The case* $\tau_2(0)=0$. For simplicity, we assume $\tau_1(s)=\tau_1$ (constant), $\tau_2(s)=s^l$ $(l>0)$, $\tau_3(s)\equiv 0$. We set

(5.5) $$\Theta(s, z, w) = \sum s^{p(m^2-m)/2} e(n^2\tau_1/2+nms^l+nz)w^m.$$

With the same notation as in (1), Y_0 is given by

(5.6) $$\begin{cases} \theta(\tau_1, z)(1+w) = 0 & \text{on } Q_0, \\ \theta(\tau_1, z) = 0 & \text{on } Q_k \quad (0<k<p). \end{cases}$$

$Y_0 \cap Q_0$ consists of two smooth curves C_0 and C_0' defined respectively by $\theta(\tau_1, z)=0$ and $1+w=0$. If we denote their closures by the same letters, C_0 is a smooth rational curve, and C_0' is a smooth elliptic curve with periods $1, \tau_1$. They meet at only one point $\mathfrak{p}$: $(s, z, w_0)=(0, (1+\tau_1)/2, -1)$. C_k and C_{k+1} meet at $\mathfrak{p}_k$: $(s, w_k, w_{k+1}^{-1})=(0, 0, 0)$ transversally. C_0' does not meet C_k $(k\neq 0)$. The defining equations of Y are given in the same way as in (1) and the smoothness of it except at $\mathfrak{p}$ can be similarly proven.

Now we shall examine the singularity of Y at $\mathfrak{p}$. We have

$$(\partial\Theta/\partial s)(0, z, w) = \begin{cases} -\theta'(\tau_1, z) \neq 0, & (l=1) \\ 0, & (l>1) \end{cases}$$
$$(\partial\Theta/\partial z)(0, z, w) = (\partial\theta/\partial w)(0, z, w) = 0 \quad \text{at } \mathfrak{p}.$$

Near $\mathfrak{p}$, Θ has an expansion of the following form:

$$\Theta = \theta(\tau_1, z)\vartheta(s, w) + s^l w\theta'(\tau_1, z)\vartheta'(s, w) + s^{2l}(\cdots)$$

where $\vartheta(s, w) = \sum_m s^{p(m^2-m)} w^m$, $\vartheta'(s, w) = \partial\vartheta(s, w)/\partial w$. We put $u = \theta(\tau_1, z)$, $v = \vartheta(s, w)$. Then s, u and v form a system of local coordinates of $P_{\mathscr{C}}$ at $\mathfrak{p}$ because $\theta'(\tau_1, (1+\tau_1)/2) \neq 0$, $\vartheta'(0, -1) \neq 0$. By modifying s slightly, Y becomes analytically isomorphic to $uv - s^l = 0$ at $\mathfrak{p}$. If $l > 1$ we obtain a nonsingular model $\tilde{X}$ of $X = Y/\Gamma$ by replacing $\mathfrak{p}$ by a chain of $(l-1)$ rational curves. It is easy to check $\tilde{X}$ is minimal.

In conclusion $\tilde{X}_0$ has a configuration of type (5. 7), and Y_0 has a configuration of type (5. 8) in $A_{\mathscr{C}}$.

(5. 7)

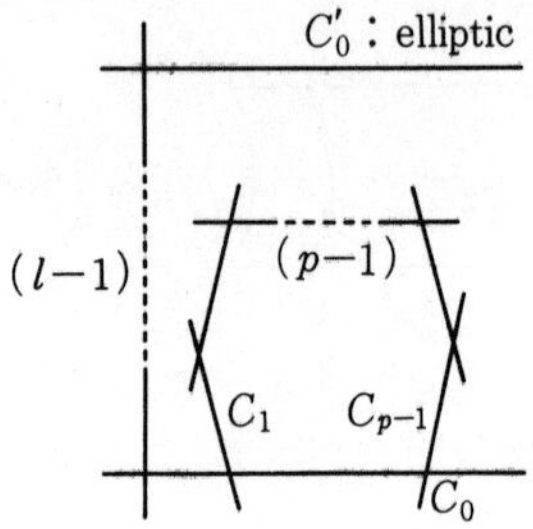

(5. 8)

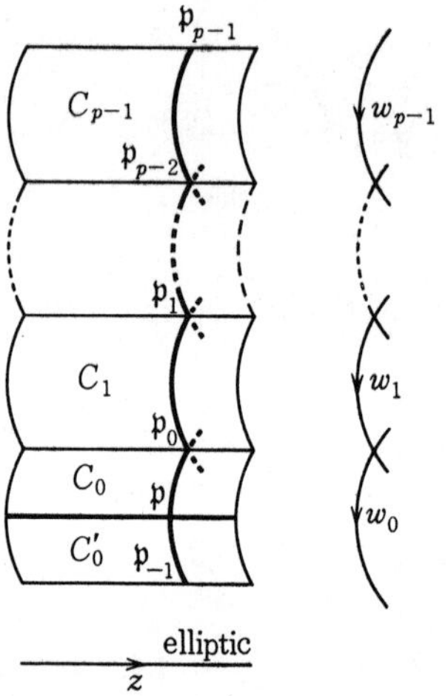

6. The Case $B = \begin{bmatrix} p & 0 \\ 0 & q \end{bmatrix}$ $(p, q > 0)$

Let $\tau(s) = \begin{bmatrix} \tau_1(s) & \tau_2(s) \\ \tau_2(s) & \tau_3(s) \end{bmatrix} + \begin{bmatrix} p & 0 \\ 0 & q \end{bmatrix} \log s/2\pi i$, $\tau_i(s)$ being holomorphic on D, $p \geqq q$. We distinguish two cases according as $\tau_2(0) \neq 0$ or $\tau_2(0) = 0$.

As an r. p. p. decomposition $\mathscr{C}$ we take the following:

$$\mathscr{C} = \{\sigma_k\ (k \in \boldsymbol{Z}^2) \text{ and their faces}\}$$

$$\sigma_k = \{(u, y_1, y_2) \in \boldsymbol{R}^3;\ (1+k_i)u - y_i \geqq 0,\ -k_i u + y_i \geqq 0 \quad (i=1, 2)\}.$$

$\mathscr{C}$ satisfies the conditions (1. 5) (1)-(3). $(P_{\mathscr{C}})_0$ consists of Δ_k ($k \in \boldsymbol{Z}^2$), each of which is by definition the closure of $(Q_k)_0$ and isomorphic to $\boldsymbol{P}_1 \times \boldsymbol{P}_1$.

(1) *The case* $\tau_2(0) \neq 0$. For simplicity we assume $\tau_1 = \tau_3 = 0$, $\tau_2(s) = \tau_2$ (constant). We set

(6. 1)
$$\begin{aligned} \Theta(s, w_1, w_2) &= \Theta(s, z + a \log s/4\pi i) \\ &= \sum s^{p(n^2-n)/2+q(m^2-m)/2} \boldsymbol{e}(nm\tau_2) w_1^n w_2^m, \end{aligned}$$

where $w_i = \boldsymbol{e}(z_i)$, and $a = (-p, -q)$.

Then we have

(6. 2)
$$\lim_{s \to 0} \Theta(s, w_1, w_2) = \begin{cases} 1 + w_1 + w_2 + \boldsymbol{e}(\tau_2) w_1 w_2, & k_1 = k_2 = 0 \\ 1 + w_2, & 0 < k_1 < p,\ k_2 = 0 \\ 1 + w_1, & k_1 = 0,\ 0 < k_2 < q \\ 1, & 0 < k_1 < p,\ 0 < k_2 < q. \end{cases}$$

We define Y' by the equation $\Theta(s, w_1, w_2) = 0$ on P', and let Y be its closure in $P_{\mathscr{C}}$. The equations (6. 2) show $Y_0 \cap \Delta_k$ is empty for $0 < k_1 < p$, $0 < k_2 < q$. Let $C_k = Y_0 \cap \Delta_k$. Then $C_{i,0}$ $(0 \leqq i < p)$ and $C_{0,j}$ $(0 \leqq j < q)$ are smooth rational curves. $C_{i,0}$ and $C_{i+1,0}$ (or $C_{0,j}$ and $C_{0,j+1}$) meet transversally at $\mathfrak{p}_i$ (or $\mathfrak{q}_j$) where $\mathfrak{p}_i$ is the point $(s, u_i, u_{i+1}^{-1}, v_0) = (0, 0, 0, -1)$, $(0 \leqq i < p)$, and $\mathfrak{q}_j$ is the point $(s, u_0, v_j, v_{j+1}^{-1}) = (0, -1, 0, 0)$, $(0 \leqq j < q)$, $(u_i = s^{-i} w_1,\ v_j = s^{-j} w_2)$.

On the other hand,

$$\lim_{s \to 0} s\Theta(s, w_1, w_2) = \begin{cases} \text{if } p = q = 1 & \\ \quad (1 + u_{-1} + \boldsymbol{e}(\tau_2) v_0 + \boldsymbol{e}(2\tau_2) u_{-1} v_0) u_{-1} & \text{on } Q_{-1,0} \\ \quad (1 + \boldsymbol{e}(\tau_2) u_0 + v_{-1} + \boldsymbol{e}(2\tau_2) u_0 v_{-1}) v_{-1} & \text{on } Q_{0,-1} \\ \text{if } p > 1,\ q = 1 & \\ \quad (1 + \boldsymbol{e}(\tau_2) v_0) u_{-1} & \text{on } Q_{-1,0} \\ \quad (1 + \boldsymbol{e}(\tau_1) u_0 + v_{-1} + \boldsymbol{e}(2\tau_2) u_0 v_{-1}) v_{-1} & \text{on } Q_{0,-1} \\ \text{if } p > 1,\ q > 1 & \\ \quad (1 + \boldsymbol{e}(\tau_2) v_0) u_{-1} & \text{on } Q_{-1,0} \\ \quad (1 + \boldsymbol{e}(\tau_2) u_0) v_{-1} & \text{on } Q_{0,-1} \end{cases}$$

(6. 3)

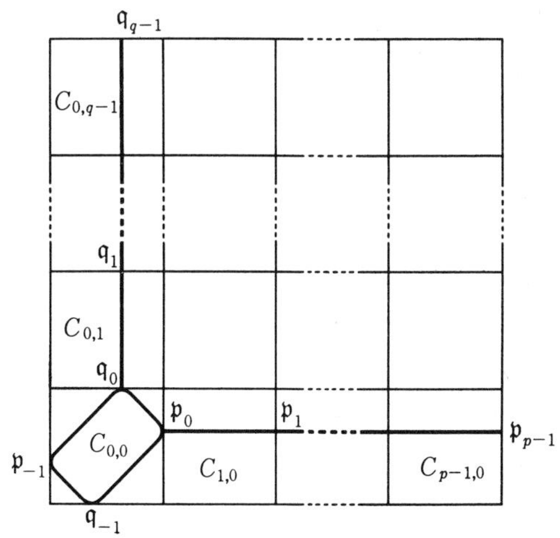

Hence $C_{-1,0}$ (or $C_{0,-1}$) and $C_{0,0}$ meet transversally at $\mathfrak{p}_{-1}$ (or $\mathfrak{q}_{-1}$) where $\mathfrak{p}_{-1}:(s, u_{-1}, u_0^{-1}, v_0)=(0, 0, 0, -\boldsymbol{e}(-\tau_2))$, $\mathfrak{q}_{-1}:(s, u_0, v_{-1}, v_0^{-1})=(0, -\boldsymbol{e}(-\tau_2), 0, 0)$. Now Y is smooth and minimal. The configuration of Y_0 in A_6 is shown in (6. 3).

(2) *The case* $\tau_2(0)=0$. For simplicity we assume $\tau_1=\tau_3=0$, $\tau_2(s)=s^l(l>0)$. Then we set

$$\Theta(s, w_1, w_2) = \sum s^{p(n^2-n)/2+q(m^2-m)/2}\boldsymbol{e}(nms^l)w_1^n w_2^m.$$

Hence we have on Q_k

$$\lim_{s\to 0}\Theta(s, w_1, w_2) = \begin{cases} (1+w_1)(1+w_2), & k_1=k_2=0 \\ 1+w_2, & 0<k_1<p,\ k_2=0 \\ 1+w_1, & k_1=0,\ 0<k_2<q \\ 1, & 0<k_1<p,\ 0<k_2<q \end{cases}$$

Let $C_k=Y_0\cap\Delta_k$. $C_{0,0}$ consists of two smooth rational curves $C'_{0,0}$ and $C''_{0,0}$ defined by $1+w_1=0$, and by $1+w_2=0$ respectively. $C'_{0,0}$ and $C''_{0,0}$ meet at $\mathfrak{p}:(s, w_1\ w_2)=(0, -1, -1)$. $C_{i,0}$ and $C_{i+1,0}$ meet transversally at $\mathfrak{p}_i(0<i<p-1)$, while $C_{0,j}$ and $C_{0,j+1}$ meet transversally at $\mathfrak{q}_i$ $(0<i<q-1)$ (see case (1)). $C'_{0,0}$ meets $C_{-1,0}$ at $\mathfrak{p}_{-1}$: $(s, u_{-1}, u_0^{-1}, v_0)=(0, -1, 0, 0)$ and $C_{0,1}$ at $\mathfrak{p}_0$. Similarly $C''_{0,0}$ meets $C_{0,-1}$ at $\mathfrak{q}_{-1}$: $(s, u_0, v_{-1}, v_0^{-1})=(0, -1, 0, 0)$ and $C_{0,1}$ at $\mathfrak{q}_0$.

The singularity of Y at $\mathfrak{p}$ is analytically isomorphic to that of the analytic set : $uv-s^l=0$ at the origin. Hence we can resolve this singularity by replacing $\mathfrak{p}$ by $(l-1)$ rational curves, so we get a configuration I_{p-q}-l for the singular fiber of $\tilde{X}$, the nonsingular model of $X=Y/\Gamma$, [8].

7. The Case $B=\begin{bmatrix} p+r & r \\ r & q+r \end{bmatrix}$ (p, q, r>0)

Assume $\tau(s)=\begin{bmatrix} p+r & r \\ r & q+r \end{bmatrix}\log s/2\pi i$ for simplicity.

We put

(7. 1)
$$\begin{aligned}\Theta(s, w_1, w_2) &= \Theta(\tau, z+a\log s/4\pi i) \\ &= \sum_{n,m} s^{A(n,m)}w_1^n w_2^m\end{aligned}$$

where $a=(-(p+r), -(q+r))$, $w_i=\boldsymbol{e}(z_i)$ and $A(n, m)=(p+r)(n^2-n)/2+(q+r)(m^2-m)/2+rnm$. We take an r. p. p. decomposition $\mathcal{D}$ (1. 6) and consider $P=P_{\mathcal{D}}$ and $A=A_{\mathcal{D}}$. A_0 consists of $pq+pr+qr$ ($=\det B$) irreducible components, each of which is isomorphic to a projective plane blown up at three distinct points.

Let N be a finite $\boldsymbol{Z}$-module $\boldsymbol{Z}^2/\boldsymbol{Z}(p+r, r)+\boldsymbol{Z}(r, q+r)$. Then N has representatives $k=(k_1, k_2)$ satisfying the following inequalities :

(7. 2)
$$\begin{aligned}&k_1\geqq 0,\quad k_2\geqq 0,\\ &p-k_1+k_2\geqq 0,\quad q+k_1-k_2\geqq 0,\\ &p+r-k_1\geqq 0,\quad q+r-k_2\geqq 0.\end{aligned}$$

For k satisfying (7. 2), $\mu(k)=0$ (4. 9). Let $u_i=s^{-i}w_1$, $v_i=s^{-i}w_2$.

On Q_k we have

$$
(7.\ 3)\quad \lim_{s\to 0}\Theta(s, w_1, w_2)=\begin{cases}
1+u_0+v_0, & k=(0, 0)\\
1+v_0, & k=(i, 0),\quad 0<i<p\\
1+v_0+u_p^{-1}v_0, & k=(p, 0)\\
1+u_{p+i}^{-1}v_i, & k=(p+i, i),\quad 0<i<r\\
1+u_{p+r}^{-1}v_r+u_{p+r}^{-1}, & k=(p+r, r)\\
1+u_{p+r}^{-1}, & k=(p+r, i+r),\quad 0<i<q\\
1+u_{p+r}^{-1}+v_{q+r}^{-1}, & k=(p+r, q+r)\\
1+v_{q+r}^{-1}, & k=(p+r-i, q+r),\quad 0<i<p\\
1+v_{q+r}^{-1}+v_{q+r}^{-1}u_r, & k=(r, q+r)\\
1+v_{q+r-i}^{-1}u_{r-i}, & k=(r-i, q+r-i),\quad 0<i<r\\
1+v_q^{-1}u_0+u_0, & k=(0, q)\\
1+u_0, & k=(0, q-i),\quad 0<i<q\\
1, & \text{otherwise.}
\end{cases}
$$

Let Y' be the divisor on P' defined by $\Theta(s, w_1, w_2)=0$, and Y its closure. Let Δ_k be the closure of $(Q_k)_0$, and $C_k=\Delta_k\cap Y_0$. C_k is a smooth rational curve. By the action of Γ, $C_{i,0}$, $C_{p+i,i}$ and $C_{p+r,r+i}$ are transformed onto $C_{r+i,q+r}$, $C_{i,q+i}$ and $C_{0,i}$ respectively. Two of them meet at most at one point and there transversally. The smoothness of Y can be proven in the same way as before. Y/Γ has a singular fiber of type I_{p-q-r} in the terminology of [8]. The configuration of Y_0 in $(P)_0$ is described as follows.

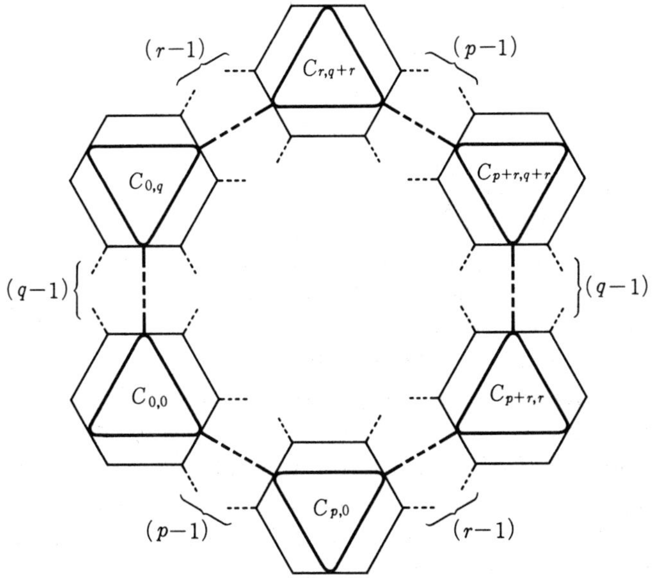

(7. 4) *Finally, the mapping*: $(u_{k_1}, v_{k_2})\to(w_1, w_2, s^p w_1^{-1}w_2, s^q w_1 w_2^{-1}, s^{p+r}w_1^{-1}, s^{q+r}w_2^{-1})$ *on* Δ_k, *for* k *satisfying* (7. 2), *gives rise to a new family of principally polarized abelian varieties, which coincides with* Proj $\tilde{R}(\tau)/\Gamma$ (cf. [7] § 2 *Remark*). *Its special fiber is a union of two projective planes, independent, surprisingly, of the choice of* p, q *and* r.

Therefore it seems to be natural to expect that this variety have some special property. Similar considerations and computations provide us with similar kinds of varieties in the three-dimensional case. The author has obtained thus the concept of stable quasi-abelian varieties. As consequence of it we can compactify the moduli space of principally polarized abelian varieties with a fixed level structure in the two and three-dimensional cases [7].

8. Final Remarks

(8. 1) Now we shall construct a family of parabolic type, not included among the types in the preceding sections. From [9], we have the case I^*_{p-q-r}, in which the monodromy is given by

$$M=\begin{bmatrix} -1 & 0 & -p-r & -r \\ 0 & -1 & -r & -q-r \\ 0 & 0 & -1 & 0 \\ 0 & 0 & 0 & -1 \end{bmatrix}.$$

Then we take a period matrix $\tau(s)=\begin{bmatrix} p+r & r \\ r & q+r \end{bmatrix} \log s/2\pi i$. If we put $s=t^2$, the pull back $\tau_1(t)=\tau(t^2)$ is a stable matrix, and $M^2=\begin{bmatrix} 1 & 0 & 2p+2r & 2r \\ 0 & 1 & 2r & 2q+2r \\ 0 & 0 & 1 & 0 \\ 0 & 0 & 0 & 1 \end{bmatrix}$.

We set $\Theta(t, z)=\sum e(n\tau_1(t)^t n/2+n^t(z+a))$, where $z=(z_1, z_2)$, and $a=((p+r)/4, (q+r)/4)$. Then we define a divisor Y' on $A'_{\mathfrak{D}}$ by $\Theta(t, z)=0$.

The automorphism σ of $A'_{\mathfrak{D}}$, $\sigma:(t, z_1, z_2)\to(-t, -z_1, -z_2)$, can be extended to a bimeromorphic mapping of $A_{\mathfrak{D}}$. It satisfies $\sigma^*\Theta=\Theta$, and induces an automorphism of the closure Y of Y'. Denote the closure of $Y_0\cap Q_k$ by C_k. Any irreducible component C_k is mapped by σ onto C_{-k}, in particular $C_{p,-q}$, $C_{p+r,r}$ and $C_{r,q+r}$ are mapped onto themselves mod Γ.

Moreover σ has two fixed points of type N_2 ([4] p. 583) on $C_{p,-q}$ mod Γ, $C_{p+r,r}$ mod Γ and $C_{r,q+r}$ mod Γ respectively. At the image of a fixed point of σ, the quotient space Y/σ has an ordinary double singularity, which we can resolve by replacing the point by one rational curve. Notice that Y is smooth.

Let X be the nonsingular model of Y/σ. Then X_0 has a configuration of type I^*_{p-q-r} [8]. Combining the results in [8] with those here one could construct a family of curves of genus two in the same way as above.

(8. 2) A smooth curve of genus three can be also realized by theta functions. In fact, a complete intersection of a theta divisor of level 1 and its translate in the jacobian variety of C, if they are in good position, consists of two smooth curves, isomorphic to C, meeting transversally at two distinct points [15]. Thanks to this fact, we can construct a family of reduced curves of genus three with at worst ordinary double points. The details will appear elsewhere.

(8. 3) Given a family X of curves we assume X_0 to be reduced and to acquire only ordinary double points as singularities. Then the local monodromy is of the form $\begin{bmatrix} 1_g & B \\ 0 & 1_g \end{bmatrix}$, ${}^tB=B\geqq 0$ [1]. Let X^* be the union of X' and X_0 with its singular

points deleted. Then a canonical morphism from X' to A' can be extended to a morphism from X^* to $\mathcal{X}$.

References

[1] Clemens, C. H. et al : Seminar on degeneration of algebraic varieties, Institute for Advanced Study, Princeton, (1969–70).

[2] Iitaka, S. : On the degenerates of a normally polarized abelian variety of dimension 2 and algebraic curve of genus two (in Japanese), Master degree thesis, University of Tokyo, (1967).

[3] Kempf, G. et al : Toroidal embeddings I, Lect. Notes in Math., 339, Springer, Berlin, (1973).

[4] Kodaira, K. : On compact analytic surfaces II-III, Ann. of Math., **77** (1963), 563–626, **78** (1963), 1–40.

[5] Miyake, K. and Oda, T. : Almost homogeneous algebraic varieties under algebraic torus action, Manifolds-Tokyo 1973, Proceedings of the International Conference on Manifolds and Related Topics in Topology, 1973, Univ. Tokyo Press, (1975).

[6] Nakamura, I. : On degeneration of abelian varieties. (unpublished)

[7] —— : On moduli of stable quasi abelian varieties, Nagoya Math. J., **58** (1975), 149–214.

[8] Namikawa, Y. and Ueno, K. : The complete classification of fibers in pencils of curves of genus two, Manuscripta Math., **9** (1973), 143–186.

[9] —— : On fibers in families of curves of genus two 1. Singular fibers of elliptic type, Number Theory, Algebraic Geometry and Commutative Algebra, in honor of Y. Akizuki, Kinokuniya, Tokyo, (1973), 297–371.

[10] Namikawa, Y. : A new compactification of the Siegel space and degeneration of abelian varieties I–II, Math. Ann. **221** (1976), 97–141, 201–241.

[11] Néron, A. : Modèles Minimaux des variétiés abéliennes sur les corps locaux et globaux, Publ. Math. IHES **21**, Paris, (1964).

[12] Ogg, A. P. : On pencils of curves of genus two, Topology **5**, (1966), 352–362.

[13] Ueno, K. : Degenerating fibers of families of abelian varieties of dimension 2. (unpublished)

[14] Weil, A. : Zur Beweis des Torellischen Satzes, Nach. Akad. Wiss. Göttingen, 2 (1957), 33–53.

[15] Gunning, R. C. : Lectures on Riemann Surfaces, Jacobi Varieties, Math. Notes, Princeton Univ. Press (1972).

Department of Mathematics
Nagoya University

(Received November 29, 1975)

Toroidal Degeneration of Abelian Varieties

Y. Namikawa

In this article we establish a method of constructing somewhat general degenerating abelian varieties as a partial answer to the problem raised in [8]. This subject was once treated by Mumford [4] from a different point of view. Here we make use of the theory of torus embeddings developed by Mumford and other mathematicians ([2], [3]). Our method of construction itself is a direct generalization of Nakamura's in [6]. However by virtue of this generalization we can obtain a synthetic view of late related results including the author's [9], and those of Oda-Seshadri [10] (cf. § 4). It is a very interesting problem to investigate how general our method is. It would be reasonable to expect that our method works not only in the complex analytic category but also in the algebraic or étale category with suitable modifications.

We take a torus embedding $\mathcal{X}_\Sigma$ of $\mathcal{T}=(\boldsymbol{G}_m)^n$ determined by a rational partial polyhedral decomposition (abbreviated to r. p. p. decomposition) $\Sigma=\{\sigma\}$ of $N_{\boldsymbol{R}}=N\otimes_{\boldsymbol{Z}}\boldsymbol{R}$ where $N=\mathrm{Hom}_{\mathrm{groups}}(\boldsymbol{G}_m,\mathcal{T})$ (see § 1 for the precise definition). Consider a $\mathcal{T}$-invariant closed set S' of $\mathcal{X}_\Sigma$ and an open neighbourhood $\mathcal{U}$ of S' in $\mathcal{X}_\Sigma$. We fix the triple $(S', \mathcal{U}, \mathcal{X}_\Sigma)$ and set $U=\mathcal{U}\cap\mathcal{T}$. We shall then consider a family of polarized abelian varieties of dimension g, $\varpi^\circ:\mathcal{A}^\circ\rightarrow U$. For simplicity we assume moreover the polarization to be principal. The family ϖ° determines a multiple-valued holomorphic map $T:U\rightarrow\mathfrak{S}_g$ called the period map, where $\mathfrak{S}_g$ denotes the Siegel upper-half plane of degree g (cf. § 2). We add some mild conditions so that the family can be extended.

The problem is *to construct degenerating fibres over S' to extend the family ϖ° to a family of analytic spaces $\varpi:\mathcal{A}\rightarrow\mathcal{U}$ which has some "nice" properties.*

Indeed there are many choices in constructing degenerating fibres. Among these possibilities we consider a special kind of fibres having much importance. Namely, letting L be a lattice of rank g, we take an r. p. p. decomposition K of $N_{\boldsymbol{R}}\times L_{\boldsymbol{R}}$ such that the projection $N_{\boldsymbol{R}}\times L_{\boldsymbol{R}}\rightarrow N_{\boldsymbol{R}}$ maps every cone κ in K onto a cone σ in Σ. (The surjectivity is necessary for the extended family to be equidimensional.) For the extendability of the family, K should satisfy a condition of compatibility with T. If the condition holds, we call K *toroidal degeneration data* and we can construct a family $\varpi:\mathcal{A}\rightarrow\mathcal{U}$ extending ϖ° (§ 3). Since $\mathcal{A}$ and all fibres of ϖ are toroidal (cf. [2] p. 54), we say this degeneration is *toroidal*. Our method seems to be quite general for constructing toroidal degenerations, although we do not know whether all toroidal degeneration can be constructed in this way. Several geometric properties of ϖ and the fibres can be stated in terms of those of K (§ 3).

In the last section, § 4, we give a number of examples to show that most of preceding results ([6], [9], [10]) fall into our category.

Because the number of pages is limited, we only indicate the ideas of proofs, which will appear in complete form somewhere else.

General assumption : All algebraic varieties in this article are defined over $\boldsymbol{C}$ and are identified with the associated analytic spaces.

1. Torus Embeddings

In this section we recall briefly the theory of torus embeddings and fix our notation. For details, see [2] (or [9])[1].

(1. 1) Let $\mathcal{T}=\operatorname{Spec}(\boldsymbol{C}[T_1, \cdots, T_n, T_1^{-1}, \cdots, T_n^{-1}])$ be an algebraic torus of dimension n over $\boldsymbol{C}$, which we often identify with the analytic group $(\boldsymbol{C}^*)^n$ where $\boldsymbol{C}^*=\boldsymbol{C}-\{0\}$. An algebraic torus of dimension one is denoted by $\boldsymbol{G}_m$.

Let $M=\operatorname{Hom}_{\text{alg.groups}}(\mathcal{T}, \boldsymbol{G}_m)$ be the group of characters : for $r \in M$ we denote by T^r the corresponding monomial in $\Gamma(\mathcal{T}, \mathcal{O})$. Let $N=\operatorname{Hom}_{\text{alg.groups}}(\boldsymbol{G}_m, \mathcal{T})$ be the group of 1-parameter subgroups of $\mathcal{T}$. M and N are free abelian groups of rank n and dual to each other by the pairing

$$\begin{array}{ccc} M\times N & \to & \boldsymbol{Z} \\ \Psi & & \Psi \\ (r, a) & \to & \langle r, a\rangle \end{array}$$

where $\langle r, a\rangle$ is defined by the condition :

$$T^r(a(t)) = t^{\langle r,a\rangle}, \quad t \in \boldsymbol{G}_m.$$

Denote $M\otimes_{\boldsymbol{Z}}\boldsymbol{R}$ and $N\otimes_{\boldsymbol{Z}}\boldsymbol{R}$ by $M_{\boldsymbol{R}}$ and $N_{\boldsymbol{R}}$ respectively.

Definition (1. 2). i) An *affine torus embedding* (resp. a *torus embedding*) of $\mathcal{T}$ is an affine scheme (resp. a scheme) over $\boldsymbol{C}$ containing $\mathcal{T}$ as an open subset, equipped with an action of $\mathcal{T}$ which extends the action of $\mathcal{T}$ on itself induced from the translation in $\mathcal{T}$.

ii) A *morphism of two torus embeddings* $\mathcal{X}_1$ and $\mathcal{X}_2$ is a morphism $f: \mathcal{X}_1 \to \mathcal{X}_2$ of schemes satisfying the conditions :

a) f induces a group epimorphism $g : \mathcal{T}_1 \to \mathcal{T}_2$ of the tori contained in them ;

b) the following diagram commutes :

$$\begin{array}{ccc} \mathcal{T}_1\times\mathcal{X}_1 & \longrightarrow & \mathcal{X}_1 \\ {\scriptstyle (g,f)}\downarrow & & \downarrow{\scriptstyle f} \\ \mathcal{T}_2\times\mathcal{X}_2 & \longrightarrow & \mathcal{X}_2, \end{array}$$

where the horizontal arrows denote the action of tori.

1) Torus embeddings considered here are only *locally* of finite type, while in [2] they consider only those of finite type. For the propositions used here, however, the same proofs as in [2] hold.

Definition (1. 3). i) A *convex rational polyhedral cone* σ in $M_{\boldsymbol{R}}$ (or $N_{\boldsymbol{R}}$) is a set of the form

$$\sigma = \{x \in M_{\boldsymbol{R}}(\text{or } N_{\boldsymbol{R}})\ ;\ l_i(x) \geq 0, \quad i = 1, \cdots, m\}$$

where l_i's are linear functionals defined over $\boldsymbol{Q}$. We abbreviate it to *c. r. p. cone.*

ii) A *rational partial polyhedral decomposition* (abbreviated to an *r. p. p. decomposition*) of $N_{\boldsymbol{R}}$ is a collection $\Sigma = \{\sigma_\alpha\}$ of c. r. p. cones in $N_{\boldsymbol{R}}$, each of which contains no linear subspace, such that

a) if for $\sigma_\alpha \in \Sigma$, σ is a face of σ_α, then $\sigma \in \Sigma$;

b) for $\sigma_\alpha \in \Sigma$ and $\sigma_\beta \in \Sigma$, $\sigma_\alpha \cap \sigma_\beta$ is again in Σ.

Given an r. p. p. decomposition of $N_{\boldsymbol{R}}$, we can construct a torus embedding of $\mathcal{T}$ in a canonical manner. First we fix some notation.

For a c. r. p. cone σ in $N_{\boldsymbol{R}}$, the dual cone $\hat{\sigma}$ in $M_{\boldsymbol{R}}$ is defined by

$$\hat{\sigma} = \{r \in M_{\boldsymbol{R}}\ ;\ \langle r, a\rangle \geq 0 \text{ for all } a \in \sigma\}$$

and is again a c. r. p. cone in $M_{\boldsymbol{R}}$. For a semigroup S in M we denote by $\boldsymbol{C}[S]$ the $\boldsymbol{C}$-vector subspace of $\Gamma(\mathcal{T}, \mathcal{O})$ generated by T^r, $r \in S$. For a c. r. p. cone σ in $N_{\boldsymbol{R}}$, $\hat{\sigma} \cap M$ is clearly a semigroup.

Proposition (1. 4). i) *For a c. r. p. cone σ in $N_{\boldsymbol{R}}$ which does not contain any linear subspace, $\mathcal{X}_\sigma = \mathrm{Spec}\,(\boldsymbol{C}[\hat{\sigma} \cap M])$ is a normal affine torus embedding of $\mathcal{T}$ of finite type.*

ii) *For an r. p. p. decomposition Σ of $N_{\boldsymbol{R}}$ the affine torus embeddings $\mathcal{X}_\sigma$, $\sigma \in \Sigma$, can be patched together to form a normal torus embedding $\mathcal{X}_\Sigma$ locally of finite type.*

Proposition (1. 5). *Let $\mathcal{X}_\Sigma$ be the torus embedding of $\mathcal{T}$ associated with an r. p. p. decomposition Σ. Then $\mathcal{X}_\Sigma$ has the following properties :*

a) *For a c. r. p. cone σ in Σ any element $a \in N$ in the relative interior of σ* (i. e., *in σ but not in any proper face of σ*) *gives the same limit $a(0) = \lim_{t \to 0} a(t)$, hence determines a $\mathcal{T}$-orbit $\mathcal{O}_\sigma$;*

b) *the correspondence $\sigma \to \mathcal{O}_\sigma$ gives a bijection between Σ and the set of $\mathcal{T}$-orbits in $\mathcal{X}_\Sigma$;*

c) *σ_1 is a face σ_2 if and only if $\bar{\mathcal{O}}_{\sigma_1} \supset \mathcal{O}_{\sigma_2}$;*

d) $\dim \sigma + \dim \mathcal{O}_\sigma = \dim \mathcal{T}$.

Many geometric properties of $\mathcal{X}_\Sigma$ can be described in terms of Σ. For later use we make the following

Definition (1. 6). An r. p. p. decomposition Σ of $N_{\boldsymbol{R}}$ is said to be *of projective type* if there exists a real convex continuous function on the convex hull of $\bigcup_{\sigma \in \Sigma} \sigma \subset N_{\boldsymbol{R}}$ satisfying the following conditions :

a) f has integral values on $N \cap (\bigcup_{\sigma \in \Sigma} \sigma)$;

b) for $\lambda \in \boldsymbol{R}^+$, $f(\lambda a) = \lambda f(a)$;

c) f is piecewise linear ;

d) for each $\sigma \in \Sigma$ there are a finite number of linear functionals l_i, $i=1, \cdots, m$, such that

$$\sigma = \{a \in N_{\boldsymbol{R}} ; f(a) = l_i(a) \text{ for all } i\} .$$

If an r. p. p. decomposition Σ is of projective type, one can construct an invertible sheaf $\mathcal{L}_f$ on $\mathcal{X}_\Sigma$ with the above function f. $\mathcal{L}_f$ is ample when Σ is a finite set.

Next we consider morphisms of torus embeddings.

Definition (1. 7). i) A *morphism of an r. p. p. decomposition* Σ *of* $N_{\boldsymbol{R}}$ to another Σ' *of* $N'_{\boldsymbol{R}}$ is a $\boldsymbol{Z}$-homomorphism $\rho : N \rightarrow N'$ of finite cokernel which maps each $\sigma_\alpha \in \Sigma$ into some $\sigma'_\beta \in \Sigma'$.

ii) A morphism of r. p. p. decompositions as above is said to be *of equidimensional type* if ρ maps each $\sigma_\alpha \in \Sigma$ onto $\sigma'_\beta \in \Sigma'$,

Proposition (1. 8). i) *A morphism of r. p. p. decompositions* ρ *as above determines a morphism of torus embeddings* $\rho_* : \mathcal{X}_\Sigma \rightarrow \mathcal{X}_{\Sigma'}$. *This correspondence* $\rho \rightarrow \rho_*$ *is functorial.*

ii) *If the morphism* ρ *above is of equidimensional type, the associated morphism* ρ_* *is equidimensional.*

2. Period Map and Degeneration Data

(2. 1) We take a torus embedding $\mathcal{X}_\Sigma$ of $\mathcal{T}$ defined by an r. p. p. decomposition Σ of $N_{\boldsymbol{R}}$. Let Σ' be a subset of Σ satisfying the condition : $\sigma \in \Sigma'$, $\sigma \prec \sigma_1$ implies $\sigma_1 \in \Sigma'$. Then the union of $\mathcal{T}$-orbits $S' = \bigcup_{\sigma \in \Sigma'} \mathcal{O}(\sigma)$ is closed in $\mathcal{X}_\Sigma$. Let $\mathcal{U}$ be an open neighbourhood of S' in $\mathcal{X}_\Sigma$. We henceforth fix the triple $(\mathcal{X}_\Sigma, S', \mathcal{U})$ and set $U = \mathcal{U} \cap \mathcal{T}$, $S = \mathcal{U} - U$.

(2. 2) Let

$$\varpi^\circ : \mathcal{A}^\circ \rightarrow U$$

be a family of polarized abelian varieties of dimension g. For simplicity we assume moreover that the polarization is principal and that ϖ° has a section $s : U \rightarrow \mathcal{A}^\circ$ which we consider as the set of the identity elements of abelian varieties. Then, as Ueno has shown in [12] § 1, one can define a multiple-valued holomorphic map (called the *period map* of ϖ°)

$$T : U \rightarrow \mathfrak{S}_g,$$

where $\mathfrak{S}_g = \{\tau \in M(g, \boldsymbol{C}) ; {}^t\tau = \tau, \operatorname{Im} \tau > 0\}$ denotes the Siegel upper-half plane of degree g, and a homomorphism (called the *monodromy* of ϖ°)

$$\Phi : \pi_1(U, t_0) \rightarrow \operatorname{Sp}(g, \boldsymbol{Z})$$

of the fundamental group of U into the symplectic group such that for a loop γ with base point t_0 the analytic continuation $T(\gamma t_0)$ of $T(t_0)$ along γ is subject to

$$T(\gamma t_0) = \Phi([\gamma]) \cdot T(t_0),$$

where $[\gamma]$ denotes the homotopy class of γ and, for $M \in \mathrm{Sp}(g, \boldsymbol{Z})$, $\tau \in \mathfrak{S}_g$, $M \cdot \tau$ denotes the usual action of $\mathrm{Sp}(g, \boldsymbol{Z})$ on $\mathfrak{S}_g$.

(2. 3) We here put one more assumption on the period map of ϖ° which is not as restrictive as it looks (cf. Remark (2. 4)).

In order to state the condition, we need some notation.

Let $\mathfrak{Y}_g$ be the vector space of real symmetric matrices of degree g and Y_g the lattice of integral matrices in $\mathfrak{Y}_g$. We denote by $\mathfrak{Y}_g^+$ the cone of positive matrices in $\mathfrak{Y}_g$. Note that $\mathfrak{S}_g = \mathfrak{Y}_g + \sqrt{-1}\,\mathfrak{Y}_g^+$. Denoting by $\bar{Y}_g^+$ the set of non-negative integral matrices in Y_g, we define the rational closure $\bar{\mathfrak{Y}}_g^+$ of $\mathfrak{Y}_g^+$ as the convex hull of $\bar{Y}_g^+$ in $\mathfrak{Y}_g$ which in fact contains $\mathfrak{Y}_g^+$. For $0 \leq g'' < g$ we have a natural injection $i : \bar{\mathfrak{Y}}_{g''}^+ \to \bar{\mathfrak{Y}}_g^+$ sending y'' to $\begin{bmatrix} 0 & 0 \\ 0 & y'' \end{bmatrix}$, by which we consider $\bar{\mathfrak{Y}}_{g''}^+$ as a closed subset of $\bar{\mathfrak{Y}}_g^+$.

We consider a map

$$\begin{array}{ccc} \boldsymbol{e} : \boldsymbol{C}^n & \longrightarrow & \mathcal{T} \\ \Downarrow & & \Downarrow \\ (z_1, \cdots, z_n) & \to & (\boldsymbol{e}(z_1), \cdots, \boldsymbol{e}(z_n)) \end{array}$$

where $\boldsymbol{e}(\ ) = \exp(2\pi\sqrt{-1}\,(\ ))$, and put $\tilde{U} = \boldsymbol{e}^{-1}(U)$.

In what follows we assume that the period map T satisfies the following

Assumption (**U**). *There exists a $\boldsymbol{Z}$-linear map*

$$B : N \to Y_g$$

such that $B_{\boldsymbol{R}} = B \otimes_{\boldsymbol{Z}} \boldsymbol{R} : N_{\boldsymbol{R}} \to \mathfrak{Y}_g$ has the following properties : i) *every $\sigma \in \Sigma$ is mapped into $\bar{\mathfrak{Y}}_g^+$*, ii) *for some g'' with $0 \leq g'' \leq g$, every $\sigma \in \Sigma'$ is mapped into $\mathfrak{Y}_{g''}^+$ (under the above identification)*, *and* iii) *the composition $\tilde{T} = T \cdot \boldsymbol{e}$ can be expressed in the form*

$$\tilde{T}(z) = B_{\boldsymbol{C}}(z) + S(\boldsymbol{e}(z))$$

on $\tilde{U}$, where $B_{\boldsymbol{C}} = B \otimes_{\boldsymbol{Z}} \boldsymbol{C}$, and $S(t)$ is a bounded single-valued holomorphic function on U with values in the symmetric complex matrices of degree g.

Remark (2. 4). i) The above condition i) on $B_{\boldsymbol{R}}$ is in fact superfluous (cf. [6] Lemma (2. 3)), and the condition ii) is added only for technical reason.

ii) In spite of the seeming complexity of the above condition, it holds in a quite general situation. Namely if $\mathcal{X}_\Sigma$ is non-singular and if for every element γ of $\pi_1(U)$, $\Phi(\gamma) - \mathrm{Id}$ is nilpotent, then the period map satisfies the assumption (U) at least locally. The idea of the proof of this is as follows. Since S is a divisor with at worst normal crossings, we may assume $\pi_1(U)$ is abelian by shrinking $\mathcal{U}$ if necessary. Hence the image of the monodromy is contained in a parabolic subgroup of $\mathrm{Sp}(g, \boldsymbol{Z})$. By transforming the period map and monodromy by a suitable element of $\mathrm{Sp}(g, \boldsymbol{Z})$ we may assume that for every $\gamma \in \pi_1(U)$, $\Phi(\gamma)$ can be written in the form

$$\begin{bmatrix} 1 & B \\ 0 & 1 \end{bmatrix}, \quad B = \begin{bmatrix} 0 & 0 \\ 0 & B'' \end{bmatrix}, \quad B'' \in \bar{Y}_{g''}^+.$$

Also g'' can be taken minimal. Then by the same argument as in [7] Th. 2 or [6] Lemma (2. 3) one can show that T satisfies the assumption (U). The above condition is quite general because one can take an equivariant resolution $\mathcal{X}_\Sigma$ ([2] p. 32) and then take a suitable covering of $\mathcal{T}$ sending t_i to t_i^n (which extends to a finite morphism of $\mathcal{X}_\Sigma$ into itself) so that the pullback family satisfies the assumption as claimed above by virtue of the theorem of quasi-unipotentness of monodromy (cf. [11]). This fact reflects a very special property of the period map.

(2. 5) Let $\mathcal{C}_g$ be an algebraic torus of dimension g and L the group of 1-parameter subgroups of $\mathcal{C}_g$ (cf. § 1). Put $L_{\boldsymbol{R}}=L\otimes_{\boldsymbol{Z}}\boldsymbol{R}$.

Given a period map T satisfying the assumption (U), we can define an action of a lattice X of rank g on $N_{\boldsymbol{R}}\times L_{\boldsymbol{R}}$ as follows:

$$\begin{array}{ccc} T_\chi : N_{\boldsymbol{R}}\times L_{\boldsymbol{R}} & \to & N_{\boldsymbol{R}}\times L_{\boldsymbol{R}} \\ \Cup & & \Cup \\ (a, x) & \to & (a, x+\chi B_{\boldsymbol{R}}(a)) \end{array}$$

for $\chi\in X$ where $B_{\boldsymbol{R}}=B\otimes_{\boldsymbol{Z}}\boldsymbol{R}$, and x and χ are considered as row g-vectors.

Definition (2. 6). i) An r. p. p. decomposition $K=\{\kappa\}$ of $N_{\boldsymbol{R}}\times L_{\boldsymbol{R}}$ is called *admissible* (with respect to T) if it satisfies the following conditions:

1) the first projection

$$p: N_{\boldsymbol{R}}\times L_{\boldsymbol{R}}\to N_{\boldsymbol{R}}$$

induces a morphism of r. p. p. decompositions from K to Σ of equidimensional type;

2) K is invariant under the action of X;

3) for every $\kappa\in K$ and $a\in N_{\boldsymbol{R}}$, $\kappa\cap p^{-1}(a)$ is contained in $\mathrm{Im} B_{\boldsymbol{R}}(a)$ $(\subset L_{\boldsymbol{R}})$;

4) for every $\sigma\in\Sigma$, there are only finitely many classes of c. r. p. cones in K mapped by p onto σ modulo X.

An r. p. p. decomposition K of $N_{\boldsymbol{R}}\times L_{\boldsymbol{R}}$ which is admissible with respect to T is called *toroidal degeneration data* of the family ϖ°.

Definition (2. 7). i) An admissible decomposition K is called *of complete type* if for every $\sigma\in\Sigma$ the union of c. r. p. cones in K contained in $p^{-1}(\sigma)$ is convex in $N_{\boldsymbol{R}}\times L_{\boldsymbol{R}}$.

ii) An admissible r. p. p. decomposition K is called *of projective type* if there is a real continuous function f with the properties: 1) for every $\sigma\in\Sigma$ the r. p. p. subdecomposition $K_\sigma=\{\kappa : p(\kappa)\subset\sigma\}$ becomes of projective type by means of f (cf. (1. 6)); 2) for $\chi\in X$, $d_\chi=f-f\cdot T_\chi$ is a linear function on $N_{\boldsymbol{R}}\times L_{\boldsymbol{R}}$.

3. Construction of Degenerating Fibres

(3. 1) Let $\varpi^\circ: \mathcal{A}^\circ\to U$ be a family of principally polarized abelian varieties of dimension g over U whose period map $T: U\to\mathfrak{S}_g$ satisfies Assumption (U), and

suppose that degeneration data K of ϖ° are given. We shall construct an extended family of degenerating abelian varieties over $\mathcal{U}$ by means of the given degeneration data. In what follows we employ freely the notation of the previous section.

(3. 2) First we note that the original family ϖ° is reconstructed by means of T as follows (cf. [12] § 1).

On $U\times\mathcal{C}_g$, the lattice X acts via

$$\begin{array}{ccc} T_\chi: U\times\mathcal{C}_g & \rightarrow & U\times\mathcal{C}_g \\ \Cup & & \Cup \\ (t, w) & \rightarrow & (t, w\cdot \boldsymbol{e}(\chi T(t))) \end{array}$$

for $\chi\in X$. Here $\boldsymbol{e}(\chi T(t))\in\mathcal{C}_g$ is uniquely determined in spite of the multiple-valuedness of T. The action is properly discontinuous and fixed point free. The quotient space $U\times\mathcal{C}_g/X$ is a fibre space over U and canonically isomorphic to $\mathcal{A}^\circ$ over U, hence we identify it with $\mathcal{A}^\circ$ from now on.

(3. 3) The construction is done in two steps.

In *the first step*, we construct by means of K the torus embedding $\mathcal{B}_K$, which admits a fibre structure $\rho:\mathcal{B}_K\rightarrow\mathcal{X}_\Sigma$ by virtue of the condition (2. 6) 1). Put $\rho^{-1}(\mathcal{U})=\mathcal{B}$ and consider the restriction

$$\rho:\mathcal{B}\rightarrow\mathcal{U},$$

which we called a family of semiuniversal coverings in [9]. Note that the restriction of ρ to U is isomorphic to the family of algebraic tori $\rho_U: U\times\mathcal{C}_g\rightarrow U$ considered above.

In *the second step*, we see that the action of X on $U\times\mathcal{C}_g$ extends to that on $\mathcal{B}$. In fact, for $\chi\in X$, if we define a linear endomorphism of $N\times L$ by sending (a, x) to $(a, x+\chi B(a))$, it induces an automorphism T'_χ of $\mathcal{B}_K$ by virtue of the condition (2. 6) 2) (cf. Prop. (1. 8) i)). Being bounded on U, S extends to a matrix-valued holomorphic function on $\mathcal{U}$ which we denote by the same letter, hence determines a holomorphic map $\mathcal{U}\rightarrow\mathcal{C}_g$ sending t to $\boldsymbol{e}(\chi S(t))$. We can see easily that the translation $(t, w)\rightarrow(t, w\cdot\boldsymbol{e}(\chi S(t)))$ extends to an automorphism T''_χ of $\mathcal{B}$. The composite $T''_\chi\cdot T'_\chi$ is the desired extension of T_χ, which we henceforce denote by the same letter T_χ. The most difficult is to show

Proposition (3. 4). *The action of X on $\mathcal{B}$ is properly discontinuous and fixed point free.*

The idea of the proof is essentially contained in [9] Prop. (13. 1) and [6] Th. (2. 6).

As a direct corollary of (3. 4), we obtain the following theorem which is one of the main results of this article.

Theorem (3. 5). i) *The quotient space $\mathcal{A}=\mathcal{B}/X$ admits a canonical structure of normal analytic space which is toroidal. The canonical fibre structure*

$$\varpi : \mathcal{A} \to \mathcal{U}$$

induced from ρ *is equidimensional (by* (2. 6), 1)) *and of finite type (by* (2. 6) 4)), *and its restriction over* U *is* ϖ°. *We call* ϖ *the toroidal family associated with the degeneration data* K.

ii) *Let* K_1 *and* K_2 *be two admissible decompositions. If the identity map of* $N \times L$ *induces a morphism of r. p. p. decompositions from* K_1 *to* K_2, *then there is a natural morphism*

$$\psi : \mathcal{A}_{K_1} \to \mathcal{A}_{K_2}$$

over $\mathcal{U}$ *which is identity over* U *(essentially by Prop.* (1. 8) i)).

If moreover $K_1 \subset K_2$, *then* ψ *is an open immersion.*

One can describe the degenerate fibres over S' quite explicitely.

Let X' be the subgroup of X consisting of the elements $\chi' = (\chi_1, \cdots, \chi_{g'}, 0, \cdots, 0)$. where $g' = g - g''$. As we have seen in (3. 3), X' acts on $\mathcal{U} \times \mathcal{C}_{g'}$ via

$$\begin{array}{ccc} T_{\chi'} : \mathcal{U} \times \mathcal{C}_{g'} & \to & \mathcal{U} \times \mathcal{C}_{g'} \\ \Cup & & \Cup \\ (t, w) & \to & (t, w \cdot \boldsymbol{e}(\chi' S'(t))), \end{array}$$

where $S'(t)$ is the g'-th principal matrix of $S(t)$. (Note that $S'(t) = T'(t)$ by Assumption (U).) The quotient space $\varpi_{\circ} : \mathcal{A}_{\circ} = \mathcal{U} \times \mathcal{C}_{g'} / X' \to \mathcal{U}$ is a family of abelian varieties over $\mathcal{U}$.

Take a c. r. p. cone σ in Σ' and let K_σ be the set of c. r. p. cones in K over σ, which is invariant under the action of X on $N_{\boldsymbol{R}} \times L_{\boldsymbol{R}}$. Consider the quotient set $\bar{K}_\sigma = K_\sigma / X$, denoting the residue class of $\kappa \in K_\sigma$ by $\bar{\kappa}$. Also in $\bar{K}_\sigma$ the incidence relation $\bar{\kappa} \prec \bar{\kappa}'$ makes sense. Then, by using (1. 5) etc., we have

Proposition (3. 6). *The fibre* $\varpi^{-1}(\mathcal{O}(\sigma))$ *has a stratification* $\{\mathcal{E}_{\bar{\kappa}}\}$ *such that*

i) *the set of strata corresponds bijectively to* $\bar{K}_\sigma$;

ii) $\bar{\mathcal{E}}_{\bar{\kappa}} \supset \mathcal{E}_{\bar{\kappa}'}$ *if and only if* $\bar{\kappa} \prec \bar{\kappa}'$;

iii) *each* $\mathcal{E}_{\bar{\kappa}}$ *is an equidimensional family of commutative group varieties over* $\mathcal{O}(\sigma)$ *and there is a natural morphism*

$$\phi_{\bar{\kappa}} : \mathcal{E}_{\bar{\kappa}} \to \mathcal{A}_{\circ | \mathcal{O}(\sigma)}$$

over $\mathcal{O}(\sigma)$ *whose kernel is a family of algebraic tori*[1]. *This* ϕ_κ *is in fact determined by* $S'''(t) = (S(t)_{i, g'+j})_{1 \leq i \leq g', 1 \leq j \leq g''}$,

iv) relative dim $\mathcal{E}_{\bar{\kappa}}$ + relative dim $\bar{\kappa} = g$.

Concerning geometric properties of ϖ we can say the following.

Proposition (3. 7). *If the r. p. p. decomposition* K *is non-singular* ([2] p. 14), *then* $\mathcal{A}$ *is non-singular.*

Proposition (3. 8). *If* K *is of complete type, then* ϖ *is proper.*

1) Here we need the condition (2. 6) 3).

Theorem (3. 9). *If K is of projective type, then ϖ is quasi-projective.*

Only the last claim is not trivial. We need to develop a theory of degenerate theta functions. The idea of the proof can be found already in [9] Chapter V.

Problem (3. 10). *Find a suitable condition for each component of a fibre of ϖ to be reduced and smooth in terms of K.*

4. Applications

The last section is devoted to exhibiting a number of examples which cover most of known results in this direction.

A) *Analytic Néron model.*

For simplicity we consider an affine torus embedding $\mathscr{X}_\Sigma$ determined by $\Sigma=\{$a c. r. p. cone σ and its faces$\}$ and let $S=\{\sigma\}$. Let K be an admissible r. p. p. decomposition such that 1) for $a\in\sigma$, $\bigcup_{\kappa\in K}(p^{-1}(a)\cap\kappa)$ is a discrete subgroup of $\boldsymbol{R}^g$; 2) for every $\kappa\in K$ over σ, $p(\kappa\cap(N\times L))=\sigma\cap N$. Then *the toroidal family $\varpi:\mathscr{A}\to\mathscr{U}$ associated with K has the structure of a family of commutative group varieties.* If we denote by K' the subdecomposition of K generated by $\sigma\times\{0\}$ (i. e., the smallest admissible r. p. p. decomposition containing $\sigma\times\{0\}$), then the family $\varpi':\mathscr{A}'\to\mathscr{U}$ associated with K' is an open subfamily whose fibre A'_t over $t\in\mathscr{U}$ is the connected component of the fibre A_t of $\mathscr{A}$ over t containing the unit.

In case $\Sigma=\{\boldsymbol{R}^+,\ \{0\}\}$ and $K=\{\kappa_n=\{(a,\ a(0\ n))\},\ n\in\boldsymbol{Z}^{g''},\ \kappa^0=\{(0,\ 0)\}\}$, the corresponding toroidal family $\mathscr{A}$ is the *analytic Néron model* constructed in [6]. Hence it is naturally expected that the above general $\mathscr{A}$ also has the minimality property described in ibid. Th. (3. 4).

It is a very interesting problem to ask whether any toroidal compactification of the Néron model can be constructed by our method.

B) *Delony-Voronoi decomposition and stable quasi-abelian varieties* ([9]).

Let f be a convex function on $\bar{\mathfrak{Y}}_g^+\times L_{\boldsymbol{R}}$ defined by

$$f(y,x)=\min_{\xi\in\boldsymbol{Z}^g}\{\xi y^t\xi+2\xi^t x\}.$$

Then f defines an r. p. p. decomposition K of $\bar{\mathfrak{Y}}_g^+\times L_{\boldsymbol{R}}$ whose projection to $\bar{\mathfrak{Y}}_g^+$ also is an r. p. p. decomposition Σ of $\mathfrak{Y}_g$. The latter is called the *Delony-Voronoi decomposition,* and the former the *mixed decomposition.* A c. r. p. cone in Σ (resp. in K) is called a Delony-Voronoi cone (abbreviated to D-V cone) (resp. a mixed cone).

If we denote by $\mathscr{T}_g$ the set of symmetric complex matrices of degree g none of the coefficients of which vanishes, ($\mathscr{T}_g$ is an algebraic torus by coefficientwise multiplication), $\mathscr{X}_\Sigma$ is a torus embedding of $\mathscr{T}_g$. Consider the map $\boldsymbol{e}:\mathfrak{S}_g\to\mathscr{T}_g$ sending $\tau=(\tau_{ij})$ to $\boldsymbol{e}(\tau)=(\exp(2\pi\sqrt{-1}\,\tau_{ij}))$, and put $\mathscr{T}^\circ=\mathrm{Im}\,\boldsymbol{e}$ and $\mathscr{X}_\Sigma^\circ=$the interior of the closure of $\mathscr{T}^\circ$ in $\mathscr{X}_\Sigma$. If we let Σ' denote the set of D-V cones meeting $\mathfrak{Y}_g^+$, then it turns out that $\mathscr{X}_\Sigma^\circ$ contains all $\mathcal{O}(\sigma)$ with $\sigma\in\Sigma'$, and $\mathscr{T}^\circ=\mathscr{X}_\Sigma^\circ\cap\mathscr{T}_g$.

Clearly the inverse map T of $\boldsymbol{e}$ on $\mathscr{T}^\circ$ satisfies Assumption (U), and the mixed decomposition K is admissible with respect to T. Moreover K is projective according to definition (by elementary calculation we have $f(y, x)-f(y, x+\chi y)=\chi y^t\chi+2\chi^t x$), and it turns out to be of complete type.

Therefore the toroidal family

$$\varpi_g : \mathcal{A}_g \rightarrow \mathcal{X}_g^\circ$$

associated with K is a projective family whose fibres we named *stable quasiabelian varieties* (abbreviated to *SQAV*). The study of these SQAV led to our notion of toroidal degeneration. The D-V decomposition and the mixed decomposition for $g\leq 3$ are already explicitly known ([13]). Using this result Nakamura studied closely the structure of SQAV of dimension ≤ 3 ([5]). For $g\leq 2$ see also [9] (14. 5).

C) *Compactification of the generalized Jacobian variety of a stable curve* (due to Oda and Seshadri [10]).

Let $\pi : \mathcal{C}\rightarrow D$ be a family of stable curves of genus $g>0$ (for the definition of stable curve, see [1]) over the disc $D=\{t\in \boldsymbol{C}; |t|<\varepsilon\}$ such that 1) $\mathcal{C}$ is non-singular and 2) π is smooth over $D'=D-\{0\}$. (Hence the fibre C_0 over 0 may contain a smooth rational component Γ with $\Gamma^2=-2$.)

Then we can define the period map $T: D'\rightarrow \mathfrak{S}_g$ satisfying the assumption (U) (cf. [7] § 4). The analytic Néron model $\varpi : \mathcal{A}\rightarrow D$ constructed in A) is nothing but the family of generalized Jacobian varieties of $\mathcal{C}$. Now we use the notations in [10] freely. We take a natural embedding i : $H^1(\Gamma(C_0), \boldsymbol{R})\rightarrow L_{\boldsymbol{R}}$ by $x\rightarrow(0\ x)$. For $\phi\in \partial C_1(\boldsymbol{R})$ let Del_ϕ be the associated polyhedral decomposition, which is called the Namikawa decomposition in [10]. We define an r. p. p. decomposition K of $\boldsymbol{R}^+\times L_{\boldsymbol{R}}$ consisting of $\{(a, i(a\Delta))\ ;\ a\in \boldsymbol{R}^+\}_{\Delta\in \mathrm{Del}\phi}$, and $\{0\}$. Then *the associated toroidal family* $\varpi^\phi : \bar{\mathcal{A}}^\phi\rightarrow D$ *is* (*the analytic counterpart of*) *the family constructed in* [10]. If ϕ is non-degenerate, then $\bar{\mathcal{A}}^\phi$ contains $\mathcal{A}$ as an open set by virtue of Theorem (3. 5), ii) and [10] Prop. 7. 6.

We remark finally that we can construct such toroidal families even over the local universal deformation space of C_0 by using our method. It would take too much space to go into the subject, so it will be discussed elsewhere.

References

[1] Deligne, P. and Mumford, D.: The irreducibility of the space of curves of given genus, Publ. Math. IHES, **36** (1969), 75–110.

[2] Kempf, G. et al.: Toroidal embeddings, I, Springer Lecture Notes, No. **339**, Springer, Berlin, 1973.

[3] Miyake, K. and Oda, T.: Almost homogeneous algebraic varieties under algebraic torus action, Manifolds-Tokyo 1973, University of Tokyo Press, Tokyo, (1975), 373–381.

[4] Mumford, D.: An analytic construction of degenerating abelian varieties over complete rings, Compositio Math., **24** (1972), 239–272.

[5] Nakamura, I.: On moduli of stable quasi-abelian varieties, Nagoya Math. J., **58** (1975), 149–214.

[6] Nakamura, I.: Relative compactification of the Néron model and its application, in this Volume.

[7] Namikawa, Y.: On the canonical holomorphic map from the moduli space of stable curves

to the Igusa monoidal transform, Nagoya Math. J., **52** (1973), 197–259.
[8] Namikawa, Y. : Studies on degeneration, Classification of algebraic varieties and compact complex manifolds, Proceedings 1974, Lecture Notes in Math., No. **412** (1974), 165–210.
[9] Namikawa, Y. : A new compactification of the Siegel space and the degeneration of abelian varieties, I–II, Math. Ann., **221** (1976), 97–141 and 201–241.
[10] Oda, T. and Seshadri, C. S. : Compactifications of the generalized Jacobian variety, to appear.
[11] Schmid, W. : Variation of Hodge structure : the singularities of the period mapping, Inventiones Math., **22** (1973), 211–319.
[12] Ueno, K. : On fibre spaces of normally polarized abelian varieties of dimension 2, I, J. Fac. Sci. Univ. Tokyo, **18** (1971), 37–95.
[13] Voronoi, G. : Nouvelles applications des paramètres continues à la théorie des formes quadratiques, II, J. reine u. angew. Math., **134** (1908), 198–287 and **136** (1909), 67–181.

Department of Mathematics
Nagoya University

(Received February 14, 1976)

Kodaira Dimensions of Complements of Divisors

F. Sakai

The Kodaira dimension of a compact complex manifold plays an important role in classification theory (Iitaka [5], Ueno [18]). The purpose of this paper is to introduce the notion of Kodaira dimension for a non-compact complex manifold.

We shall mainly consider a complex manifold which has a compactification. Let X be a complex manifold of dimension n which is a Zariski open set of a compact complex manifold $\bar{X}$ such that $D=\bar{X}-X$ is a divisor with at most normal crossings. In this case, the Kodaira dimension $\kappa(X)$ of X is defined as follows. Let $K_{\bar{X}}$ denote the canonical bundle of $\bar{X}$. For a positive integer m, let $\varphi_0,\cdots,\varphi_N$ be a basis of $H^0(\bar{X}, \mathcal{O}(mK_{\bar{X}}+(m-1)[D]))$. Define a meromorphic map

$$\Phi_m : \bar{X} \ni w \longrightarrow (\varphi_0(w) : \cdots : \varphi_N(w))$$

of $\bar{X}$ into $\boldsymbol{P}_N$. Then

$$\kappa(X) = \begin{cases} \max_{m\in N(X)} \{\dim \Phi_m(\bar{X})\} & \text{if } N(X)\neq\phi, \\ -\infty & \text{if } N(X)=\phi, \end{cases}$$

where $N(X)=\{m>0 \mid \dim H^0(\bar{X}, \mathcal{O}(mK_{\bar{X}}+(m-1)[D]))>0\}$.

In Section 3, we define an analog of a Bergman kernel form on X. In Section 4, 5, we shall prove that if X is quasi-projective and $\kappa(X)=n$, then X is measure-hyperbolic and the automorphism group $\mathrm{Aut}(X)$ is a finite group.

Notations

Throughout this paper, we use the following notations : $\boldsymbol{C}$=complex plane, $\boldsymbol{C}^* =\boldsymbol{C}-\{0\}$; $\Delta_r=\{z\in \boldsymbol{C} \mid |z|<r\}$, $\Delta_r^*=\Delta_r-\{0\}$. Further Δ_1, Δ_1^* are simply denoted by Δ, Δ^*, respectively.

We shall use the following definition of meromorphic maps (Remmert [15], p. 367). Let X, Y be complex manifolds. A correspondence $f: X\rightarrow Y$ is a *meromorphic map* if

(i) for every point $x\in X$, $f(x)$ is a non-empty compact subset of Y,

(ii) the graph $G_f=\{(x,y)\in X\times Y \mid y\in f(x)\}$ of f is an irreducible analytic subset of $X\times Y$,

(iii) there is a dense subset X^* of X such that for every point $x\in X^*$, $f(x)$ is a single point.

Denote by π_X (resp. π_Y) the restriction to G_f of the natural projection from $X\times Y$

onto X (resp. Y). It follows that π_X is a proper modification of X. Note that f is a proper meromorphic map if π_Y is a proper map. Moreover if π_Y is a proper modification, f^{-1} becomes a meromorphic map. In this case, we call f a *bimeromorphic* map. A bimeromorphic map $g: X \to X$ is called a *bimeromorphic transformation* of X.

Let X, Y be complex manifolds and $f: X \to Y$ a holomorphic map. The rank of f at a point $x \in X$ is defined by

$$\text{rank}_x(f) = \dim_x X - \dim_x f^{-1}(f(x)).$$

The (total) *rank* of f is defined by

$$\text{rank}(f) = \max_{x \in X} \{\text{rank}_x(f)\}.$$

In case f is a meromorphic map, rank(f) is also defined, but the maximum is taken where f is holomorphic. The image $f(X)$ is a topological space and

$$\dim_{\boldsymbol{R}} f(X) = 2\,\text{rank}(f)$$

(cf. Remmert-Stein [16]).

1. Kodaira Dimensions of Non-Compact Complex Manifolds

Let X be a complex manifold of dimension n. Let $\{U_\alpha\}$ be an open covering of X with holomorphic coordinates $(w_\alpha^1, \cdots, w_\alpha^n)$ in U_α. The canonical bundle K_X of X is defined by transition functions $k_{\alpha\beta} = \det(\partial w_\beta^i / \partial w_\alpha^j)$. For any positive integer m, a holomorphic section $\varphi \in H^0(X, \mathcal{O}(mK_X))$ is given by holomorphic functions φ_α in U_α such that $\varphi_\alpha = k_{\alpha\beta}^m \varphi_\beta$ in $U_\alpha \cap U_\beta$. So

$$(1.1) \qquad \omega = \varphi_\alpha(w_\alpha)(dw_\alpha^1 \wedge \cdots \wedge dw_\alpha^n)^m, \quad \text{in} \quad U_\alpha,$$

is a global m-ple n-form on X. Thus we can view an element of $H^0(X, \mathcal{O}(mK_X))$ as a holomorphic m-ple n-form on X. For any holomorphic m-ple n-form ω on X, written locally as (1.1), denote by $(\omega \wedge \bar{\omega})^{1/m}$ the continuous (n, n)-form given by

$$(\omega \wedge \bar{\omega})^{1/m} = |\varphi_\alpha(w_\alpha)|^{2/m} \prod_{i=1}^{n} (\sqrt{-1}/2\pi) dw_\alpha^i \wedge d\bar{w}_\alpha^i, \quad \text{in} \quad U_\alpha.$$

Moreover we set

$$\|\omega\| = \left\{ \int_X (\omega \wedge \bar{\omega})^{1/m} \right\}^{m/2}.$$

Clearly $\| \ \|$ satisfies the following properties.

(i) $\|\omega\| = 0$ if and only if $\omega = 0$,

(ii) $\|c\omega\| = |c| \, \|\omega\|$ for $c \in \boldsymbol{C}$.

(iii) $\|\omega_1 + \omega_2\| \leqq C_m(\|\omega_1\| + \|\omega_2\|)$ $\quad (C_1 = 1, \ C_m = 2^{(m/2)-1}$ for $m \geqq 2)$.

If $\|\omega\| < \infty$, we say that ω is *integrable* ($L_{2/m}$-integrable).

Definition. Define $F_m(X)$ to be the subspace of $H^0(X, \mathcal{O}(mK_X))$ consisting of all integrable holomorphic m-ple n-forms on X.

Remark. Note that $F_1(X)$ is a Hilbert space in a natural manner (See [7], [12]).

We shall see later that if X has a compactification, then $\dim F_m(X)<\infty$.

Set $N(X)=\{m>0|\ F_m(X)\neq\{0\}\}$. For a finite set of elements $\omega_0, \cdots, \omega_N$ of $F_m(X)$, $m\in N(X)$, we can define a meromorphic map $\Phi_{\{\omega_0,\cdots,\omega_N\}}$ associated to them by

$$\begin{array}{cccc} \Phi_{\{\omega_0,\cdots,\omega_N\}}: & X & \longrightarrow & \boldsymbol{P}_N \\ & \Cup & & \Cup \\ & w & \longrightarrow & (\omega_0(w):\cdots:\omega_N(w)). \end{array}$$

Define the *m-th rank* $\mathrm{rk}_m(X)$ of X by

$$\mathrm{rk}_m(X)=\max\{\mathrm{rank}\,(\Phi_{\{\omega_0,\cdots,\omega_N\}})\},$$

where the maximum is taken over all choices of finite sets $\omega_0, \cdots, \omega_N$ of $F_m(X)$, for $N=0, 1, 2,\cdots$

Definition. The *Kodaira dimension* $\kappa(X)$ of X is defined by

$$\kappa(X)=\begin{cases} \max\limits_{m\in N(X)}\{\mathrm{rk}_m(X)\} & \text{if } N(X)\neq\phi, \\ -\infty & \text{if } N(X)=\phi. \end{cases}$$

(Note that $\kappa(X)$ takes one of the values $-\infty, 0, 1, \cdots, n$.)

Theorem 1. 1. *The Kodaira dimension $\kappa(X)$ is a bimeromorphic invariant of a complex manifold X.*

Proof. Let X' be a complex manifold which is bimeromorphic to X and $f: X'\to X$ a bimeromorphic map. Then f^* induces an isomorphism of $F_m(X)$ and $F_m(X')$, which implies that $\kappa(X)=\kappa(X')$. In fact, for an element $\omega\in F_m(X)$, $f^*\omega$ is a holomorphic m-ple n-form on $X-S(f)$. Here $S(f)$ is the set of points where f is not holomorphic. Since $S(f)$ is an analytic subset of X' of codimension$\geqq 2$ (See [15], p. 333), $f^*\omega$ extends to a holomorphic m-ple n-form on X' by a theorem of Hartogs. Further

$$\int_{X'}(f^*\omega\wedge\overline{f^*\omega})^{1/m}=\int_X(\omega\wedge\bar{\omega})^{1/m}<\infty.$$

Hence $f^*\omega\in F_m(X')$. Obviously f^* is an injection. Considering f^{-1}, we get the surjectivity of f^*. q. e. d.

As a consequence, we may formulate the following

Definition. For a singular complex space X, the *Kodaira dimension* $\kappa(X)$ is defined to be $\kappa(X^*)$, where X^* is a complex manifold which is bimeromorphic to X.

A holomorphic m-ple n-form ω on Δ^* can be written as $\omega=f(z)(dz)^m$, where f is a holomorphic function on Δ^*.

Lemma 1. 1 (cf. Kobayashi-Ochiai [9], Addendum). *Let $\omega=f(z)(dz)^m$ be a holo-*

morphic m-ple n-form on Δ^*. *Let* $f(z)=\sum_{n=-\infty}^{\infty} a_n z^n$ *be the Laurent expansion of* $f(z)$. *If* $\omega \in F_m(\Delta^*)$, *then* $a_n=0$ *for* $n \leqq -m$.

Proof. We have, by definition, $(\omega\wedge\bar{\omega})^{1/m}=|f(z)|^{2/m}(\sqrt{-1}/2\pi)dz\wedge d\bar{z}$. Let p be the least integer $\geqq m$ such that $a_{-p}\neq 0$. Put $f_1(z)=\sum_{n=-(m-1)}^{\infty} a_n z^n$, and $f_2(z)=\sum_{n=-\infty}^{-p} a_n z^n$. Assume that $f_2(z)\neq 0$. Letting r_0 be small enough, we can make

$$|f_1/f_2| < b < 1 \text{ and } |f_2| \geqq |a_{-p}|\,/\,|z|^p \quad \text{for} \quad z \in \Delta^*_{r_0}.$$

It follows that

$$\int_{\Delta^*} |f(z)|^{2/m}(\sqrt{-1}/2\pi)dz\wedge d\bar{z} \geqq (1-b)^{2/m}\int_{\Delta^*_{r_0}} |f_2(z)|^{2/m}(\sqrt{-1}/2\pi)dz\wedge d\bar{z}$$
$$\geqq (1-b)^{2/m}|a_{-p}|^{2/m}\int_{\Delta^*_{r_0}} |z|^{-2p/m}(\sqrt{-1}/2\pi)dz\wedge d\bar{z}.$$

On the other hand, using polar coordinates $z=re^{\sqrt{-1}\theta}$, we have

$$\int_{\Delta^*_{r_0}} |z|^{-2p/m}(\sqrt{-1}/2\pi)dz\wedge d\bar{z} = \lim_{\varepsilon\to 0}\int_{\varepsilon}^{r_0}\int_0^{2\pi} r^{1-(2p/m)}\frac{dr\,d\theta}{\pi}$$
$$= \lim_{\varepsilon\to 0}\begin{cases} (1/m-p)[r^{2(m-p)/m}]_{\varepsilon}^{r_0} & \text{if } p > m, \\ 2[\log r]_{\varepsilon}^{r_0} & \text{if } p = m, \end{cases}$$
$$= \infty.$$

Combining these inequalities, we get $\|\omega\|=\infty$. This is a contradiction. q. e. d.

Corollary. $\kappa(\boldsymbol{C})=-\infty$, $\kappa(\boldsymbol{C}^*)=-\infty$.

Proof. Here we prove $\kappa(\boldsymbol{C}^*)=-\infty$. Let ω be an element of $F_m(\boldsymbol{C}^*)$. We can represent ω as $\omega=(\sum_{n=-\infty}^{\infty} a_n z^n)(dz)^m$. Clearly $\omega_{|\Delta^*} \in F_m(\Delta^*)$. So by the above lemma, we have $a_n=0$ for $n\leqq -m$. Let $w=1/z$. Then the domain $\{z\in \boldsymbol{C}\,|\,|z|>1\}$ is transformed onto the punctured unit disc $\Delta^*_w=\{w\in \boldsymbol{C}\,|\,0<|w|<1\}$. Then

$$\omega = \sum_{n=-(m-1)}^{\infty} a_n(1/w)^n(-dw/w^2)^m \in F_m(\Delta^*_w).$$

It follows from the above lemma that $a_n=0$ for $n\geqq -(m-1)$. Consequently, we get $\omega=0$. q. e. d.

Corollary. $\kappa(\boldsymbol{C}^*\times\Delta^{n-1})=-\infty$, $\kappa(\boldsymbol{C}\times\Delta^{n-1})=-\infty$ *and* $\kappa(\boldsymbol{C}^n)=-\infty$, $\kappa(\boldsymbol{C}^{*n})=-\infty$.

The Kodaira dimension has the following properties.

Proposition 1. 1. *Let* X, Y *be complex manifolds of dimension* n *such that* $X\subset Y$. *Then* $\kappa(X)\geqq\kappa(Y)$. *In particular, if* $\kappa(X)=-\infty$, *then* $\kappa(Y)=-\infty$.

Corollary. *If a complex manifold* X *contains* $\boldsymbol{C}^n$ *as its open subset, then* $\kappa(X)=-\infty$.

In case X is compact, this fact is obtained by Kodaira in [11], using Nevanlinna theory.

Proposition 1. 2. *Let X be a complex manifold of dimension n and Z an analytic subset of X. If codim $Z \geqq 2$, then $\kappa(X-Z)=\kappa(X)$.*

Proposition 1. 3. *Let X, Y be complex manifolds of dimension n. Suppose that there is a proper surjective meromorphic map $f\colon X \to Y$. Then we have*

$$\kappa(X) \geqq \kappa(Y).$$

Proof. First, assume that f is holomorphic. The set $E_f = \{x \in X \mid \mathrm{rank}_x(f) < n\}$ is an analytic subset of X. Set $E' = f(E_f)$, $E = f^{-1}(E')$. By Lemma 2. 2 in [1], there exist analytic subsets S (resp. S') of $X-E$ (resp. $Y-E'$) and a positive integer s such that $f\colon X-E-S \to Y-E'-S'$ is locally biholomorphic and $f^{-1}(y)$ consists of s distinct points for every $y \in Y-E'-S'$. The number s is called the sheet number of f. Take an element $\omega \in F_m(Y)$. Then $f^*\omega \in F_m(X)$, because

$$\int_X (f^*\omega \wedge \overline{f^*\omega})^{1/m} = \int_{X-E-S} (f^*\omega \wedge \overline{f^*\omega})^{1/m} = s \int_{Y-E'-S'} (\omega \wedge \bar{\omega})^{1/m}$$
$$= s \int_Y (\omega \wedge \bar{\omega})^{1/m} < \infty.$$

Clearly f^* is injective. Therefore we get $\kappa(X) \geqq \kappa(Y)$. In case f is only meromorphic, the assertion follows from the first case by using the desingularization of the graph of f (cf. Theorem 1. 1). q. e. d.

Proposition 1. 4. *Let X be a complex manifold of dimension n and g a bimeromorphic transformation of X. Then g^* induces an automorphism of $F_m(X)$ for every positive integer m.*

2. Complements of Divisors

In what follows, we shall consider a complex manifold X of dimension n which has a compactification. In other words, X is a complement of an analytic subset of a compact complex manifold. According to Hironaka, there is a smooth compactification $\bar{X}$ of X. Namely, $\bar{X}$ is a compact complex manifold of dimension n and $D = \bar{X} - X$ is a divisor which has at most normal crossings. With respect to local coordinates $(w_1, \cdots, w_n)$ of $\bar{X}$, D is given locally by $w_1 \cdots w_j = 0$, $j \leqq n$. Denote by $[D]$ the line bundle on $\bar{X}$ determined by the divisor D.

Theorem 2. 1. *Let X be a complex manifold of dimension n with a smooth compactification $\bar{X}$. Put $D = \bar{X} - X$. Then we have an isomorphism*

$$F_m(X) \cong H^0(\bar{X}, \mathcal{O}(mK_{\bar{X}} + (m-1)[D])).$$

Proof. First we prove

Lemma 2. 1. *We have an isomorphism* :

$$H^0(\bar{X}, \mathcal{O}(mK_{\bar{X}}+(m-1)[D])) \cong \left\{ \begin{array}{l} \textit{meromorphic } m\textit{-ple } n\textit{-forms on } \bar{X} \textit{ (holomorphic} \\ \textit{on } X) \textit{ with at most } (m-1)\textit{-ple poles along } D \end{array} \right\}$$

Proof. The identification is described as follows. Cover $\bar{X}$ by coordinate neighborhoods $\{U_\alpha\}$ with holomorphic coordinates $(w_\alpha^1, \cdots, w_\alpha^n)$ in U_α. Let $\{k_{\alpha\beta}\}$, $\{\delta_{\alpha\beta}\}$ be transition functions of the line bundles $K_{\bar{X}}$, $[D]$, respectively. Take a holomorphic section $\varphi \in H^0(\bar{X}, \mathcal{O}(mK_{\bar{X}}+(m-1)\ [D]))$, which is given by holomorphic functions φ_α in U_α such that $\varphi_\alpha = k_{\alpha\beta}^m \delta_{\alpha\beta}^{m-1} \varphi_\beta$ in $U_\alpha \cap U_\beta$. Let $\{\delta_\alpha\}$ be a holomorphic section of $[D]$ defining D. We associate to φ the following m-ple n-form on $\bar{X}$:

$$\frac{\varphi_\alpha}{\delta_\alpha^{m-1}} (dw_\alpha^1 \wedge \cdots \wedge dw_\alpha^n)^m \quad \text{in} \quad U_\alpha.$$

This gives the above isomorphism. q. e. d.

We proceed to the proof of Theorem 2. 1. Let $\varphi \in H^0(\bar{X}, \mathcal{O}(mK_{\bar{X}}+(m-1)\ [D]))$. Let $(w_1, \cdots, w_n)$ be local coordinates on $\bar{X}$ such that $D = \{w_1 \cdots w_j = 0\}$, $j \leqq n$, and let $U = \{|w_1| \leqq r, \cdots, |w_n| \leqq r\}$ be a small polycylindrical coordinate neighborhood. Via the identification of Lemma 2. 1, we can write φ in U as

$$\varphi = g(w) \frac{(dw_1 \wedge \cdots \wedge dw_n)^m}{(w_1)^{s_1} \cdots (w_j)^{s_j}},$$

where $g(w)$ is a holomorphic function on U and $s_i \leqq (m-1)$ for $i = 1, \cdots, j$. For convenience sake, we put $s_i = 0$ for $i = j+1, \cdots, n$. Then

$$(\varphi \wedge \bar{\varphi})^{1/m} = |g|^{2/m} \prod_{i=1}^n (\sqrt{-1}\ /2\pi) \frac{dw_i \wedge d\bar{w}_i}{|w_i|^{2s_i/m}}.$$

Using polar coordinates $w_i = r_i e^{\sqrt{-1}\theta_i}$, we get

$$(\varphi \wedge \bar{\varphi})^{1/m} = |g|^{2/m} \prod_{i=1}^n (1/\pi) r_i^{(1-(2s_i/m))} dr_i d\theta_i.$$

Let $C = \max_U |g|^{2/m}$. We obtain the estimate :

$$\int_U (\varphi \wedge \bar{\varphi})^{1/m} \leqq (C/\pi^n) \int_0^{2\pi} \cdots \int_0^{2\pi} \left\{ \prod_{i=1}^n \int_0^r r_i^{(1-(2s_i/m))} dr_i \right\} d\theta_1 \cdots d\theta_n,$$

$$\leqq C \prod_{i=1}^n (m/(m-s_i)) r^{2(m-s_i)/m} < \infty,$$

because $1-(2s_i/m) > -1$. Since $\bar{X}$ is compact, we can cover $\bar{X}$ by a finite number of polycylindrical coordinate neighborhoods. It follows that $\|\varphi\| < \infty$ and hence we get $\varphi \in F_m(X)$.

On the other hand, given $\omega \in F_m(X)$, we must prove that ω has at most $(m-1)$-ple poles along D. We examine ω in a neighborhood of a point in which D is non-singular. We can find local coordinates $(w_1, \cdots, w_n)$ such that $D = \{w_1 = 0\}$. Let $U = \{|w_1| < r, \cdots, |w_n| < r\}$ be a small polycylindrical coordinate neighborhood. Clearly $\omega_{|U-D} \in F_m(U-D)$ (Proposition 1. 1). Noting that $U-D \cong \Delta_r^* \times \Delta_r^{n-1}$, it follows from Lemma 1. 1, that $\omega_{|U}$ has at most $(m-1)$-ple poles along D. Hence ω can be extended to a meromorphic m-ple n-form on $\bar{X}-S$, where S is the singular locus of

D. Since codim $S \geqq 2$, ω extends meromorphically to $\bar{X}$. Moreover ω has at most $(m-1)$-ple poles along D. q. e. d.

Remark. In particular, we have $F_1(X) \cong F_1(\bar{X}) \cong H^0(\bar{X}, \mathcal{O}(K_{\bar{X}}))$.

Definition. Let X, $\bar{X}$ and D be as above. Set

$$\gamma_m = \gamma_m(X) = \dim F_m(X) = \dim H^0(\bar{X}, \mathcal{O}(mK_{\bar{X}}+(m-1)[D])) < \infty.$$

Corollary. *Let X, $\bar{X}$ and D be the same as in Theorem 2. 1. Let $\varphi_0, \cdots, \varphi_N$ be a basis of $H^0(\bar{X}, \mathcal{O}(mK_{\bar{X}}+(m-1)[D]))$, with $N+1=\gamma_m$. Define a meromorphic map*

$$\Phi_m : \bar{X} \ni w \rightarrow \Phi_m(w) = (\varphi_0(w) : \cdots : \varphi_N(w)) \tag{2. 1}$$

of $\bar{X}$ into $\boldsymbol{P}_N$. Then we have

$$\kappa(X) = \begin{cases} \max\limits_{m \in N(X)} \{\dim \Phi_m(\bar{X})\} & \text{if } N(X) \neq \phi, \\ -\infty & \text{if } N(X) = \phi, \end{cases}$$

where $N(X) = \{m>0 | \gamma_m>0\}$.

Proof. The assertion follows easily from Theorem 2. 1.

Remark. In view of the above corollary, in case X is compact, our Kodaira dimension $\kappa(X)$ agrees with the original one in Iitaka [5] (cf. Ueno [18]).

Example. We present a classification of algebraic curves:

κ	$\gamma_1=g$	$\gamma_m \quad (m \geqq 2)$	structure
$-\infty$	0	0	$\boldsymbol{P}_1$, $\boldsymbol{P}_1-\{a_1\}$, $\boldsymbol{P}_1-\{a_1\}-\{a_2\}$
0	1	1	elliptic curves
1	0	$m(k-2)-k+1$(except $k=3$, $m=2$)	$\boldsymbol{P}_1-\bigcup_{i=1}^{k}\{a_i\}$, $k \geqq 3$
	1	$mk-k$	elliptic curves$-\bigcup_{i=1}^{k}\{a_i\}$, $k \geqq 1$
	$g \geqq 2$	$m(k+2g-2)-k+1-g$	curves of genus $\geqq 2-\bigcup_{i=1}^{k}\{a_i\}$, $k \geqq 0$

Example. Let D be a hypersurface of degree d in the projective space $\boldsymbol{P}_n$. If D has at most normal crossings, then

$$\kappa(\boldsymbol{P}_n - D) = \begin{cases} n & \text{if } d>n+1, \\ -\infty & \text{if } d \leqq n+1. \end{cases}$$

For a compact complex manifold $\bar{X}$, the *L-dimension* of $\bar{X}$, written $\kappa(L, \bar{X})$, of a line bundle L on $\bar{X}$ is defined as follows ([5], [18]). For a positive integer m, let Φ_{mL} be a meromorphic map defined by a basis of $H^0(\bar{X}, \mathcal{O}(mL))$. Then

$$\kappa(L, \bar{X}) = \begin{cases} \max\limits_{m \in N(L, \bar{X})} \{\Phi_{mL}(\bar{X})\} & \text{if } N(L, \bar{X}) \neq \phi, \\ -\infty & \text{if } N(L, \bar{X}) = \phi, \end{cases}$$

where $N(L, \bar{X}) = \{m>0 | \dim H^0(\bar{X}, \mathcal{O}(mL))>0\}$. For a divisor D on $\bar{X}$, we put $\kappa(D, \bar{X}) = \kappa([D], \bar{X})$.

Our Kodaira dimension has the following relation with the $K_{\bar{X}}+D$-dimension of $\bar{X}$.

Proposition 2. 1. *Let X, $\bar{X}$ and D be the same as in Theorem 2. 1. Then we have $\kappa(X) \leqq \kappa(K_{\bar{X}}+D, \bar{X})$. Further if $\kappa(X) \geqq 0$, then $\kappa(X) = \kappa(K_{\bar{X}}+D, \bar{X})$.*

Proof. Since $H^0(\bar{X}, \mathcal{O}(mK_{\bar{X}}+(m-1)[D]))$ can be regarded as a subspace of $H^0(\bar{X}, \mathcal{O}(mK_{\bar{X}}+m[D]))$, we have by definition, $\kappa(X) \leqq \kappa(K_{\bar{X}}+D, \bar{X})$. If $\kappa(X) \geqq 0$, there is a non-zero element $\varphi \in H^0(\bar{X}, \mathcal{O}(m_0K_{\bar{X}}+(m_0-1)[D]))$, for some positive integer m_0. Let $\psi_0, \cdots, \psi_N$ be a basis of $H^0(\bar{X}, \mathcal{O}(mK_{\bar{X}}+m[D]))$. Then $\varphi\psi_i$ belongs to $H^0(\bar{X}, \mathcal{O}((m+m_0)K_{\bar{X}}+(m+m_0-1)[D]))$, from which we readily infer that $\kappa(X) \geqq \kappa(K_{\bar{X}}+D, \bar{X})$. q. e. d.

Remark. The above proposition is valid without the assumption that D has normal crossings (See Appendix, Theorem A. 1). Iitaka in [6] calls $\kappa(K_{\bar{X}}+D, \bar{X})$ the logarithmic Kodaira dimension of X and denotes it by $\bar{\kappa}(X)$.

Proposition 2. 2. *Let X, $\bar{X}$ and D be the same as in Theorem 2. 1. Then $\kappa(X)=n$ if and only if $\kappa(K_{\bar{X}}+D, \bar{X})=n$.*

Proof. In view of Proposition 2. 1, it suffices to prove that $\kappa(K_{\bar{X}}+D, \bar{X})=n$ implies $\kappa(X)=n$. Suppose that $\kappa(K_{\bar{X}}+D, \bar{X})=n$. Then $\bar{X}$ is a Moišezon manifold[1]. So we may assume that $\bar{X}$ is projective algebraic, because there exists a proper modification $\pi : \tilde{X} \to \bar{X}$ such that $\tilde{X}$ is projective algebraic. Let H be a hyperplane section of $\bar{X}$. Since $\kappa(K_{\bar{X}}+D, \bar{X})=n$, there exists an effective divisor Z in the complete linear system $|m(K_{\bar{X}}+D)-D-H|$ for a large integer m. Then $mK_{\bar{X}}+(m-1)[D]=[H]+[Z]$. Hence it follows easily that $\kappa(X)=n$. q. e. d.

Remark. In particular, if $K_{\bar{X}}+D$ is ample on $\bar{X}$, then the map Φ_m defined by (2. 1) is an imbedding of X for a large integer m. This fact is a corollary of the proof of Theorem 4 in Kodaira [10]. Further $H^i(\bar{X}, \mathcal{O}(mK_{\bar{X}}+(m-1)[D]))=0$ for $i \geqq 1$ and for every positive integer $m \geqq 2$.

Theorem 2. 2. *Let X be a complex manifold of dimension n which has a compactification. Then there exist positive numbers α, β such that the estimate*

$$\alpha m^{\kappa(X)} \leqq \gamma_m \leqq \beta m^{\kappa(X)}$$

1) A compact complex manifold is called a Moišezon manifold if the transcendence degree of its field of meromorphic functions is equal to its dimension.

holds for every sufficiently large integer m. In particular, $\kappa(X)=n$ *if and only if*

$$\limsup_{m\to\infty} \{\gamma_m/m^n\} > 0.$$

Proof. This follows easily from Proposition 2. 1 and Theorem 1 in [5].

Proposition 2. 3. *Let X, Y de complex manifolds having smooth compactifications* $\bar{X}$, $\bar{Y}$, *respectively. Then* $\kappa(X\times Y)=\kappa(X)+\kappa(Y)$.

Proof. Put $D=\bar{X}-X$, $C=\bar{Y}-Y$. Let π_1, π_2 be the projections from $\bar{X}\times\bar{Y}$ to $\bar{X}$, $\bar{Y}$, respectively. It follows that $\bar{X}\times\bar{Y}$ is a smooth compactification of $X\times Y$ and $\bar{X}\times\bar{Y}-X\times Y=D\times\bar{Y}+\bar{X}\times C$. Obviously $D\times\bar{Y}=\pi_1^*(D)$, $\bar{X}\times C=\pi_2^*(C)$. It suffices to prove that $F_m(X\times Y)=F_m(X)\otimes F_m(Y)$ for every m. Using Theorem 2. 1, we get

$$\begin{aligned} F_m(X\times Y) &\cong H^0(\bar{X}\times\bar{Y}, \mathcal{O}(mK_{\bar{X}\times\bar{Y}}+(m-1)[D\times\bar{Y}+\bar{X}\times C])) \\ &= H^0(\bar{X}\times\bar{Y}, \mathcal{O}(\pi_1^*(mK_{\bar{X}}+(m-1)[D])+\pi_2^*(mK_{\bar{Y}}+(m-1)[C]))) \\ &= H^0(\bar{X}, \mathcal{O}(mK_{\bar{X}}+(m-1)[D]))\otimes H^0(\bar{Y}, \mathcal{O}(mK_{\bar{Y}}+(m-1)[C])) \\ &\qquad\qquad\qquad\qquad\qquad\qquad \text{(by Künneth formula)} \\ &\cong F_m(X)\otimes F_m(Y). \end{aligned}$$

q. e. d.

Proposition 2. 4. *Let X be a complex manifold with a smooth compactification* $\bar{X}$. *Let* $f: \bar{X}\to\bar{Y}$ *be a surjective holomorphic map, where* $\bar{Y}$ *is a compact complex manifold. Then for a general point y in* $f(X)$, *we have*

$$\kappa(X) \leqq \kappa(X_y)+\dim \bar{Y},$$

where $X_y=X\cap f^{-1}(y)$.

Proof. This can be proven in a manner similar to that in Ueno [18], Theorem 5. 11.

We have the following structure theorem (cf. Iitaka [5]).

Theorem 2. 3. *Let X be a complex manifold of dimension n with a smooth compactification* $\bar{X}$. *Assume that* $\kappa(X)\geqq 1$. *Then there exist a complex manifold* X^* *and its smooth compactification* $\bar{X}^*$, *a projective algebraic manifold* $\bar{Y}^*$ *and a surjective holomorphic map* $f: \bar{X}^*\to\bar{Y}^*$ *which satisfy the following conditions.*

(i) X^* (*resp.* $\bar{X}^*$) *is bimeromorphic to* X (*resp.* $\bar{X}$),

(ii) $\dim \bar{Y}^*=\kappa(X)$,

(iii) *for a general point* $y\in f(X^*)$, $\bar{X}_y^*=f^{-1}(y)$ *is a smooth compactification of* $X_y^*=X^*\cap f^{-1}(y)$ *and* $\kappa(X_y^*)=0$.

3. Bergman Kernel Forms

Let X be a complex manifold of dimension n. Let $\{U_\alpha\}$ be an open covering of X with holomorphic coordinates $(w_\alpha^1, \cdots, w_\alpha^n)$ in U_α. An *m-ple quasi-volume form* Ω on X is an m-ple (n, n)-form defined by a continuous section $\{\xi_\alpha\}$ of the line bundle $|mK_X|^2$ which is C^∞ and positive outside an analytic subset of X. We write

$$\Omega=\xi_\alpha\left\{\prod_{i=1}^{n}(\sqrt{-1}/2\pi)dw_\alpha^i\wedge d\bar{w}_\alpha^i\right\}^m,\quad \text{in } U_\alpha.$$

Further, if $\{\xi_\alpha\}$ is an everywhere C^∞ and positive section of $|mK_X|^2$, i. e., if $\{\xi_\alpha\}$ is a metric of mK_X, we call Ω an *m-ple volume form.* Denote by $\Omega^{1/m}$ the quasi-volume form defined by

$$\Omega^{1/m}=\xi_\alpha^{1/m}\prod_{i=1}^{n}(\sqrt{-1}/2\pi)dw_\alpha^i\wedge d\bar{w}_\alpha^i,\quad \text{in}\quad U_\alpha.$$

The Ricci form Ric Ω of Ω is the real (1, 1)-form, given locally by

$$\operatorname{Ric}\Omega=dd^c\log\xi_\alpha,\quad \text{in}\quad U_\alpha,$$

where $d^c=(\sqrt{-1}/4\pi)(\bar{\partial}-\partial)$. Let ω be a holomorphic m-ple n-form on X. Then $\mu_m\omega\wedge\bar{\omega}$ is an m-ple quasi-volume form, where

$$\mu_m=\{(\sqrt{-1})^{n^2}/(2\pi)^n\}^m.$$

Using the above notation, we see that $(\omega\wedge\bar{\omega})^{1/m}=\{\mu_m\omega\wedge\bar{\omega}\}^{1/m}$.

Assume now that X has a smooth compactification $\bar{X}$. Put $D=\bar{X}-X$.

Definition. For $m\in N(X)$, we define an m-ple quasi-volume form V_m by

$$V_m=V_{m,X}=\sup_{\|\omega\|=1}\{\mu_m\omega\wedge\bar{\omega}\},$$

where the supremum is taken over all elements $\omega\in F_m(X)$ such that $\|\omega\|=1$. Further we put $v_m=\{V_m\}^{1/m}$.

Lemma 3. 1. $\int_X v_m<\infty$.

Proof. Cover $\bar{X}$ by coordinate neighborhoods $\{U_\alpha\}$ with holomorphic coordinates $(w_\alpha^1,\cdots,w_\alpha^n)$ in U_α. Let $\{\delta_\alpha\}$ be a holomorphic section of $[D]$ defining D and $\{a_\alpha\}$ a metric in $[D]$. A length $\|\delta\|$ of δ is given by $\|\delta\|^2=|\delta_\alpha|^2/a_\alpha$ in U_α. By Lemma 2. 1, $\|\delta\|^{2(m-1)/m}(\omega\wedge\bar{\omega})^{1/m}$ is a continuous m-ple (n,n)-form on $\bar{X}$, for $\omega\in F_m(X)$. Hence

$$u_m=\sup_{\|\omega\|=1}\{\|\delta\|^{2(m-1)/m}(\omega\wedge\bar{\omega})^{1/m}\}$$

is a continuous (n,n)-form on $\bar{X}$ (for a proof, see Narasimhan-Simha [14], Appendix). Clearly $v_m=\|\delta\|^{-2(m-1)/m}u_m$. Thus the assertion follows from the proof of Theorem 2. 1. q. e. d.

Definition. We define a hermitian form on $F_m(X)$ by

$$(\omega_1,\omega_2)=\int_X\frac{\mu_m\omega_1\wedge\bar{\omega}_2}{(v_m)^{m-1}},\quad \text{for}\quad \omega_1,\omega_2\in F_m(X).$$

Since $|\mu_m\omega_1\wedge\bar{\omega}_2|\leqq(v_m)^m$, we get $(\omega_1,\omega_2)<\infty$, because of Lemma 3. 1. This form is positive-definite.

Remark. In general, the product (ω_1,ω_2) can be defined for $\omega_1\in F_m(X)\cong H^0(\bar{X},\mathcal{O}(mK_{\bar{X}}+(m-1)[D]))$, $\omega_2\in H^0(\bar{X},\mathcal{O}(mK_{\bar{X}}+m[D]))$.

Let $\omega_0,\cdots,\omega_N$ be an orthonormal basis of $F_m(X)$ with respect to the above hermitian form, where $N+1=\gamma_m$. That is, $(\omega_i, \omega_j)=\delta_{ij}$. Denote by $X^\dagger$ the complex manifold which is conjugate to X. Then

$$B_{m,X}(w, \bar{\zeta}) = \sum_{i=0}^{N} \mu_m \omega_i(w) \wedge \overline{\omega_i(\zeta)}, \quad \text{for} \quad w, \zeta \in X,$$

is a holomorphic m-ple $2n$-form on $X \times X^\dagger$ (cf. [7]). Note that $B_{m,X}(w, \bar{\zeta})$ is independent of the choice of an orthonormal basis of $F_m(X)$. Letting $w=\zeta$, we obtain an m-ple quasi-volume form B_m on X:

$$B_m = B_{m,X} = B_{m,X}(w, \bar{w}).$$

We call it the *m-th Bergman kernel form* of X. Put $b_m = \{B_m\}^{1/m}$.

Lemma 3. 2. $B_m = \sup_{(\omega,\omega)=1} \{\mu_m \omega \wedge \bar{\omega}\}$, *for* $\omega \in F_m(X)$.

Proof. See [7], [12].

Theorem 3. 1. *Let g be a bimeromorphic transformation of X. Then we have*

$$g^* V_m = V_m, \quad g^* B_m = B_m,$$
$$(g^*\omega_1, g^*\omega_2) = (\omega_1, \omega_2), \quad \textit{for} \quad \omega_1, \omega_2 \in F_m(X).$$

Proof. From the fact that $\|g^*\omega\| = \|\omega\|$ for $\omega \in F_m(X)$, it follows that $g^* V_m = V_m$, which implies the rest of the above equalities. q. e. d.

Let $S = \{\omega \in F_m(X) | (\omega, \omega)=1\}$ be the unit sphere in $F_m(X)$. Since $\| \ \|$ is continuous on $F_m(X)$, we can define

$$\eta = \max_{\omega \in S} \|\omega\|, \quad \tau = \min_{\omega \in S} \|\omega\|.$$

We have, by definition, $\eta \geqq \tau > 0$.

Lemma 3. 3. $\tau\sqrt{(\omega, \omega)} \leqq \|\omega\| \leqq \eta\sqrt{(\omega, \omega)}$, *for* $\omega \in F_m(X)$.

Proof. Take an element $\omega \in F_m(X)$. Then $\omega/\sqrt{(\omega, \omega)} \in S$. By definition, we get $\tau \leqq \|\omega/\sqrt{(\omega, \omega)}\| \leqq \eta$, which implies the above inequality. q. e. d.

Proposition 3. 1. $\tau^2 V_m \leqq B_m \leqq \eta^2 V_m$.

Proof. By definition, if $(\omega, \omega)=1$, then $\|\omega\| \leqq \eta$. Hence

$$B_m = \sup_{(\omega,\omega)=1} \{\mu_m \omega \wedge \bar{\omega}\} \leqq \sup_{\|\omega\| \leqq \eta} \{\mu_m \omega \wedge \bar{\omega}\}$$
$$= \eta^2 \sup_{\|\omega'\|=1} \{\mu_m \omega' \wedge \bar{\omega}'\} = \eta^2 V_m.$$

We can similarly prove the other side of the above inequality. q. e. d.

Remark. We say that $F_m(X)$ is base point free if for every point $x \in X$, there exists an element $\omega \in F_m(X)$ such that $\omega(x) \neq 0$. If $F_m(X)$ is base point free, then V_m and B_m become m-ple volume forms on X, and define metrics in the line bundle mK_X. For $\omega \in H^0(X, \mathcal{O}(mK_X))$, denote by $\|\omega\|_V$, $\|\omega\|_B$ the lengths of ω with respect to the metrics defined by V_m, B_m, respectively. Namely, we have

$$\|\omega\|_V^2 = \frac{\mu_m \omega \wedge \bar{\omega}}{V_m}, \quad \|\omega\|_B^2 = \frac{\mu_m \omega \wedge \bar{\omega}}{B_m},$$

and then

$$\|\omega\| = \left\{\int_X \|\omega\|_V^{2/m} v_m\right\}^{m/2} = \left\{\int_X \|\omega\|_B^{2/m} b_m\right\}^{m/2} \tag{3. 1}$$

We conclude that $\omega \in F_m(X)$ if and only if $\|\omega\|_V$ (or $\|\omega\|_B$) is uniformly bounded on X. In fact, if $\|\omega\|_V < C$, for a constant C on X, then by (3. 1), we get $\|\omega\| \leq C \int_X v_m < \infty$. On the other hand, if $\omega \in F_m(X)$, then by definition, $\|\omega\|_V \leq \|\omega\|$. In view of Proposition 3. 1, the assertion for $\|\omega\|_B$ follows from this.

In this case, the map Φ_m (cf. (2. 1)) is holomorphic and $\mathrm{Ric}(B_m)$ is induced from a standard Kähler form of $\boldsymbol{P}_N$ by Φ_m.

4. Extension of Holomorphic Maps

Let X be a complex manifold of dimension n. We say that X is *quasi-projective* if X is given as a complement of an analytic subset of a projective algebraic manifold. We shall consider the case in which $\kappa(X)=n$. First we recall the following

Proposition 4. 1. *Let X be a quasi-projective manifold of dimension n such that $\kappa(X)=n$. Then there exists a quasi-volume form Ψ on X such that* $(\mathrm{Ric}\ \Psi)^n \geq \Psi$.

Proof. See [7], Proposition 1. For later application, we define the form Ψ in the following way, which is a slight modification of that in [17]. Let $\bar{X}$ be a smooth compactification of X, which is projective algebraic. We may assume that $D=\bar{X}-X$ is a union of non-singular divisors D_i : $D=D_1+\cdots+D_k$, and that D has normal crossings. Cover $\bar{X}$ by coordinate neighborhoods $\{U_\alpha\}$ with holomorphic coordinates $(w_\alpha^1, \cdots, w_\alpha^n)$ in U_α. Let $\delta_i = \{\delta_{i,\alpha}\}$ be a holomorphic section of $[D_i]$ defining D_i and $\{a_{i,\alpha}\}$ a metric in $[D_i]$, for each i, and let $\{a_\alpha\} = \left\{\prod_{i=1}^k a_{i,\alpha}\right\}$ be a metric of $[D]$. Then $\|\delta_i\|^2 = |\delta_{i,\alpha}|^2/a_{i,\alpha}$ in U_α defines a length of δ_i. Let H be a very ample line bundle on $\bar{X}$. Further we assume that the line bundle $H+[D]$ is ample on $\bar{X}$. Since $\kappa(X)=n$, we have, by Proposition 2. 2, $\kappa(K_{\bar{X}}+D, \bar{X})=n$, and then there exists an effective divisor E in the complete linear system $|m(K_{\bar{X}}+D)-H-D|$ for a large integer m. Let $\{\sigma_\alpha\}$ be a holomorphic section of $[E]$ defining E. Let $\{h_\alpha\}$ be a metric of H such that $dd^c \log(h_\alpha a_\alpha)$ is a positive (1, 1) form on $\bar{X}$. Define Ψ by

$$\Psi = \frac{c\,\{a_\alpha h_\alpha |\sigma_\alpha|^2\}^{1/m}}{\prod_{i=1}^k (\log\|\delta_i\|^2)^2 |\delta_{i,\alpha}|^2} \prod_{i=1}^n (\sqrt{-1}/2\pi)\, dw_\alpha^i \wedge d\bar{w}_\alpha^i, \quad \text{in} \quad U_\alpha.$$

We obtain the desired form, by letting the constant c be small enough and multiplying the metrics $\{a_{i,\alpha}\}$ by suitable constants (See [17], for details). q. e. d.

Definition. The Poincaré volume forms on Δ^n and $\Delta^* \times \Delta^{n-1}$ are defined by

$$V_{\Delta^n} = \prod_{i=1}^{n} \frac{2}{(1-|z_i|^2)^2} \cdot \Phi,$$
$$V_{\Delta^* \times \Delta^{n-1}} = \frac{2}{(\log|z_1|^2)^2|z_1|^2} \prod_{i=2}^{n} \frac{2}{(1-|z_i|^2)^2} \cdot \Phi,$$

where $\Phi = \prod_{i=1}^{n} (\sqrt{-1}/2\pi) dz_i \wedge d\bar{z}_i$. Note that $(\mathrm{Ric}\, V)^n = V$, in both cases.

Lemma 4. 1. *Let X have the same meaning as in Proposition* 4. 1. *Let $f: \Delta^n \to X$ (resp. $f: \Delta^* \times \Delta^{n-1} \to X$) be a non-degenerate holomorphic map, i. e., a holomorphic map such that* $\mathrm{rank}(f) = \dim X = n$. *Then*

$$f^*\Psi \leqq V,$$

where V is the Poincaré volume form of Δ^n (resp. $\Delta^* \times \Delta^{n-1}$).

Proof. See the proof of Proposition 2 in [17].

As a consequence of the above lemma (Schwarz' lemma), we obtain the following

Proposition 4. 2 (cf. Proposition 3 in [17]). *Let X be the same as in Proposition* 4. 1. *Let $f: \Delta^* \times \Delta^{n-1} \to X$ be a non-degenerate holomorphic map. Then f extends to a meromorphic map from Δ^n to any compactification of X.*

Proof. Let $\bar{X}$ be a smooth compactification of X as in the proof of Proposition 4. 1. Let $\{U_\alpha\}$, D, δ_i, H, h_α, E and σ_α be the same as in Proposition 4. 1. Since H is very ample, there exists a basis $s_0, \cdots, s_N$ of $H^0(\bar{X}, \mathcal{O}(H))$, with $N+1 = \dim H^0(\bar{X}, \mathcal{O}(H))$ such that the map $\Phi_H : w \to (s_0(w) : \cdots : s_N(w))$ is an imbedding of $\bar{X}$ into $\boldsymbol{P}_N$. Therefore, it suffices to prove that the map $\Phi_H \circ f$ extends to a meromorphic map from Δ^n. Set $\delta = \prod_{i=1}^{k} \delta_i$. Let

$$\omega_i = \frac{s_{i,\alpha}\sigma_\alpha}{\delta_\alpha^{m-1}} (dw_\alpha^1 \wedge \cdots \wedge dw_\alpha^n)^m, \quad \text{in} \quad U_\alpha$$

be an m-ple n-form on X for each i. By Lemma 2. 1, each $\omega_i \in F_m(X)$.

$$(\omega_i \wedge \bar{\omega}_i)^{1/m} = \frac{|s_{i,\alpha}\sigma_\alpha|^{2/m}}{|\delta_\alpha|^{2(m-1)/m}} \prod_{i=1}^{n} (\sqrt{-1}/2\pi) dw_\alpha^i \wedge d\bar{w}_\alpha^i, \quad \text{in} \quad U_\alpha.$$

Letting $\|\delta_i\|$ be small enough by multiplying the metrics $\{a_{i,\alpha}\}$ by constants, we can make

$$\prod_{i=1}^{k} (\log\|\delta_i\|^2)^2 \|\delta_i\|^{2/m} \leqq 1.$$

Hence

$$(\omega_i \wedge \bar{\omega}_i)^{1/m} \leqq \frac{1}{\prod_{i=1}^{k} (\log\|\delta_i\|^2)^2 \|\delta_i\|^{2/m}} (\omega_i \wedge \bar{\omega}_i)^{1/m} \leqq (1/c)\Psi,$$

by multiplyng the metric $\{h_\alpha\}$ by a constant as $\|s_i\| \leqq 1$. Combining this with Lemma 3. 1, we get

$$(f^*\omega_i \wedge \overline{f^*\omega_i})^{1/m} \leqq (1/c) V,$$

which implies that

$$\int_{\Delta_r^* \times \Delta_r^{n-1}} (f^*\omega_i \wedge \overline{f^*\omega_i})^{1/m} < \infty \quad \text{for} \quad 0<r<1.$$

By Lemma 1. 1, $f^*\omega_i$ extends meromorphically to Δ_r^n, and then to Δ^n by letting $r \to 1$. Noting that $(f^*\omega_0 : \cdots : f^*\omega_N) = (f^*s_0 : \cdots : f^*s_N)$, we infer that $\Phi_H \circ f$ extends to a meromorphic map from Δ^n into $\boldsymbol{P}_N$. q. e. d.

Proposition 4. 3. *Let X be as in Proposition 4. 1. Let $\bar{X}$ be a compactification of X. Then every biholomorphic transformation of X extends to a bimeromorphic transformation of $\bar{X}$.*

Let X be a complex manifold. Given a Borel subset Ξ in X, choose holomorphic maps $f_i : \Delta^n \to X$ and Borel subsets Ξ_i in Δ^n, such that $\Xi \subset \bigcup_i f_i(\Xi_i)$. Define

$$\mu_X(\Xi) = \inf \sum_i \int_{\Xi_i} V,$$

where the infimum is taken over all possible choices of f_i, Ξ_i, and V is the Poincaré volume form on Δ^n. We say that X is *measure-hyperbolic* if $\mu_X(\Xi) > 0$ for all non-empty open subsets Ξ in X (cf. [8]).

Theorem 4. 1. *Let X be a quasi-projective manifold of dimension n such that $\kappa(X) = n$. Then X is measure-hyperbolic.*

Proof. Let Ξ be a nonempty open set in X and choose holomorphic maps f_i : $\Delta^n \to X$ and Borel subsets Ξ_i such that $\Xi \subset \bigcup_i f_i(\Xi_i)$. Then, we have by Lemma 4. 1,

$$\sum_i \int_{\Xi_i} V \geqq \sum \int_{\Xi_i} f_i^* \Psi = \sum_i \int_{f_i(\Xi_i)} \Psi \geqq \int_{\Xi} \Psi,$$

which implies that $\mu_X(\Xi) \geqq \int_\Xi \Psi > 0$. q. e. d.

Corollary. *Let X be as above. Then there exists no non-degenerate holomorphic map $f : \boldsymbol{C} \times \Delta^{n-1} \to X$.*

5. Representation of the Group of Bimeromorphic Transformations

Let X be a complex manifold. Denote by $\mathrm{Bim}\,(X)$ (resp. $\mathrm{Aut}(X)$) the group of bimeromorphic transformations (resp. biholomorphic transformations) of X. Moreover in case X has a compactification $\bar{X}$, we denote by $\mathrm{Bim}_{\mathrm{alg}}(X)$ the group of bimeromorphic transformations of X which can be extended to bimeromorphic transformations of $\bar{X}$. So $\mathrm{Bim}_{\mathrm{alg}}(X) = \mathrm{Bim}(\bar{X}) \cap \mathrm{Bim}(X)$.

Now assume that X is a complex manifold of dimension n having a smooth compactification $\bar{X}$. In view of Proposition 1. 4 and Theorem 3. 1, we have a unitary representation

$$\rho_m : \mathrm{Bim}(X) \to \mathrm{U}(F_m(X)),$$

where $U(F_m(X)) \subset GL(F_m(X))$ is the group of unitary matrices with respect to the hermitian form of $F_m(X)$ defined in Section 3. We shall prove the following

Theorem 5. 1. *Let X be a complex manifold of dimension n having a smooth compactification $\bar{X}$. Assume that $\bar{X}$ is a Moišezon manifold. Then $\rho_m(\mathrm{Bim}_{\mathrm{alg}}(X))$ is a finite group.*

As a corollary, we obtain

Theorem 5. 2. *Let X be a quasi-projective manifold of dimension n such that $\kappa(X)=n$. Then* $\mathrm{Aut}(X)$ *is a finite group.*

Proof. By Proposition 4. 3, every element of $\mathrm{Aut}(X)$ extends to an element of $\mathrm{Bim}(\bar{X})$. So the above theorem implies that $\rho_m(\mathrm{Aut}(X))$ is a finite group. Since $\kappa(X)=n$, the map Φ_m (defined by (2. 1)) becomes a bimeromorphic map for a large integer m (cf. the proof of Proposition 4. 3.). Hence the representation ρ_m is faithful for this m. It follows that $\mathrm{Aut}(X)$ is a finite group. q. e. d.

Corollary. *Let X be a complex manifold of dimension n with a smooth compactification $\bar{X}$. Put $D=\bar{X}-X$. If $K_{\bar{X}}+[D]$ is ample, then* $\mathrm{Bim}(X)=\mathrm{Aut}(X)$ *is a finite group.*

Example. Let D be a hypersurface of degree d in $\boldsymbol{P}_n$ which has at most normal crossings. If $d>n+1$, then $\mathrm{Bim}(\boldsymbol{P}_n-D)=\mathrm{Aut}(\boldsymbol{P}_n-D)$ is a finite group.

Remark. Some partial results are obtained in [3], [4]. See also [6].

Proof of Theorem 5. 1. We follow the argument of Ueno [18], Nakamura-Ueno [13].

Lemma 5. 1 (cf. [18], Proposition 14. 4.). *For an element $g \in \mathrm{Bim}_{\mathrm{alg}}(X)$, if $g^*\omega = \lambda\omega$ holds for some non-zero element $\omega \in F_m(X)$, then λ is an algebraic integer.*

Proof. By hypothesis, g extends to an element $\bar{g} \in \mathrm{Bim}(\bar{X})$. If $m=1$, then we have seen that $F_1(X) \cong H^0(\bar{X}, \mathcal{O}(K_{\bar{X}}))$. The assertion is the same as in [18]. In fact, $\bar{g}$ induces an endomorphism of $H^n(\bar{X}, \boldsymbol{Z})_0$ and then λ is an algebraic integer and the degree of the minimal equation of λ is bounded by the n-th Betti number $b_n(\bar{X})$ of $\bar{X}$.

Consider the case $m \geqq 2$. Let $\{U_\alpha\}$ be coordinate neighborhoods of $\bar{X}$ with holomorphic coordinates $(w_\alpha^1, \cdots, w_\alpha^n)$ in U_α. Put $D=\bar{X}-X$. Let $\{\delta_\alpha\}$ be a holomorphic section of $[D]$ defining D and let $\{k_{\alpha\beta}\}$ be transition functions of $K_{\bar{X}}$. By Lemma 2. 1, we can write ω locally as

$$\omega = \frac{\varphi_\alpha}{\delta_\alpha^{m-1}}(dw_\alpha^1 \wedge \cdots \wedge dw_\alpha^n)^m, \quad \text{in} \quad U_\alpha,$$

where $\{\varphi_\alpha\}$ is a holomorphic section of $H^0(\bar{X}, \mathcal{O}(mK_{\bar{X}}+(m-1)[D]))$. Let $\pi: M \to \bar{X}$ be the $\boldsymbol{P}_1$-bundle over $\bar{X}$ corresponding to the canonical bundle $K_{\bar{X}}$ of $\bar{X}$. We may assume that $M_{|U_\alpha} \cong U_\alpha \times \boldsymbol{P}_1$. We choose coordinates (w_α, ζ_α) on $U_\alpha \times \boldsymbol{P}_1$, with inhomogeneous coordinates ζ_α of $\boldsymbol{P}_1$, such that $\zeta_\alpha = k_{\alpha\beta}\zeta_\beta$ in $M_{|U_\alpha \cap U_\beta}$. We see that $\bar{g}$ induces a bimeromorphic transformation $\hat{g}$ of M by

$$\hat{g} : (w_\alpha, \zeta_\alpha) \to (\bar{g}(w_\alpha), \det(\partial \bar{g}/\partial w_\alpha)^{-1}\zeta_\alpha).$$

Let m_ν be the automorphism of M defined by

$$m_\nu : (w_\alpha, \zeta_\alpha) \to (w_\alpha, \nu\zeta_\alpha), \quad \text{in } M_{|U_\alpha}, \text{ with } \nu^m = \lambda.$$

Let $\bar{Y}$ be the subvariety of M defined by

$$\zeta_\alpha^m \delta_\alpha^{m-1}(w_\alpha) = \varphi_\alpha(w_\alpha), \quad \text{in} \quad M_{|U_\alpha}.$$

Then $\bar{Y}$ is an m-fold covering of $\bar{X}$. By a sequence of monoidal transformations of M, we can desingularize $\bar{Y}$. Let $\pi_1 : M^* \to M$ be such a desingularization and let $\bar{Y}^*$ be the strict transform of $\bar{Y}$ by π_1. Put $f = \pi \circ \pi_1$ and $Y^* = f^{-1}_{|\bar{Y}^*}(X)$. Then Y^* is a generically m-fold covering of X. Moreover, $m_\nu \circ \hat{g}$ induces a bimeromorphic transformation h of $\bar{Y}^*$. Define a meromorphic n-form Θ on M by

$$\Theta = \zeta_\alpha dw_\alpha^1 \wedge \cdots \wedge dw_\alpha^n \quad \text{in} \quad M_{|U_\alpha}.$$

Then Θ can be lifted to a meromorphic n-form on M^* and induces a meromorphic n-form θ on $\bar{Y}^*$. By definition, we get

$$(\theta)^m = f^*(\omega).$$

Further, since $m_\nu \hat{g}^* \Theta = \nu\Theta$, we have $h^*(\theta) = \nu\theta$. By the proof of Proposition 1. 3, we see that $f^*(\omega)_{|Y^*} \in F_m(Y^*)$, from which follows that $\theta \in F_1(Y^*)$. Hence θ is a holomorphic n-form on $\bar{Y}^*$. Therefore ν is an algebraic integer, by the case in which $m=1$. Moreover $[\boldsymbol{Q}(\lambda) : \boldsymbol{Q}] \leqq b_n(\bar{Y}^*)$. q. e. d.

Lemma 5. 2. *Let X, g, ω and λ be the same as in Lemma* 5. 1. *Then λ is a root of unity.*

Proof. For the proof of this fact, we need the hypothesis that $\bar{X}$ is a Moišezon manifold and then we can assume that $\bar{X}$ is projective algebraic. Because ρ_m is a unitary representation, we see that $|\lambda| = 1$. The assertion follows from the proof of Proposition 14. 5 in [18]. q. e. d.

We proceed to the proof of Theorem 5. 1. For $g \in \mathrm{Bim}_{\mathrm{alg}}(X)$, $\rho_m(g)$ is of finite order by Lemma 5. 2. To prove the theorem, it suffices to show that the order of $\rho_m(g)$ is uniformly bounded for all $g \in \mathrm{Bim}_{\mathrm{alg}}(X)$ (Burnside). This can be established in a manner similar to that in [18], Theorem 14. 10. q. e. d.

Appendix

Let W be a compact complex manifold of dimension n and D an effective (reduced) divisor on W. As we have seen in Section 2, to calculate the Kodaira dimension $\kappa(W-D)$, we need to desingularize D. Here we examine the process of desingularization more precisely. According to Hironaka, we can find a sequence of monoidal transformations $\pi_i : W_i \to W_{i-1}$ with non-singular centers C_{i-1} for $i=1, \cdots, l$ such that

(i) $W_0 = W$, $W_l = W^*$,

(ii) $D_i =$ the support of $\pi_i^*(D_{i-1})$,

(iii) $D_l = D^*$ has at most normal crossings,

(iv) $\pi : W^* - D^* \to W - D$ is biholomorphic, where $\pi = \pi_l \circ \cdots \circ \pi_1$.

Thus W^* is a smooth compactification of $W-D$. We use the following notations : $\bar{D}_i$=the strict transform of D_{i-1} by π_i ; E_i=the exceptional locus of π_i, i. e., $\pi_i^{-1}(C_{i-1})$; δ_i=the codimension of C_{i-1} in W_{i-1} ; ν_i=the multiplicity of the singular locus of D_{i-1} along C_{i-1}. Then we have

$$D_i = \bar{D}_i + E_i, \quad \pi_i^*(D_{i-1}) = \bar{D}_i + \nu_i E_i, \quad K_{W_i} = \pi_i^*(K_{W_{i-1}}) + (\delta_i - 1)[E_i].$$

It follows that

$$\text{(A. 1)} \qquad \begin{aligned} K_{W_i} + [D_i] &= \pi_i^*(K_{W_{i-1}} + [D_{i-1}]) + (\delta_i - \nu_i)\,[E_i], \\ mK_{W_i} + (m-1)[D_i] &= \pi_i^*(mK_{W_{i-1}} + (m-1)[D_{i-1}]) \\ &\quad + \{m(\delta_i - \nu_i) + (\nu_i - 1)\}\,[E_i]. \end{aligned}$$

Proposition A. 1. *We have*

$$\begin{aligned} \gamma_m(W-D) &= \dim H^0(W^*, \mathcal{O}(mK_{W^*} + (m-1)[D^*])) \\ &\leqq \dim H^0(W, \mathcal{O}(mK_W + (m-1)[D])). \end{aligned}$$

Proof. Set $L_i = mK_{W_i} + (m-1)\,[D_i]$ and $b_i = m(\delta_i - \nu_i) + (\nu_i - 1)$. By (A. 1), we see that $L_i = \pi_i^*(L_{i-1}) + b_i\,[E_i]$. It suffices to prove that

$$\text{(A. 2)} \qquad \dim H^0(W_i, \mathcal{O}(L_i)) \leqq \dim H^0(W_{i-1}, \mathcal{O}(L_{i-1})),$$

for each i. If $b_i \leqq 0$, (A. 2) is clear. In case $b_i > 0$, for any effective divisor Z belonging to the complete linear system $|L_i|$, the direct image $T = \pi_{i*}(Z)$ belongs to $|L_{i-1}|$, which implies that $Z - \pi_i^*(T)$ is linearly equivalent to $b_i E_i$ and then $Z = \pi_i^*(T) + b_i E_i$, because E_i is exceptional. Therefore the map $\pi_{i*} : |L_i| \to |L_{i-1}|$ is an isomorphism. Thus we obatin the equality $\dim H^0(W_i, \mathcal{O}(L_i)) = \dim H^0(W_{i-1}, \mathcal{O}(L_{i-1}))$.

q. e. d.

Lemma A. 1. *If* $\dim H^0(W_i, \mathcal{O}(mK_{W_i} + (m-1)\,[D_i])) > 0$ *for some positive integer* m, *then we have*

$$\kappa(K_{W_i} + D_i, W_i) = \kappa(K_{W_{i-1}} + D_{i-1}, W_{i-1}).$$

Proof. Let $\Gamma_i = K_{W_i} + [D_i]$ for each i. By (A. 1), we obtain the equality $\Gamma_i = \pi_i^*(\Gamma_{i-1}) + (\delta_i - \nu_i)[E_i]$. In case $(\delta_i - \nu_i) \geqq 0$, we can show that $\dim H^0(W_i, \mathcal{O}(\Gamma_i)) = \dim H^0(W_{i-1}, \mathcal{O}(\Gamma_{i-1}))$ in a similar manner as in the proof of the above proposition. Consider the case in which $(\delta_i - \nu_i) < 0$. Using (A. 1), we get

$$\pi_i^*(m\Gamma_{i-1}) = mK_{W_i} + (m-1)[D_i] + [\bar{D}_i] + \{m(\nu_i - \delta_i) + 1\}\,[E_i].$$

By hypothesis, there is an effective divisor $Z \in |mK_{W_i} + (m-1)D_i|$, and then we get

$$\begin{aligned} \kappa(\Gamma_{i-1}, W_{i-1}) &= \kappa(\pi_i^*(\Gamma_{i-1}), W_{i-1}) \\ &= \kappa(Z + \bar{D}_i + \{m(\nu_i - \delta_i) + 1\}\,E_i, W_i) \\ &= \kappa(Z + \bar{D}_i + E_i, W_i) \qquad \text{(cf. Lemma 5, in [17])} \\ &= \kappa(\Gamma_i, W_i). \end{aligned}$$

q. e. d.

As a consequence of this lemma, we obtain

Proposition A. 2 (Iitaka). *If* $\kappa(W-D)\geqq 0$, *then we have*

$$\kappa(K_{W*}+D^*, W^*) = \kappa(K_W+D, W).$$

Proof. Since $\kappa(W-D)\geqq 0$, we have $\dim H^0(W^*, \mathcal{O}(mK_{W*}+(m-1)D^*))>0$, for some positive integer m. By Lemma A. 1, we get

$$\kappa(K_{W*}+D^*, W^*) = \kappa(K_{W_{l-1}}+D_{l-1}, W_{l-1})$$

and by (A. 2), we have

$$\dim H^0(W_{l-1}, \mathcal{O}(mK_{W_{l-1}}+(m-1)[D_{l-1}])) \geqq \dim H^0(W^*, \mathcal{O}(mK_{W*}+(m-1)[D^*])) > 0.$$

Therefore by induction, we can prove Proposition A. 2. q. e. d.

Combining this with Proposition 2. 1, we have the following

Theorem A. 1. *Let W be a compact complex manifold and D an effective divisor on W. If* $\kappa(W-D)\geqq 0$, *then*

$$\kappa(W-D) = \kappa(K_W+D, W).$$

Example. Let H_1, H_2, H_3, H_4 be four lines in $\boldsymbol{P}_2$ which meet as in the diagram below. Set $D=H_1+H_2+H_3+H_4$. In this case, we have

$$\begin{cases} \kappa(K_{\boldsymbol{P}_2}+D, \boldsymbol{P}_2) = 2, \\ \bar{\kappa}(\boldsymbol{P}_2-D) = 1, \\ \kappa(\boldsymbol{P}_2-D) = -\infty. \end{cases}$$

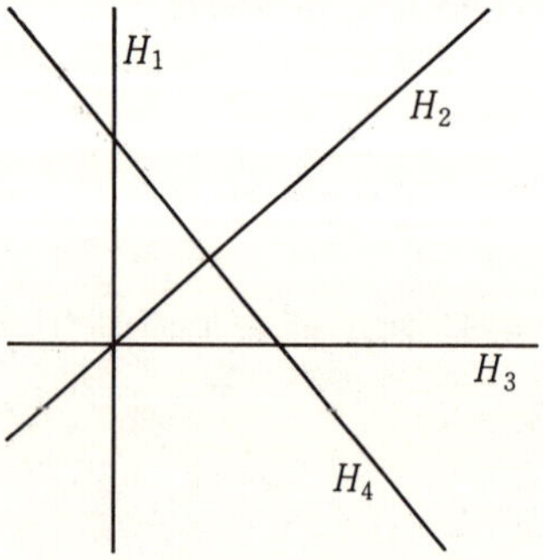

References

[1] Andreotti, A. and Stoll, W.: Analytic and algebraic dependence of meromorphic functions. Lecture Notes in Mathematics 234, Springer, Berlin-Heidelberg-New York, 1971.

[2] Griffiths, P.: Holomorphic mappings into canonical algebraic varieties. Ann. of Math. **93**

(1971), 439–458.
[3] Carlson, J.: Some degeneracy theorems for entire functions in an algebraic varieties. Trans. Amer. Math. Soc. **168** (1972), 273–301.
[4] Fujimoto, H.: Families of holomorphic maps into the projective space omitting hyperplanes. J. Math. Soc. Japan **25** (1973), 235–249.
[5] Iitaka, S.: On *D*-dimensions of algebraic varieties. J. Math. Soc. Japan **23** (1971), 356–373.
[6] ——: On logarithmic Kodaira dimension of algebraic varieties, in this Volume.
[7] Kobayashi, S.: Geometry of bounded domains. Trans. Amer. Math. Soc. **92** (1959), 267–290.
[8] ——: Hyperbolic manifolds and holomorphic mappings. Marcel Dekker, New York, 1970.
[9] —— and Ochiai, T.: Mappings into compact complex manifolds with negative first Chern class. J. Math. Soc. Japan **23** (1971), 137–148.
[10] Kodaira, K.: On Kähler varieties of restricted type. Ann. of Math. **60** (1954), 28–48.
[11] ——: On holomorphic mappings of polydiscs into compact complex manifolds. J. Diff. Geometry **6** (1971), 31–46.
[12] Lichnérowicz, A.: Varietés complexes et tenseur de Bergman. Ann. Ins. Fourier **15** (1965), 345–408.
[13] Nakamura, I. and Ueno, K.: An addition formula for Kodaira dimensions of analytic fibre bundles whose fibre are Moišezon manifolds. J. Math. Soc. Japan **25** (1973), 363–371.
[14] Narasimhan, M. S. and Simha, R. R.: Manifolds with ample canonical class. Invent. Math. **5** (1968), 120–128.
[15] Remmert, R.: Holomorphe und meromorphe Abbildungen komplexer Räume. Math. Ann. **133** (1957), 328–370.
[16] —— and Stein, K.: Eigentliche holomorphe Abbildungen. Math. Zeitschr. **73** (1960), 159–189.
[17] Sakai, F.: Degeneracy of holomorphic maps with ramification. Invent. Math. **26** (1974), 213–229.
[18] Ueno, K.: Classification theory of algebraic varieties and compact complex spaces. Lecture Notes in Mathematics 439, Springer, Berlin-Heidelberg-New York, 1975.

Department of Mathematics
Kochi University

(Received February 9, 1976)

Compact Quotients of C^3 by Affine Transformation Groups, II

T. Suwa[1)]

Let G denote a group of affine transformations of the three-dimensional complex vector space $\boldsymbol{C}^3$. Assume that the action of G is free and properly discontinuous and that the quotient $\boldsymbol{C}^3/G$ is compact. This is the second part of a study of compact complex threefolds of the form $\boldsymbol{C}^3/G$. The main results were announced in [4].

Each element g in G is represented by a 4×4 marix

$$g=\begin{bmatrix} A(g) & b(g) \\ 0 & 1 \end{bmatrix},$$

where $A(g)$ (the holonomy part) is in $GL(3, \boldsymbol{C})$ and $b(g)$ (the translation part) is a column vector in $\boldsymbol{C}^3$. The action of g on $\boldsymbol{C}^3=\{z|z={}^t(z_1, z_2, z_3)\}$ is given by

$$\begin{bmatrix} z \\ 1 \end{bmatrix} \to \begin{bmatrix} A(g) & b(g) \\ 0 & 1 \end{bmatrix} \begin{bmatrix} z \\ 1 \end{bmatrix}.$$

In the first part [5], it has been shown that, if G contains no elements whose holonomy parts have three different eigenvalues, then G contains a nilpotent subgroup N of finite index. Thus if G satisfies the condition given above, the quotient $\boldsymbol{C}^3/G$ is finitely covered by $\boldsymbol{C}^3/N$, where N is nilpotent. In the present paper, the quotients $\boldsymbol{C}^3/N$ are classified. In section 1, the number of linearly independent closed holomorphic 1-forms is computed in each case. We determine, in section 2, the structure of $\boldsymbol{C}^3/N$ by analyzing the Albanese map, which is defined by Blanchard [1]. The quotients $\boldsymbol{C}^3/N$ are classified in Tables I and II. They include some new non-Kähler manifolds. In section 3, we study holomorphic forms on $\boldsymbol{C}^3/N$.

1. Closed Holomorphic 1-Forms on C^3/N

We may assume ([4]) that, for every element g in N,

$$(1) \qquad A(g)=\begin{bmatrix} 1 & a_{12}(g) & a_{13}(g) \\ 0 & 1 & a_{23}(g) \\ 0 & 0 & 1 \end{bmatrix}.$$

Also let ${}^tb(g)=(b_1(g), b_2(g), b_3(g))$. For a pair (g_1, g_2) of elements of N, we let $[g_1, g_2]=g_1g_2g_1^{-1}g_2^{-1}$. For functions $a_i: N\to\boldsymbol{C}$, $1\le i\le r$, we define $[a_1, \cdots, a_r]: N^r=\overbrace{N\times\cdots\times N}^{r}$

1) Research supported in part by NSF grant GP 43614 X.

$\to \boldsymbol{C}$ by $[a_1, \cdots, a_r]\,(g_1, \cdots, g_r) = \det\,(a_i(g_j))$. For each fixed r, the functions : $N^r \to \boldsymbol{C}$ form a $\boldsymbol{C}$-algebra in a natural manner. We have

$$(2)\qquad [g_1, g_2] = \begin{bmatrix} 1 & 0 & [a_{12}, a_{23}]\,(g_1, g_2) & * \\ 0 & 1 & 0 & [a_{23}, b_3]\,(g_1, g_2) \\ 0 & 0 & 1 & 0 \\ 0 & 0 & 0 & 1 \end{bmatrix},$$

where $* = ([a_{12}, b_2] + [a_{13}, b_3])(g_1, g_2) - [a_{12}, a_{23}]\,(g_1, g_2) \cdot (b_3(g_1) + b_3(g_2))$. Let $U_i = \{z_i\}$, $1 \leq i \leq 3$, denote the i-th factor of $\boldsymbol{C}^3 = \boldsymbol{C} \times \boldsymbol{C} \times \boldsymbol{C}$. For each fixed z_3, the commutator group $N^{(1)} = [N, N]$ of N acts on $U_1 \times U_2 \times \{z_3\}$ effectively and properly discontinuously as a group of translations. Hence we have $N^{(1)} \subset \boldsymbol{Z}^4 = \boldsymbol{Z} \oplus \boldsymbol{Z} \oplus \boldsymbol{Z} \oplus \boldsymbol{Z}$, where $\boldsymbol{Z}$ denotes the ring of integers (Wolf [6], 2. 5. 4. Lemma). Note that N, being the fundamental group of a compact space, is finitely generated. Let $\mathcal{O}$ denote the structure sheaf of a complex manifold.

In this section, we compute the number $d = \dim H^0(\boldsymbol{C}^3/N, d\mathcal{O})$ of linearly independent closed holomorphic 1-forms on the quotient $\boldsymbol{C}^3/N$. We may identify $H^0(\boldsymbol{C}^3/N, d\mathcal{O})$ with the space $H^0(\boldsymbol{C}^3, d\mathcal{O})^N$ of N-invariant closed holomorphic 1-forms on $\boldsymbol{C}^3$. Take $\omega \in H^0(\boldsymbol{C}^3, d\mathcal{O})^N$ and let

$$\omega = \sum_{i=1}^{3} f_i(z) dz_i, \quad f_i \in H^0(\boldsymbol{C}^3, \mathcal{O}), 1 \leq i \leq 3.$$

Since the pullback $g^*\omega$ of ω by $g \in N$ is given by $g^*\omega = f_1(gz)dz_1 + (f_1(gz)a_{12}(g) + f_2(gz))dz_2 + (f_1(gz)a_{13}(g) + f_2(gz)a_{23}(g) + f_3(gz))dz_3$, the N-invariance conditions are

$$(3)\qquad f_1(z) = f_1(gz),$$

$$(4)\qquad f_2(z) = f_1(gz)a_{12}(g) + f_2(gz), \text{ and}$$

$$(5)\qquad f_3(z) = f_1(gz)a_{13}(g) + f_2(gz)a_{23}(g) + f_3(gz), \text{ for all } g \in N.$$

Also, since ω is closed, we have

$$(6)\qquad \frac{\partial f_2}{\partial z_1} = \frac{\partial f_3}{\partial z_1} = 0, \quad \frac{\partial f_2}{\partial z_3} = \frac{\partial f_3}{\partial z_2}.$$

By (3), we may think of f_1 as a holomorphic function on $\boldsymbol{C}^3/N$, which is compact. Hence $f_1(z) = c_1$ is a constant. From (4), we have $\frac{\partial f_2}{\partial z_2}(z) = \frac{\partial f_2}{\partial z_2}(gz)$. Hence $\frac{\partial f_2}{\partial z_2} = c_2$ is a constant and we have $f_2(z) = c_2 z_2 + \varphi_1(z_3)$, where φ_1 is a holomorphic function of z_3. From (4), we infer that φ_1 must be a polynomial in z_3 of degree at most 2. Therefore we have

$$(7)\qquad f_2(z) = c_2 z_2 + c_3 + c_4 z_3 + c_5 z_3^2.$$

From (6), we have $\partial f_3/\partial z_2 = \partial f_2/\partial z_3 = c_4 + 2c_5 z_3$. Hence $f_3(z) = (c_4 + 2c_5 z_3) z_2 + \varphi_2(z_3)$, where φ_2 is a holomorphic function of z_3. From (5), we infer that φ_2 must be a polynomial of degree at most 3. Therefore we have

$$(8)\qquad f_3(z) = (c_4 + 2c_5 z_3) z_2 + c_6 + c_7 z_3 + c_8 z_3^2 + c_9 z_3^3.$$

The N-invariance conditions (3), (4) and (5) become

$$(9)\qquad c_1 a_{12} + c_2 b_2 + c_4 b_3 + c_5 b_3^2 = 0,$$

$$(10)\qquad c_2 a_{23} + 2c_5 b_3 = 0,$$

(11) $$c_5a_{23}+c_9b_3=0,$$

(12) $$c_1a_{13}+c_3a_{23}+c_4(a_{23}b_3+b_2)+c_7b_3+c_8b_3^2=0,$$

(13) $$2c_4a_{23}+c_5(2b_2-a_{23}b_3)+2c_8b_3=0.$$

Definition 1. A *permissible coordinate transformation* (p. c. t., hereafter) of $\boldsymbol{C}^3$ is a coordinate transformation (analytic automorphism) φ of $\boldsymbol{C}^3$ such that (1) for any $g\in N$, $\varphi g\varphi^{-1}$ is also an affine transformation, (2) if the (i,j)-entry of g is 0, so is the (i,j)-entry of $\varphi g\varphi^{-1}$, and (3) the diagonals of $\varphi g\varphi^{-1}$ are all 1.

Lemma 1. *If* $[a_{23}, b_3]=0$, *by a suitable p. c. t. of* $\boldsymbol{C}^3$, *we may assume that* $a_{23}=0$.

Proof. If $b_3=0$, the quotient $\boldsymbol{C}^3/N$ would be a fiber space over $U_3=\boldsymbol{C}$. Hence we have $b_3(g_0)\neq 0$, for some $g_0\in N$. If $[a_{23}, b_3]=0$, letting $\alpha=a_{23}(g_0)/b_3(g_0)$, we have $a_{23}=\alpha b_3$. Apply the p. c. t. $\varphi(z_1, z_2, z_3)=(z_1, z_2-\frac{\alpha}{2}z_3^2, z_3)$, then the (2, 3)-entry of $\varphi g\varphi^{-1}$ becomes zero for every $g\in N$, q. e. d.

Case I) Suppose $N^{(1)}=0$.

Lemma 2. *In this case, by a suitable coordinate transformation of* $\boldsymbol{C}^3$, *every element of* N *is reduced to a translation* (cf. [2] Theorem 2. 1).

Proof. By Lemma 1, we may assume that $a_{23}=0$. Since $\boldsymbol{C}^3/N$ is compact, there is an element g_0 with $b_2(g_0)\neq 0$, and $b_3(g_0)\neq 0$. Letting $\alpha=a_{12}(g_0)/b_2(g_0)$ and $\beta=a_{13}(g_0)/b_3(g_0)$, consider the coordinate transformation $\varphi(z_1, z_2, z_3)=(z_1-\frac{\alpha}{2}z_2^2-\frac{\beta}{2}z_3^2, z_2, z_3)$. Then φ is permissible and g_0 is reduced to a translation. By a suitable linear transformation on $U_2\times U_3$, we may assume that $A(g_0)=I$, ${}^tb(g_0)=(b_1(g_0), 0, 1)$. Then we have $b_1([g, g_0])=a_{13}(g)$ for all $g\in N$. Since $N^{(1)}=0$, we have $a_{13}=0$. Then we have $b_1([g_1, g_2])=[a_{12}, b_2]\,(g_1, g_2)$, for all $g_1, g_2\in N$. Since $N^{(1)}=0$, we have $a_{12}=\gamma b_2$ for some constant γ. Finally every element is reduced to a translation by the p. c. t. $\phi(z_1, z_2, z_3)=(z_1-\frac{\gamma}{2}z_2^2, z_2, z_3)$, q. e. d.

By the above lemma, the quotient $\boldsymbol{C}^3/N$ is a complex 3-torus. Therefore $d=3$ and dz_1, dz_2, dz_3 form a basis for $H^0(\boldsymbol{C}^3/N, d\mathcal{O})=H^0(\boldsymbol{C}^3, d\mathcal{O})^N$.

Case II) Suppose $N^{(1)}\neq 0$ and $[a_{23}, b_3]=0$. By Lemma 1, we may assume $a_{23}=0$. Thus the functions a_{12}, a_{13}, b_2, and b_3 are homomorphisms of N into the additive group $\boldsymbol{C}$. Let N_1 and Δ be, respectively, the kernel and the image of the homomorphism $g\to(b_2(g), b_3(g))$ of N into $\boldsymbol{C}^2=U_2\times U_3$. We have $N^{(1)}\subset N_1$. Moreover, since N has no fixed points on $\boldsymbol{C}^3$, we have $N_1=\{g\in N|a_{12}(g)=a_{13}(g)=b_2(g)=b_3(g)=0\}$. Therefore, N_1 acts on $U_1=\boldsymbol{C}$ effectively and properly discontinuously as a group of translations. Thus $N_1\subset\boldsymbol{Z}^2$. Let V denote the real vector subspace of $U_2\times U_3$ spanned by Δ. Note that Δ is a finitely generated free abelian group.

Lemma 3. $\dim_{\mathbf{R}} V=4$ *and hence* rank $\Delta\geq 4$.

Proof. We have $\mathbf{C}^3/N=(\mathbf{C}^3/N_1)/(N/N_1)=(U_1/N_1\times U_2\times U_3)/(N/N_1)$. If $\dim_{\mathbf{R}} V<4$, we would have $U_2\times U_3=V\times W$ for some positive dimensional real vector subspace W of $U_2\times U_3$. Then, since $N/N_1\simeq\Delta$ acts trivially on W, $\mathbf{C}^3/N=((U_1/N_1\times V)/(N/N_1))\times W$, which is not compact, q. e. d.

Lemma 4. $[b_2, b_3]\neq 0$.

Proof. If $[b_2, b_3]=0$, then $b_2=\alpha b_3$ for some $\alpha\in\mathbf{C}$, contradicting Lemma 3, q. e. d.

Since $a_{23}=0$ and $b_3\neq 0$, we have, from (10) and (11), $c_5=c_9=0$. Then (13) implies $c_8=0$. The N-invariance conditions reduce to

$$(9)' \qquad c_1a_{12}+c_2b_2+c_4b_3=0,$$

$$(12)' \qquad c_1a_{13}+c_4b_2+c_7b_3=0.$$

From these we get $c_1([a_{12}, b_2]+[a_{13}, b_3])=0$. Since $N^{(1)}\neq 0$, we have $c_1=0$. Lemma 4, (9)′, and (12)′ imply $c_2=c_4=c_7=0$. The constants c_3 and c_6 are arbitrary. Therefore $d=2$ and dz_2, dz_3 form a basis for $H^0(\mathbf{C}^3/N, d\mathcal{O})$.

Case III) Suppose $[a_{23}, b_3]\neq 0$, and $a_{12}=0$. From (10), (11), and (13), we get $c_2=c_4=c_5=c_8=c_9=0$. The invariance condition is

$$(12)'' \qquad c_1a_{13}+c_3a_{23}+c_7b_3=0.$$

Since $a_{12}=0$, the functions a_{13}, a_{23}, and b_3 are homomorphisms. Take a set $g_1, g_2, \cdots, g_r$ of generators of N and set $D_{ijk}=[a_{13}, a_{23}, b_3]\,(g_i, g_j, g_k)$, $1\leq i, j, k\leq r$.

Lemma 5. *If* $D_{ijk}=0$ *for all* i, j, k, *then the situation is reduced to that of case II.*

Proof. From the condition, we infer that the equation (12)″ has a non-trivial solution for (c_1, c_3, c_7). Since $[a_{23}, b_3]\neq 0$, we have $c_1\neq 0$. Thus we have $a_{13}=\alpha a_{23}+\beta b_3$ for some $\alpha, \beta\in\mathbf{C}$. Applying the p. c. t. $\varphi(z_1, z_2, z_3)=(z_1-\alpha z_2-\frac{\beta}{2}z_3^2, z_2, z_3)$ of $\mathbf{C}^3$, we see that we may assume $a_{13}=0$. Then, by interchanging z_1 and z_2, the situation is reduced to that of case II, q. e. d.

In view of Lemma 5, we assume that $D_{ijk}\neq 0$ for some i, j, k. Then from (12)″, we have $c_1=c_3=c_7=0$. Therefore $d=1$, and dz_3 is a basis for $H^0(\mathbf{C}^3/N, d\mathcal{O})$.

Case IV) Suppose $[a_{23}, b_3]\neq 0$ and $a_{12}\neq 0$. From (10), (11), and (13), we get $c_2=c_4=c_5=c_8=c_9=0$. Also, from (9), we have $c_1=0$. Then (12) implies $c_3=c_7=0$. Therefore, $d=1$ and dz_3 is a basis for $H^0(\mathbf{C}^3/N, d\mathcal{O})$.

Summarizing the above we have the following

Theorem 1. i) *By a suitable coordinate transformation of* $\mathbf{C}^3$, *every case is reduced to one of the followings* :

(I) $N^{(1)}=0$,
(II) $N^{(1)}\neq 0$, $a_{23}=0$ *and* $[b_2, b_3]\neq 0$,
(III) $a_{12}=0$ *and* $[a_{13}, a_{23}, b_3]\neq 0$,
(IV) $a_{12}\neq 0$, *and* $[a_{23}, b_3]\neq 0$.

ii) *In each case, we can choose the following set as a basis for* $H^0(\boldsymbol{C}^3/N, d\mathcal{O})=H^0(\boldsymbol{C}^3, d\mathcal{O})^N$:

(I) dz_1, dz_2, dz_3; (II) dz_2, dz_3; (III) *and* (IV) dz_3.

2. Structures

In this scetion, we determine the structure of the quotient $\boldsymbol{C}^3/N$ by analyzing the Albanese map of $\boldsymbol{C}^3/N$. Recall that the Albanese variety $\mathrm{Alb}(M)$ of a compact complex manifold M is defined as follows (see Blanchard [1]). Consider the space D dual to $H^0(M, d\mathcal{O})$, and let Δ denote the image of the homomorphism $F: H_1(M, \boldsymbol{Z})\to D$ given by the integration of 1-forms on the 1-cycle γ, i. e. $F(\gamma)=\int_\gamma$. Also let $\bar{\Delta}$ denote the smallest closed subgroup of D containing Δ whose component of 0 is a complex vector subspace of D. Then $\mathrm{Alb}(M)$ is the complex torus $D/\bar{\Delta}$. Fix a point z_0 on M. The Albanese map $M\to\mathrm{Alb}(M)$ is defined by $z\to\int_{z_0}^{z}$ (mod. $\bar{\Delta}$).

Let $\eta: N=\pi_1(\boldsymbol{C}^3/N)\to H_1(\boldsymbol{C}^3/N, \boldsymbol{Z})$ be the canonical surjection (Hurewicz homomorphism), and let $N_1=\ker F\circ\eta$. We have the exact sequences:

$$(*)\qquad \begin{array}{ccccccccc} 1 \to & N^{(1)} & \to & N & \xrightarrow{\eta} & H_1(\boldsymbol{C}^3/N, \boldsymbol{Z}) & \to 0 \\ & \cap & & \| & & \downarrow & \\ 1 \to & N_1 & \to & N & \xrightarrow{F\circ\eta} & \Delta & \to 0. \end{array}$$

Note that Δ is a finitely generated free abelian group. We will see below that $N_1\subset\boldsymbol{Z}^4$ and that rank N_1+rank $\Delta=6$. Thus the group N has at most 6 generators. We denote by T^k a k-dimensional complex torus.

Case I) $N^{(1)}=0$. As was proved in section 1, the quotient $\boldsymbol{C}^3/N$ is T^3, and $\mathrm{Alb}(\boldsymbol{C}^3/N)=\boldsymbol{C}^3/N$.

Case II) $N^{(1)}\neq 0$, $a_{23}=0$, and $[b_2, b_3]\neq 0$. We identify D with $U_2\times U_3$ by taking the basis dual to (dz_2, dz_3) for D. Then we have $F\circ\eta(g)=\left(\int_{\eta(g)}dz_2, \int_{\eta(g)}dz_3\right)=(b_2(g), b_3(g))$. Hence N_1 coincides with the kernel of the natural action of N on $U_2\times U_3$. Moreover, if $h\in N_1$, then $a_{12}(h)=a_{13}(h)=0$, since otherwise h would have a fixed point on $\boldsymbol{C}^3$.

Lemma 6. $N_1\subset\boldsymbol{Z}^2$.
Proof. See section 1, Case II).

Lemma 7. *We may assume that there are elements* g_1 *and* g_2 *in* N *which are represented* *as follows* :

$$g_1=\begin{bmatrix}1&0&\alpha&\beta_1\\0&1&0&1\\0&0&1&0\\0&0&0&1\end{bmatrix},\quad g_2=\begin{bmatrix}1&0&0&\beta_2\\0&1&0&0\\0&0&1&1\\0&0&0&1\end{bmatrix}.$$

Proof. We have $[b_2, b_3]\,(g_1, g_2)\neq 0$ for some g_1, g_2 in N. By a suitable linear transformation on $U_2\times U_3$, we may assume $(b_2(g_1), b_3(g_1))=(1, 0)$ and $(b_2(g_2), b_3(g_2))=(0, 1)$. Consider the p. c. t. $\psi(z_1, z_2, z_3)=(z_1-\frac{a_{12}(g_1)}{2}z_2^2-\frac{a_{13}(g_2)}{2}z_3^2, z_2, z_3)$ of $\boldsymbol{C}^3$. In terms of the new coordinate system, g_1 and g_2 are represented by

$$g_1=\begin{bmatrix}1&0&\alpha_1'&\beta_1'\\0&1&0&1\\0&0&1&0\\0&0&0&1\end{bmatrix}\quad\text{and}\quad g_2=\begin{bmatrix}1&\alpha_2'&0&\beta_2'\\0&1&0&0\\0&0&1&1\\0&0&0&1\end{bmatrix}.$$

Finally if we apply the p. c. t. $\psi(z_1, z_2, z_3)=(z_1-\alpha_2' z_2 z_3, z_2, z_3)$, then g_1 and g_2 are represented as in the statement, q. e. d.

Let V denote the real vector subspace of $U_2\times U_3$ spanned by Δ as in § 1. By Lemma 3, $\dim_{\boldsymbol{R}} V=4$ and rank $\Delta\geq 4$. Let $h_1, \cdots, h_s (s\leq 2)$ be a $\boldsymbol{Z}$-basis for N_1 and choose $g_3, \cdots, g_r (r\geq 4)$ so that $(b_2(g_1), b_3(g_1)), \cdots, (b_2(g_r), b_3(g_r))$ form a $\boldsymbol{Z}$-basis for $\Delta=\operatorname{Im} F\circ\eta=\operatorname{Im} F$. Then N is generated by $h_1, \cdots, h_s, g_1, \cdots, g_r$. Finally let $k_1, \cdots, k_t$ $(t\leq s)$ be a $\boldsymbol{Z}$-basis for $N^{(1)}$. Lettig m_i, $1\leq i\leq t$, be the invariant factors of $N^{(1)}$ in N, we may assume $k_i=h_i^{m_i}$, $1\leq i\leq t$.

II i) Suppose $N_1=\boldsymbol{Z}$. In this case, $N^{(1)}=\boldsymbol{Z}$. We may assume that $A(h_1)=I$ and ${}^t b(h_1)=(1, 0, 0)$.

Lemma 8. $r=\operatorname{rank}\Delta>4$.

Proof. Suppose rank $\Delta=4$. Then Δ is a lattice in $U_2\times U_3$, and $N/N_1\simeq\Delta$ acts properly discontinuously on $U_2\times U_3$. Since $N_1=\boldsymbol{Z}$, the quotient $\boldsymbol{C}^3/N=(\boldsymbol{C}^3/N_1)/(N/N_1)$ $=((\boldsymbol{C}/\boldsymbol{Z})\times U_2\times U_3)/(N/N_1)$ is a $\boldsymbol{C}^*$-bundle over T^2, which is a contradiction,

q. e. d.

We have $[g_i, g_j]=h_1^{n_{ij}}$, for some $n_{ij}\in\boldsymbol{Z}$, $1\leq i, j\leq r$. Hence we have

(14) $$([a_{12}, b_2]+[a_{13}, b_3])(g_i, g_j)=n_{ij},\ 1\leq i, j\leq r.$$

In particular, we have

(15) $$\alpha=n_{12},$$

(16) $$a_{12}(g_i)-\alpha b_3(g_i)=n_{i1},$$

(17) $$a_{13}(g_i)=n_{i2},\quad 1\leq i\leq r.$$

Lemma 9. *For some* $\lambda\in\boldsymbol{C}$, *we have*

(18) $$a_{12}+\lambda(a_{13}-nb_2)=0,\quad n=n_{12}.$$

Proof. Let n_1 and n_2 be defined by $n_1(g_i)=n_{i1}$ and $n_2(g_i)=n_{i2}$. Substituting (15), (16), and (17) in (14), we have

$$(19)\qquad ([n_1, b_2]+[n_2, b_3]-n[b_2, b_3])(g_i, g_j) = n_{ij}.$$

From (19), we have

$$(20)\qquad [n_1, b_2, b_3](g_i, g_j, g_k) = n_{ij}b_3(g_k)+n_{jk}b_3(g_i)+n_{ki}b_3(g_j),$$

$$(21)\qquad [n_2, b_2, b_3](g_i, g_j, g_k) = -(n_{ij}b_2(g_k)+n_{jk}b_2(g_i)+n_{kj}b_2(g_j)),$$

$$(22)\qquad ([n_1, n_2, b_3]-n[n_1, b_2, b_3])(g_i, g_j, g_k) = n^{(1)}_{ijk},$$

$$(23)\qquad ([n_1, n_2, b_2]+n[n_2, b_2, b_3])(g_i, g_j, g_k) = -n^{(2)}_{ijk},$$

where $n^{(l)}_{ijk}=n_{ij}n_{kl}+n_{jk}n_{il}+n_{ki}n_{jl}$, $1\leq i, j, k\leq r$, $l=1, 2$. Letting $m_{ij}=[n_1, n_2]\ (g_i, g_j) -nn_{ij}(\in \boldsymbol{Z})$, we have from (20)–(23),

$$(24)\qquad n^{(2)}_{ijk}\begin{bmatrix}1\\0\end{bmatrix}-n^{(1)}_{ijk}\begin{bmatrix}0\\1\end{bmatrix}+m_{ij}\begin{bmatrix}b_2(g_k)\\b_3(g_k)\end{bmatrix}+m_{jk}\begin{bmatrix}b_2(g_i)\\b_3(g_i)\end{bmatrix}+m_{ki}\begin{bmatrix}b_2(g_j)\\b_3(g_j)\end{bmatrix}=0.$$

Recall $(b_2(g_1), b_3(g_1))=(1, 0)$ and $(b_2(g_2), b_3(g_2))=(0, 1)$ (Lemma 7). From Lemma 8, we must have

$$(25)\qquad n^{(1)}_{ijk} = n^{(2)}_{ijk} = m_{ij} = 0,\quad 1\leq i, j, k\leq r.$$

Since $n_1=a_{12}-nb_3$ and $n_2=a_{13}$, we have from (19)

$$\begin{aligned}0 = m_{ij}&=[n_1, n_2]\ (g_i, g_j)-nn_{ij}\\ &= ([a_{12}-nb_3, a_{13}]-n([a_{12}-nb_3, b_2]+[a_{13}, b_3]-n[b_2, b_3]))(g_i, g_j)\\ &= ([a_{12}, a_{13}]-n[a_{12}, b_2])(g_i, g_j) = [a_{12}, a_{13}-nb_2](g_i, g_j).\end{aligned}$$

Hence we infer that the equations

$$(26)\qquad c_1a_{12}(g_i)+c_2(a_{13}(g_i)-nb_2(g_i)) = 0,\quad 1\leq i\leq r,$$

have a non-trivial solution for (c_1, c_2). If $c_1=0$, then $nb_2(g_i)=a_{13}(g_i)=n_{i2}$. Hence $\dim_{\boldsymbol{R}} V\leq 3$, which is a contradiction. Thus $c_1\neq 0$. Hence for some $\lambda\in \boldsymbol{C}$, $a_{12}(g_i)+\lambda(a_{13}(g_i)-nb_2(g_i))=0$, $1\leq i\leq r$. Since N is generated by $h_1=k_1^{m_1}$, $g_1, \cdots, g_r$ and a_{12}, a_{13}, b_2 are homomorphisms, we have $a_{12}(g)+\lambda(a_{13}(g)-nb_2(g))=0$ for all $g\in N$, q. e. d.

Lemma 10. *We may assume* $a_{12}=0$.

Proof. If $\lambda=0$, we have $a_{12}=0$. Suppose $\lambda\neq 0$. Consider the coordinate transformation $\varphi(z_1, z_2, z_3)=(z_1', z_2', z_3')=(z_1, z_2, z_2-\lambda^{-1}z_3)$. Then, in terms of the new coordinate system, every element $g\in N$ is represented by

$$(27)\qquad g=\begin{bmatrix}1 & a_{12}'(g) & a_{13}'(g) & b_1'(g)\\ 0 & 1 & 0 & b_2'(g)\\ 0 & 0 & 1 & b_3'(g)\\ 0 & 0 & 0 & 1\end{bmatrix}=\begin{bmatrix}1 & a_{12}(g)+\lambda a_{13}(g) & -\lambda a_{13}(g) & b_1(g)\\ 0 & 1 & 0 & b_2(g)\\ 0 & 0 & 1 & b_2(g)-\lambda^{-1}b_3(g)\\ 0 & 0 & 0 & 1\end{bmatrix}.$$

From (18), we have $a_{12}'(g)=n\lambda b_2'(g)$. Applying the p. c. t. $\psi(z_1', z_2', z_3')=(z_1'-\frac{n\lambda}{2}z_2'^2, z_2', z_3')$, we see that we may assume $a_{12}'=0$, q. e. d.

Let N_2 and Δ_1 denote, respectively, the kernel and the image of the homomorphism $g \to b_3(g)$ of N into the additive group $U_3 = \boldsymbol{C}$. Also let V_1 denote the real vector subspace of U_3 spanned by Δ_1. Since $\boldsymbol{C}^3/N$ is compact, $\dim_{\boldsymbol{R}} V_1 = 2$ and rank $\Delta_1 \geq 2$.

Lemma 11. *If $g \in N_2$, then $a_{13}(g) = 0$.*

Proof. Take $g \in N_2$. Then, for any $h \in N$, we have $b_1([g, h]) = a_{13}(g) b_3(h)$. If $a_{13}(g) \neq 0$, then, since rank $\Delta_1 \geq 2$, we have rank $N^{(1)} \geq 2$, which is a contradiction,

q. e. d.

By Lemma 11, the group N_2 acts on $U_1 \times U_2$ effectively and properly discontinuously as a group of translations. Hence $N_2 \subset \boldsymbol{Z}^4$. Let rank $\Delta_1 = p$ and choose $f_1, \ldots, f_p \in N$ so that $b_3(f_1), \cdots, b_3(f_p)$ form a $\boldsymbol{Z}$-basis for Δ_1. Also let $l_1, \cdots, l_q$ be a $\boldsymbol{Z}$-basis for N_2. Clearly the group N is generated by $f_1, \cdots, f_p, l_1, \cdots, l_q$.

Lemma 12. *$N/N_2 \simeq \Delta_1$ acts properly discontinuously on U_3.*

Proof. We have $[f_i, f_j] = h_1^{m_{ij}}$, for some $m_{ij} \in \boldsymbol{Z}$. Since $a_{12} = 0$, we have $[a_{13}, b_3](f_i, f_j) = m_{ij}$. Therefore, $m_{ij} b_3(f_k) + m_{jk} b_3(f_i) + m_{ki} b_3(f_j) = 0$. Suppose $p \geq 3$. Then we have $m_{ij} = 0$, $1 \leq i, j \leq p$. Since the group N is generated by $f_1, \cdots, f_p, l_1, \cdots, l_q$, and a_{13}, and b_3 are homomorphisms, we have $[a_{13}, b_3] = 0$ and $N^{(1)} = 0$, which is a contradiction. Thus $p \leq 2$ and the group N/N_2 acts properly discontinuously on U_3,

q. e. d.

Theorem 2. *In Case* II, *if $N_1 = \boldsymbol{Z}$,*

1) *the Albanese map of $\boldsymbol{C}^3/N$ gives a T^2-bundle structure on $\boldsymbol{C}^3/N$ over $Alb(\boldsymbol{C}^3/N) = T^1$,*

2) *the group N is generated by*

$$l_i = \begin{bmatrix} 1 & 0 & 0 & b_1(l_i) \\ 0 & 1 & 0 & b_2(l_i) \\ 0 & 0 & 1 & 0 \\ 0 & 0 & 0 & 1 \end{bmatrix}, \quad f_j = \begin{bmatrix} 1 & 0 & a_{13}(f_j) & b_1(f_j) \\ 0 & 1 & 0 & b_2(f_j) \\ 0 & 0 & 1 & b_3(f_j) \\ 0 & 0 & 0 & 1 \end{bmatrix}, \quad 1 \leq i \leq 4, j = 1, 2,$$

where a) $(b_1(l_1), b_2(l_1)) = (1, 0)$, $(b_1(l_2), b_2(l_2)) = (0, 1)$, $b_3(f_1) = 1$, b) $(b_1(l_i), b_2(l_i))$, $1 \leq i \leq 4$, *are linearly independent over* $\boldsymbol{R}$, *and* $b_3(f_j)$, $j = 1, 2$, *are linearly independent over* $\boldsymbol{R}$, c) $(b_2(l_i), 0)$, $2 \leq i \leq 4$, *and* $(b_2(f_j), b_3(f_j))$, $j = 1, 2$, *are linearly independent over* $\boldsymbol{Z}$, d) $[a_{13}, b_3](f_1, f_2) = m \in \boldsymbol{Z} - 0$,

3) $H_1(\boldsymbol{C}^3/N, \boldsymbol{Z}) = \boldsymbol{Z}^5 \oplus \boldsymbol{Z}_m$.

Proof. 1) We have $\boldsymbol{C}^3/N = (\boldsymbol{C}^3/N_2)/(N/N_2) = (((U_1 \times U_2)/N_2) \times U_3)/(N/N_2)$. By Lemma 12, the canonical projection $\boldsymbol{C}^3/N \to U_3/\Delta_1$ is a bundle map onto the manifold U_3/Δ_1 with fiber $(U_1 \times U_2)/N_2$. Since $\boldsymbol{C}^3/N$ is compact, and Δ_1 and N_2 are groups of translations, $U_3/\Delta_1 = T^1$ and $(U_2 \times U_3)/N_2 = T^2$ are complex tori. Clearly $\tilde{\Delta} = U_2 \times \Delta_1$, Alb $(\boldsymbol{C}^3/N) = U_2 \times U_3/\tilde{\Delta} = U_3/\Delta_1$ and the Albanese map is the bundle map. 2) From 1), we have $p = 2$, and $q = 4$. The assertion b) also follows from 1). By suitable linear transformations of $U_2 \times U_3$ and U_3, we may assume the situation is as in a). Lemma 8 implies c). Since $a_{12} = a_{23} = 0$, $N^{(1)}$ is generated by $[f_1, f_2]$. Moreover, since $N^{(1)} \subset N_2$, we may let $[f_1, f_2] = l_1^m$, for some $m \in \boldsymbol{Z} - 0$, which yields d). 3) In view of the exact sequence (*), it suffices to show that $\eta(l_2)$, $\eta(l_3)$, $\eta(l_4)$, $\eta(f_1)$, $\eta(f_2)$ are linearly independent over $\boldsymbol{Z}$. Suppose we have $\sum_{i=2}^{4} m_i\eta(l_i)+\sum_{j=1}^{2} n_j\eta(f_j)=0$. Then we have $g=\prod_{i=2}^{4} l_i^{m_i}\cdot\prod_{j=1}^{2} f_j^{n_j}\in N^{(1)}$. Hence $0=b_3(g)=\sum_{j=1}^{2} n_j b_3(f_j)$, which yields $n_j=0$, $j=1, 2$. Then we have $m_i=0$, $i=2, 3, 4$, q. e. d.

Remark. From 2a, c), we have rank $\Delta=5$.

II ii) Suppose $N_1=\boldsymbol{Z}^2$. In this case we have the following

Theorem 3. *In case* II, *if* $N_1=\boldsymbol{Z}^2$,

1) *the Albanese map of* $\boldsymbol{C}^3/N$ *gives a* T^1*-bundle structure on* $\boldsymbol{C}^3/N$ *over* $\mathrm{Alb}(\boldsymbol{C}^3/N)=T^2$,

2) *the group* N *is generated by*

$$h_i=\begin{bmatrix}1 & 0 & 0 & b_1(h_i)\\ 0 & 1 & 0 & 0\\ 0 & 0 & 1 & 0\\ 0 & 0 & 0 & 1\end{bmatrix},\quad g_i=\begin{bmatrix}1 & a_{12}(g_j) & a_{13}(g_j) & b_1(g_j)\\ 0 & 1 & 0 & b_2(g_j)\\ 0 & 0 & 1 & b_3(g_j)\\ 0 & 0 & 0 & 1\end{bmatrix},\quad i=1, 2,\quad 1\leq j\leq 4,$$

where a) $b_1(h_i)$, $i=1, 2$, *are linearly independent over* $\boldsymbol{R}$, $(b_2(g_j), b_3(g_j))$, $1\leq j\leq 4$, *are linearly independent over* $\boldsymbol{R}$, b) $[g_i, g_j]$, $1\leq i, j\leq 4$, *generate* $N^{(1)}$,

3) $$H_1(\boldsymbol{C}^3/N, \boldsymbol{Z})=\begin{cases}\boldsymbol{Z}^5\oplus\boldsymbol{Z}_{m_1}, & \text{if } N^{(1)}=\boldsymbol{Z},\\ \boldsymbol{Z}^4\oplus\boldsymbol{Z}_{m_1}\oplus\boldsymbol{Z}_{m_2}, & \text{if } N^{(1)}=\boldsymbol{Z}^2.\end{cases}$$

Proof. 1) Since $N_1=\boldsymbol{Z}^2$, $U_1/N_1=T^1$ is a complex torus. We have $\boldsymbol{C}^3/N=(\boldsymbol{C}^3/N_1)/(N/N_1)=(T^1\times U_2\times U_3)/(N/N_1)$, and the projection $\boldsymbol{C}^3/N_1\to U_2\times U_3$ is proper. Hence $N/N_1\simeq\Delta$ acts properly discontinuously on $U_2\times U_3$, and $\boldsymbol{C}^3/N$ is a T^1-bundle over $T^2=U_2\times U_3/\Delta$. Clearly $\mathrm{Alb}(\boldsymbol{C}^3/N)=(U_2\times U_3)/\Delta$, and the bundle map is the Albanese map. 2) From 1), we have $s=2$, and $r=4$. The assertion a) also follows from 1). Since $a_{23}=0$, a_{12}, a_{13}, b_2, and b_3 are homomorphisms. Hence we have b). 3) if $N^{(1)}=\boldsymbol{Z}$, we may assume $h_1^{m_1}=k_1$, $m_1\in\boldsymbol{Z}-0$. In view of the exact sequence (*), it suffices to prove that $\eta(h_2)$, $\eta(g_1)$, $\cdots$, $\eta(g_4)$ are linearly independent over $\boldsymbol{Z}$. Suppose $n\eta(h_2)+\sum_{j=1}^{4} n_j\eta(g_j)=0$. Then $g=h_2^n\prod_{j=1}^{4} g_j^{n_j}\in N^{(1)}\subset N_1$. Hence $(0, 0)=(b_2(g), b_3(g))=\sum_{j=1}^{4} n_j(b_2(g_j), b_3(g_j))$. Therefore $n_j=0$, $1\leq j\leq 4$. Then we have also $n=0$. If $N^{(1)}=\boldsymbol{Z}^2$, we may assume $h_i^{m_i}=k_i$, $m_i\in\boldsymbol{Z}-0$, $i=1, 2$. We can similarly prove that $\eta(g_j)$, $1\leq j\leq 4$, are linearly independent over $\boldsymbol{Z}$, q. e. d.

Remarks. 1. By Lemma 6, we may let $a_{12}(g_1)=a_{12}(g_2)=a_{13}(g_2)=0$.

2. By suitable linear transformations of U_1 and $U_2\times U_3$, we may let $b_1(h_1)=1$, $(b_2(g_1), b_3(g_1))=(1, 0)$, and $(b_2(g_2), b_3(g_2))=(0, 1)$. Also let $b_1(h_2)=\omega$. Then, in terms of the entries, the condition 2b) is given as follows. If $N^{(1)}=\boldsymbol{Z}$, $a_{12}(g_1)$, $a_{12}(g_i)-a_{12}(g_1)b_3(g_i)$, $a_{13}(g_i)$, $i=3, 4$, and $([a_{12}, b_2]+[a_{13}, b_3])$ (g_3, g_4) are integers. The integer m_1 in 3) is the g. c. d. of those integers. If $N^{(1)}=\boldsymbol{Z}^2$, those complex numbers generate a subgroup of rank 2 in the free abelian group generated by 1 and ω. The integers m_1 and m_2 in 3) are the invariant factors of this subgroup.

Case III) $a_{12}=0$ and $[a_{13}, a_{23}, b_3]\neq 0$. We identify U_3 with D by taking the basis dual to dz_3 for D. Then we have $F\circ\eta(g)=\int_{\eta(g)} dz_3=b_3(g)$. Hence N_1 coincides with the kernel of the natural action of N on U_3. For each fixed z_3, N_1 acts on $U_1\times U_2\times\{z_3\}$ effectively and properly discontinuously as a group of translations. Hence we have the following lemma (cf. [5], 2. 5. 4. Lemma).

Lemma 13. $N_1\subset \boldsymbol{Z}^4$.

Let V denote the real vector subspace of U_3 spanned by $\varDelta$. Since $\boldsymbol{C}^3/N$ is compact, we have $\dim_{\boldsymbol{R}} V=2$ (cf. the proof of Lemma 3). Hence rank $\varDelta\geq 2$. Let $h_1, \cdots, h_s$ and $k_1, \cdots, k_t$ be, respectively, $\boldsymbol{Z}$-bases of N_1 and $N^{(1)}(\subset N_1)$ such that $k_i=h_i^{m_i}$, $m_i\in \boldsymbol{Z}-0$, $1\leq i\leq t\leq s\leq 4$. Also choose $g_1,\cdots, g_r$, $r\geq 2$, so that $b_3(g_1), \cdots, b_3(g_r)$ form a $\boldsymbol{Z}$-basis for $\varDelta$. From the equation (2), we have $A([g, h])=I$, ${}^tb([g, h])=([a_{13}, b_3]\,(g, h), [a_{23}, b_3]\,(g, h), 0)$, for all $g, h\in N$.

III i) $N^{(1)}=\boldsymbol{Z}$. By a suitable coordinate transformation, we may assume $A(k_1)=I$, ${}^tb(k_1)=(b_1(k_1), 0, 0)$. Then we have $[a_{23}, b_3]=0$, which contradicts the condition $[a_{13}, a_{23}, b_3]\neq 0$. Hence $N^{(1)}$ cannot be $\boldsymbol{Z}$.

III ii) $N^{(1)}=\boldsymbol{Z}^2$. Since $N^{(1)}$ acts on $U_1\times U_2$ properly discontinuously as a group of translations, the vectors $v_i=(b_1(k_i), b_2(k_i))$, $i=1, 2$, are linearly indepedent over $\boldsymbol{R}$. If they are linearly dependent over $\boldsymbol{C}$, we may assume ${}^tb(k_i)=(b_1(k_i), 0, 0)$, $i=1, 2$. Then we have $[a_{23}, b_3]=0$, which is a contradiction. Hence v_1 and v_2 are linearly independent over $\boldsymbol{C}$, and we may assume ${}^tb(k_1)=(1, 0, 0)$ and ${}^tb(k_2)=(0, 1, 0)$.

Lemma 14. *If* $h\in N_1$, *then* $a_{13}(h)=a_{23}(h)=0$.

Proof. Take $h\in N_1$. Then $[h, g_i]=k_1^{n_i}k_2^{m_i}$ for some n_i, $m_i\in\boldsymbol{Z}$. Hence we have $a_{13}(h)b_3(g_i)=n_i$, $a_{23}(h)b_3(g_i)=m_i$. Since rank $\varDelta\geq 2$, we have $a_{13}(h)=a_{23}(h)=0$, q. e. d.

Lemma 15. rank $\varDelta=2$ *and* rank $N_1=4$.

Proof. If rank $N_1=4$, N/N_1 acts properly discontinuously on U_3. Hence rank $\varDelta=2$. Suppose rank $N_1\leq 3$. Then, since $\boldsymbol{C}^3/N$ is compact, we must have $r=\text{rank }\varDelta\geq 3$. We have $[g_i, g_j]=k_1^{n_{ij}}k_2^{m_{ij}}$ for some n_{ij}, $m_{ij}\in\boldsymbol{Z}$. This implies $[a_{13}, b_3](g_i, g_j)=n_{ij}$ and $[a_{23}, b_3](g_i, g_j)=m_{ij}$ and we have $n_{ij}b_3(g_k)+n_{jk}b_3(g_i)+n_{ki}b_3(g_j)=0$ and $m_{ij}b_3(g_k)+m_{jk}b_3(g_i)+m_{ki}b_3(g_j)=0$. Hence $n_{ij}=0$, $1\leq i,j\leq r$. For any $g\in N$, we have $b_3(g)=\sum_{i=1}^{r} n_ib_3(g_i)$ with $n_i\in\boldsymbol{Z}$. The element $g^{-1}\prod_{i=1}^{r} g_i^{n_i}$ is in N_1 so by Lemma 14, we have $a_{13}(g)=\sum_{i=1}^{r} n_ia_{13}(g_i)$ and $a_{23}(g)=\sum_{i=1}^{r} n_ia_{23}(g_i)$. Thus we have $[a_{13}, b_3]=[a_{23}, b_3]=0$, which is a contradiction, q. e. d.

Thus N is generated by $g_1, g_2, h_1, \cdots, h_4$. Since $[h_i, h_j]=[g_i, h_j]=1$, $N^{(1)}$ cannot be $\boldsymbol{Z}^2$.

III iii) $N^{(1)}=\boldsymbol{Z}^3$, $N_1=\boldsymbol{Z}^3$. In this case, $a_{13}(h)=a_{23}(h)=0$ for all $h\in N_1$. We may assume that $b_3(g_1)=1$. By the p. c. t. $\varphi(z_1, z_2, z_3)=(z_1-\frac{a_{13}(g_1)}{2}z_3^2, z_2-\frac{a_{23}(g_1)}{2}z_3^2, z_3)$, g_1 is reduced to a translation. Also we may assume that ${}^t b(h_1)=(1, 0, 0)$, ${}^t b(h_2)=(0, 1, 0)$, and ${}^t b(h_3)=(\omega_1, \omega_2, 0)$. For $g, h \in N$, we have $[g, g_1]=h_1^{n(g)}h_2^{m(g)}h_3^{l(g)}$, and $[g, h]=h_1^{n(g,h)}h_2^{m(g,h)}h_3^{l(g,h)}$, for some $n(g)$, $m(g)$, $l(g)$, $n(g, h)$, $m(g, h)$ and $l(g, h) \in \boldsymbol{Z}$. Hence we have

$$\begin{bmatrix} a_{13}(g) \\ a_{23}(g) \end{bmatrix} = n(g)\begin{bmatrix} 1 \\ 0 \end{bmatrix} + m(g)\begin{bmatrix} 0 \\ 1 \end{bmatrix} + l(g)\begin{bmatrix} \omega_1 \\ \omega_2 \end{bmatrix}, \tag{28}$$

$$\begin{bmatrix} [a_{13}, b_3](g, h) \\ [a_{23}, b_3](g, h) \end{bmatrix} = n(g, h)\begin{bmatrix} 1 \\ 0 \end{bmatrix} + m(g, h)\begin{bmatrix} 0 \\ 1 \end{bmatrix} + l(g, h)\begin{bmatrix} \omega_1 \\ \omega_2 \end{bmatrix}. \tag{29}$$

Let $z_\nu=x_\nu+iy_\nu$, $\omega_\mu=p_\mu+iq_\mu$, $\mu=1, 2, 3$, $\nu=1, 2$, $i=\sqrt{-1}$. We may assume $q_1 \neq 0$, then the vectors $v_1=(1, 0)$, $v_2=(0, 1)$, $v_3=(\omega_1, \omega_2)$, and $v_4=(0, i)$ are linearly independent over $\boldsymbol{R}$. Choose these vectors as a real basis for $U_1 \times U_2$. Let r_i, $1 \leq i \leq 4$, be defined by $(z_1, z_2)=\sum_{i=1}^{4} r_i v_i$. Moreover, let $a_{13}=r_{13}+is_{13}$, $a_{23}=r_{23}+is_{23}$, $b_\nu=t_\nu+iu_\nu$, $1 \leq \nu \leq 3$. Then, on the $(r_1, \cdots, r_4, x_3, y_3)$ space, $g \in N$ is represented by

$$g=\begin{bmatrix} 1 & 0 & 0 & 0 & * & * & * \\ 0 & 1 & 0 & 0 & * & * & * \\ 0 & 0 & 1 & 0 & * & * & * \\ 0 & 0 & 0 & 1 & \alpha(g) & \beta(g) & \gamma(g) \\ 0 & 0 & 0 & 0 & 1 & 0 & t_3(g) \\ 0 & 0 & 0 & 0 & 0 & 1 & u_3(g) \\ 0 & 0 & 0 & 0 & 0 & 0 & 1 \end{bmatrix},$$

where $\alpha=-\frac{q_2}{q_1}s_{13}+s_{23}$, $\beta=-\frac{q_2}{q_1}r_{13}+r_{23}$, $\gamma=-\frac{q_2}{q_1}u_1+u_2$. From (28) and (29), we have $\alpha=0$, $\beta=cu_3$ for some $c \in \boldsymbol{R}$. Consider the coordinate transformation $\varphi(r_1, \cdots, r_4, x_3, y_3)=(\xi_1, \cdots, \xi_6)=(r_1, r_2, r_3, r_4-\frac{c}{2}y_3^2, x_3, y_3)$. Then, on the ξ-space, g is represented by

$$g=\begin{bmatrix} 1 & 0 & 0 & 0 & \alpha_{15}(g) & \alpha_{16}(g) & \beta_1(g) \\ 0 & 1 & 0 & 0 & \alpha_{25}(g) & \alpha_{26}(g) & \beta_2(g) \\ 0 & 0 & 1 & 0 & \alpha_{35}(g) & \alpha_{36}(g) & \beta_3(g) \\ 0 & 0 & 0 & 1 & 0 & 0 & \beta_4(g) \\ 0 & 0 & 0 & 0 & 1 & 0 & \beta_5(g) \\ 0 & 0 & 0 & 0 & 0 & 1 & \beta_6(g) \\ 0 & 0 & 0 & 0 & 0 & 0 & 1 \end{bmatrix}, \tag{30}$$

where $\beta_4=-\frac{q_2}{q_1}u_1+u_2-\frac{c}{2}u_3^2$, $\beta_5=t_3$, $\beta_6=u_3$, $\beta_1=t_1-\frac{p_1}{q_1}u_1$, $\beta_2=-\frac{p_2}{q_1}u_1+t_2$, $\beta_3=\frac{1}{q_1}\cdot u_1$. Let $N_1'=\{g \in N | \beta_4(g)=\beta_5(g)=\beta_6(g)=0\}$.

Lemma 16. *$N_1'=N_1$. Thus, if $g \in N_1'$, $\alpha_{ij}(g)=0$, $1 \leq i \leq 3$, $j=5, 6$.*

Proof. We have $\beta_4(h_i)=\beta_5(h_i)=\beta_6(h_i)=0$. Hence $N_1 \subset N_1'$. Conversely, if $g \in N_1'$, $t_3(g)=u_3(g)=0$. Hence $N_1' \subset N_1$, q. e. d.

Theorem 4. *In Case* III, rank $N^{(1)} \geq 3$. *If* $N_1 = \boldsymbol{Z}^3$, *then* $N^{(1)} = \boldsymbol{Z}^3$, *and*

1) $\mathrm{Alb}(\boldsymbol{C}^3/N) = 0$,

2) *differentiably,* $\boldsymbol{C}^3/N$ *is a real* 3-*torus bundle over the real* 3-*torus,*

3) *the group* N *is generated by*

$$h_i = \begin{bmatrix} 1 & 0 & 0 & b_1(h_i) \\ 0 & 1 & 0 & b_2(h_i) \\ 0 & 0 & 1 & 0 \\ 0 & 0 & 0 & 1 \end{bmatrix}, \quad g_j = \begin{bmatrix} 1 & 0 & a_{13}(g_j) & b_1(g_j) \\ 0 & 1 & a_{23}(g_j) & b_2(g_j) \\ 0 & 0 & 1 & b_3(g_j) \\ 0 & 0 & 0 & 1 \end{bmatrix}, \quad 1 \leq i, j \leq 3,$$

where a) $(b_1(h_i), b_2(h_i))$, $1 \leq i \leq 3$, *are linearly independent over* $\boldsymbol{R}$, b) $b_3(g_j)$, $1 \leq j \leq 3$, *are linearly independent over* $\boldsymbol{Z}$, c) $[g_i, g_j]$, $1 \leq i, j \leq 3$, *generate* $N^{(1)}$,

4) $H_1(\boldsymbol{C}_3/N, \boldsymbol{Z}) = \boldsymbol{Z}^3 \oplus \boldsymbol{Z}_{m_1} \oplus \boldsymbol{Z}_{m_2} \oplus \boldsymbol{Z}_{m_3}$.

Proof. First we prove 2). Introduce, on $\boldsymbol{C}^3 = \boldsymbol{R}^6$, the ξ-coordinate system described above, and let $W_i = \{\xi_i\}$ denote the i-th factor of $\boldsymbol{R}^6$. From Lemma 16, $N_1' = N_1 = \boldsymbol{Z}^3$ acts on $W_1 \times W_2 \times W_3$ properly discontinuously as a group of translations. Hence we have $\boldsymbol{C}^3/N = (\boldsymbol{C}^3/N_1)/(N/N_1) = (\boldsymbol{R}^6/N_1')/(N/N_1') = (T^3_{\boldsymbol{R}} \times W_4 \times W_5 \times W_6)/(N/N_1')$, where $T^3_{\boldsymbol{R}} = W_1 \times W_2 \times W_3/N_1'$ is the real 3-torus. Since the projection $\boldsymbol{C}^3 \to W_4 \times W_5 \times W_6/(N/N_1')$ is proper, N/N_1' acts on $W_4 \times W_5 \times W_6$ properly discontinuously as a group of translations (see (30)). Since $\boldsymbol{C}^3/N$ is compact, $(W_4 \times W_5 \times W_6)/(N/N_1')$ must also be the real 3-torus. Therefore, $\boldsymbol{C}^3/N$ is (the total space of) a $T^3_{\boldsymbol{R}}$-bundle over $T^3_{\boldsymbol{R}}$. 1) We have $\Delta \simeq N/N_1 = N/N_1' = \boldsymbol{Z}^3$. Therefore, $\mathrm{Alb}(\boldsymbol{C}^3/N) = U_3/\bar{\Delta} = 0$. 3) From 2), we see that N is generated by $h_1, h_2, h_3, g_1, g_2, g_3$. The assertions a) and b) also follow from 2). Finally, since $a_{12} = 0$, a_{13}, a_{23}, and b_3 are homomorphisms. Hence $N^{(1)}$ is generated by $[g_i, g_j]$, $1 \leq i, j \leq 3$. 4) We have $h_i^{m_i} = k_i$, $m_i \in \boldsymbol{Z} - 0$, $1 \leq i \leq 3$. It is easy to show that $\eta(g_j)$, $1 \leq j \leq 3$, are linearly independent over $\boldsymbol{Z}$,

q. e. d.

Remarks. 1. By a suitable linear transformation of $U_1 \times U_2$, we may assume $(b_1(h_1), b_2(h_1)) = (1, 0)$, $(b_1(h_2), b_2(h_2)) = (0, 1)$.

2. In terms of the entries, the condition 3c) is given as follows. The pairs $([a_{13}, b_3](g_i, g_j), [a_{23}, b_3](g_i, g_j))$, $1 \leq i, j \leq 3$, generate a subgroup of rank 3 in the free abelian group generated by $(b_1(h_i), b_2(h_i))$, $1 \leq i \leq 3$. The integers m_i, $1 \leq i \leq 3$, in 4) are the invariant factors of this subgroup.

Definition 2. Let X and Y be complex manifolds. We say X is a *regular fiber space* over Y if there is a proper holomorphic map f of X onto Y such that the Jacobian matrix of f is everywhere of maximal rank. Thus $f: X \to Y$ is a complex analytic family of compact complex manifolds, and is differentiably a fiber bundle.

III iv) Suppose $N_1 = \boldsymbol{Z}^4$. In this case, we have the following

Theorem 5. *In Case* III, *if* $N_1 = \boldsymbol{Z}^4$, *then* $N^{(1)} = \boldsymbol{Z}^3$, *and*

1) *the Albanese map gives a regular fiber space structure on* $\boldsymbol{C}^3/N$ *over* $\mathrm{Alb}(\boldsymbol{C}^3/N) = T^1$. *Each fiber is a complex* 2-*torus, but the complex structure may vary,*

2) *the group* N *is generated by*

$$h_i=\begin{bmatrix}1 & 0 & a_{13}(h_i) & b_1(h_i)\\ 0 & 1 & a_{23}(h_i) & b_2(h_i)\\ 0 & 0 & 1 & 0\\ 0 & 0 & 0 & 1\end{bmatrix},\quad g_j=\begin{bmatrix}1 & 0 & a_{13}(g_j) & b_1(g_j)\\ 0 & 1 & a_{23}(g_j) & b_2(g_j)\\ 0 & 0 & 1 & b_3(g_j)\\ 0 & 0 & 0 & 1\end{bmatrix},\quad 1\leq i\leq 4,\ j=1,2,$$

where a) $a_{13}(h_i)=a_{23}(h_i)=0$, $1\leq i\leq 3$, b) $(b_1(h_i), b_2(h_i))$, $1\leq i\leq 3$, *and* $(a_{13}(h_4)z_3+b_1(h_4)$, $a_{23}(h_4)z_3+b_2(h_4))$ *are linearly independent over* $\boldsymbol{R}$, *for any* $z_3\in U_3=\boldsymbol{C}$, c) $b_3(g_j), j=1, 2$, *are linearly independent over* $\boldsymbol{R}$, d) $[h_4, g_i]$, $i=1, 2$, *and* $[g_1, g_2]$ *generate* $N^{(1)}$,

3) $H_1(\boldsymbol{C}^3/N, \boldsymbol{Z})=\boldsymbol{Z}^3\oplus\boldsymbol{Z}_{m_1}\oplus\boldsymbol{Z}_{m_2}\oplus\boldsymbol{Z}_{m_3}$.

Proof. For each fixed z_3, N_1 acts on $U_1\times U_2\times\{z_3\}$ properly discontinuously as a group of translations. Since $N_1=\boldsymbol{Z}^4$, the canonical projection $\boldsymbol{C}^3/N_1\to U_3$ gives a regular fiber space structure on $\boldsymbol{C}^3/N_1$ whose fibers are complex 2-tori $T^2_{z_3}$, $z_3\in U_3$. The group N/N_1 acts on $\boldsymbol{C}/N_1\to U_3$ as a fiber preserving automorphism group, and the projection is proper. Hence $N/N_1\simeq\Delta$ acts on U_3 properly discontinuously as a group of translations. Therefore, $\boldsymbol{C}^3/N=(\boldsymbol{C}^3/N_1)/(N/N_1)$ is a regular fiber space of complex 2-tori over $T^1=U_3/\Delta$. Clearly Alb $(\boldsymbol{C}^3/N)=U_3/\Delta$, and the map $\boldsymbol{C}^3/N\to U_3/\Delta$ is the Albanese map. 2) From 1), we see that N is generated by h_1, h_2, h_3, h_4, g_1, g_2. We claim $N^{(1)}=\boldsymbol{Z}^3$. If $N^{(1)}=\boldsymbol{Z}^4$, we may assume $h_i^{m_i}=k_i$, $m_i\in\boldsymbol{Z}-0$, $1\leq i\leq 4$. Hence $a_{13}(h_i)=a_{23}(h_i)=0$, $1\leq i\leq 4$. Since $a_{12}=0$, a_{13}, a_{23}, b_3 are homomorphisms. Hence $N^{(1)}$ is generated by $[h_i, h_j]$, $1\leq i, j\leq 4$, $[h_i, g_j]$, $1\leq i\leq 4, j=1$, 2, and $[g_1, g_2]$. But since $[h_i, h_j]=[h_i, g_j]=1$, $N^{(1)}$ cannot be $\boldsymbol{Z}^4$. Therefore $N^{(1)}=\boldsymbol{Z}^3$, and we may assume $h_i^{m_i}=k_i$, $1\leq i\leq 3$, which yields a). The assertions b) and c) follow from 1). Finally the condition $a_{12}=0$ implies d). 3) We have $h_i^{m_i}=k_i$, $1\leq i\leq 3$. Hence it suffices to show that $\eta(h_4)$, $\eta(g_1)$, and $\eta(g_2)$ are linearly independent over $\boldsymbol{Z}$. Suppose $n\eta(h_4)+\sum_{i=1}^{2}n_i\eta(g_i)=0$, n, $n_i\in\boldsymbol{Z}$, then the element $g=h_4^n g_1^{n_1}g_2^{n_2}$ is in $N^{(1)}$. Hence $0=b_3(g)=\sum_{i=1}^{2}n_ib_3(g_i)$, which yields $n_1=n_2=0$. Since any non-zero power of h_4 is not in $N^{(1)}$, we have $n=0$, q. e. d.

Remarks. 1. Since h_4 has no fixed points on $\boldsymbol{C}^3$, if $a_{13}(h_4)\neq 0$ or $a_{23}(h_4)\neq 0$, then $a_{13}(h_4)b_2(h_4)-a_{23}(h_4)b_1(h_4)\neq 0$.

2. We may assume $(b_1(h_1), b_2(h_1))=(1, 0)$, $(b_1(h_2), b_2(h_2))=(0, 1)$. The period matrix of the torus $T^2_{z_3}$ is $\begin{pmatrix}1 & 0 & \omega_1 & \alpha_1z_3+\omega_3\\ 0 & 1 & \omega_2 & \alpha_2z_3+\omega_4\end{pmatrix}$, where $\omega_1=b_1(h_3)$, $\omega_2=b_2(h_3)$, $\omega_3=b_1(h_4)$, $\omega_4=b_2(h_4)$, $\alpha_i=a_{i3}(h_4)$, $i=1$, 2, 3. In terms of the entries, the condition 2d) is given as follows. The pairs $(a_{13}(h_4)b_3(g_i), a_{23}(h_4)b_3(g_i))$, $i=1$, 2, and $([a_{13}, b_3](g_1, g_2), [a_{23}, b_3](g_1, g_2))$ generate a subgroup of rank 3 in the free abelian group generated by $(b_1(h_i), b_2(h_i))$, $1\leq i\leq 3$. The integers m_i, $1\leq i\leq 3$, in 3) are the invariant factors of this subgroup.

Case IV) $a_{12}\neq 0$, and $[a_{23}, b_3]\neq 0$. We identify U_3 with D by taking the basis dual to dz_3 for D. Then we have $F\circ\eta(g)=\int_{\eta(g)}dz_3=b_3(g)$. Hence N_1 coincides with the kernel of the natural action of N on U_3. Let V denote the real vector subspace spanned by Δ. We have $\dim_{\boldsymbol{R}} V=2$ and rank $\Delta \geq 2$.

Lemma 17. *If $h \in N_1$, then $a_{12}(h)=0$.*

Proof. Suppose $a_{12}(h) \neq 0$. Then we have $a_{23}(h)=0$, since otherwise h would have a fixed point on $\boldsymbol{C}^3$. Thus for any $g \in N$,

$$[h, g]=\begin{bmatrix} 1 & 0 & a_{12}(h)a_{23}(g) & * \\ 0 & 1 & 0 & 0 \\ 0 & 0 & 1 & 0 \\ 0 & 0 & 0 & 1 \end{bmatrix}.$$

Since $[h, g]$ has no fixed points on $\boldsymbol{C}^3$, we get $a_{12}(h)a_{23}(g)=0$. Hence we have $a_{23}=0$, which is a contradiction, q. e. d.

Lemma 18. $N_1 \subset \boldsymbol{Z}^4$.

Proof. By Lemma 17, for each fixed $z_3 \in U_3$, N_1 acts on $U_1 \times U_2 \times \{z_3\}$ properly discontinuously as a group of translations. Hence $N_1 \subset \boldsymbol{Z}^4$, q. e. d.

Let $h_1, \cdots, h_s$ and $k_1, \cdots, k_t$ be, respectively, $\boldsymbol{Z}$-bases of N_1 and $N^{(1)}(\subset N_1)$ such that $k_i=h_i^{m_i}$, $m_i \in \boldsymbol{Z}-0$, $1 \leq i \leq t \leq s \leq 4$. Also choose $g_1, \cdots, g_r$, $r \geq 2$, so that $b_3(g_1), \cdots, b_3(g_r)$ form a $\boldsymbol{Z}$-basis for Δ. Finally let $N^{(2)}=[N, N^{(1)}]$ be the commutator group of N and $N^{(1)}$. It is easy to see that for any $g \in N^{(2)}$, $a_{12}(g)=a_{13}(g)=a_{23}(g)=b_2(g)=b_3(g)=0$. Hence $N^{(2)}$ acts on U_1 properly discontinuously as a group of translations. Thus $N^{(2)} \subset \boldsymbol{Z}^2$.

Lemma 19. $N^{(2)} \neq 0$.

Proof. Since $[a_{23}, b_3] \neq 0$, we have $b_2(k) \neq 0$ for some $k \in N^{(1)}$. Then for any $g \in N$, $A([g, k])=I$, ${}^t b([g, k])=(a_{12}(g)b_2(k)-a_{13}(k)b_3(g), 0, 0)$. Suppose $N^{(2)}=0$. Then we would have $a_{12}=\alpha b_3$, for some $\alpha \in \boldsymbol{C}^*$. This implies $A([g, [g, g']])=I$, ${}^t b([g, [g, g']])=(2\alpha b_3(g)[a_{23}, b_3](g, g'), 0, 0)$, for all g, g' in N. Since $N^{(2)}=0$, we have $[a_{23}, b_3]=0$, which is a contradiction, q. e. d.

Lemma 20. rank $N^{(1)} \geq 2$.

Proof. Since $N^{(2)} \subset N^{(1)}$, if $N^{(1)}=\boldsymbol{Z}$, we would have $b_2(k)=0$, for all k in $N^{(1)}$. This yields $[a_{23}, b_3]=0$, q. e. d.

Lemma 21. rank $N^{(1)} \leq 3$.

Proof. Suppose $N^{(1)}=\boldsymbol{Z}^4$. Then $N_1=\boldsymbol{Z}^4$, and we may assume $k_i=h_i^{m_i}$, $m_i \in \boldsymbol{Z}-0$, $1 \leq i \leq 4$. Hence $a_{23}(h)=0$, for all $h \in N_1$. Moreover, if $N_1=\boldsymbol{Z}^4$, noting Lemma 17, we can show that $\boldsymbol{C}^3/N$ is a regular fiber space of complex 2-tori over $T^1=U_3/\Delta$ by the same argument as in the proof of Theorem 5, 1). Thus $r=\text{rank } \Delta=2$, and the group N is generated by $h_1, \cdots, h_4$, g_1 and g_2. Since a_{23} and b_3 are homomorphisms, we have $b_2(k_i)=n_i[a_{23}, b_3]\ (g_1, g_2)$, $n_i \in \boldsymbol{Z}$, $1 \leq i \leq 4$. Then $b_2(h_i)=\dfrac{1}{m_i}b_2(k_i)$ $=\dfrac{n_i}{m_i}[a_{23}, b_3]\ (g_1, g_2)$. This means that $(b_1(h_i), b_2(h_i))$, $1 \leq i \leq 4$, cannot be linearly independent over $\boldsymbol{R}$, and hence N_1 fails to act properly discontinuously on $U_1 \times U_2 \times \{0\}$, a contradiction, q. e. d.

Lemma 22. *If* rank $N_1 \leq 3$, *then* rank $\Delta \geq 3$.

Proof. If rank $N_1 \leq 3$, the projection $\boldsymbol{C}^3/N_1 \to U_3$ gives a fiber space structure on $\boldsymbol{C}^3/N_1$ with noncompact fibers $((U_1 \times U_2)/N_1) \times \{z_3\}$, $z_3 \in U_3$. Suppose rank $\Delta \leq 2$. Then the fiber preserving automorphism group $N/N_1 \simeq \Delta$ would also act properly discontinuously on U_3, and $\boldsymbol{C}^3/N=(\boldsymbol{C}^3/N_1)/(N/N_1) \to U_3/\Delta$ would be a fiber space over the manifold U_3/Δ with noncompact fibers, a contradiction, q. e. d.

Lemma 23. *If* $N^{(1)}=\boldsymbol{Z}^2$, *then* $N_1=\boldsymbol{Z}^4$.

Proof. Since $0 \neq N^{(2)} \subset N^{(1)}$, we may assume $b_2(k_1)=0$. Then we have $[a_{23}, b_3] \cdot (g_i, g_j)=n_{ij}b_2(k_2)$, $n_{ij} \in \boldsymbol{Z}$, $1 \leq i, j \leq r$, as $N^{(1)}$ is generated by k_1 and k_2. This yields $n_{ij}b_3(g_k)+n_{jk}b_3(g_i)+n_{ki}b_3(g_j)=0$, $1 \leq i, j, k \leq r$. Suppose rank $N_1 \leq 3$. Then by Lemma 22, rank $\Delta \geq 3$, and hence $n_{ij}=0$, $1 \leq i, j \leq r$. Take $h \in N_1$. Then for any $g \in N$, $a_{23}(h)b_3(g)=[a_{23}, b_3](h, g)=n(g, h)b_2(k_2)$, for some $n(g, h) \in \boldsymbol{Z}$. Since rank $\Delta \geq 3$, this implies $a_{23}(h)=0$. The group N is generated by $h_1, \cdots, h_s, g_1, \cdots, g_r$, and a_{23} and b_3 are homomorphisms. Thus we conclude $[a_{23}, b_3]=0$, which is a contradiction, q. e. d.

Lemma 24. *If* $N^{(1)}=\boldsymbol{Z}^3$, *then* $N_1=\boldsymbol{Z}^4$.

Proof. Suppose $N_1=\boldsymbol{Z}^3$. Then, from $k_i=h_i^{m_i}$, $1 \leq i \leq 3$, we have $a_{23}(h)=0$, for all $h \in N_1$. Moreover, by Lemma 22, rank $\Delta \geq 3$. We claim $N^{(2)}=\boldsymbol{Z}$. Suppose $N^{(2)}=\boldsymbol{Z}^2$. Then, since $N^{(2)} \subset N^{(1)}$, we may assume that $A(k_i)=I$, $b_2(k_i)=0$, $i=1, 2$. Thus we get $[a_{23}, b_3](g_i, g_j)=n_{ij}b_2(k_3)$, $n_{ij} \in \boldsymbol{Z}$, $1 \leq i, j \leq r$. This yields $n_{ij}=0$, and $[a_{23}, b_3]=0$ as in Lemma 23. Hence $N^{(2)}=\boldsymbol{Z}$. Let l be a generator of $N^{(2)}$. We have $A([g_i, k_p])$ $=I$, ${}^tb([g_i, k_p])=(a_{12}(g_i)b_2(k_p)-a_{13}(k_p)b_3(g_i), 0, 0)$, $1 \leq i \leq r$, $1 \leq p \leq 3$. Since $[g_i, k_p]$ $=l^{n_{ip}}$, for some $n_{ip} \in \boldsymbol{Z}$, we get

$$(31) \qquad a_{12}(g_i)b_2(k_p)-a_{13}(k_p)b_3(g_i) = n_{ip}b_1(l).$$

This yields $[b_2, a_{13}](k_p, k_q) \cdot b_3(g_i)=(n_{ip}b_2(k_q)-n_{iq}b_2(k_p))b_1(l)$. Since rank $\Delta \geq 3$, we have $[b_2, a_{13}](k_p, k_q)=0$, $1 \leq p, q \leq 3$. Hence $a_{13}(k_p)=\alpha b_2(k_p)$, $1 \leq p \leq 3$, for some $\alpha \in \boldsymbol{C}$, and we have, from (31), $b_2(k_p)(a_{12}(g_i)-\alpha b_3(g_i))=n_{ip}b_1(l)$. We claim $a_{12}(g_i)=\alpha b_3(g_i)$, $1 \leq i \leq r$. Suppose $a_{12}(g_{i_0}) \neq \alpha b_3(g_{i_0})$, for some i_0. Then, letting $\beta=b_1(l)(a_{12}(g_{i_0})-\alpha b_3(g_{i_0}))^{-1}$, $n_p=n_{i_0p}$, we would have $b_2(k_p)=n_p\beta$, $1 \leq p \leq 3$. On the other hand, $[a_{23}, b_3](g_i, g_j)=\sum_{p=1}^{3} n_{ijp}b_2(k_p)$, for some $n_{ijp} \in \boldsymbol{Z}$. Thus we get $[a_{23}, b_3](g_i, g_j)=m_{ij}\beta$, where $m_{ij}=\sum_{p=1}^{3} n_{ijp}n_p$. Therefore we conclude $[a_{23}, b_3]=0$ as before. Hence $a_{12}(g_i)=\alpha b_3(g_i)$, $1 \leq i \leq r$. Moreover, since $a_{12} \neq 0$, $\alpha \neq 0$. We have $[g_i, [g_i, g_j]]=l^{e_{ij}}$, for some

$e_{ij}\in \boldsymbol{Z}$, $1\leq i, j\leq r$. Using $a_{12}(g_i)=\alpha b_3(g_i)$, we get

$$2\alpha b_3(g_i)\cdot[a_{23}, b_3](g_i, g_j) = e_{ij}b_1(l). \tag{32}$$

This yields $e_{ij}b_3(g_j)+e_{ji}b_3(g_i)=0$. Since rank $\Delta\geq 3$, we have $e_{ij}=0$. Then from (32), $[a_{23}, b_3]$ $(g_i, g_j)=0$, $1\leq i, j\leq r$. In view of $a_{23}(h_i)=b_3(h_i)=0$, $1\leq i\leq 3$, we conclude $[a_{23}, b_3]=0$, a contradiction, q. e. d.

Theorem 6. *In Case* IV, *we have* $N_1=\boldsymbol{Z}^4$, $N^{(1)}=\boldsymbol{Z}^2$ *or* $\boldsymbol{Z}^3$, *and*

1) *the Albanese map gives a regular fiber space structure on* $\boldsymbol{C}^3/N$ *over* $\mathrm{Alb}(\boldsymbol{C}^3/N)=T^1$,

2) *the group* N *is generated by* h_i, $1\leq i\leq 4$, *and* g_j, $j=1, 2$, *where* a) $A(h_i)$ *and* $A(g_j)$ *are upper triangular matrices whose diagonal entries are all* 1, b) $a_{13}(h_1)=b_2(h_1)=0$, $a_{23}(h_i)=0$, $1\leq i\leq t=\mathrm{rank}\ N^{(1)}$, c) $(a_{13}(h_i)z_3+b_1(h_i),\ a_{23}(h_i)z_3+b_2(h_i))$, $1\leq i\leq 4$, *are linearly independent over* $\boldsymbol{R}$, *for any* $z_3\in U_3=\boldsymbol{C}$, d) $b_3(g_j)$, $j=1, 2$, *are linearly independent over* $\boldsymbol{R}$,

3) $$H_1(\boldsymbol{C}^3/N, \boldsymbol{Z}) = \begin{cases} \boldsymbol{Z}^4\oplus\boldsymbol{Z}_{m_1}\oplus\boldsymbol{Z}_{m_2}, & \textit{if}\ \ N^{(1)} = \boldsymbol{Z}^2, \\ \boldsymbol{Z}^3\oplus\boldsymbol{Z}_{m_1}\oplus\boldsymbol{Z}_{m_2}\oplus\boldsymbol{Z}_{m_3}, & \textit{if}\ \ N^{(1)} = \boldsymbol{Z}^3. \end{cases}$$

Proof. From Lemmas 20–24, we have $N_1=\boldsymbol{Z}^4$, and $N^{(1)}=\boldsymbol{Z}^2$ or $\boldsymbol{Z}^3$. The same argument as in the proof of Theorem 5, 1), shows 1) and 2c, d). Since $0\neq N^{(2)}\subset N^{(1)}\subset N_1$, we have 2b). The assertion 3) is also proved as in Theorem 5, q. e. d.

Remark. It does not seem easy to find the integers m_i from $g_1, g_2, h_1, \cdots, h_4$. The group $N^{(2)}$ is isomorphic to $\boldsymbol{Z}$ or $\boldsymbol{Z}^2$ and is generated by $[g_i, [g_1, g_2]]$, $i=1, 2$.

3. Holomorphic Forms

Let Ω^p denote the sheaf of germs of holomorphic p-forms on the quotient $\boldsymbol{C}^3/N$. In this section we compute $h^{p,0}=\dim H^0(\boldsymbol{C}^3/N, \Omega^p)$, the number of linearly independent holomorphic p-forms on $\boldsymbol{C}^3/N$. We may identify $H^0(\boldsymbol{C}^3/N, \Omega^p)$ with the space $H^0(\boldsymbol{C}^3, \Omega^p)^N$ of N-invariant holomorphic p-forms on $\boldsymbol{C}^3$. Note that the pullbacks g^*dz_i, $i=1, 2, 3$, are given by

$$\begin{bmatrix} g^*dz_1 \\ g^*dz_2 \\ g^*dz_3 \end{bmatrix} = A(g)\begin{bmatrix} dz_1 \\ dz_2 \\ dz_3 \end{bmatrix}. \tag{33}$$

Lemma 25. $h^{3,0}=1$.

Proof. Take $\omega\in H^0(\boldsymbol{C}^3, \Omega^3)^N$ and let $\omega=f(z)dz_1\wedge dz_2\wedge dz_3$, where $f\in H^0(\boldsymbol{C}^3, \mathcal{O})$. Then we have $g^*\omega=f(gz)dz_1\wedge dz_2\wedge dz_3$. Thus we have $f(z)=f(gz)$. Hence f reduces to a constant, q. e. d.

Take $\omega\in H^0(\boldsymbol{C}^3/N, \Omega^1)=H^0(\boldsymbol{C}^3, \Omega^1)^N$, and let $\omega=\sum_{i=1}^{3} f_i(z)dz_i$, $f_i\in H^0(\boldsymbol{C}^3, \mathcal{O})$, $i=1, 2, 3$. The N-invariance conditions for ω are given by (3), (4), (5) in § 1. Also take $\theta\in H^0(\boldsymbol{C}^3/N, \Omega^2)=H^0(\boldsymbol{C}^3, \Omega^2)^N$ and let $\theta=h_1(z)dz_2\wedge dz_3+h_2(z)dz_3\wedge dz_1+h_3(z)dz_1\wedge dz_2$, $h_i\in H^0(\boldsymbol{C}^3, \mathcal{O})$. The N-invariance conditions for θ are

$$h_1(z)=h_1(gz)-a_{12}(g)h_2(gz)+(a_{12}(g)a_{23}(g)-a_{13}(g))h_3(gz), \tag{34}$$
$$h_2(z)=h_2(gz)-a_{23}(g)h_3(gz), \tag{35}$$
$$h_3(z)=h_3(gz). \tag{36}$$

Case I) Clearly, $h^{1,0}=h^{2,0}=3$.

Case II) In this case, we have

Lemma 26. *Let $f\in H^0(\boldsymbol{C}^3, \mathcal{O})$. If we have*

$$f(z)=a(g)+f(gz), \quad \text{for all } g\in N, \tag{37}$$

for some function $a: N\to \boldsymbol{C}$, then $f(z)=\alpha z_2+\beta z_3+\gamma$, for some $\alpha, \beta, \gamma\in \boldsymbol{C}$.

Proof. We have $\dfrac{\partial f}{\partial z_1}(z)=\dfrac{\partial f}{\partial z_1}(gz)$. Hence we have $f(z)=c_1z_1+\varphi(z_2, z_3)$, where φ is holomorphic in (z_2, z_3). From (37), we get $\dfrac{\partial\varphi}{\partial z_2}(z_2, z_3)=c_1a_{12}(g)+\dfrac{\partial\varphi}{\partial z_2}(g(z_2, z_3))$. Hence we get $\varphi(z_2, z_3)=c_2z_2^2+\psi_1(z_3)z_2+\psi_2(z_3)$, where ψ_i, $i=1, 2$, are holomorphic in z_3. Also from (37), we infer that ψ_1 and ψ_2 are polynomials of degree at most 1 and 3, respectively. Therefore

$$f(z)=c_1z_1+c_2z_2^2+(c_3z_3+c_4)z_2+c_5z_3^3+c_6z_3^2+c_7z_3+c_8. \tag{38}$$

Substituting (38) in (37), we get $c_5b_3=c_1a_{13}+c_3b_2+3c_5b_3^2+2c_6b_3=c_1a_{12}+2c_2b_2+c_3b_3=0$. Hence we have $c_5=0$ and $c_1([a_{12}, b_2]+[a_{13}, b_3])=0$. Thus $c_1=0$. Finally, since $[b_2, b_3]\neq 0$, we have $c_2=c_3=c_6=0$, q. e. d.

Take $\omega=\sum_{i=1}^{3}f_i(z)dz_i\in H^0(\boldsymbol{C}^3, \Omega^1)^N$. Since $a_{23}=0$, we have from (3), (4), (5) and Lemma 26, $f_1=c_1$, $f_2(z)=c_2z_2+c_3z_3+c_4$, $f_3(z)=c_5z_2+c_6z_3+c_7$. The N-invariance conditions are

$$c_1a_{12}+c_2b_2+c_3b_3=0, \tag{39}$$
$$c_1a_{13}+c_5b_2+c_6b_3=0. \tag{40}$$

Letting $g=g_1$ or g_2 (see Lemma 7), we conclude $c_2=0$, $\alpha c_1+c_5=0$, $c_3=c_6=0$. Hence $h^{1,0}=2$ or 3. Note that $h^{1,0}=3$ if and only if $a_{12}=a_{13}-\alpha b_2=0$.

Lemma 27. *If $N^{(1)}=\boldsymbol{Z}$, then $h^{1,0}=2$.*

Proof. We have $c_1([a_{12}, b_2]+[a_{13}, b_3])=c_1[a_{13}, b_3]=-c_5[b_2, b_3]$. Also we have $[g_i, g_j]=k_1^{n_{ij}}$, for some $n_{ij}\in\boldsymbol{Z}$. Hence $c_1([a_{12}, b_2]+[a_{13}, b_3])(g_i, g_j)=-c_5n_{ij}b_1(k_1)$. Thus we have

$$c_5\left(n_{ij}\begin{bmatrix}b_2(g_k)\\ b_3(g_k)\end{bmatrix}+n_{jk}\begin{bmatrix}b_2(g_i)\\ b_3(g_i)\end{bmatrix}+n_{ki}\begin{bmatrix}b_2(g_j)\\ b_3(g_j)\end{bmatrix}\right)=0.$$

Since rank $\Delta\geq 4$, $c_5=0$. Therefore $c_1=0$, q. e. d.

Take $\theta\in H^0(\boldsymbol{C}^3, \Omega^2)^N$. Since $a_{23}=0$, from (34), (35), (36), and Lemma 26, we have $h_1(z)=c_1+c_4z_2+c_5z_3$, $h_2(z)=c_2$, $h_3(z)=c_3$. The N-invariance condition is

(41) $$c_2a_{12}+c_3a_{13}-c_4b_2-c_5b_3 = 0.$$

Letting $g=g_1$ or g_2, we have $c_5=\alpha c_3-c_4=0$. Hence $h^{2,0}=1$, 2 or 3.

Lemma 28. *$h^{2,0}=3$ if and only if $h^{1,0}=3$.*

Proof. $h^{1,0}=3$ if and only if $a_{12}=a_{13}-\alpha b_2=0$, q. e. d.

Lemma 29. *If $N_1=\boldsymbol{Z}$, $h^{2,0}=2$.*

Proof. By Lemmas 10 and 11. q. e. d.

Case III) In this case we have

Lemma 30. *Let $f\in H^0(\boldsymbol{C}^3, \mathcal{O})$. If we have*

(42) $$f(z) = a(g)+f(gz), \textit{for all } g\in N,$$

for some function $a: N\to \boldsymbol{C}$, then $f(z)=\alpha+\beta z_3$ for some α, $\beta\in \boldsymbol{C}$, *and $a=-\beta b_3$.*

Proof. Since $a_{12}=0$, from (42), we have $f(z)=c_1z_1+c_2z_2+\varphi(z_3)$, where φ is holomorphic in z_3. We infer readily that φ is a polynomial of degree at most 2. Hence we have $f(z)=c_1z_1+c_2z_2+c_3+c_4z_3+c_5z_3^2$. Comparing the z_3 terms in (42), we get $c_1a_{13}+c_2a_{23}+2c_5b_3=0$. Since $[a_{13}, a_{23}, b_3]\neq 0$, we have $c_1=c_2=c_5=0$. Finally comparing the constant terms in (42), we get $a+c_4b_3=0$, q. e. d.

Take $\omega=\sum_{i=1}^{3} f_i(z)dz_i\in H^0(\boldsymbol{C}^3, \Omega^1)^N$. Since $a_{12}=0$, from (3), (4), (5) and Lemma 30, we get $f_1=c_1$, $f_2=c_2$, $f_3(z)=c_3+c_4z_3$, $c_1a_{13}+c_2a_{23}+c_4b_3=0$. Since $[a_{13}, a_{23}, b_3]\neq 0$, we have $c_1=c_2=c_4=0$. Hence $h^{1,0}=1$. Take $\theta\in H^0(\boldsymbol{C}^3, \Omega^2)^N$. From (34), (35), (36) and Lemma 30, we have $h_1(z)=c_1+c_4z_3$, $h_2(z)=c_2+c_5z_3$, $h_3=c_3$, $-c_3a_{13}+c_4b_3=0$, $-c_3a_{23}+c_5b_3=0$. Since $[a_{23}, b_3]\neq 0$, we get $c_3=c_5=0$. Hence $c_4=0$. Therefore $h^{2,0}=2$, and $dz_2\wedge dz_3$, $dz_3\wedge dz_1$ form a basis for $H^0(\boldsymbol{C}^3/N, \Omega^2)$.

Case IV) In this case, we have

Lemma 31. *Let $f\in H^0(\boldsymbol{C}^3, \mathcal{O})$. If*

(43) $$f(z) = a(g)+b(g)z_3+f(gz), \textit{for all } g\in N,$$

for some functions a, $b: N\to \boldsymbol{C}$, then $f(z)=\alpha+\beta z_2+\gamma z_3+\delta z_3^2$ for some α, β, γ, $\delta\in \boldsymbol{C}$, where $b+\beta a_{23}+2\delta b_3=0$ and $a+\beta b_2+\gamma b_3+\delta b_3^2=0$.

Proof. From (43), we have $f(z)=c_1z_1+\varphi(z_2, z_3)$, where φ is holomorphic in (z_2, z_3). Let $\varphi(z_2, z_3)=\sum_{n\geq 0}\varphi_n(z_3)z_2^n$ be the Taylor series expansion in z_2. (43) implies $\partial^2\varphi/\partial z_2^2=$constant. Hence we have $\varphi(z_2, z_3)=c_2z_2^2+\varphi_1(z_3)z_2+\varphi_2(z_3)$. From (43), we infer that φ_1 and φ_2 are polynomials of degree at most 2 and 4, respectively. Hence we have

(44) $$f(z) = c_1z_1+c_2z_2^2+(c_3+c_4z_3+c_5z_3^2)z_2+c_6+c_7z_3+c_8z_3^2+c_9z_3^3+c_{10}z_3^4.$$

Comparing the z_3^3 terms in (43), we get $c_5a_{23}+4c_{10}b_3=0$. Since $[a_{23}, b_3]\neq 0$, we have

$c_5=c_{10}=0$. Then, comparing the z_2z_3 terms, we get $c_2=0$. Also comparing the z_3^2 terms, we have $c_4a_{23}+3c_9b_3=0$. Hence $c_4=c_9=0$. Since $a_{12}\neq 0$, comparing the z_2 terms, we get $c_1=0$. Then, comparing the z_3 and constant terms, we get $a+c_3b_2+c_7b_3+c_8b_3^2=0$, $b+c_3a_{23}+2c_8b_3=0$, q. e. d.

Take $\omega=\sum_{i=1}^{3} f_i(z)dz_i \in H^0(\boldsymbol{C}^3, \Omega^1)^N$. Then from (3), (4) and Lemma 31, we have $f_1=c_1$, $f_2(z)=c_2+c_3z_2+c_4z_3+c_5z_3^2$, $c_3a_{23}+2c_5b_3=0$, $c_1a_{12}+c_3b_2+c_4b_3+c_5b_3^2=0$. Since $[a_{23}, b_3]\neq 0$, $c_3=c_5=0$. Hence

$$f_2(z) = c_2+c_4z_3,\ c_1a_{12}+c_4b_3 = 0. \tag{45}$$

From (5) and Lemma 31, we have $f_3(z)=c_6+c_7z_2+c_8z_3+c_9z_3^2$, $c_4a_{23}+c_7a_{23}+2c_9b_3=0$, $c_1a_{13}+c_2a_{23}+c_4a_{23}b_3+c_7b_2+c_8b_3+c_9b_3^2=0$. Since $[a_{23}, b_3]\neq 0$, $c_4+c_7=c_9=0$. Hence we have

$$f_3(z) = c_6+c_7z_2+c_8z_3,\ c_1a_{13}+c_2a_{23}+c_4(a_{23}b_3-b_2)+c_8b_3 = 0. \tag{46}$$

Hence we see that $h^{1,0}=1$ or 2.

Take $\theta=h_1(z)dz_2\wedge dz_3+h_2(z)dz_3\wedge dz_1+h_3(z)dz_1\wedge dz_2$. From (35), (36) and Lemma 31, we have $h_3=c_3$, $h_2(z)=c_2+c_4z_2+c_5z_3+c_6z_3^2$, $c_4a_{23}+2c_6b_3=0$, $-c_3a_{23}+c_4b_2+c_5b_3+c_6b_3^2=0$. Since $[a_{23}, b_3]\neq 0$, we have $c_4=c_6=0$. Hence $c_3=c_5=0$. From (34) and Lemma 31, we have $h_1(z)=c_1+c_7z_2+c_8z_3+c_9z_3^2$, $c_7a_{23}+2c_9b_3=0$, $-c_2a_{12}+c_7b_2+c_8b_3+c_9b_3^2=0$. Hence we have $c_7=c_9=0$, and

$$h_1(z) = c_1+c_8z_3,\ -c_2a_{12}+c_8b_3 = 0. \tag{47}$$

Hence $h^{2,0}=1$ or 2.

Summarising the results in sections 2 and 3, we obtain the following tables.

Table I

	d	$h^{1,0}$	$h^{2,0}$	$H_1(\boldsymbol{C}^3/N, \boldsymbol{Z})$	structure (Albanese map)
I	3	3	3	$\boldsymbol{Z}^6$	(i) T^3
II	2	2	2	$\boldsymbol{Z}^5\oplus\boldsymbol{Z}_m$	(ii) T^2-bundle over T^1 (Th. 2)
			1, 2	$\boldsymbol{Z}^5\oplus\boldsymbol{Z}_m$	(iii) T^1-bundle over T^2 (Th. 3)
		2, 3	1, 2, 3	$\boldsymbol{Z}^4\oplus\boldsymbol{Z}_{m_1}\oplus\boldsymbol{Z}_{m_2}$	(iv) T^1-bundle over T^2 (Th. 3)
III	1	1	2	$\boldsymbol{Z}^3\oplus\boldsymbol{Z}_{m_1}\oplus\boldsymbol{Z}_{m_2}\oplus\boldsymbol{Z}_{m_3}$	(v) Alb$=0$ (Th. 4)
					(vi) regular fiber space of complex 2-tori over T^1 (Th. 5)
IV	1	1, 2	1, 2	$\boldsymbol{Z}^4\oplus\boldsymbol{Z}_{m_1}\oplus\boldsymbol{Z}_{m_2}$ or $\boldsymbol{Z}^3\oplus\boldsymbol{Z}_{m_1}\oplus\boldsymbol{Z}_{m_2}\oplus\boldsymbol{Z}_{m_3}$	(vii) regular fiber space of complex 2-tori over T^1 (Th. 6)

where 1) T^k denotes a complex k-torus, 2) d and $h^{p,0}$ denote, respectively, the num-

bers of linearly independent closed holomorphic 1-forms and (not necessarily closed) holomorphic p-forms. In every case, $h^{3,0}=1$. For the definition of a regular fiber space, see Definition 2.

Table II

Class	Δ	N_1	$N^{(1)}$	$N^{(2)}$
i	$\boldsymbol{Z}^6$	0	0	0
ii	$\boldsymbol{Z}^5$	$\boldsymbol{Z}$	$\boldsymbol{Z}$	0
iii	$\boldsymbol{Z}^4$	$\boldsymbol{Z}^2$	$\boldsymbol{Z}$	0
iv	$\boldsymbol{Z}^4$	$\boldsymbol{Z}^2$	$\boldsymbol{Z}^2$	0
v	$\boldsymbol{Z}^3$	$\boldsymbol{Z}^3$	$\boldsymbol{Z}^3$	0
vi	$\boldsymbol{Z}^2$	$\boldsymbol{Z}^4$	$\boldsymbol{Z}^3$	0
vii	$\boldsymbol{Z}^2$	$\boldsymbol{Z}^4$	$\boldsymbol{Z}^2$ or $\boldsymbol{Z}^3$	$\boldsymbol{Z}^1$ or $\boldsymbol{Z}^2$

where 1) Δ and N_1 denote, respectively, the image and the kernel of the homomorphism $F\circ\eta : N=\pi_1(\boldsymbol{C}^3/N)\to D=H^0(\boldsymbol{C}^3/N, d\mathcal{O})^*$, 2) $N^{(1)}=[N, N]$, $N^{(2)}=[N, N^{(1)}]$.

Remark. It is not difficult to construct all manifolds in each of the classes (i)–(vi). Simply choose generators of N so that the conditions in the corresponding theorems are satisfied. For examples, see [4].

References

[1] Blanchard, A.: Sur les variétés analytiques complexes, Ann. Sci. Éc. Norm. Sup., **73** (1956), 157–202.
[2] Sakane, Y.: On compact complex affine manifolds, to appear.
[3] Suwa, T.: Compact quotient spaces of $\boldsymbol{C}^2$ by affine transformation groups, J. Diff. Geom. **10** (1975), 239–252.
[4] Suwa, T.: Compact quotients of $\boldsymbol{C}^3$ by affine transformation groups, to appear in the Proc. AMS Summer Institute, Williamstown 1975.
[5] Suwa, T.: Compact quotients of $\boldsymbol{C}^3$ by affine transformation groups I, to appear.
[6] Wolf, J.: "Spaces of Constant Curvature", Publish or Perish, Inc. 1974.

Department of Mathematics
The University of Michigan
(Current address)
Department of Mathematics
Hokkaido University

(Received January 19, 1976)

Kodaira Dimensions for Certain Fibre Spaces

K. Ueno

In what follows, we mean by an algebraic manifold V a complete irreducible non-singular algebraic variety defined over $\boldsymbol{C}$, and by K_V the canonical line bundle of V. For any algebraic manifold (or more generally any irreducible compact complex space) V, the notion of Kodaira dimension $\kappa(V)$ of V has been introduced by Iitaka [3] and its properties have been studied by Iitaka [3] and Ueno [11], [12] from the viewpoint of the classification theory of algebraic manifolds. There are many unsolved problems on Kodaira dimensions. Among others, the following conjecture is most important (see Ueno [12] § 11).

Conjecture C_n. *Let $\varphi: V \to W$ be a surjective morphism of an n-dimensional algebraic manifold V onto an algebraic manifold W with connected fibres. Then we have*

$$\kappa(V) \geqq \kappa(V_x) + \kappa(W)$$

where V_x is a general fibre of φ.

Very little is known about this conjecture. In the present paper we give an affirmative answer to the conjecture for certain cases. Nakamura and Ueno [7] have solved the conjecture when $\varphi: V \to W$ is an analytic fibre bundle (in this case, the total space V need not be algebraic). Ueno [11] has given an affirmative answer when a general fibre of φ is an elliptic curve and φ has locally meromorphic sections at every point of W. On the other hand it has been known that Conjecture C_2 is true as a corollary of Enriques and Kodaira's classification theory of surfaces. (See for example Ueno [12] Theorem 11. 5. 2). Since Conjecture C_n is one of the key steps for building the classification theory of algebraic manifolds, it is desirable to find a proof of Conjecture C_2 without using the classification theory of surfaces.

In the first part of the present paper we shall give a new proof of Conjecture C_2. The proof is a corollary to the following theorem which has its own rightful interest.

Theorem. *Let $\varphi: S \to C$ be a surjective morphism of a non-singular surface S onto a non-singular curve C. Suppose that a general fibre of φ is a curve of genus $g \geqq 2$. We set $K_{S/C} = K_S \otimes \varphi^*(K_C^{-1})$. Then the sheaf $\mathcal{F} = \varphi_*(\mathcal{O}(K_{S/C}))$ is a locally free sheaf of rank g and its degree $d(\mathcal{F})$ is non-negative. Moreover $d(\mathcal{F}) = 0$ if and only if $\varphi: S \to C$ is birationally equivalent to a fibre bundle over C with respect to the etale topology on C.*

A similar result holds when the general fibre is an elliptic curve. This will be treated in the appendix. The proof of the theorem is based on a deep analysis of the singular fibres of a family of curves. In § 1 we shall construct a family of Jacobian varieties $\varpi : \mathcal{J}(S/C) \to C$ with a section o associated to a fibre space $\varphi : S \to C$. It will be shown that the sheaf $\mathcal{F}$ is isomorphic to the sheaf $\mathcal{O}(N^*)$ where N^* is the dual of the normal bundle of $o(C)$ in $\mathcal{J}(S/C)$.

In § 2 using this isomorphism we shall prove the above theorem and Conjecture C_2. Here we use a cusp form on the Siegel upper half plane S_g to construct a holomorphic section of a certain line bundle.

We also note that $d(\mathcal{F})$ is related to numerical invariants of singular fibres. When a general fibre is a curve of genus 2, $d(\mathcal{F})$ is explicitly calculated in Ueno [13].

By the Grothendieck duality theorem, it is known that the sheaf $\mathcal{F}$ is dual to the sheaf $R^1\varphi_*\mathcal{O}_S$. Hence the above theorem can be considered as a result on the degree of the locally free sheaf $R^1\varphi_*\mathcal{O}_S$ on C. This leads to the following theorem.

Theorem. *Let $\varphi : V \to C$ be a surjective morphism of an algebraic manifold V onto a non-singular curve C. If $p_g(V_x)$ is one or two for a regular fibre V_x of φ and the genus of C is strictly greater than one, we have*

$$\kappa(V) \geqq 1.$$

From this theorem we infer that in the following cases Conjecture C_n is valid (the base space is assumed to be a curve of genus $\geqq 2$).

1) A general fibre of φ is a surface S with $p_g(S)=1$ or 2.

2) A general fibre V_x of φ is of Kodaira dimension zero and $p_g(V_x)=1$. (For example, this is the case, if V_x is birationally equivalent to an abelian variety or a $K3$ surface.)

In the appendix we shall give a simple proof of the canonical bundle formula for elliptic surfaces.

Notations

In what follows we shall use freely the following notations.

$e_m = \exp(2\pi\sqrt{-1}\,/m)$.

$\boldsymbol{e}(x) = \exp(2\pi\sqrt{-1}\,x)$

$K_{V/W} = K_V \otimes \varphi^*(K_W^{-1})$ for a fibre space $\varphi : V \to W$.

$\pi(C)$: the genus of a non-singular curve C.

1. Families of Jacobian Varieties

By a fibre space of curves of genus g over a non-singular curve C, we mean a surjective holomorphic mapping $\varphi : S \to C$ of a non-singular surface S onto the curve C whose general fibre is a curve of genus g. Moreover, we assume that no fibre of φ contains exceptional curves of the first kind.

There exists a set of points $\{p_1, p_2, \cdots, p_l\}$ on the curve C such that the scheme-theoretic fibre $S_{p_i}=\varphi^{-1}(p_i)$ is singular and the that fibre $S_x=\varphi^{-1}(x)$, $x\in C'=C-\{p_1, \cdots, p_l\}$, is non-singular. We put $S'=\varphi^{-1}(C')$, $\varphi'=\varphi|_{C'}$. The fibre $S_{p_i}=\varphi^{-1}(p_i)$ is called a singular fibre over the point p_i.

Let $\tilde{C}'$ be the universal covering of C' with the covering mapping $\pi:\tilde{C}'\to C'$. We can construct a period mapping $T:\tilde{C}'\to\mathfrak{S}_g$ of $\tilde{C}'$ into the Siegel upper half plane of degree g and a group representation $\Phi:\pi_1(C')\to Sp(g, \boldsymbol{Z})$ such that

$$T(\gamma\cdot\tilde{u}) = (A_\gamma T(\tilde{u})+B_\gamma)(C_\gamma T(\tilde{u})+D_\gamma)^{-1}, \qquad (1.1)$$
$$\Phi(\gamma) = \begin{bmatrix} A_\gamma & B_\gamma \\ C_\gamma & D_\gamma \end{bmatrix}, \quad \gamma\in\pi_1(C'),$$

(see, for example, Namikawa and Ueno [9]). From these data (T, Φ), we can construct a polarized bundle $\varpi:\mathcal{J}(S'/C')\to C'$ (see Ueno [10], § 1). For each point $x\in C'$, the fibre J_x of ϖ is the Jacobian variety of the curve S_x. Moreover $\varpi:\mathcal{J}(S'/C')\to C'$ has a holomorphic section (zero section) $o':C'\to\mathcal{J}(S'/C')$. $\mathcal{J}(S'/C')$ is constructed as follows. The abelian group $\boldsymbol{Z}^{2g}$ acts on $\tilde{C}'\times\boldsymbol{C}^g$ properly discontinuously and freely as

$$\tilde{C}'\times\boldsymbol{C}^g \to \tilde{C}'\times\boldsymbol{C}^g$$
$$(\tilde{u}, \zeta)\mapsto\left(\tilde{u}, \zeta+\nu\begin{pmatrix} T(\tilde{u}) \\ I_g \end{pmatrix}\right), \quad \nu\in\boldsymbol{Z}^{2g}.$$

We set $\tilde{\mathcal{J}}=\tilde{C}'\times\boldsymbol{C}^g/\boldsymbol{Z}^{2g}$ and by the symbol $(\tilde{u}, [\zeta])$ we denote the point of $\tilde{\mathcal{J}}$ corresponding to a point $(\tilde{u}, \zeta)\in\tilde{C}'\times\boldsymbol{C}^g$. Then the fundamental group $\pi_1(C')$ acts on $\tilde{\mathcal{J}}$ freely and properly discontinuously as follows.

$$\gamma:\tilde{\mathcal{J}}\to\tilde{\mathcal{J}}$$
$$(\tilde{u}, [\zeta])\mapsto(\gamma\tilde{u}, [\zeta\cdot f_\gamma(\tilde{u})]), \ \gamma\in\pi_1(C')$$

where

$$f_\gamma(\tilde{u}) = (C_\gamma T(\tilde{u})+D_\gamma)^{-1}, \quad \Phi(\gamma) = \begin{bmatrix} A_\gamma & B_\gamma \\ C_\gamma & D_\gamma \end{bmatrix}. \qquad (1.2)$$

$\mathcal{J}(S'/C')$ is, by definition, the quotient space $\tilde{\mathcal{J}}/\pi_1(C')$. By $[\tilde{u}, [\zeta]]$ we denote the point on $\mathcal{J}(S'/C')$ corresponding to a point $(\tilde{u}, [\zeta])$. The zero section o' is defined by

$$o'(u) = [\tilde{u}, [0]],$$

where $\pi(\tilde{u})=u$. From our construction it follows that the normal bundle N' of $o'(C')$ in $\mathcal{J}(C')$ is analytically isomorphic to the quotient $\tilde{C}'\times\boldsymbol{C}^g/\pi_1(C')$ of the trivial bundle $\tilde{C}'\times\boldsymbol{C}^g$ on $\tilde{C}'$ by the fundamental group $\pi_1(C')$ acting on $\tilde{C}'\times\boldsymbol{C}^g$ as

$$\tilde{C}'\times\boldsymbol{C}^g \to \tilde{C}'\times\boldsymbol{C}^g$$
$$(\tilde{u}, \zeta)\mapsto(\gamma\tilde{u}, \zeta\cdot f_\gamma(\tilde{u})), \ \gamma\in\pi_1(C')$$

where $f_\gamma(\tilde{u})$ is defined in (1.2). Hence if we set $K_{S'/C'}=K_{S'}\otimes\varphi^*(K_{C'}^{-1})$, we have a natural isomorphism

$$\varphi_*\mathcal{O}(K_{S'/C'}) \simeq \mathcal{O}(N'^*), \qquad (1.3)$$

where N'^* is the dual bundle of N'.

Next we extend the polarized bundle $\varpi' : \mathcal{J}(S'/C') \to C'$ to a fibre space $\varpi : \mathcal{J}(S/C) \to C$ of generalized Jacobians over the curve C. For our purposes it is sufficient to construct $\varpi : \mathcal{J}(S/C) \to C$ as a group scheme on C. (It is possible to construct a "compactification" $\varpi : \overline{\mathcal{J}(S/C)} \to C$ of $\varpi : \mathcal{J}(S/C) \to C$. See Nakamura [6], Namikawa [8] and Ueno [10].). Since the problem is local, we consider a small coordinate neighbourhood $D = \{t \mid |t| < \varepsilon\}$ in C with a local coordinate t with center p_i. We set $X = S|_D$, $\psi = \varphi|_D$, $X^* = \psi^{-1}(D^*)$, $D^* = D - \{0\}$. Let $M \in S_p(g, \mathbf{Z})$ be the Picard-Lefschetz transformation of the family $\psi : X \to D$ around the singular fibre $X_0 = S_{p_i}$. Let m be the smallest positive integer such that M^m is unipotent. We set

$$\mu : \hat{D} = \{s \mid |s| < \varepsilon^{1/m}\} \to D$$
$$s \mapsto t = s^m.$$

We let $\hat{X}$ be the minimal non-singular model of $X \underset{D}{\times} \hat{D}$ with a natural holomorphic mapping $\hat{\psi} : \hat{X} \to \hat{D}$. Then the scheme-theoretic fibre $\hat{X}_0$ of $\hat{\psi}$ over the origin is reduced and has only ordinary double points $\{q_1, \cdots, q_k\}$ as its singularities (see Mayer [1]). Moreover there is an analytic automorphism g of order m of $\hat{X}$ which is the lift of the automorphism $s \to e_m s$ such that X is the minimal non-singular model of $\hat{X}/\langle g \rangle$. We write $\hat{X}_0 = \bigcup_{\lambda \in \Lambda} C_\lambda$ where C_λ are irreducible components of $\hat{X}_0$. Let $\nu : \tilde{X}_0 \to \hat{X}_0$ be the normalization of $\hat{X}_0$. Hence $\tilde{X}_0 = \bigcup_{\lambda \in \Lambda} \tilde{C}_\lambda$ where $\nu_\lambda : \tilde{C}_\lambda \to C_\lambda$ is the normalization of C_λ. Then a holomorphic section $\omega \in H^0(\hat{X}_0, \mathcal{O}(K_{\hat{X}_0}))$, where $K_{\hat{X}_0} = K_{\hat{X}}|_{\hat{X}_0}$, corresponds to a meromorphic 1-form ϖ on $\tilde{X}_0$ with at most simple poles at q_i^+ and q_i^- such that

$$\text{Res}_{q_i^+}(\varpi) + \text{Res}_{q_i^-}(\varpi) = 0, \quad \nu^{-1}(q_i) = \{q_i^+, q_i^-\},$$

and this correspondence is bijective (see, for example, Jambois [1]). Moreover we can prove that $\dim_{\mathbf{C}} H^0(\hat{X}_0, \mathcal{O}(K_{\hat{X}_0})) = g$. A basis of $H^0(\hat{X}_0, \mathcal{O}(K_{\hat{X}_0}))$ can be chosen as follows. First we choose double points $q_{i_1}, \cdots, q_{i_d}$ of $\hat{X}_0$ in such a way that $\hat{X}_0 - \{q_{i_1}, \cdots, q_{i_d}\}$ is connected but $\hat{X}_0 - \{q_{i_1}, \cdots, q_{i_d}, q_j\}$, $q_j \neq q_{i_l}$, $l = 1, 2, \cdots, d$ is not connected. We may assume that $q_{i_l} = q_l$, $l = 1, 2, \cdots, d$. For a fixed point $s \in \hat{D}^*$, we let $\delta_{s,i}$ be the vanishing cycle of $H_1(\hat{X}_s, \mathbf{Z})$ corresponding to the point q_i, $i = 1, \cdots, d$. Moreover since we choose ε small enough, there exists a retraction $r : \hat{X} \to \hat{X}_0$. We set $r_s = r|_{\hat{X}_s}$. A basis $\{\alpha_1, \cdots, \alpha_{g'}, \beta_1, \cdots, \beta_{g'}\}$ of $H_1(\tilde{X}_0, \mathbf{Z})$ can be chosen so that

$$\alpha_i \cdot \alpha_j = 0, \quad \beta_i \cdot \beta_j = 0, \quad \alpha_i \cdot \beta_j = \delta_{ij}.$$

We can choose cycles $\alpha_{s,i}$, $\beta_{s,i}$, $i = 1, \cdots, g'$ on $\hat{X}_s$ in such a way that

$$(r_s)_*(\alpha_{s,i}) = \alpha_i, \quad (r_s)_*(\beta_{s,i}) = \beta_i,$$
$$\alpha_{s,i} \cdot \delta_{s,j} = \beta_{s,i} \cdot \delta_{s,j} = 0.$$

Then there exist cycles $\gamma_{s,1}, \cdots, \gamma_{s,d}$ on $\hat{X}_s$ which satisfy

$$\alpha_{s,i} \cdot \gamma_{s,j} = \beta_{s,i} \cdot \gamma_{s,j} = 0$$
$$\gamma_{s,i} \cdot \gamma_{s,j} = 0, \quad \gamma_{s,i} \cdot \delta_{s,j} = \delta_{ij}.$$

The monodromy $\hat{M}=M^m$ of $\hat{\varphi}:\hat{X}\to\hat{D}$ around the origin has the form

$$(1.4)\qquad \begin{cases} \hat{M}(\alpha_{s,i}) = \alpha_{s,i} \\ \hat{M}(\beta_{s,i}) = \beta_{s,i} \\ \hat{M}(\gamma_{s,i}) = \gamma_{s,i}+\sum a_{i,j}\delta_{s,j} \\ \hat{M}(\delta_{s,i}) = \delta_{s,i}. \end{cases}$$

Therefore we can find 2-chains β_i, $i=1, 2, \cdots, g'$, δ_j, on $\hat{X}$ such that the restrictions $\beta_i(s^*)$, $\delta_j(s^*)$ of β_i, δ_j on $\hat{X}_{s^*}$, $s^*\in\hat{D}^*$, are elements of $H_1(\hat{X}_{s^*}, \boldsymbol{Z})$ and $\beta_i(s)=\beta_{s,i}$, $\delta_j(s)=\delta_{s,j}$. Then we can find a basis $\{\omega_1(0), \cdots, \omega_g(0)\}$ of $H^0(\hat{X}_0, \mathcal{O}(K_{\hat{X}}))$ and elements $\omega_1, \cdots, \omega_g$ of $H^0(\hat{X}, \mathcal{O}(K_{\hat{X}/\hat{D}}))$ such that $\{\omega_1, \cdots, \omega_g\}$ is a basis of $H^0(\hat{X}, \mathcal{O}(K_{\hat{X}/\hat{D}}))$ as an $H^0(\hat{D}, \mathcal{O}_{\hat{D}})$-module, $\omega_i|\hat{X}_0=\omega_i(0)$ and that

$$(1.5)\qquad \begin{bmatrix} \int_{\beta_i(s)}\omega_k(s) & \int_{\delta_j(s)}\omega_k(s) \\ \int_{\beta_i(s)}\omega_{g'+l}(s) & \int_{\delta_j(s)}\omega_{g'+l}(s) \end{bmatrix} = I_g, \qquad \begin{matrix} i, k = 1, 2, \cdots, g' \\ j, l = 1, 2, \cdots, d, \end{matrix}$$

for each point $s\in\hat{D}^*$ where $\omega_j(s)=\omega_j|\hat{X}_s$. From the definition of the period mapping and (1.4) (1.5), it follows that

$$(1.6)\qquad T(s) = \frac{1}{2\pi\sqrt{-1}}\begin{bmatrix} 0 & 0 \\ 0 & \Delta \end{bmatrix}\log s+T_0(s), \quad s\in\hat{D}^*,$$

where $T_0(s)$ is holomorphic on $\hat{D}$ and $\Delta=(a_{ij})$, where the a_{ij} appear in (1.4). The abelian group $\boldsymbol{Z}^g$ acts on $\hat{D}^*\times\boldsymbol{C}^{*g}$:

$$\begin{aligned} \nu:\hat{D}^*\times\boldsymbol{C}^{*g} &\to \hat{D}^*\times\boldsymbol{C}^{*g} \\ (s, (w_j)) &\to (s, (w_j\boldsymbol{e}(\textstyle\sum\nu_j t_{ij}(s)))) \end{aligned}$$

where $T(s)=(t_{ij}(s))$, $\nu=(\nu_1, \cdots, \nu_g)\in\boldsymbol{Z}^g$. This automorphism cannot be extended to one of $\hat{D}\times\boldsymbol{C}^{*g}$, but we can construct the *quotient manifold* $\hat{\mathcal{J}}=\hat{D}\times\boldsymbol{C}^{*g}/\boldsymbol{Z}^g$ (see Kodaira [4] II, p. 597). There is a surjective morphism $\hat{\varpi}:\hat{\mathcal{J}}\to\hat{D}$ such that the fibre over the origin is an analytic Lie group which is an extension of the Jacobian variety of $\Pi\,\tilde{C}_i$ by $\boldsymbol{C}^{*d}$, that is, the generalized Jacobian variety of the curve $\hat{X}_0$. By the symbol $(s, [w_j])$ we denote the point on $\hat{\mathcal{J}}$ corresponding to a point $(s, (w_j))$. Note that $\varpi:\hat{\mathcal{J}}\to\hat{D}$ has a holomorphic section $\hat{o}$ defined by $\hat{o}(s)=(s, [1, \cdots, 1])$.

Let G be the cyclic group of order m of analytic automorphisms of $\hat{\mathcal{J}}$ generated by the automorphism

$$\begin{aligned} \hat{g}:\hat{\mathcal{J}} &\to \hat{\mathcal{J}} \\ (s, [w_j]) &\mapsto (e_m s, [w'_j]), \\ w'_j &= \boldsymbol{e}(\sum_i f_{ij}(s)\log w_i) \end{aligned}$$

where

$$(f_{ij}(s)) \equiv (CT(s)+D)^{-1}, \quad M = \begin{bmatrix} A & B \\ C & D \end{bmatrix}.$$

Let $\mathcal{J}(X/D)$ be a minimal non-singular model of the quotient space $\hat{\mathcal{J}}/G$. There is a natural surjective morphism $\varpi_D:\mathcal{J}(X/D)\to D$ and the restriction $\varpi_D|_{D^*}$: $\mathcal{J}(X/D)|_{D^*}\to D^*$ is naturally isomorphic to

$$\varpi'|_{D^*}:\mathcal{J}(S'/C')|_{D^*}\to D^*.$$

Note that $F(s)=(CT(s)+D)^{-1}$ is a $g\times g$ matrix valued holomorphic function on $\hat{D}$, since we have

$$F(s)\cdot F(e_m s)\cdot \cdots \cdot F(e_m^{m-1}s) = I_g.$$

Moreover from this relation we infer that there exists a $g\times g$ matrix valued holomorphic function $G(s)$ on $\hat{D}$ such that

$$F(s) = G(s)HG(e_m s)^{-1}, \tag{1.7}$$

where

$$H = \begin{bmatrix} e_m{}^{a_1} & & 0 \\ & \ddots & \\ 0 & & e_m{}^{a_g} \end{bmatrix}, \quad 0 \leqq a_1 \leqq \cdots \leqq a_g < m$$

is a Jordan normal form of the matrix $F(0)$.

We study the behaviour of the automorphism $\hat{g}$ in a neighbourhood of the subvariety $\hat{o}(\hat{D})$ so that we can describe a neighbourhood of the image of the zero section in $\mathcal{J}(X/D)$. We choose local coordinates $s, \eta_1, \cdots, \eta_g$ in a neighbourhood U of $\hat{o}(\hat{D})$ in $\hat{\mathcal{J}}$ in such a way that

$$\begin{cases} w_j = \boldsymbol{e}(\zeta_j), \quad j = 1, 2, \cdots, g \\ (\zeta_1, \cdots, \zeta_g) = (\eta_1, \cdots, \eta_s)G(s)^{-1}, \quad |\eta_i| < \varepsilon'. \end{cases} \tag{1.8}$$

From 1. 7 we infer that, in these local coordinates, the automorphism $\hat{g}$ is expressed in the form

$$\hat{g} : (s, \eta_1, \cdots, \eta_g) \mapsto (e_m s, e_m^{a_1}\eta_1, \cdots, e_m^{a_g}\eta_g).$$

We may assume

$$a_1 = \cdots = a_j = 0, \quad 1 \leqq a_{j+1} \leqq \cdots \leqq a_g.$$

We consider the meromorphic mapping

$$\begin{aligned} h : (s, \eta_1, \cdots, \eta_g) &\mapsto \left(s^m, \eta_1, \cdots, \eta_j, \frac{\eta_{j+1}}{s^{a_{j+1}}}, \cdots, \frac{\eta_g}{s^{a_g}}\right) \\ &= (t, \eta_1, \cdots, \eta_j, z_{j+1}, \cdots, z_g) \end{aligned} \tag{1.9}$$

of U into $D\times \boldsymbol{C}^g$. Let V be the open set in $D\times \boldsymbol{C}^g$ defined by the inequalities

$$\begin{aligned} &|t| < \varepsilon, \qquad |\eta_i| < \varepsilon', \quad i = 1, 2, \cdots, j, \\ &|z_k^m t^{a_k}| < (\varepsilon')^m, \quad k = j+1, \cdots, g. \end{aligned}$$

Then $h(U)=V$ and V is contained in $\mathcal{J}(X/D)$ as an open set, and the above meromorphic mapping h induces a part of blowing-ups of the singularities of the quotient space $\hat{\mathcal{J}}/G$. On V the mapping ϖ_D is expressed as the projection to the first factor t. The zero section o_D of $\varpi_D : \mathcal{J}(X/D)\to D$ is given by

$$o_D : t \mapsto (t, 0, 0, \cdots, 0).$$

It is easy to show that on D^* we have $o'|_{D^*}=o_D|_{D^*}$. Since $\varpi_D : \mathcal{J}(X/D)\to D$ and $\varpi' : \mathcal{J}(S'/C')\to C'$ is isomorphic when restricted to D^*, we can patch together $\mathcal{J}(S'/C')$ and all $\mathcal{J}(X/D)$' s, and obtain a fiber space $\varpi : \mathcal{J}(S/C)\to C$ and a holomorphic section o.

On the other hand, the automorphism $\hat{g}$ induces an automorphism $\hat{g}^*$ on $H^0(\hat{D}, \hat{\varphi}_*\mathcal{O}(K_{\hat{X}/\hat{D}}))$; which has the form

$$\begin{cases} \hat{g}^*(\omega) = b_1(e_m s)\hat{g}^*(\omega_1) + \cdots + b_g(e_m s)\hat{g}^*(\omega_g) \\ (\hat{g}^*(\omega_1), \cdots, \hat{g}^*(\omega_g)) = (\omega_1, \cdots, \omega_g)F(s) \end{cases}$$

where

$$\omega \in H^0(\hat{D}, \hat{\varphi}_*\mathcal{O}(K_{\hat{X}/\hat{D}})), \quad \omega = \sum b_i(s)\omega_i, \quad b_i(s) \in H^0(\hat{D}, \mathcal{O}_{\hat{D}}).$$

We set

$$(\omega'_1, \cdots, \omega'_g) = (\omega_1, \cdots, \omega_g)G(s)^{-1}.$$

Then $\{\omega'_1, \cdots, \omega'_g\}$ is a basis of $H^0(\hat{D}, \hat{\varphi}_*\mathcal{O}(K_{\hat{X}/\hat{D}}))$ as an $H^0(\hat{D}, \mathcal{O}_{\hat{D}})$-module. Let us consider $\omega'_1, \cdots, \omega'_g$ as elements of $H^0(\hat{X}, \mathcal{O}(K_{\hat{X}/\hat{D}}))$. Then $ds\wedge\omega'_1, \cdots, ds\wedge\omega'_g$ are holomorphic 2-forms on $\hat{X}$ and we have

$$(1.\ 10) \qquad \hat{g}^*(ds \wedge \omega'_i) = e_m^{a_i+1} ds \wedge \omega'_i.$$

Put $b_i = m - a_i - 1$. Then $s^{b_i}ds\wedge\omega'_i$ is invariant under the action of $\hat{g}^*$. Hence, by a theorem of Freitag [2], $s^{b_i}ds\wedge\omega'_i$ induces a holomorphic 2-form ω^*_i on X. Then we can write

$$\omega^*_i = \frac{1}{m} dt \wedge \tau_i,$$

where $\tau_i \in H^0(D, \varphi_*\mathcal{O}(K_{X/D}))$. Therefore $\{\tau_1, \cdots, \tau_g\}$ is a basis of $H^0(D, \varphi_*\mathcal{O}(K_{X/D}))$ as an $H^0(D, \mathcal{O}_D)$-module. Let $\psi : \hat{X} \to X$ be the natural meromorphic mapping induced by the quotient morphism. Then we infer that

$$\psi^*(\tau_i) = \omega'_i/s^{a_i}.$$

Combining this with 1. 9 and 1. 10 we infer that there is a natural isomorphism between $\varphi_*\mathcal{O}(K_{X/D})$ and $\mathcal{O}(N^*_D)$, where N^*_D is the dual bundle of the normal bundle N_D of $o_D(D)$ in $\mathcal{J}(X/D)$. Moreover on D^* this isomorphism is compatible with the isomorphism 1. 3. Therefore we obtain the following theorem.

Theorem 1. 11. *The sheaf* $\mathcal{F} = \varphi_*\mathcal{O}(K_{S/C})$ *is locally free, of rank g, and is isomorphic to* $\mathcal{O}(N^*)$, *where* N^* *is the dual of the normal bundle N of the curve* $o(C)$ *in* $\mathcal{J}(S/C)$.

2. Conjecture C_2

Let $\varphi : S \to C$ be the same as above and we use the same notations as above. We set $\mathcal{J} = \mathcal{J}(S/C)$, $\mathcal{J}' = \mathcal{J}(S'/C')$. We now prove the following theorem.

Theorem 2. 1. *The degree* $d(\mathcal{F})$ *of the locally free sheaf* $\mathcal{F}$ *is always non-negative. Moreover* $d(\mathcal{F}) = 0$ *if and only if* $\varphi : S \to C$ *is an analytic fibre bundle.*

Proof. From the above theorem 1. 11, it is sufficient to consider the vector bundle N^* instead of $\mathcal{F}$. We set $L = \overset{g}{\wedge} N^*$. By the adjunction formula we have

$$K_{\mathcal{J}}|_{o(C)} = L \otimes K_{o(C)}.$$

We prove the theorem by constructing a certain meromorphic section of $K_{\mathscr{J}}^{\otimes n}$ in a neighbourhood of $o(C)$ in $\mathscr{J}(S/C)$ for a suitable positive integer n. For a sufficiently large positive even integer n, there exists a cusp form $f(z)$ of weight n on the Siegel upper half plane S_g of degree g. Let τ be a meromorphic 1-form on C. We set

$$\tilde{\Omega}=f(T(\tilde{u}))(\tilde{\tau}\wedge d\zeta_1\wedge\cdots\wedge d\zeta_g)^n$$

where $\tilde{\tau}=\pi^*(\tau)$, $\pi:\tilde{C}'\to C'$. $\tilde{\Omega}$ is a meromorphic section of $K_{\tilde{\mathscr{J}}}^{\otimes n}$ and for every $\gamma\in\pi_1(C')$ we have $\gamma^*\tilde{\Omega}=\tilde{\Omega}$. Hence $\tilde{\Omega}$ induces a meromorphic section Ω' of $K_{\mathscr{J}'}^n$. Now we construct the extension Ω of Ω' in a neighoburhood of $o(C)$ in $\mathscr{J}$. For that purpose it is enough to consider the problem locally on $\varpi_D:\mathscr{J}(X/D)\to D$. We set $\hat{\Omega}'=\hat{\phi}^*(\Omega')$ where $\hat{\phi}:\mathscr{J}(\hat{X}/\hat{D})\to\mathscr{J}(X/D)$ is the natural meromorphic mapping induced by the meromorphic mapping 1. 9. On $\mathscr{J}(\hat{X}/\hat{D})|_{\hat{D}^*}$, $\hat{\Omega}'$ may be written in the form

$$\hat{\Omega}'=f(T(s))\,a(t)^n\left(dt\wedge\frac{dw_1}{w_1}\wedge\cdots\wedge\frac{dw_g}{w_g}\right)$$

where $\tau=a(t)dt$. Since $f(Z)$ is a cusp form, it has a Fourier expansion

$$f(Z)=\sum_S\beta(S)\boldsymbol{e}(tr(S\cdot Z))$$

where the sum is taken over all half-integer positive-definite symmetric matrices of size g. Hence we have

$$f(T(s))=\sum_{S>0}\beta(S)\boldsymbol{e}\left(tr\left(\frac{\log s}{2\pi\sqrt{-1}}\cdot\begin{bmatrix}0&0\\0&\Delta\end{bmatrix}\right)+tr(S\cdot T_0(s))\right).$$

Since Δ is positive-definite, we infer that

$$f(T(s))=s^a g(s)$$

where $g(s)$ is holomorphic on $\hat{D}$ with $g(0)\neq 0$, and a is an integer with

$$a\geqq \operatorname{tr}\Delta>0.$$

Hence, using 1. 8 and 1. 9, we can write in a neighbourhood V of the section $o_D(D)$ in $\mathscr{J}(X/D)$

$$\begin{aligned}\hat{\Omega}'&=f(T(s))\,a(t)^n b(s)^n(dt\wedge d\eta_1\wedge\cdots\wedge d\eta_g)^n\\&=s^\alpha g(s)a(t)^n b(s)^n(dt\wedge d\eta_1\wedge\cdots\wedge d\eta_j\wedge dz_{j+1}\wedge\cdots\wedge dz_g)^n\end{aligned}$$

where

$$\begin{aligned}\alpha&=a+a_1+a_2+\cdots+a_g\\b(s)&=\det(F(s)).\end{aligned}$$

Since $\hat{\Omega}'=\hat{\phi}^*\hat{\Omega}'$, we infer that $e_m^\alpha s^\alpha g(e_m s)b(e_m s)^n=s^\alpha g(s)b(s)^n$. Hence we can write

$$\Omega=t^\beta a(t)^n h(t)(dt\wedge d\eta_1\wedge\cdots\wedge d\eta_j\wedge dz_{j+1}\wedge\cdots\wedge dz_g)^n$$

where

$$\begin{aligned}\alpha&=m\beta\\h(s^m)&=g(s)b(s)^n.\end{aligned}$$

Therefore Ω is meromorphic in a neighbourhood of $o_D(D)$ in $\mathscr{J}(X/D)$. Thus we obtain the desired extension Ω of Ω' in a neighbourhood of $o(C)$ in $\mathscr{J}$. Since Ω does not vanish identically, it follows that $\Omega/(\tau)^n$ is a well-defined holomorphic section of

L^n on $o(C)$. Hence deg $L \geqq 0$. Moreover if $\alpha > 0$, then $\beta > 0$ and this implies that deg $L > 0$. Therefore assume that all α 's are zero. This occurs only when the local monodromy around each singular fibre is trivial. In this case the period mapping T induces a holomorphic mapping $\bar{T}: C \to S_g^* = S_g/Sp(g, z)$. Moreover if φ contains a singular fibre, the image $\bar{T}(C)$ of the holomorphic mapping $\bar{T}$ is of positive dimension. Then $\bar{T}(C)$ intersects the subvariety $\bar{F}$ of the Satake compactification $\overline{S_g^*}$ defined by $f(z)=0$, because the divisor $\bar{F}$ is an ample Cartier divisor by construction of the Satake compactification. Then $f(T(\tilde{u}))$ has zeros on $\tilde{C}'$, and hence Ω has zeros on C. Hence deg $L > 0$. Therefore if deg $L=0$, then $\varphi: S \to C$ is a smooth morphism and $\bar{T}(C)$ is a point. That is, $\varphi: S \to C$ is an analytic fibre bundle. The converse is trivial. q. e. d.

Corollary 2. 2. *Let $\varphi: S \to C$ be the same as above. Then we have*

$$p_g(S) \geqq g(\pi(C)-1) + \deg \mathcal{F}.$$

Corollary 2. 3. *Let $\varphi: S \to C$ be the same as above. Then*

$$\kappa(S) \geqq \kappa(S_x) + \kappa(C)$$

where S_x is a general fibre of φ.

Proof. If $\pi(C) \geqq 2$ and deg $\mathcal{F} > 0$, or deg $\mathcal{F} \geqq 2$, by the above corollary, $p_g(S) \geqq 2$, hence a fortiori $\kappa(S) > 0$. From the fundamental theorem on pluricanonical fibrations (see Iitaka [3], Ueno [12] Theorem 6. 11), it is easy to show that the above inequality holds. If deg $\mathcal{F}=0$, then from Theorem 2. 1 and Nakamura-Ueno [7] we obtain the desired result. Therefore it remains to consider the case where $\pi(C) = 1$, deg $\mathcal{F}=1$, $p_g(S)=1$. In this case we need the detailed study of curves which are components of the effective canonical divisor of S. Since the proof is complicated we omit it. The proof can be found in Ueno [14] § 9, Theorem 9. 6.

q. e. d.

Corollary 2. 4. *Conjecture C_2 is true.*

For the proof we should only consider the case where $\varphi: S \to C$ is an elliptic surface. This will be done in the appendix.

3. Fibre Space over a Curve

In this section we consider a surjective morphism $\varphi: V \to C$ of an algebraic manifold V of dimension $n+1$ to a non-singular curve C of genus g with connected fibres. Let us consider the spectral sequence

$$E_2^{p,q} = H^p(C, R^q\varphi_*\mathcal{O}_V) \Rightarrow H^{p+q}(V, \mathcal{O}_V).$$

This spectral sequence degenerates at E_2-terms and we have an isomorphism

$$H^{n+1}(V, \mathcal{O}_V) \simeq H^1(C, R^n\varphi_*\mathcal{O}_V). \tag{3. 1}$$

Moreover we have a canonical exact sequence

$$(3.2)\qquad 0 \to H^1(C, R^{n-1}\varphi_*\mathcal{O}_V) \to H^n(V, \mathcal{O}_V) \to H^0(C, R^n\varphi_*\mathcal{O}_V) \to 0.$$

Since $R^n\varphi_*\mathcal{O}_V$ is a coherent sheaf we can write

$$R^n\varphi_*\mathcal{O}_V = \mathcal{L} \oplus T$$

where $\mathcal{L}$ is a locally free sheaf and T is a torsion sheaf. As V is algebraic we have

$$p_g(V) = \dim H^1(C, \mathcal{L}).$$

The rank of $\mathcal{L}$ is equal to the geometric genus $p_g(V_x)$ of a general fibre V_x of φ. (Note that p_g is invariant under algebraic deformations.) We put $h = p_g(V_x)$. Suppose that the degree $d(\mathcal{L})$ of $\mathcal{L}$ is strictly greater than $h(g-1)$. Then by the Riemann-Roch theorem, we have

$$\dim H^0(C, \mathcal{L}) > 0.$$

From the exact sequence 3. 2 it follows that there exists an element ω of $H^n(V, \mathcal{O}_V)$ which induces a non-zero element of $H^0(C, \mathcal{L})$. By the Dolbeault isomorphism, ω can be considered as a $\bar{\partial}$-closed $(0, n)$-form. Moreover by the Hodge theory on an algebraic manifold, we can choose ω in such a way that the complex conjugate $\bar{\omega}$ is a holomorphic n-form (see, for example, Ueno [12] § 9). Since ω induces a non-zero element of $H^0(C, \mathcal{L})$, $\bar{\omega}$ induces a non-zero holomorphic n-form on a general fibre of φ. Therefore for a holomorphic 1-form π on C, $\varphi^*(\pi) \wedge \bar{\omega}$ is a non-zero holomorphic $(n+1)$-form. Thus we obtain the following Lemma.

Lemma. *Suppose that $d(\mathcal{L}) > h(g-1)$ and $g \geqq 2$. Then $p_g(V) \geqq 2$. Hence, a fortiori $\kappa(V) \geqq 1$.*

Now suppose that $d(\mathcal{L}) \leqq h(g-1)$. By the Grothendieck duality theorem we have an isomorphism

$$\varphi_*\mathcal{O}(K_{V/C}) \simeq \mathcal{H}om_{\mathcal{O}_C}(R^n\varphi_*\mathcal{O}_V, \mathcal{O}_C) \simeq \mathcal{H}om_{\mathcal{O}_C}(\mathcal{L}, \mathcal{O}_C).$$

Therefore the sheaf $\mathcal{F} = \varphi_*\mathcal{O}(K_{V/C})$ is a locally free sheaf of rank h. Hence, $\deg(\mathcal{F}) \geqq -h(g-1)$, by our assumption. Hence from the Riemann-Roch theorem we obtain

$$\dim H^0(C, \mathcal{S}^m(\mathcal{F} \otimes K_C)) - \dim H^1(C, \mathcal{S}^m(\mathcal{F} \otimes K_C))$$
$$\geqq \left\{p(h, m)h - \binom{m+h-1}{m}\right\}(g-1)$$

where $p(h, m)$ is defined by

$$p(0, m) = 1, \quad p(l, m) = \sum_{k=1}^{m} p(l-1, k).$$

It follows that

$$(3.3)\qquad \dim H^0(C, \mathcal{S}^3(\mathcal{F} \otimes K_C)) \geqq \frac{h(h+1)(h+2)}{3}(g-1).$$

On the otherhand there is a natural $\mathcal{O}_C$-module homomorphism

$$\theta : \mathcal{S}^3(\mathcal{F} \otimes K_C) = \mathcal{S}^3(\varphi_*\mathcal{O}(K_V)) \to \varphi_*(\mathcal{O}(K_V^{\otimes 3})),$$

which induces a homomorphism

$$\theta_* : H^0(C, \mathcal{S}^3(\mathcal{F} \otimes K_C)) \to H^0(C, \varphi_*(\mathcal{O}(K_V^{\otimes 3}))) \cong H^0(V, \mathcal{O}(K_V^{\otimes 3})).$$

Suppose that Ker θ is a torsion sheaf. (This is the case if $h=1$, 2 or more generally $\Phi_{K(V_x)}(V_x)=\boldsymbol{P}^{h-1}$ for almost all points x of C). If $h\geqq 1$ and $g\geqq 2$, it follows from 3. 3 that

$$\dim \theta_*(H^0(C, \mathcal{S}^3(\mathcal{F}\otimes K_C))) \geqq 2.$$

Therefore we obtain the following proposition.

Proposition 3. 4. *Suppose that* Ker θ *is a torsion sheaf. Then* $p_3(V)\geqq 2$. *Hence a fortiori* $\kappa(V)\geqq 1$.

Corollary 3. 5. Let $\varphi : V\rightarrow C$ be the same as above. *Suppose that for a general fibre* V_x *of* φ *we have* $p_g(V_x)=1$, *or* $p_g(V_x)=2$ *and* $g\geqq 2$. *Then we have* $p_3(V)\geqq 2$. *Hence a fortiori* $\kappa(V)\geqq 1$.

Corollary 3. 6. *Let* $\varphi : V\rightarrow C$ *be a surjective morphism from an algebraic threefold* V *onto a non-singular curve* C *with connected fibres. Suppose that* $\pi(C)\geqq 2$ *and* $p_g(V_x)=1$ *or* $p_g(V_x)=2$, *for a general fibre* V_x *of* φ. *Then we have*

$$p_3(V) \geqq 2,$$
$$\kappa(V) \geqq \kappa(V_x)+\kappa(C).$$

Proof. From 3. 5 we have $p_3(V)\geqq 2$ and $\kappa(V)\geqq 1$. Hence, if $\kappa(V_x)=0$, or $\kappa(V_x)=1$ and $\pi(C)=1$, there is nothing to prove. Therefore first we assume that $\kappa(V_x)=1$ and $\pi(C)\geqq 2$. Suppose that $\kappa(V)=1$. Then by the fundamental theorem due to Iitaka (see Iitaka [3] and Ueno [12], § 6) there exist an algebraic threefold $V^\natural$ birationally equivalent to V, a non-singular curve Δ and a surjective morphism $\psi : V^\natural\rightarrow\Delta$ with connected fibres such that for a general fibre $V^\natural_\alpha$, $\alpha\in\Delta$, of ψ, $\kappa(V^\natural_\alpha)=0$. Let V^*_α be the image of $V^\natural_\alpha$ in V under the birational correspondence. V^*_α cannot be contained in a fibre of φ, since $\kappa(V^*_\alpha)=0$ and $\kappa(V_x)=1$ for a general fibre V_x of φ. Therefore φ induces a surjective morphism $\varphi^* : V^*_\alpha\rightarrow C$. Then by Corollary 2. 4, $\kappa(V^*_\alpha)\geqq\kappa(C)=1$. This is a contradiction. Therefore we must have $\kappa(V)\geqq 2$.

Next suppose that $\kappa(V_x)=2$ and $\pi(C)=1$. If $\kappa(V)=1$, we have the same surjective morphism $\psi : V^\natural\rightarrow\Delta$ as above. Let V^*_α be the same as above. Then φ induces a surjective morphism $\varphi^* : V^*_\alpha\rightarrow C$. Since $\kappa(V^*_\alpha)=0$ and C is a non-singular elliptic curve, by Corollary 2. 4 and the canonical bundle formula for an elliptic surface, $\varphi^*: V^*_\alpha\rightarrow C$ is birationally equivalent to an elliptic bundle over C. Thus the intersection $V^*_\alpha\cap V_x$ is birationally equivalent to an elliptic curve. Therefore V_x has an (irrational) pencil of (singular) elliptic curves. Hence $\kappa(V_x)\leqq 1$. This is a contradiction. Hence we have $\kappa(V)\geqq 2$.

Finally suppose that $\kappa(V_x)=2$ and $\pi(C)\geqq 2$. If $\kappa(V)=1$, by an argument similar to that above we have a contradiction. Therefore suppose that $\kappa(V)=2$. Then there exist an algebraic threefold $V^\natural$ birationally equivalent to V, a non-singular surface W and a surjective morphism $\psi : V^\natural\rightarrow W$ with connected fibres such that a general fibre is an elliptic curve. Let H be a general hypersurface section of W of high degree and set $V^\natural_H=\psi^{-1}(H)$. We let V^*_H be the corresponding subvariety in V under the above birational correspondence. Since $V_H^{\natural}$ is an elliptic surface over a curve H, we have $\kappa(V_H^*)=\kappa(V_H^{\natural})=\kappa(H)=1$. Therefore V_H^* cannot be contained in a fibre of φ. Hence φ induces a surjective morphism $\varphi^* : V_H^* \to C$. As $\kappa(V_H^*)=\kappa(C)=1$, a general fibre of φ^* is birationally equivalent to the union of a finite number of elliptic curves. Therefore $V_H^* \cap V_x$ is birationally equivalent to the union of a finite number of elliptic curves. Hence V_x has an (irrational) pencil of (singular) elliptic curves. This implies that $\kappa(V_x) \leqq 1$. This is a contradiction. q. e. d.

Appendix

Let $\varphi : S \to \Delta$ be an elliptic surface, that is, a general fibre of φ is a non-singular elliptic curve. We assume furthermore that *no fibre of φ contains exceptional curves of the first kind.* In this appendix we will provide a simple proof of the following theorem due to Kodaira [5].

Theorem. *The canonical divisor K of an elliptic surface $\varphi : S \to \Delta$ has the following form:*

$$K = \varphi^*(K_\Delta - \mathfrak{f}) + \sum_{j=1}^{l} (m_j - 1)E_j$$

where

1) $\mathfrak{f}$ *is a divisor on Δ such that*

$$\deg(\mathfrak{f}) = -(p_g(S) - q(S) + 1)$$

2) $m_j E_j$, $j=1, 2, \cdots, l$ *are all multiple singular fibres of $\varphi : S \to \Delta$ with multiplicities m_j.*

Proof. For sufficiently large N we let $x_1, x_2, \cdots, x_N$ be points on Δ such that the scheme-theoretic fibre S_{x_i} is non-singular. Let us consider the spectral sequence

$$E_2^{p,q} = H^p(\Delta, R^q\varphi_*\mathcal{O}(-\sum_{i=1}^{N} C_i)) \Rightarrow H^{p+q}(S, \mathcal{O}(-\sum_{i=1}^{N} C_i)),$$

which degenerates at E_2-terms. We have isomorphisms

$$R^q\varphi_*\mathcal{O}(-\sum_{i=1}^{N} C_i) \simeq (R^q\varphi_*\mathcal{O}_S)(-\sum_{i=1}^{N} x_i)$$

$$\varphi_*\mathcal{O}_S \simeq \mathcal{O}_C.$$

Since N is sufficiently large it follows that

$$E_2^{0,1} = 0.$$

Thus we obtain

$$\dim_C H^1(S, \mathcal{O}(-\sum_{i=1}^{N} C_i)) = N + \pi(\Delta) - 1.$$

Therefore from the Riemann-Roch theorem we infer that

$$\dim_C H^2(S, \mathcal{O}(-\sum_{i=1}^{N} C_i)) = p_g(S) - q(S) + N + \pi(\Delta).$$

By the Serre duality we have

(A. 1) $$\dim|K+\sum_{i=1}^{N} C_i| = p_g(S)-q(S)+N+\pi(\Delta)-1.$$

It follows that the complete linear system $|K+\sum_{i=1}^{N} C_i|$ contains an effective divisor D. Since we have $D\cdot C_i=0$, the support of the divisor D is contained in a finite union of fibres of φ. Since we have assumed that each fibre of φ does not contain exceptional curves of the first kind, we infer that $D^2=0$. Therefore we can write

$$K = D-\sum_{i=1}^{N} C_i = \sum m_y S_y+\sum_{j=1}^{l} k_j E_j, \quad 0 \leqq k_j < m_i,$$

where m_y are integers. Thus we have

(A. 2) $$K = \varphi^*(K_\Delta-\mathfrak{f})+\sum_{j=1}^{l} k_j E_j$$

where $\mathfrak{f}$ is a suitable divisor on Δ. It is easily seen that $\sum_{j=1}^{l} k_j E_j$ is the fixed component of the linear system $|K+\sum_{j=1}^{N} C_j|$. Hence from A. 1 and A. 2 we infer that

$$\deg \mathfrak{f} = -(p_g(S)-q(S)+1).$$

Let us show that $(m_j-1)E_j$ is a fixed component of $|K+\sum_{j=1}^{N} C_j|$. On Δ we can choose points $y_1, \cdots, y_{N-1}$ which are different from $a_i=\varphi(E_i)$, $i=1, 2, \cdots, l$ such that the divisor $\sum_{i=1}^{N} x_i$ is linearly equivalent to $\sum_{i=1}^{N-1} y_i+a_i$. Then we have

$$|K+\sum_{i=1}^{N} C_i| = |K+\sum_{i=1}^{N-1} S_{y_i}+m_j E_j|.$$

Hence it is enough to show that

$$\dim|K+\sum_{i=1}^{N} C_i| = \dim|K+\sum_{i=1}^{N-1} S_{y_i}+E_j|.$$

From the exact sequence

$$0\to \mathcal{O}(-\sum_{i=1}^{N-1} S_{y_i}-E_j)\to \mathcal{O}_S\to \bigoplus_{i=1}^{N-1} \mathcal{O}_{S_{y_i}}\oplus \mathcal{O}_{E_j}\to 0,$$

we have the long exact sequence

$$H^1(S, \mathcal{O}_S)\xrightarrow{r} \bigoplus_{i=1}^{N-1} H^1(S_{y_i}, \mathcal{O}_{S_{y_i}})\oplus H^1(E_j, \mathcal{O}_{E_j})\to$$
$$\to H^2(S, \mathcal{O}(-\sum_{i=1}^{N-1} S_{y_i}-E_j))\to H^2(S, \mathcal{O}_S)\to 0.$$

Hence we have

$$\dim_C H^2(S, \mathcal{O}(-\sum_{i=1}^{N-1} S_{y_i}-E_j)) = p_g(S)+\dim_C \operatorname{Coker} r.$$

On the other hand, from the spectral sequence

$$E_2^{p,q} = H^p(\Delta, R^q\varphi_*\mathcal{O}_S) \Rightarrow H^{p+q}(S, \mathcal{O}_S),$$

we obtain the exact sequence

$$0\to H^1(\Delta, \mathcal{O}_\Delta)\xrightarrow{t} H^1(S, \mathcal{O}_S)\to H^0(\Delta, R^1\varphi_*\mathcal{O}_S)\to 0.$$

As $r \cdot \iota = 0$, the homomorphism r can be factored through the homomorphism

$$\bar{r} : H^0(\Delta, R^1\varphi_* \mathcal{O}_S) \to \bigoplus_{i=1}^{N-1} H^1(S_{y_i}, \mathcal{O}_{S_{y_i}}) \oplus H^1(E_j, \mathcal{O}_{E_j}).$$

Since N is sufficiently large, we have Ker $\bar{r}=0$, that is, dim Ker $r=\pi(\Delta)$. Thus we obtain

$$\dim H^2(S, \mathcal{O}(-\sum_{i=1}^{N-1} S_{y_i} - E_j)) = p_g(S) - q(S) + N + \pi(\Delta).$$

By the Serre duality we obtain the desired result. q. e. d.

Remark. Let $\varphi : S \to \Delta$ be the same as above. It is easy to show that

$$\varphi_*(K_{S/\Delta}) \simeq \mathcal{O}(-\mathfrak{f}).$$

Moreover deg $(\mathfrak{f})=0$ if and only if the moduli of regular fibres of φ are constant and φ has only multiple singular fibres.

References

[1] Clemens, H., Griffiths, P. A., Jambois, T. F. and Mayer, A. L. : Seminar on degenerations of algebraic varieties, Institute for Advanced Study, Princeton (1969–70).

[2] Freitag, E. : Über die Struktur der Funktionenkörper zu hyperabelschen Gruppen, I. J. Rein Angew. Math., **247** (1971), 91–117.

[3] Iitaka, S. : On D-dimensions of algebraic varieties, J. Math. Soc. Japan, **23** (1971), 356–373.

[4] Kodaira, K. : On compact analytic surface II, Ann. of Math., **77** (1963), 563–626.

[5] ——: On the structure of compact complex analytic surfaces I, Amer. J. Math., **86** (1964), 751–798.

[6] Nakamura, I. : On moduli of stable quasi-abelian varieties, Nagoya Math. J., **58** (1975), 149–214.

[7] Nakamura, I. and Ueno, K. : An addition formula for Kodaira dimensions of analytic fibre bundles whose fibres are Moišezon manifolds, J. Math. Soc. Japan, **25** (1973), 363–371.

[8] Namikawa, Y. : A new compactification of Siegel space and degeneration of abelian varieties, Math. Ann. **221** (1976), 97–141, 201–214.

[9] Namikawa, Y. and Ueno, K. : The complete classification of fibres in pencil of curves of genus two, Manuscripta Math., **9** (1973), 143–186.

[10] Ueno, K. : On fibre spaces of normally polarized abelian varieties of dimension 2, I, J. Fac. Sci. Univ. Tokyo, Sec. IA., **17** (1971), 37–95.

[11] ——: Classification of algebraic varieties I, Compositio Math., **27** (1973), 277–342.

[12] ——: Classification theory of algebraic varieties and compact complex spaces, Lecture Notes in Math. **439** (1975), Springer.

[13] ——: Algebraic surfaces which have pencils of curves of genus 2, To appear.

[14] ——: Introduction to the classification theory of algebraic surfaces, Lecture Notes, University of Amsterdam, 1975.

Department of Mathematics
Kyoto University

(Received January 17, 1976)

Part III

Some Remarks on Formal Poincaré Lemma

A. Andreotti and M. Nacinovich

Given a complex of differential operators it is a basic question to decide if the complex admits the so called Poincaré lemma. This lemma, which seems to have been established, for exterior differentiation, for the first time by Volterra, plays an essential role in trying to pass, via cohomology, from local to global problems.

For operators with constant coefficients reasonable criteria essentially due to Malgrange, Ehrenpreis and Lojasiewicz are available. For operators with variable coefficients the problem is still open. In this paper we deal only with the formal side of the question and we establish two criteria for the validity of the formal Poincaré lemma. The first is intended to show that in some cases this is a consequence of the Poincaré lemma and the second to cover some cases in which, although the Poincaré lemma may fail to be true, the formal side of it is preserved.

1. Preliminaries

a) Let Ω be an open set in $\boldsymbol{R}^n$ where $x=(x_1, \cdots, x_n)$ denote the coordinates. By $\mathcal{E}(\Omega)$ we denote the space of C^∞ functions on Ω endowed with the Schwartz topology. Given $x_0 \in \boldsymbol{R}^n$ we denote by ϕ_{x_0} the ring of formal power series in $x-x_0$, i. e. centered at x_0.

Let

$$\mathcal{E}^{p_0}(\Omega) \xrightarrow{A^0(x,D)} \mathcal{E}^{p_1}(\Omega) \xrightarrow{A^1(x,D)} \mathcal{E}^{p_2}(\Omega) \xrightarrow{A^2(x,D)} \cdots \tag{1}$$

be a complex of differential operators on Ω. Here $\mathcal{E}^{p_i}(\Omega)$ denotes the space $\mathcal{E}(\Omega) \times \cdots \times \mathcal{E}(\Omega)$ p_i times and $A^i(x, D) = (a^{(i)}_{\alpha\beta}(x, D))$ denotes a matrix of type $p_{i+1} \times p_i$ whose entries $a^{(i)}_{\alpha\beta}(x, D)$ are differential operators with complex valued C^∞ coefficients:

$$a^{(i)}_{\alpha\beta}(x, D) = \sum_{|\mu| \leq k} a^{(i)}_{\alpha\beta\mu}(x) D^\mu$$

where $\mu = (\mu_1, \cdots, \mu_n) \in \boldsymbol{N}^n$ is a multi-index, $|\mu| = \sum \mu_i$, and where $D_i = \dfrac{\partial}{\partial x_i}$ and D^μ stands for $\partial^{|\mu|}/\partial x_1^{\mu_1} \cdots \partial x_n^{\mu_n}$.

b) We say that the complex (1) *admits the Poincaré lemma at* $x_0 \in \Omega$ if the sequence

$$\mathcal{E}^{p_0}_{x_0} \xrightarrow{A^0(x,D)} \mathcal{E}^{p_1}_{x_0} \xrightarrow{A^1(x,D)} \mathcal{E}^{p_2}_{x_0} \xrightarrow{A^2(x,D)} \cdots \tag{2}$$

where $\mathcal{E}_{x_0}$ denotes the ring of germs of C^∞ functions at x_0, is an exact sequence.

We say that the complex (1) *admits the strict Poincaré lemma* at the point $x_0 \in \Omega$ if we can find a fundamental sequence of neighborhoods $\{w_\nu\}_{\nu \in N}$ of x_0, open and contained in Ω such that, for every $\nu \in \boldsymbol{N}$, we have an exact sequence

$$\mathcal{E}^{p_0}(w_\nu) \xrightarrow{A^0(x,D)} \mathcal{E}^{p_1}(w_\nu) \xrightarrow{A^1(x,D)} \mathcal{E}^{p_2}(w_\nu) \xrightarrow{A^2(x,D)} \cdots. \tag{3}$$

We say that the complex (1) *admits the formal Poincaré lemma* at $x_0 \in \Omega$ if we have an exact sequence

$$\phi^{p_0}_{x_0} \xrightarrow{A^0(x,D)} \phi^{p_1}_{x_0} \xrightarrow{A^1(x,D)} \phi^{p_2}_{x_0} \xrightarrow{A^2(x,D)} \cdots. \tag{4}$$

Clearly the strict Poincaré lemma implies the Poincaré lemma but all the other possible implications may fail to be true for some particular complexes.

2. Honest Differential Operators

a) Let

$$\mathcal{E}^p(\Omega) \xrightarrow{A(x,D)} \mathcal{E}^q(\Omega)$$

be a differential operator with C^∞ coefficients in Ω. Let $\mathcal{E}'(\Omega)$ denote the space of distributions with compact support in Ω

$$\mathcal{E}'(\Omega) = \text{Hom cont } (\mathcal{E}(\Omega), \boldsymbol{C}),$$

and let

$$\mathcal{E}'(\Omega)^p \xleftarrow{{}^tA(x,D)} \mathcal{E}'(\Omega)^q$$

be the transposed differential operator. If

$$A(x, D) = \sum_{|\alpha| \leq k} a_\alpha(x) D^\alpha,$$

one usually takes

$${}^tA(x, D) = \sum (-1)^{|\alpha|} D^\alpha({}^t a_\alpha(x)).$$

Let $\mathcal{E}'_{x_0}$ be the space of distributions supported at the point $x_0 \in \Omega$; every element $T \in \mathcal{E}'_{x_0}$ is of the form

$$T = \sum a_\alpha D^\alpha \delta_{x_0},$$

where the sum is finite and when δ_{x_0} denotes the Dirac measure.

We say that the operator $A(x, D)$ is a honest operator at $x_0 \in \Omega$ if the following condition is satisfied.

(H) *for any distribution* $T_{x_0} = \sum a_\alpha D^\alpha \delta_{x_0} \in (\mathcal{E}'_{x_0})^p$ *such that*

$$T_{x_0} \in \bigcap_{w \ni x_0} {}^tA(x, D)(\mathcal{E}'(w))^q$$

(w *runs through a fundamental sequence of neighborhoods of* x_0 *in* Ω) *we can find a distribution* $\theta_{x_0} = \sum b_\beta D^\beta \delta_{x_0} \in (\mathcal{E}'_{x_0})^q$ *with the property*

$$T_{x_0} = {}^tA(x, D)\theta_{x_0}.$$

b) Differential operators with constant coefficients are honest operators. This is a consequence of the following

Proposition 1. *Let* $A(D) : \mathcal{E}^p(\boldsymbol{R}^n) \rightarrow \mathcal{E}^q(\boldsymbol{R}^n)$ *be a differential operator with constant coefficients. Let* $x_0 \in \boldsymbol{R}^n$ *and let* $T_{x_0} = \sum a_\alpha D^\alpha \delta_{x_0} \in (\mathcal{E}'_{x_0})^p$. *If the equation*

$$T_{x_0} = {}^tA(D)\theta$$

admits a solution $\theta \in (\mathcal{E}'(\boldsymbol{R}^n))^q$, *then there exists an element* $\theta_{x_0} \in (\mathcal{E}'_{x_0})^q$ *such that*

$$T_{x_0} = {}^tA(D)\theta_{x_0}$$

Proof. Let $A = \sum_{|\alpha| \le k} c_\alpha D^\alpha$ with $c_\alpha \in \boldsymbol{C}$ so that ${}^tA = \sum_{|\alpha| \le k} (-1)^{|\alpha|}\, {}^tc_\alpha D^\alpha$ is also an operator with constant coefficients. By Fourier transform it is then enough to prove the following statement (taking x_0 at the origin, as we can assume).

Let $B(\xi)$ be a $p \times q$ matrix with polynomial entries and let $p(\xi) \in \mathcal{P}^p$ be a vector with components in the ring $\mathcal{P}$ of polynomials in n variables. Assume that the equation

$$p(\xi) = B(\xi)\mu$$

admits a solution $\mu \in \mathcal{H}(\boldsymbol{C}^n)^q$, where $\mathcal{H}(\boldsymbol{C}^n)$ denotes the ring of entire functions in $\boldsymbol{C}^n$. Then the same equation admits a polynomial solution $g(\xi) \in \mathcal{P}^q$.

But this is a consequence of the fact that the ring $\mathcal{H}(\boldsymbol{C}^n)$ is faithfully flat over the ring $\mathcal{P}$ (cf. [1] lemma 2, § 1, section 3).

c) We can now prove the following

Proposition 2. *Let the complex* (1) *be a complex of honest differential operators at a point* $x_0 \in \Omega$.

If the complex (1) *admits the strict Poincaré lemma at* x_0 *then it also admits the formal Poincaré lemma at* x_0.

Proof.

(α) Under the specified assumptions we have to show that the sequence (4) is exact.

We have

$$\phi_{x_0} \simeq \prod_{\alpha \in N^n} \boldsymbol{C}_\alpha$$

where $\boldsymbol{C}_\alpha$ stands for a copy of $\boldsymbol{C}$ with index $\alpha \in \boldsymbol{N}^n$. We take on ϕ_{x_0} the product topology of $\prod \boldsymbol{C}_\alpha$. This is a Fréchet space as product of a countable set of Fréchet spaces. The dual space is $\phi'_{x_0} = \text{Hom cont}\,(\phi_{x_0}, \boldsymbol{C})$

$$\phi'_{x_0} \simeq \prod_{\alpha \in N^n} \boldsymbol{C}_\alpha$$

with the locally convex direct sum topology. It can be identified with the space of distributions supported at x_0.

The dual sequence of (4) is the sequence

$$(4') \qquad (\phi'_{x_0})^{p_0} \xleftarrow{{}^tA^0(x,D)} (\phi'_{x_0})^{p_1} \xleftarrow{{}^tA^1(x,D)} (\phi'_{x_0})^{p_2} \xleftarrow{{}^tA^2(x,D)} \cdots.$$

We claim that the exactness of (4) is equivalent to the exactness of (4′). This is a consequence of the following facts :

i) any formal differential operator (i. e., with coefficients in $\phi=\boldsymbol{C}[[x_1, \cdots, x_n]]$)

$$\phi^p \xrightarrow{C(x,D)} \phi^q$$

has a closed image.

ii) for any complex of formal differential operators

$$\phi^p \xrightarrow{A(x,D)} \phi^q \xrightarrow{B(x,D)} \phi^t$$

and its dual

$$(\phi')^p \xleftarrow{{}^tA(x,D)} (\phi')^q \xleftarrow{{}^tB(x,D)} (\phi')^t$$

we do have

$$\frac{\operatorname{Ker}{}^tA}{\operatorname{Im}{}^tB} \simeq \text{Hom cont.}\left(\frac{\operatorname{Ker} B}{\operatorname{Im} A}, \boldsymbol{C}\right).$$

In view of i) the second statement is the content of the duality lemma for a complex of Fréchet spaces (See [4]).

(β) To establish the above statement i) we use an argument of Banach.

Let

$$(*) \qquad \sum_{\substack{\beta\in N\\ \alpha\in N}} c_{\alpha\beta}g_\beta = u_\alpha$$

be a countable system of linear equations with the following property :

For every $\alpha \in \boldsymbol{N}$ there exists an integer $\nu(\alpha) \in \boldsymbol{N}$ such that $c_{\alpha\beta}=0$ if $\beta>\nu(\alpha)$.

Because of this, in each equation the sum involves only finite many terms.

Now the necessary and sufficient condition for the system (*) to have a solution $g \in \prod_{\beta\in N} \boldsymbol{C}_\beta$ is that :

for any finite set of constants k_α such that

$$\sum k_\alpha c_{\alpha\beta} = 0, \quad \forall \beta \in \boldsymbol{N}$$

we should also have

$$\sum k_\alpha u_\alpha = 0$$

(Banach [3] theorem 12, pg. 51).

The system (*) defines a linear map

$$C: \prod_{\beta\in N} \boldsymbol{C}_\beta \to \prod_{\alpha\in N} \boldsymbol{C}_\alpha.$$

We claim that $\operatorname{Im} C$ is closed. Indeed let $\{g^\nu\}_{\nu\in N}$ in $\prod_{\beta\in N} \boldsymbol{C}_\beta$ be a sequence of points such that $Cg^\nu=u^\nu\to u$. We do have for any finite set of constants k_α such that $\sum k_\alpha c_{\alpha\beta}=0$ that also $\sum k_\alpha u^\nu_\alpha=0$. Therefore, at the limit we do also have $\sum k_\alpha u_\alpha=0$. The theorem of Banach shows that $u \in \operatorname{Im} C$ and the claim is proved.

To establish point i) we have, therefore, only to show that a formal differential operator leads to a system of linear equations of the above type.

Now if

$$C(x, D) = \sum_{\substack{|\alpha| \leq k \\ \beta \in N^n}} c_{\alpha\beta} x^\beta D^\alpha,$$

setting $g = \sum g_\gamma x^\gamma / \gamma! \in \phi^p$ and $u = \sum u_\sigma x^\sigma / \sigma! \in \phi^q$ the equation $C(x, D)g = u$ is equivalent to the infinite system

$$\sum_{\substack{|\alpha| \leq k \\ \beta \leq \sigma}} c_{\alpha\beta} \binom{\sigma}{\beta} g_{\sigma+\alpha-\beta} = u_\sigma \qquad \text{for all} \quad \sigma \in \boldsymbol{N}^n$$

(with the standard notations). This is a system of the type we have considered.

(γ) We now make use of the fact that the complex (1) admits the strict Poincaré lemma at x_0. To simplify the notations we will choose coordinates in $\boldsymbol{R}^n$ so that x_0 is at the origin.

For a fundamental sequence of neighborhoods w_ν of $0 \in \boldsymbol{R}^n$ we have an exact sequence of the type

$$\mathcal{E}^p(w_\nu) \xrightarrow{A(x,D)} \mathcal{E}^q(w_\nu) \xrightarrow{B(x,D)} \mathcal{E}^r(w_\nu).$$

We want to show (in view of the above remarks) that the sequence

$$(\phi')^p \xleftarrow{{}^tA(x,D)} (\phi')^q \xleftarrow{{}^tB(x,D)} (\phi')^r$$

is exact. Now this is a complex, thus we need only to show that if $T_0 = \sum c_\alpha D^\alpha \delta_0 \in (\phi')^q$ is such that ${}^tA(x, D)\, T_0 = 0$, then there exists $\theta_0 = \sum b_\beta D^\beta \delta_0 \in (\phi')^r$ such that

$$T_0 = {}^tB(x, D)\theta_0.$$

Now for every $\nu \in \boldsymbol{N}$ and every $v \in \mathcal{E}^p(w_\nu)$ we have

$$T_0[Av] = 0.$$

As $A(x, D)\mathcal{E}^p(w_\nu) = \operatorname{Ker} \{\mathcal{E}^q(w_\nu) \xrightarrow{B(x,D)} \mathcal{E}^r(w_\nu)\}$ by assumption, we have

$$T_0[w] = 0, \quad \forall\, w \in \mathcal{E}^q(w_\nu) \quad \text{with} \quad B(x, D)w = 0.$$

Therefore, T_0 defines a linear continuous map

$$t: \frac{\mathcal{E}^q(w_\nu)}{\operatorname{Ker} B} \to \boldsymbol{C}.$$

As B has a closed image (because of the assumption that the complex admits the strict Poincaré lemma), the injection

$$\frac{\mathcal{E}^q(w_\nu)}{\operatorname{Ker} B} \to B(x, D)\mathcal{E}^q(w_\nu) \subset \mathcal{E}^r(w_\nu)$$

is a topological homomorphism. Therefore t is continuous also for the topology induced on it by $\mathcal{E}^r(w_\nu)$. We can, therefore, extend t to a linear continuous functional

$$\theta_\nu : \mathcal{E}^r(w_\nu) \to \boldsymbol{C}.$$

Thus, $\theta_\nu \in (\mathcal{E}'(w_\nu))^r$ is a distribution compactly supported in w_ν and such that

$$T_0 = {}^tB(x, D)\theta_\nu.$$

Since B is an honest differential operator at 0 by assumption, we can find $\theta_0 \in (\phi')^r$ such that

$$T_0 = {}^tB(x, D)\theta_0,$$

d) A complex of differential operators with constant coefficients obtained by Fourier transform from a Hilbert resolution (cf. [1]) is called a *Hilbert complex*. It is characterized by the property of being a complex of differential operators with constant coefficients which is exact on every open convex set ([1]). Therefore a Hilbert complex is a particular complex of honest differential operators admitting the strict Poincaré lemma. Therefore

Corollary. *A Hilbert complex admits the formal Poincaré lemma at any point* $x_0 \in \boldsymbol{R}^n$.

3. The Symbolic Complex

a) Let

$$A^j(x, D) = \sum_{|\alpha| \leq k_j} a^j_\alpha(x) D^\alpha$$

be the j-th differential operator in the complex (1). Let $x_0 \in \Omega$ and let $k_j(x_0)$ be largest integer such that for some $\alpha \in \boldsymbol{N}^n$ with $|\alpha| = k_j(x_0)$ we have $a^j_\alpha(x_0) \neq 0$. Let us consider the symbol of $A_j(x, D)$ at x_0

$$A^j_0(x_0, \xi) = \sum_{|\alpha| = k_j(x_0)} a^j_\alpha(x_0)\xi^\alpha$$

where $\xi = (\xi_1, \cdots, \xi_n)$. Because (1) is a complex we do have for any $j \geq 0$

$$A^{j+1}_0(x_0, \xi) A^j_0(x_0, \xi) \equiv 0.$$

For any fixed $x_0 \in \Omega$ we can thus consider the complex of differential operators with constant coefficients

$$\mathcal{E}^{p_0}(\boldsymbol{R}^n) \xrightarrow{A_0{}^0(x_0, D)} \mathcal{E}^{p_1}(\boldsymbol{R}^n) \xrightarrow{A_0{}^1(x_0, D)} \mathcal{E}^{p_2}(\boldsymbol{R}^n) \xrightarrow{A_0{}^2(x_0, D)} \cdots. \tag{5}$$

This complex will be called the *symbolic complex for* (1) *at the point* $x_0 \in \Omega$.

Note that in the complex (5) every operator $A^j_0(x_0, D)$ is represented by a matrix of differential operators all homogeneous and of the same order $k_j(x_0)$.

In particular we can consider at the point $x_0 \in \Omega$ the complex of formal differential operators,

$$\phi^{p_0}_{x_0} \xrightarrow{A_0{}^0(x_0, D)} \phi^{p_1}_{x_0} \xrightarrow{A_0{}^1(x_0, D)} \phi^{p_2}_{x_0} \xrightarrow{A_0{}^2(x_0, D)} \cdots. \tag{6}$$

The exactness of (6) means that the symbolic complex (5) admits the formal Poincaré lemma at x_0.

b) We want to prove a criterion for the validity of the formal Poincaré lemma for the original complex (1). *We will assume that the operators $A_j(x, D)$ have a constant order $k_j(x) = k_j$ in a sufficiently small neighborhood of x_0.* The criterion is the following :

Proposition 3. *Suppose that the symbolic complex* (5) *admits the formal Poincaré lemma at* $x_0 \in \Omega$. *Then also the original complex* (1) *admits the formal Poincaré lemma at* x_0, *i. e.,* *the sequence*

(7) $$\phi_{x_0}^{p_0} \xrightarrow{A^0(x,D)} \phi_{x_0}^{p_1} \xrightarrow{A^1(x,D)} \phi_{x_0}^{p_2} \xrightarrow{A^2(x,D)} \cdots$$

is also exact.

We first prove the following

Lemma. *Let $A(D): \phi^p \to \phi^q$ be the operator defined by a $q \times p$ matrix of differential operators with constant coefficients all homogeneous and of the same order k.*

Let $g \in \phi^q$ be such that the equation

$$A(D)u = g$$

admits a solution $u \in \phi^p$. Let $\mathfrak{M}$ denote the maximal ideal of ϕ.

If $g \in \mathfrak{M}^l \phi^q$ then the above equation admits also a solution $u \in \mathfrak{M}^{l+k}\phi^p$.

Proof. Set $g = g_0 + g_1 + g_2 + \cdots$ where the g_i's are homogeneous polynomials of degree i. Similarly set $u = u_0 + u_1 + u_2 + \cdots$.

We have :

$$A(D)u_i = 0 \quad \text{for} \quad i < k$$
$$A(D)u_{k+s} = g_s \quad \text{for} \quad s = 0, 1, 2, \cdots.$$

Therefore, if $g_0 = g_1 = \cdots = g_{l-1} = 0$, setting

$$v = u_{k+l} + u_{k+l+1} + \cdots,$$

we have $v \in \mathfrak{M}^{k+l}\phi^p$ and $A(D)v = g$.

Proof of proposition 3. (α) We have to show that if $g \in \phi_{x_0}^{p_j}$ and if $A^j(x, D)g = 0$, then the equation

$$A^{j-1}(x, D)u = g$$

admits a solution $u \in \phi_{x_0}^{p_{j-1}}$, and this for $j = 1, 2, 3, \cdots$.

To simplify the notations we may assume x_0 at the origin of the coordinates of $\boldsymbol{R}^n$ and set

$$A^j(x, D) = B(x, D) \qquad \text{order of } A^j = \beta \quad \text{near } 0$$
$$A^{j-1}(x, D) = A(x, D) \quad \text{order of } A^{j-1} = \alpha \quad \text{near } 0.$$

Also we will denote by $B_0(D)$ and $A_0(D)$ the operators with constant coefficients corresponding to the symbols respectively of B and A at the origin. Therefore, we can write

$$A(x, D) = A_0(D) + A_1(x, D)$$
$$B(x, D) = B_0(D) + B_1(x, D)$$

where $A_1(x, D)$ (respectively $B_1(x, D)$) is a differential operator of order $\leq \alpha$ (respectively $\leq \beta$) with coefficients of terms of maximal order α (respectively β) vanishing at the origin.

Let

$$\phi^p \xrightarrow{A(x,D)} \phi^q \xrightarrow{B(x,D)} \phi^r$$

be the part of the formal complex at the origin we are considering and which we want to show to be acyclic.

(β) Let $g \in \phi^q$ with $B(x, D)g=0$. We first show that the equation

$$A(x, D)u = g$$

has a solution $u \in \phi^p$ under the additional assumption

$$g \in \mathfrak{M}^\beta \phi^q \quad (\mathfrak{M} = \text{maximal ideal of } \phi).$$

From $Bg=0$ we derive that

$$B_0(D)g = -B_1(x, D)g.$$

As $g \in \mathfrak{M}^\beta \phi^q$ we get that $B_1(x, D)g \in \mathfrak{M}\phi^r$. Therefore, by the previous lemma we can find $g_0 \in \mathfrak{M}^{\beta+1}\phi^q$ such that

(*) $$B_0(D)g_0 = -B_1(x, D)g.$$

Therefore,

$$B_0(D)(g-g_0) = 0$$

and $g-g_0 \in \mathfrak{M}^\beta \phi^q$. By the hypothesis of the exactness of the formal symbolic complex (6) we can find $u_0 \in \mathfrak{M}^{\alpha+\beta}\phi^p$ such that

$$A_0(D)u_0 = g-g_0.$$

Now the equation (*) gives

$$\begin{aligned} B_0(D)g_0 &= -B_1(x, D)(g_0+(g-g_0)) \\ &= -B_1(x, D)(g_0+A_0(D)u_0) \\ &= -B_1(x, D)g_0+(B_0(D)A_1(x, D)+B_1(x, D)A_1(x, D))u_0 \end{aligned}$$

since, from $BA\equiv 0$ and $B_0A_0=0$, we derive that

$$B_0A_1+B_1A_0+B_1A_1 = 0$$

so that

$$-B_1A_0 = B_0A_1+B_1A_1.$$

Hence,

$$B_0(g_0-A_1u_0) = -B_1(g_0-A_1u_0).$$

Now $g_0 \in \mathfrak{M}^{\beta+1}\phi^q$ and $u_0 \in \mathfrak{M}^{\alpha+\beta}\phi^p$, therefore $A_1u_0 \in \mathfrak{M}^{\beta+1}\phi^q$ and thus $g_0-A_1u_0 \in \mathfrak{M}^{\beta+1}\phi^q$, so that

$$B_1(g_0-A_1u_0) \in \mathfrak{M}^2\phi^r$$

Therefore, by the previous lemma we can find $g_1 \in \mathfrak{M}^{\beta+2}\phi^q$ such that

(**) $$B_0g_1 = -B_1(g_0-A_1u_0).$$

We then obtain

$$B_0(g_0-A_1u_0-g_1) = 0.$$

Now $g_0-A_1u_0-g_1 \in \mathfrak{M}^{\beta+1}\phi^q$ thus by the hypothesis and the lemma we can find $u_1 \in \mathfrak{M}^{\alpha+\beta+1}\phi^p$ with

$$A_0u_1 = -A_1u_0+g_0-g_1.$$

The equation (**) gives

$$\begin{aligned} B_0 g_1 &= -B_1(g_0 - A_1 u_0) \\ &= -B_1(A_0 u_1 + g_1) \\ &= -B_1 g_1 + (B_0 A_1 + B_1 A_1) u_1. \end{aligned}$$

i. e.,

$$B_0(g_1 - A_1 u_1) = -B_1(g_1 - A_1 u_1).$$

Now $g_1 - A_1 u_1 \in \mathfrak{M}^{\beta+2}\phi^q$, therefore,

$$B_1(g_1 - A_1 u_1) \in \mathfrak{M}^3 \phi^r.$$

Therefore, we can find $g_2 \in \mathfrak{M}^{\beta+3}\phi^q$ such that

$$(***) \qquad B_0 g_2 = -B_1(g_1 - A_1 u_1).$$

Consequently we obtain

$$B_0(g_1 - A_1 u_1 - g_2) = 0$$

and we can solve

$$A_0 u_2 = g_1 - A_1 u_1 - g_2$$

with $u_2 \in \mathfrak{M}^{\alpha+\beta+2}\phi^p$.

Now (***) gives

$$\begin{aligned} B_0 g_2 &= -B_1(g_1 - A_1 u_2) \\ &= -B_1(A_0 u_2 + g_2) \\ &= -B_1 g_2 + (B_0 A_1 + B_1 A_1) u_2 \end{aligned}$$

i. e.,

$$B_0(g_2 - A_1 u_2) = -B_1(g_2 - A_1 u_2).$$

As $g_2 - A_1 u_2 \in \mathfrak{M}^{\beta+3}\phi^q$ we have $B_1(g_2 - A_1 u_2) \in \mathfrak{M}^4\phi^r$ so that we can find $g_3 \in \mathfrak{M}^{\beta+4}\phi^q$ such that

$$(****) \qquad B_0 g_3 = -B_1(g_2 - A_1 u_2)$$

and thus,

$$B_0(g_2 - A_1 u_2 - g_3) = 0$$

so that we can solve

$$A_0 u_3 = g_2 - A_1 u_2 - g_3$$

with $u_3 \in \mathfrak{M}^{\alpha+\beta+3}\phi^p$.

Proceeding in this way we have successively

$$\begin{aligned} A_0 u_0 &= g - g_0 & & u_0 \in \mathfrak{M}^{\alpha+\beta}\phi^p,\ g \in \mathfrak{M}^{\beta}\phi^q,\ g_0 \in \mathfrak{M}^{\beta+1}\phi^q \\ A_0 u_1 + A_1 u_0 &= g_0 - g_1 & & u_1 \in \mathfrak{M}^{\alpha+\beta+1}\phi^p,\ g_1 \in \mathfrak{M}^{\beta+2}\phi^q \\ A_0 u_2 + A_1 u_1 &= g_1 - g_2 & & u_2 \in \mathfrak{M}^{\alpha+\beta+2}\phi^p,\ g_2 \in \mathfrak{M}^{\beta+3}\phi^q \\ A_0 u_m + A_1 u_{m-1} &= g_{m-1} - g_m & & u_m \in \mathfrak{M}^{\alpha+\beta+m}\phi^p,\ g_m \in \mathfrak{M}^{\beta+m+1}\phi^q. \end{aligned}$$

Therefore, the series

$$\begin{aligned} u &= u_0 + u_1 + u_2 + u_3 + \cdots \\ g &= (g - g_0) + (g_0 - g_1) + (g_1 - g_2) + \cdots \end{aligned}$$

are $\mathfrak{M}$-adically convergent and the second has for sum g.

Adding the above equations we thus obtain

$$A_0 u + A_1 u = g \quad \text{or} \quad Au = g.$$

(γ) We now show that we can drop the assumption

$$g \in \mathfrak{M}^{\beta} \phi^q.$$

Let $g \in \mathfrak{M}^{\beta-1} \phi^q$ with $Bg = 0$. We can write

$$g = g_0 + g_1$$

with g_0 homogeneous of degree $\beta - 1$ and $g_1 \in \mathfrak{M}^{\beta} \phi^q$. As B_0 is homogeneous of order β, we do have $B_0 g_0 = 0$ so that we can solve

$$A_0 u = g_0 \quad \text{with} \quad u \in \mathfrak{M}^{\alpha+\beta-1} \phi^p.$$

Then

$$Au = g_0 + A_1 u = g + (A_1 u - g_1).$$

Now $A_1 u - g_1 \in \mathfrak{M}^{\beta} \phi^q$. As $Bg = 0$ and $BA \equiv 0$, we have

$$B(A_1 u - g_1) = 0.$$

Therefore, by the previous argument, we can find $v \in \phi^p$ such that

$$Av = A_1 u - g_1.$$

Hence

$$A(u - v) = g.$$

Let $g \in \mathfrak{M}^{\beta-2} \phi^q$ with $Bg = 0$ and set

$$g = g_0 + g_1$$

with g_0 homogeneous of degree $\beta - 2$ and $g_1 \in \mathfrak{M}^{\beta-1} \phi^q$. As $B_0 g_0 = 0$, we can solve $A_0 u = g_0$ with $u \in \mathfrak{M}^{\alpha+\beta-2} \phi^p$. Then

$$Au = g_0 + A_1 u = g + (A_1 u - g_1)$$

and $A_1 u - g_1 \in \mathfrak{M}^{\beta-1} \phi^q$. Moreover $B(A_1 u - g_1) = 0$ and, therefore, we can find $v \in \phi^p$ with

$$Av = A_1 u - g_1.$$

Consequently

$$A(u - v) = g.$$

The inductive argument is now obvious.

c) As an example of the application of the previous proposition we can show that, *given a hypersurface S in $\boldsymbol{C}^n$ the tangential Cauchy-Riemann complex along S admits the formal Poincaré lemma.* It is known [2] that this complex does not admit in general the Poincaré lemma.

The question being local at a point $x_0 \in S$ we may assume x_0 at the origin of $\boldsymbol{C}^n$ and S given by an equation

$$y_n = g(z_1, \cdots, z_{n-1}, x_n)$$

where $z_j=x_j+iy_j$ are the holomorphic coordinates in $\boldsymbol{C}^n$ and where g vanishes at the origin with its first partial derivatives.

If $\rho=y_n-g=\frac{1}{2i}(z_n-\bar{z}_n)-g$, we get

$$\bar{\partial}\rho=-\frac{1}{2i}d\bar{z}_n+\sum_1^n\alpha_j d\bar{z}_j$$

where the functions α_j vanish at the origin.

For a C^∞ function f on S we set

$$\frac{\partial f}{\partial w_j}=\frac{\partial f}{\partial \bar{z}_j}+2i\beta'_j\frac{\partial f}{\partial x_n},\quad \beta_j=\frac{\alpha_j}{1+\alpha_n},\quad 1\leq j\leq n-1.$$

Set $w_j=d\bar{z}_j$ for $1\leq j\leq n-1$. Forms of type $(0, s)$ on S are, by definition, the elements of the quotient space of forms of type $(0, s)$ in $\boldsymbol{C}^n$ modulo those of type

$$\rho\alpha+\bar{\partial}\rho\wedge\beta$$

where α is of type $(0, s)$ and β of type $(0, s-1)$. We see then that, near the origin on S, the space $\mathscr{A}^j$ of forms of type $(0, s)$ is isomorphic to the space

$$\mathscr{A}^s=\{w=\sum_{j_1<\cdots<j_s}a_{j,\cdots,j_s}(x)w_{j_1}\wedge\cdots\wedge w_{j_s}\}$$

w having the coefficients $a_{j_1\cdots j_s}(x)$ C^∞ on S. The operator

$$\bar{\partial}_S\colon \mathscr{A}^j\longrightarrow\mathscr{A}^{j+1}$$

is then defined by

$$\bar{\partial}_S w=\sum_{j_0<\cdots<j_s}(\sum(-1)^h\frac{\partial a_{j_0\cdots\hat{j}_h\cdots j_s}}{\partial w_{j_h}})w_{j_0}\wedge\cdots\wedge w_{j_s}.$$

If follows then (as the α_j's vanish at the origin) that the symbolic complex at the origin reduces to the complex of the $\bar{\partial}$-operator in the variables $z_1, \cdots, z_{n-1}$. The symbolic complex is thus a Hilbert complex and has the formal Poincaré lemma. The same is, therefore, true for the tangential Cauchy-Riemann complex.

References

[1] Andreotti, A. and Nacinovich, M. : Complexes of differential operators with constant coefficients, Ann. Scuola Norm. Sup. Pisa (to appear).
[2] Andreotti, A. and Hill, C. D. : E. E. Levi convexity and Hans Lewy problems, Part I and II. Ann. Scuola Norm. Sup. Pisa, **26** (1972), 325–363, 767–806.
[3] Banach, S. : Théorie des Opérations Linéaires, Warszawa (1932).
[4] Serre, J.-P. : Un théorème de dualité, Comm. Math. Helvetici, **29** (1955), 9–26.

Istituto Matematico,
Università di Pisa

(Received April 6, 1976)

Special Arithmetic Groups and Eisenstein Series

W. L. Baily, Jr.[1)]

In the course of earlier investigations [1] on Eisenstein series, it became evident that Eisenstein series on tube domains were in a natural way automorphic forms for certain particular types of arithmetic groups. It seemed appropriate to designate these as *special arithmetic groups* because of the connection of their localizations with what Bruhat and Tits have called special maximal compact subgroups of simply-connected, semi-simple, p-adic algebraic groups. The present note represents some preliminary investigations on these. We hope to treat further some of the questions raised here on another occasion.

Many of the results here are straightforward consequences of known facts, but we have not seen them assembled in print elsewhere. The known facts include results on linear, semisimple algebraic groups over an arbitrary field, including their classification, and results of Bruhat and Tits on the buildings associated with reductive p-adic linear groups. The main references are [5, 7, 8, 11].

We wish to acknowledge here our debt to conversations with H. Hijikata and G. Shimura, as well as to a written communication from J. Tits. All of these have proved most helpful in bringing together the conclusions developed here. The author trusts each of the aforementioned understands his thanks for their help.

Let k be an algebraic number field of finite degree and let G be a connected, semi-simple, linear algebraic group defined over k. We generally assume G is simply connected, though for the most elementary definitions this is unnecessary. Let $\mathfrak{o}$ be the ring of integers in k. If $\mathfrak{p}$ is a prime ideal of $\mathfrak{o}$, let $\mathfrak{o}_\mathfrak{p}$ and $k_\mathfrak{p}$ denote the respective completions of $\mathfrak{o}$ and k at $\mathfrak{p}$, and extend this notation, as usual, to include the infinite primes of k. If H is any topological group, let H^0 denote its identity component.

Let $\rho: G \to GL(V)$ be a strictly faithful, rational, linear representation of G defined over k, where V is a finite-dimensional vector space defined over k. Let Λ be a k-lattice in V; i. e., Λ is an $\mathfrak{o}$-submodule of finite type in V_k such that for every extension field K of k, $\Lambda \otimes_\mathfrak{o} K \simeq V_K$. Let $\mathfrak{p}$ be a prime ideal of $\mathfrak{o}$ and put

$$G_\Lambda = \{g \in G | \rho(g).\Lambda = \Lambda\}, \quad \Lambda_\mathfrak{p} = \Lambda \otimes_\mathfrak{o} \mathfrak{o}_\mathfrak{p},$$
$$G_{\Lambda\mathfrak{p}} = \{g \in G | \rho(g).\Lambda_\mathfrak{p} = \Lambda_\mathfrak{p}\}.$$

Clearly, $G_\Lambda \subset G_k$, $G_{\Lambda\mathfrak{p}} \subset G_{k\mathfrak{p}}$. If Λ and Λ' are two k-lattices in V, there exists a finite set

1) Research done in part with support from National Science Foundation grant MCS 75-06336.

$S=S(\Lambda, \Lambda')$ of primes such that $\Lambda'_{\mathfrak{p}}=\Lambda_{\mathfrak{p}}$ for all $\mathfrak{p}\notin S$. $\Lambda_{\mathfrak{p}}$ is called a special $k_{\mathfrak{p}}$-lattice if $G_{\Lambda_{\mathfrak{p}}}$ is a special (or "good") maximal compact (SMC) subgroup of $G_{k_{\mathfrak{p}}}$ (in the sense of [8, § 4]). Now assume G is simply connected. If Λ is any k-lattice in V, it is known (unpublished proof of H. Hijikata) that $\Lambda_{\mathfrak{p}}$ is a special lattice for all but a finite number of primes $\mathfrak{p}$. Λ is called special if $\Lambda_{\mathfrak{p}}$ is a special $k_{\mathfrak{p}}$-lattice for *all* $\mathfrak{p}$. As special $k_{\mathfrak{p}}$-lattices exist for each finite prime $\mathfrak{p}$, it follows from a fundamental theorem on lattices that special lattices always exist. The stabilizer G_{Λ} of a special lattice Λ is called a special arithmetic subgroup (SAG) of G_k. We shall prove a theorem concerning the possible finiteness of the number of isomorphism classes of SAG's in G_k when G is simply-connected. If K is any extension field of k, let ${}_KT$ denote a maximal K-split torus of G, and let T be a maximal torus of G defined over k. One has (cf. [3], § 2. 5)

Proposition 1. *Let G be a semi-simple, connected, linear algebraic group defined over an algebraic number field k and let P be a finite set of (finite or infinite) places of k. Then there exists in G a maximal k-split torus ${}_kT$, a maximal torus T defined over k, and for each $\mathfrak{p}\in P$ a maximal $k_{\mathfrak{p}}$-split torus ${}_{k\mathfrak{p}}T$ satisfying ${}_kT\subset{}_{k\mathfrak{p}}T\subset T$.*

Proof. We assume first that G is simply-connected, so that strong approximation holds [12].

Let ${}_kT$ be any maximal k-split torus of G and let $Z=Z_G({}_kT)$ be its centralizer in G. If L is a field containing k, ${}_kT$ is contained in some ${}_LT$, and any ${}_LT$ containing ${}_kT$ is contained in Z. We may write $Z=H.S$, where S is a central torus of Z, H is semisimple, H and S are both defined over k, and $H\cap S$ is finite. Moreover, H is k-anisotropic and simply-connected [6, 4. 4]. Thus we may assume G is k-anisotropic, and then we only need show there is a maximal torus T of G defined over k such that for each $\mathfrak{p}\in P$ there exists a maximal $k_{\mathfrak{p}}$-split torus ${}_{k\mathfrak{p}}T\subset T$.

To prove this, if $\mathfrak{p}\in P$, let $T^{(\mathfrak{p})}$ be a maximal torus of G defined over $k_{\mathfrak{p}}$ and containing a maximal $k_{\mathfrak{p}}$-split torus of G, and let $T^{(\mathfrak{p})\prime}$ be its set of regular points. The mapping $(x, y)\rightarrow xyx^{-1}$ of $G\times T^{(\mathfrak{p})\prime}$ into G is everywhere submersive (within $T^{(\mathfrak{p})\prime}$), hence its restriction to the set of $\mathfrak{p}$-adic points is an open mapping in the $\mathfrak{p}$-adic topology; let $A_{\mathfrak{p}}$ be the image of that set of $\mathfrak{p}$-adic points. Then $A_{\mathfrak{p}}$ is open in $G_{k\mathfrak{p}}$ and every point of $A_{\mathfrak{p}}$ is a regular point $x_{\mathfrak{p}}$ such that $Z(x_{\mathfrak{p}})^0$ is a maximal torus of G containing a maximal $k_{\mathfrak{p}}$-split torus. By strong approximation, there exists a point $x\in G_k\cap(\prod_{\mathfrak{p}\in P}A_{\mathfrak{p}})$. Let $T=Z(x)^0$. Then T is a maximal torus of G defined over k and contains a maximal $k_{\mathfrak{p}}$-split torus for each $\mathfrak{p}\in P$.

Now, if G is not simply-connected, let $\tilde{G}$ be its universal covering and let $\Psi:\tilde{G}\rightarrow G$ be the covering isogeny, all defined over k. Then [5] Ψ establishes a one-to-one correspondence between maximal L-split tori of $\tilde{G}$ and those of G for any extension field L of k. This completes the proof.

Let ${}_L\Phi=\Phi({}_LT, G)$, where L is any extension field of k, denote the set of roots of G with respect to ${}_LT$, and omit L when $T={}_LT$ is a maximal torus in G. Let $\tau({}_L\Phi)$

be the type of ${}_L\Phi$ in the Killing-Cartan-Bourbaki [7] classification of root systems.

Let $\mathcal{A}(G)$ be the algebraic group of all rational automorphisms of G and let $\mathcal{I}(G)=\mathcal{A}(G)^0\simeq \mathrm{Ad}\, G$, the subgroup of inner automorphisms. All these objects *and the isomorphism identifying* $\mathcal{I}(G)$ *with* Ad G *are defined over* k. The factor group $\mathcal{A}(G)/\mathcal{I}(G)$ is isomorphic to the group of symmetries of the (ordinary) Dynkin diagram of a simple root system of Φ.

Now let $L=k_\mathfrak{p}$ and $Z=Z_G({}_LT)$. Let $a\in {}_L\Phi$. We define the subgroup L_a of G_L and the coset M_a of Z_L as in § 6 of [8] (where, of course, what we denote by G_L is denoted simply by G), so that in the definitions of § 6, *loc. cit.*, $(Z_L, (U_{La,}, M_a)_{a\in_L\Phi})$ is a generating system of root data in G_L. We assume the system of root data is supplied with a discrete valuation $\varphi=\{\varphi_a\}_{a\in_L\Phi}$ according to the axioms of § 6. 2 of *loc. cit.*, and such that φ is suitably compatible with the existing valuation ω on $L=k_\mathfrak{p}$, so normalized that $\omega(\pi)=1$ for a prime element π of $\mathfrak{p}$. The relation between ω and a naturally associated valuation $\varphi^\omega=\{\varphi_a^\omega\}_{a\in_L\Phi}$ can be made quite explicit for groups of Chevalley type (*loc. cit.* § 6. 2. 2–b)) and for groups of classical type (*loc. cit.* § 10) ; $\varphi=\varphi^\omega$ is unique up to equivalence for "most" G. Further, we define the subgroup $U_{a,k}$ of $U_{a,L}$ and the subset $M_{a,k}$ of M_a as in *loc. cit.* § 6. 2. Taking the affine root system Σ and the laddering (échelonnage) $\mathcal{E}_L$ of Σ by ${}_L\Phi$ as defined in § 6. 2. 6, *loc. cit.*, one has a natural homomorphism ν of the group N', generated by all the sets $M_{a,k}$, onto the affine Weyl group $\tilde{W}$ of Σ. Fix a chamber $\boldsymbol{C}$ for $\tilde{W}$, and let S be the set of reflections $r_{a,k}$ associated to its walls. The elements of S are the vertices of the Dynkin diagram $\mathcal{D}(\mathcal{E}_L)$ of $\mathcal{E}_L$, which is completed by joining certain pairs of vertices and by indicating certain metric data by suitable markings. We now assume ${}_L\Phi$ is irreducible. Then $r=r_{a,k}\in S$ is called special if the set $S'=S-\{r\}$ generates a finite subgroup $W_{S'}$ of $\tilde{W}$ isomorphic to the ordinary L-Weyl group ${}^vW$ of ${}_L\Phi$. Let $B=P_{fc}$ be as defined in § 6. 5, *loc. cit.* Then, when G is simply-connected, r is a special vertex if and only if $BW_{S'}B$ is an SMC subgroup of G_L. More generally, $r\in S$ is called special if it is a special point of an irreducible component of ${}_L\Phi$.

The group G_L operates transitively on the chambers of the building $\mathcal{I}$ associated to the Tits system (G_L, B, N, S). Moreover, any $g\in\mathcal{A}(G)_L$ operates on the building and permutes its chambers in a natural way, for the action of such g is continuous in the natural, locally compact topology on G_L. (*loc. cit.* § 2. 7, 3, 5). So if $\alpha\in\mathcal{A}(G)_L$, there exists $g\in G_L$ such that $g\cdot\alpha(\boldsymbol{C})=\boldsymbol{C}$. Thus, $g\cdot\alpha$ defines an automorphism of $\boldsymbol{C}$, hence also one of $\mathcal{D}(\mathcal{E}_L)$, which we denote by $\mu(\alpha)$. Thus, μ is a homomorphism of $\mathcal{A}(G)_L$ into $\mathrm{Aut}(\mathcal{D}(\mathcal{E}_L))$ and G_L is in the kernel of μ ; we denote the image of $\mathcal{A}(G)_L$ under μ by A_L. Our problem is to find for what $\mathfrak{p}$ the group $A_{k_\mathfrak{p}}$ acts transitively on the special vertices of $\mathcal{D}(\mathcal{E}_{k_\mathfrak{p}})$. At times, we may speak, simply, of the action of $\mathcal{A}(G)_{k_\mathfrak{p}}$ on $\mathcal{D}(\mathcal{E}_{k_\mathfrak{p}})$, understanding thereby its action via μ.

We say that G has the property $P_\mathfrak{p}$ *if* $\mathcal{A}(G)_{k_\mathfrak{p}}$ *operates transitively on the special vertices of* $\mathcal{D}(\mathcal{E}_{k_\mathfrak{p}})$.

Henceforth, assume G to be almost k-simple (i. e., to have no proper, connected, normal subgroups defined over k). We know then that there exists a finite algebraic extension k' of k and an absolutely almost simple, semisimple, connected, linear algebraic group G' defined over k' such that $G=R_{k'/k}G'$. For the definition and categorical properties of the "groundfield reduction functor" $R_{k'/k}$, the reader may consult Chapter 1 of [14]. Among such properties are :

1. There is a natural isomorphism of $G'_{k'}$ with G_k, which is bi-continuous if k' and k are topological fields.

2. If $\mathfrak{p}$ is a prime of k, we have

$$G_{k_{\mathfrak{p}}} \simeq \prod_{\mathfrak{p}'|\mathfrak{p}} G'_{k'_{\mathfrak{p}'}}, \tag{1}$$

where $\mathfrak{p}'|\mathfrak{p}$ means $\mathfrak{p}'$ runs over the primes of k' dividing $\mathfrak{p}$.

3. It follows from 1., in the notation of 2., that

$$G'_{k'_{\mathfrak{p}'}} \simeq (R_{k'_{\mathfrak{p}'}/k_{\mathfrak{p}}}G')_{k_{\mathfrak{p}}}.$$

Moreover, $\mathscr{A}(G)$, as an algebraic group defined over k, is an extension, by a finite permutation group, of $R_{k'/k}\mathscr{A}(G')$. If ${}_{k'_{\mathfrak{p}'}}T'$ is a maximal $k'_{\mathfrak{p}'}$-split torus in G', then its group $D_{\mathfrak{p}'}$ of $k'_{\mathfrak{p}'}$-rational points is "diagonally" imbedded in $R_{k'_{\mathfrak{p}'}/k_{\mathfrak{p}}}(G')$ and the Zariski closure of $D_{\mathfrak{p}'}$ is a maximal $k_{\mathfrak{p}}$-split torus ${}_{k_{\mathfrak{p}}}T'$ of $R_{k'_{\mathfrak{p}'}/k_{\mathfrak{p}}}(G')$; therefore, $\mathrm{rank}_{k_{\mathfrak{p}}}(G)=\sum_{\mathfrak{p}'|\mathfrak{p}} \mathrm{rank}_{k'_{\mathfrak{p}'}}(G')$, $\Phi({}_{k_{\mathfrak{p}}}T, G)=\sum_{\mathfrak{p}'|\mathfrak{p}} \Phi({}_{k'_{\mathfrak{p}'}}T, G')$, and if $\mathscr{E}_{k'_{\mathfrak{p}'}}$ is the laddering associated in like manner to $G'_{k'_{\mathfrak{p}'}}$ and $\mathscr{D}(\mathscr{E}_{k'_{\mathfrak{p}'}})$ is its Dynkin diagram, then

$$\mathscr{D}(\mathscr{E}_{k_{\mathfrak{p}}}) = \bigcup_{\mathfrak{p}'|\mathfrak{p}} \mathscr{D}(\mathscr{E}_{k'_{\mathfrak{p}'}}), \tag{2}$$

and the set of special points on the left is also the set of special points on the right, so that a good, maximal compact subgroup of $G_{k_{\mathfrak{p}}}$ is the product of SMC subgroups of $G_{k'_{\mathfrak{p}'}}$ over all $\mathfrak{p}'|\mathfrak{p}$. Hence, if G' has property $P_{\mathfrak{p}'}$ for all $\mathfrak{p}'|\mathfrak{p}$, then G has property $P_{\mathfrak{p}}$.

We want to show that when G is simply-connected, there exists a finite set $S=S(G, \rho, V)$ of primes of k with the property that if $\mathfrak{p}$ is a prime of k not in S, and if no simple component ${}_{k_{\mathfrak{p}}}\Phi$ is of type BC (in symbols, ${}_{k_{\mathfrak{p}}}\Phi \cap BC=\phi$), then G has property $P_{\mathfrak{p}}$. From the foregoing, it is evidently enough to prove this when G is absolutely almost simple.

If G is absolutely almost simple, then Φ is one of the Cartan-Killing types A_n, B_n, C_n, D_n, $E_{6,7,8}$, F_4, or G_2. If G is of Chevalley type over $k_{\mathfrak{p}}$, it is known that G has property $P_{\mathfrak{p}}$. More precisely, recalling the identification of $\mathscr{A}(G)^0$ with Ad G, which is defined over k, it may be shown that any two special maximal compact subgroups of $G_{k_{\mathfrak{p}}}$ are conjugate under $(\mathrm{Ad}\, G)_{k_{\mathfrak{p}}}$ [11]. In fact, the group Ω defined there [11, p. 15] is transitive on the set of special vertices of the extended Dynkin diagram of $\Phi={}_L\Phi$ [11, pp. 18–20], and Ω is contained in the homomorphic image of the subgroup $\mathfrak{W}$ of $(\mathrm{Ad}\, G)_{k_{\mathfrak{p}}}$ in the group of affine transformations normalizing the affine Weyl group $\tilde{W}=DW$ [11, pp. 6, 28]. If $\tau(\Phi) \in B_n$, C_n, $E_{7,8}$, F_4, or G_2, then G is of Chevalley type over $k_{\mathfrak{p}}$, hence has property $P_{\mathfrak{p}}$, for almost all primes $\mathfrak{p}$ of k. This is a consequence of the fact that, in any event, G is quasi-split (i. e., has a Borel subgroup defined) over $k_{\mathfrak{p}}$ for almost all $\mathfrak{p}$, as is known. Moreover, it is well-known [5] that there exists a finite normal extension k' of k such that G is split over k' ; of course, the number of primes of k which ramify in k' is finite. Hence, it suffices to prove :

Theorem 1. *Let G be a connected, semi-simple, simply-connected, almost absolutely simple linear algebraic group defined over a complete* $\mathfrak{p}$*-adic number field* $k_{\mathfrak{p}}$*. Suppose that G is quasi-split over* $k_{\mathfrak{p}}$*, that G splits over an unramified extension of* $k_{\mathfrak{p}}$*, that* $\tau({}_{k_{\mathfrak{p}}}\Phi) \notin BC$*, and that* $\mathfrak{p}$ *does not divide* 2. *Then G has property* $P_{\mathfrak{p}}$.

Proof. We have seen that this is true when G actually splits over $k_{\mathfrak{p}}$. We shall use this fact to complete the proof of the theorem when G is only quasi-split over $k_{\mathfrak{p}}$. Let $k_{\mathfrak{p}}=L$.

Let ${}^v\Sigma$ be the reduced root system associated as described in [8, p. 22, § 1. 3. 8], to the affine root system Σ of G_L, and let $\mathcal{E}_L$ be the laddering of Σ by ${}_L\Phi$. The type $\tau(\mathcal{E}_L)$ of $\mathcal{E}_L$ is, by definition, the hyphenated pair $\tau({}^v\Sigma)-\tau({}_L\Phi)$; if $\tau({}^v\Sigma)=\tau({}_L\Phi)=X$ *and if* ${}^v\Sigma$ is proportional to ${}_L\Phi$, then we write $\tau(\mathcal{E}_L)=X$, as in [8, § 1. 4].

We first show, under the hypotheses of the theorem, that if $\tau(\mathcal{E}_L)=X-Y$, then $X=Y$. To begin with, if $X\neq Y$, then $\tau(\mathcal{E}_L)$ must be one of the types $B-C$, $C-B$, $B-BC$, or $C-BC$, according to the table on pp. 29–30 of [8]. Since $\tau({}_L\Phi)\notin BC$, only types $B-C$ or $C-B$ might occur, thus, G must be one of the classical types discussed in [8, § 10. 1]. We now refer to the notation and enumeration of cases of §§ 10. 1. 25–26 of [8]. Since G is quasi-split over L, the division algebra K there must be a commutative extension of degree one or two of L. Then either $K=L$, $\sigma=$id., and the anisotropic kernel X_0 must have dimension one or two (since we are supposing $\tau({}^v\Sigma)\neq\tau({}_L\Phi)$, which excludes the case $\tau({}^v\Sigma)=\tau({}_L\Phi)=C$ occurring when $X_0=0$) ; or else $[K:L]=2$, $\sigma\neq$id., and $\dim_K X_0$ equals one or zero. Suppose the latter occurs. Then K, which must be contained in any splitting field for G containing L, must be unramified over L, and this fact eliminates case $B-C$. Since, by hypothesis, $\tau({}_L\Phi)\notin BC$, only the case $\tau(\mathcal{E}_L)=C_r$ remains when $K\neq L$. If $K=L$, $\sigma=$id., then $\tau(\mathcal{E}_L)\in B=B-B$ if $\dim X_0=1$; while if $\dim X_0=2$, then $\tau({}^v\Sigma)\neq\tau({}_L\Phi)$ or ${}^v\Sigma$ is not proportional to ${}_L\Phi$ (type C_2-B_2) only if $\omega(q(X_0-\{0\}))=\mathbf{Z}$; if the latter holds and if $\mathfrak{p}\nmid 2$, we may assume the quadratic form $q|_{X_0}$ is expressible as x^2+dy^2 in a suitable coordinate system, where d is ($\mathfrak{p}$-adic) integral. If x and y are such that $\omega(x^2+dy^2)=1$, and if d were a $\mathfrak{p}$-adic unit, then Hensel's lemma would imply $q|_{X_0}$ had a non-trivial isotropic vector. Thus we may assume $d=\pi$, a prime, but then x^2+dy^2 splits only in a ramified extension of L containing $\sqrt{-\pi}$. In conclusion, we see that under the hypotheses of the theorem, $\tau({}^v\Sigma)=\tau({}_L\Phi)$, and if $a\in{}_L\Phi$, then $2a\notin{}_L\Phi$. There are still two cases where $\tau({}^v\Sigma)=\tau({}_L\Phi)$ but yet ${}^v\Sigma$ and ${}_L\Phi$ are not proportional ; these are the cases (*loc. cit.*) F_4^I and G_2^I, but as the Dynkin diagram of the laddering of each of them has but a single special vertex, the theorem is obviously true when these arise. Thus, without any loss, we may assume that ${}^v\Sigma={}_L\Phi$.

For each $a\in{}_L\Phi$, then, U_a is Abelian and if $\mathfrak{U}_a$ is its Lie algebra, the mapping *exp* is a continuous isomorphism of $\mathfrak{U}_a$ onto U_a. Moreover, under the adjoint action of the L-split torus ${}_LT$, each $t\in{}_LT$ acts on $\mathfrak{U}_a$ through scalar multiplication by the character value $a(t)$. Let the system of root data $(Z_L, (U_{a,L}, M_a)_{a\in_L\Phi})$ be supplied with a valuation $\varphi=\{\varphi_a\}_{a\in_L\Phi}$ as before, and for each a in the simple root system Δ let $\mathfrak{U}_a'$ be a one-dimensional subgroup of U_a containing an element x of $U_{a,L}$ of minimum positive value $\varphi_a(x)$, and we may assume φ so normalized that $0 \in \Gamma_a = \Gamma'_a$ for all $a \in \Delta$. Then the groups U'_a, $a \in \Delta$, and ${}_LT$ are contained in a uniquely determined L-split reductive subgroup G' of G with root system $\Phi' = {}_L\Phi$ [5, § 7], and, by restriction, φ determines a valuation $\varphi' = \{\varphi'_a\}_{a \in {}_L\Phi}$ of the root data $({}_LT_L, (U'_{a,L}, M'_a)_{a \in {}_L\Phi})$, where M'_a is a subset of M_a and is determined from the groups ${}_LT_L$, $U'_{\pm a,L}$ in the same way as M_a is from Z_L and $U_{\pm a,L}$. One may easily see that the value groups of φ and φ' are the same for each $a \in {}_L\Phi$, and that $M'_{a,k} \subset M_{a,k}$ for each $a \in {}_L\Phi$, $k \in \Gamma_a$. It is known [6, § 4. 6] that G' is simply-connected, and it is easy to see that representatives of the set S of generators of the affine Weyl group $\tilde{W}$ of G_L may be chosen to coincide with representatives of the set S' of generators of the affine Weyl group $\tilde{W}'$ of G'_L. Of course, $(\mathrm{Ad}\ G')_L \subset (\mathrm{Ad}\ G)_L$, because, as may be readily verified, the center of G' is contained in the center of G. Applying what we know from the split case, we obtain the full assertion of the theorem.

Now let G be a connected, simply-connected, semisimple, linear algebraic group defined over $\boldsymbol{Q}$ and almost $\boldsymbol{Q}$-simple. Let Γ be an arithmetic subgroup of $G_{\boldsymbol{Q}}$. As G is simply-connected, the $\mathfrak{p}$-adic closure $\Gamma_{\mathfrak{p}}$ of Γ in $G_{\boldsymbol{Q}_{\mathfrak{p}}}$ is open and compact for all finite primes $\mathfrak{p}$ and the product

$$\prod_{\mathfrak{p} | \infty} G_{\boldsymbol{Q}_{\mathfrak{p}}} \times \prod_{\mathfrak{p} < \infty} \Gamma_{\mathfrak{p}}$$

is open in the adele group G_A of G. Moreover, from what we have said earlier, $\Gamma_{\mathfrak{p}}$ is a *special* maximal compact subgroup of $G_{\boldsymbol{Q}_{\mathfrak{p}}}$ for all but, at most, a finite number of finite primes $\mathfrak{p}$. In accordance with what we have said earlier, if $\Gamma_{\mathfrak{p}}$ is a special maximal compact subgroup of $G_{\boldsymbol{Q}_{\mathfrak{p}}}$ for *all* finite primes $\mathfrak{p}$, then Γ is called a special arithmetic group (SAG). By changing a lattice Λ whose stabilizer is Γ in the representation space of G for a finite number of primes, one obtains a lattice whose stabilizer is an SAG. Our problem is, broadly speaking, to determine the isomorphism classes of SAG's in $G_{\boldsymbol{Q}}$ with respect to the operation of $\mathrm{Aut}(G)_{\boldsymbol{Q}}$. Thus, we propose to investigate the orbits of $\mathrm{Aut}\ (G)_{\boldsymbol{Q}}$ in the set of SAG's. As a first step, we shall exhibit necessary and sufficient conditions for the number of orbits to be finite.

There exists a number field k of finite degree and a connected, simply-connected, semisimple, and absolutely almost simple linear algebraic group G' defined over k such that $G = R_{k/\boldsymbol{Q}} G'$. There is a natural isomorphism of $G_{\boldsymbol{Q}}$ with G'_k which establishes a one-to-one correspondence between arithmetic subgroups of $G_{\boldsymbol{Q}}$ and those of G'_k, as we have defined them, and in particular, in virtue of (2), between special arithmetic subgroups of $G_{\boldsymbol{Q}}$ and those of G'_k. Now, $\mathcal{A}(G)$ is not, in general, equal to $R_{k/\boldsymbol{Q}}\mathcal{A}(G')$, but contains the latter as a subgroup of finite index. However, for certain purposes, including determination of when the number of isomorphism classes of SAG's is finite, it will suffice to consider arithmetic subgroups Γ' of G'_k.

So now assume G' is a subgroup of $GL(V')$, where V' is a finite-dimensional vector space defined over k, and for each prime $\mathfrak{p}$ of k, let $V'_{\mathfrak{p}} = V'_k \otimes_k k_{\mathfrak{p}}$, and if Λ' is a k-lattice in V', let $\Lambda'_{\mathfrak{p}}$ be, as before, its closure in $V'_{k_{\mathfrak{p}}}$.

Let $H'=\mathcal{A}(G')$, $\mathcal{G}'=\mathcal{G}(G')\simeq\mathrm{Ad}(G')$. For any linear algebraic group defined over k, denote by affixation of the subscript A (or A_k) formation of its group of adeles. Thus, for example,

$$G'_A = \prod_{\mathfrak{p}}{}' G'_{k_\mathfrak{p}},$$

where $\prod_{\mathfrak{p}}'$ means the direct product over all primes $\mathfrak{p}$ of k, restricted by the condition that for any given $g=(g_\mathfrak{p}) \in G'_A$, we have $g_\mathfrak{p} \in \Gamma'_{k_\mathfrak{p}}$, the stabilizer of the $\mathfrak{p}$-adic component of a fixed k-lattice in V', for all but a finite number of $\mathfrak{p}$.

We now fix G' for the following discussion, let V'' be the k-vector space underlying the Lie algebra of G', let H' operate on V'' in the usual way, and let G' operate on V'' via the adjoint operation.

Let Λ', Λ'' be k-lattices in V''_k. We say that Λ' and Λ'' are in the same genus if for each finite prime $\mathfrak{p}$ there exists $h_\mathfrak{p} \in H'_{k_\mathfrak{p}}$ such that $h_\mathfrak{p}.\Lambda'_\mathfrak{p}=\Lambda''_\mathfrak{p}$. Since $\Lambda'_\mathfrak{p}=\Lambda''_\mathfrak{p}$ for almost all $\mathfrak{p}$, we may take $h_\mathfrak{p}=1$ for almost all $\mathfrak{p}$, hence $h=(h_\mathfrak{p}) \in H'_A$. Thus, H'_A is transitive on the set of lattices in a given genus, and the genera are just the orbits of H'_A in this sense. Let Γ' and Γ'' be arithmetic subgroups of G'_k (i. e., subgroups commensurable with G'_0). If there exist k-lattices Λ', Λ'' in V''_k such that Γ' (resp. Γ'') is the stabilizer of Λ' (resp. of Λ'') in G'_k and such that Λ' and Λ'' are in the same genus, then we say that Γ' and Γ'' are in the same genus. In other words, Γ' and Γ'' are in the same genus if and only if for each finite prime $\mathfrak{p}$ there exists $h_\mathfrak{p} \in H'_{k_\mathfrak{p}}$ such that $h_\mathfrak{p}(\Gamma'_\mathfrak{p})=\Gamma''_\mathfrak{p}$. We say that Γ' and Γ'' are in the same class if there exists $h \in H'_k$ such that $h(\Gamma')=\Gamma''$. Clearly, all elements of the same class are in the same genus.

We prove [4] that the number of classes in a genus is finite. Let $H'_\infty=\prod_{\mathfrak{p}|\infty} H'_{k_\mathfrak{p}}$. Let Λ' be a k-lattice in V''_k such that Γ' is the stabilizer of Λ' in G'_k, and let, for each finite prime $\mathfrak{p}$,

$$\Gamma'_{1_\mathfrak{p}} = \{h \in H'_{k_\mathfrak{p}} | h.\Lambda'_\mathfrak{p} = \Lambda'_\mathfrak{p}\}.$$

Let U'' be the open subgroup $H'_\infty \times \prod_{\mathfrak{p}<\infty} \Gamma'_{1_\mathfrak{p}}$ of H'_A. It follows from [4] that

$$H'_A = \bigcup_{\alpha \in E} H'_k \alpha U'' \tag{3}$$

where E is some finite subset of H'_A. We know that H'_A operates on the set of k-lattices in V''_k (if $h=(h_\mathfrak{p}) \in H'_A$, let $h.\Lambda'$ be the lattice such that $(h.\Lambda')_\mathfrak{p}=h_\mathfrak{p}.\Lambda'_\mathfrak{p}$, $\mathfrak{p}<\infty$), so that U'' is exactly the stabilizer in H'_A of the adelic closure of Λ', and

$$L = \{h.\Lambda' | h \in H'_A\}$$

is the set of lattices in the genus of Λ'. Now (3) implies that the number of orbits of H'_k in L is finite, hence the number of classes in the genus of Λ' or in the genus of Γ' is finite.

Now let $\gamma(G')$ (resp. $\gamma(G)$) be the number of genera of special arithmetic subgroups of G'_k (resp. of G_Q). As for the finiteness of these, we have

Proposition 2. *Let $G=R_{k/Q}G'$ as above. Let $S(G', k, BC)$ be the set of finite primes $\mathfrak{p}$ of k such that ${}_{k_\mathfrak{p}}\Phi$ is of type BC. Then the following statements are equivalent:*

i) $\gamma(G)$ *is finite* ;
ii) $\gamma(G')$ *is finite* ; *and*
iii) $S(G', k, BC)$ *is a finite set.*

Proof. From earlier observations, it is clear that $\mathscr{A}(G')_k$ may be viewed as a subgroup of finite index in $\mathscr{A}(G)_{\boldsymbol{Q}}$. This shows that i) and ii) are equivalent.

Now we prove the equivalence of ii) and iii). If $\mathfrak{p}$ is a finite prime of k, let $\sigma_{\mathfrak{p}}$ be the number of special points in $\mathscr{D}(\mathscr{E}_{k\mathfrak{p}})$. Then

$$\gamma(G') \leq \prod_{\mathfrak{p} \in S} {}_{(G', k, BC) \cup S} \sigma_{\mathfrak{p}}$$

where $S-S(G', \rho', V')$ (here $\rho' : G' \to GL(V')$ is the inclusion homomorphism). This shows that iii) implies ii).

Suppose $S(G', k, BC)$ is infinite. Since G' is quasi-split over $k_{\mathfrak{p}}$ for almost all $\mathfrak{p}$, it follows from an examination of the classification of indices that this can happen only if Φ is of type A_{2n} and G'_k is the special unitary group of a Hermitian sesquilinear form q on a finite-dimensional vector space X over a central division algebra D over a quadratic extension l of k, having an involution ι of the second kind of which k is the fixed point set in l. In this case, for infinitely many $\mathfrak{p}$, $G'_{k_{\mathfrak{p}}}$ is the special unitary group of a Hermitian form on an odd-dimensional vector space over a quadratic unramified extension K of $k_{\mathfrak{p}}$, and then $\tau(\mathscr{E}_{k_{\mathfrak{p}}}) \in C-BC^{IV}$, thus $\mathscr{D}(\mathscr{E}_{k_{\mathfrak{p}}})$ has two special points which are not symmetrically situated, hence cannot be interchanged by an element of $\mathscr{A}(G')_{k_{\mathfrak{p}}}$. For each $\mathfrak{p}$, let $N_{\mathfrak{p}}$ be the number of orbits of $\mathscr{A}(G')_{k_{\mathfrak{p}}}$ in the set of special points of $\mathscr{D}(\mathscr{E}_{k_{\mathfrak{p}}})$. It is clear from the definition of "genus" that $\gamma(G') = \prod_{\mathfrak{p} < \infty} N_{\mathfrak{p}}$. In the present case, this implies $\gamma(G')$ is infinite. q. e. d. (Consideration of the asymmetry properties of $\mathscr{D}(\mathscr{E}_{k_{\mathfrak{p}}})$ was suggested by J. Tits.)

In many cases, at least, the determination of $N_{\mathfrak{p}}$ for each finite prime $\mathfrak{p}$ appears to be a straightforward computation. Moreover, except in the case of certain groups G' of (absolute) type A_{2n}, we have $N_{\mathfrak{p}}=1$ for almost all $\mathfrak{p}$. However, calculation of the class number h of a given genus appears more difficult (and interesting). Simple examples will be given in the following.

Now we make further assumptions about G. Let K be a maximal compact subgroup of $G_{\boldsymbol{R}}$ (which is connected because G is supposed simply-connected). Assume that $K \backslash G_{\boldsymbol{R}} = X$ is a Hermitian symmetric space, that $\operatorname{rank}_{\boldsymbol{Q}}(G) > 0$, and that ${}_{\boldsymbol{Q}}\Phi$ is of type C. This implies [3, §§ 2. 8–10] that ${}_{\boldsymbol{R}}\Phi$ is of type C, hence that X is isomorphic as a complex manifold to a tube domain

$$\mathscr{T} = \{x+iy = z \in \boldsymbol{C}^n \mid y \in \mathfrak{R}\} \subset \boldsymbol{C}^n,$$

where $\mathfrak{R}$ is a certain type of cone in $\boldsymbol{R}^n$. Moreover, it follows as a corollary to Proposition 1. that the set of circled vertices (i. e., roots with non-trivial restriction to a maximal split torus of G' over the given field) of (the Dynkin diagram of) the k-index of G' is a subset of the set of circled vertices of the $k_{\mathfrak{p}}$-index of G'. This sometimes affords useful information about the behaviour of G' over the various $k_{\mathfrak{p}}$; for example, it implies that if, in addition to the above assumptions, $\Phi \in E_7$, then G'

splits over $k_{\mathfrak{p}}$ for all finite $\mathfrak{p}$.

With the tube domain $\mathcal{T}$ as above, let $\mathrm{Aff}(\mathcal{T})$ be the Lie group of linear affine transformations of $\mathcal{T}$ onto itself and let P be the Zariski closure in G of the inverse image of $\mathrm{Aff}(\mathcal{T})^0$ in $G_{\boldsymbol{R}}$. By appropriate choice of coordinates, we may assume P is defined over $\boldsymbol{Q}$ (since $\mathrm{rank}_{\boldsymbol{Q}}(G)=\mathrm{rank}_k(G')>0$, and a longest $\boldsymbol{R}$-root restricts to a longest k-root [3, § 2]) ; then P is a maximal $\boldsymbol{Q}$-parabolic subgroup of G. With these relationships being as stated, the pair (G, T) is called a rational tube domain. If Γ is an arithmetic subgroup of $G_{\boldsymbol{Q}}$, then Γ operates on $\mathcal{T}$. Let Γ be a special arithmetic subgroup of $G_{\boldsymbol{Q}}$ and let K^{Γ} be the subgroup $K.\prod_{\mathfrak{p}<\infty}\Gamma_{\mathfrak{p}}$ of G_A. We have $G_A=K^{\Gamma}.P_A$ (since $G_{\boldsymbol{Q}_{\mathfrak{p}}}=\Gamma_{\mathfrak{p}}.P_{\boldsymbol{Q}_{\mathfrak{p}}}$ for all finite $\mathfrak{p}$, because Γ is special, and $G_{\boldsymbol{R}}=K.P_{\boldsymbol{R}}$ (Iwasawa decomposition)). Let $P_{\boldsymbol{R}}=P_\infty$ and let P_A^0 be the set of $g=(g_{\mathfrak{p}})\in P_A$ such that $g_\infty=1$. Then $P_A=P_A^0.P_\infty$ and $G_A=K^{\Gamma}.P_A^0.P_\infty$. Now denote by ν a multiplicative holomorphic one-cocycle on $G_{\boldsymbol{R}}$ with values in the group of nowhere zero holomorphic functions on $\mathcal{T}:\nu(z, g_1, g_2)=\nu(z, g_1)\nu(z, g_1, g_2)$, $z\in\mathcal{T}$, $g_1, g_2\in G_{\boldsymbol{R}}$. We assume for every $g\in P_\infty$ that $\nu(z, g)$ is independent of $z\in\mathcal{T}$, $\nu(z, g)=\chi(g)$, where $\chi: P\to\boldsymbol{C}^\times$ is a $\boldsymbol{Q}$-rational character. The product formula then implies that $|\chi(g)|_A=1$ for $g\in P_{\boldsymbol{Q}}$, where $|\ |_A$ is the adelic norm. Moreover, we may identify the complexification K_C of K with a subgroup of P_C, so that $\chi(k)$ is defined for $k\in K$. If $g=(g_{\mathfrak{p}})\in G_A$, $g_{\mathfrak{p}}=k_{\mathfrak{p}}.\pi_{\mathfrak{p}}$, $\pi_{\mathfrak{p}}\in P_{\boldsymbol{Q}_{\mathfrak{p}}}$, $k_{\mathfrak{p}}\in\Gamma_{\mathfrak{p}}$, with $\Gamma_\infty=K$, put $\pi(g)=(\pi_{\mathfrak{p}})\in P_A$ and define $\varphi(g)=\chi(k_\infty).|\chi(\pi(g))|_A$ (this is well defined since $|\chi(\pi_{\mathfrak{p}})|_{\mathfrak{p}}=1$ if $\pi_{\mathfrak{p}}\in\Gamma_{\mathfrak{p}}\cap P_{\boldsymbol{Q}_{\mathfrak{p}}}$). Then φ is a complex-valued function on G_A and we define an Eisenstein series (briefly, E-series) on G_A by ([1])

$$\tilde{E}_\nu(g)=\sum_{\gamma\in G_{\boldsymbol{Q}}/P_{\boldsymbol{Q}}}\varphi(g\gamma). \tag{4}$$

Let z_0 be the unique fixed point of K and if $z\in\mathcal{T}$, let $g\in G_{\boldsymbol{R}}$ be such that $z=z_0.g$. Define $E_\nu(z)=\nu(z_0, g)^{-1}.\tilde{E}_\nu(g)$; this is a well defined complex-valued function on $\mathcal{T}$ and we have

$$E_\nu(z)=\sum_{\gamma\in G_{\boldsymbol{Q}}/P_{\boldsymbol{Q}}}\nu(z, \gamma)|\chi(p_\gamma)|, \tag{5}$$

where $p_\gamma\in P_A^0$ is such that $\gamma\in K^{\Gamma}.p_\gamma.P_\infty$. When ν has been chosen, holomorphic in $z\in\mathcal{T}$, so as to ensure uniform convergence of the series (5) on compact sets, then E_ν is a holomorphic function on $\mathcal{T}$, and is an automorphic form on $\mathcal{T}$ with respect to Γ. In particular, we may let $\nu(z, g)=j(z, g)^n$, where j is the functional (or Jacobian) determinant of g at z and n is a sufficiently large and divisible positive integer. Tsao and I [13, 1] have shown that (for such ν) the Fourier coefficients of E_ν are rational numbers. Moreover, these E_ν generate the field of automorphic functions with respect to a certain discontinuous group $\Gamma^\natural$ acting on $\mathcal{T}$ such that $\Gamma\subset\Gamma^\natural$ and $[\Gamma^\natural:\Gamma]<\infty$. Thus, the Satake compactification of $X_\natural=\mathcal{T}/\Gamma^\natural$ is defined over $\boldsymbol{Q}$. (The existence of $\Gamma^\natural$ may be demonstrated using ideas of [2].) In many cases, $X_\natural$ or $X=\mathcal{T}/\Gamma$ or some other related quotient space is associated to some moduli problem. It would be interesting to see how general this phenomenon might be.

Examples

1. Let k be a totally real algebraic number field (of finite degree), $G'_k = SL(2, k)$, $G = R_{k/Q}G'$, $H'_{1k} = PGL(2, k)$, $H_1 = R_{k/Q}H'_1$, and Γ be the subgroup $SL(2, \mathfrak{o})$ ($\mathfrak{o}=$ integers of k) of $G'_k \simeq G_Q$. Let $\mathfrak{p}$ be a prime of k. Every maximal compact subgroup of $G'_{k_\mathfrak{p}}$ is conjugate *in* $G'_{k_\mathfrak{p}}$ either to $SL(2, \mathfrak{o}_\mathfrak{p})$ or to $\mathrm{Ad}\, \alpha_\pi(SL(2, \mathfrak{o}_\mathfrak{p}))$, where $\alpha_\pi = \begin{bmatrix} \pi & 0 \\ 0 & 1 \end{bmatrix}$, π being a prime of $\mathfrak{o}_\mathfrak{p}$, and these are conjugate to each other in $H'_{1k_\mathfrak{p}}$ to which α_π belongs. It follows, by strong approximation that every special arithmetic subgroup of G_Q is, when viewed as a subgroup of G'_k, conjugate to one of the form

$$\Gamma_\mathfrak{a} = SL(2, k) \cap \begin{bmatrix} \mathfrak{o} & \mathfrak{a} \\ \mathfrak{a}^{-1} & \mathfrak{o} \end{bmatrix},$$

for some ideal $\mathfrak{a}$ in $\mathfrak{o}$. Two such groups $\Gamma_\mathfrak{a}$ and $\Gamma_\mathfrak{b}$ are certainly conjugate if $\mathfrak{a}\mathfrak{b}^{-1}$ is a principal ideal ; hence, the number of outer isomorphism classes of SAG's is $\leq h = h(k)$ (the class number of k). As G. Shimura has observed, it follows from [10] that $\Gamma_\mathfrak{a}$ is conjugate to $\Gamma_\mathfrak{b}$ by an element of $GL(2, k)$ if and only if $\mathfrak{a}\mathfrak{b}^{-1}$ is equivalent to the square of an ideal (a fact also obtained in [9]). As the (usual) Dynkin diagram of $SL_2 \in A_1$ has a trivial group of symmetries, the algebraic group $\mathcal{A}(G')$ of *rational* automorphisms of G' is connected and may be identified with Ad G'. It is easy to see that for any field K containing k we have $(\mathrm{Ad}\, G')_K = \mathrm{Ad}(GL(2, K))$. Hence, the number of outer isomorphism classes of SAG's in G'_k is equal to 2^{c_2}, where c_2 is the number of 2-primary cyclic summands of the ideal class group of k.

2. Let $G'_k = \mathrm{Sp}(n, k)$, the symplectic group over k contained in $GL(2n, k)$, where k is assume to be totally real. Defining c_2 as before, one may establish in essentially the same way that the number of outer isomorphism classes of SAG's in $\mathrm{Sp}(n, k) \simeq G_Q$ is equal to 2^{c_2}. In particular, every SAG in $\mathrm{Sp}(n, \boldsymbol{Q})$ is *outer* isomorphic to the Siegel modular group, $\mathrm{Sp}(n, \boldsymbol{Z})$.

3. More generally, let $(G, \mathcal{T})$ be a rational tube domain with G simply-connected. In this case it can be shown that the number of primes $\mathfrak{p}$ for which an irreducible component of ${}_{Q_\mathfrak{p}}\Phi$ is of type BC is finite, for the full root system Φ of G cannot in this case be of type A_{2n}. Hence, the number of outer isomorphism classes of SAG's in G_Q is always finite in this case.

References

[1] Baily, W. L., Jr. : On the Fourier coefficients of certain Eisenstein series on the adele group, in Number Theory, Algebraic Geometry and Commutative Algebra in honor of Y. Akizuki, Kinokuniya, Tokyo, (1973), 23–43.

[2] ——: Eisenstein Series on Tube Domains, in Problems in Analysis, A Symposium in Honor of Salomon Bochner, Ed. R. C. Gunning, Princeton University Press, Princeton (1970), 139–156.

[3] ——, and Borel, A. : Compactification of arithmetic quotients of bounded symmetric domains, Ann. of Math., **84** (1966), 442–528.

[4] Borel, A. : Some finiteness properties of adele groups over number fields, Publ. Math. IHES, **16** (1963), 5–30.

[5] ——, and Tits, J. : Groupes réductifs, Publ. Math. IHES, **27** (1965), 55–150.

[6] —— : Compléments a l'Article "Groupes réductifs", Publ. Math. IHES, **41** (1972), 253–276.

[7] Bourbaki, N. : Groupes et algèbres de Lie, Chapitres 4, 5 and 6, Hermann, Paris, 1968.

[8] Bruhat, F. and Tits, J. : Groupes réductifs sur un corps local, I., Données radicielles valuées, Publ. Math. IHES, **41** (1972), 5–252.

[9] Chevalley, C. : L'Arithmétique dans les Algèbres de Matrices, Act. Sci. et Ind. No. 323, Hermann, Paris, 1936.

[10] Eichler, M. : Über die Idealklassenzahl hyperkomplexer Systeme, Math. Z. **43** (1937/38) 481–494.

[11] Iwahori, N. and Matsumoto, H. : On some Bruhat decomposition and the structure of the Hecke rings of p-adic Chevalley groups, Publ. Math. IHES, **25** (1965), 5–48.

[12] Platonov, V. P. : The problem of strong approximation and the Kneser-Tits conjecture for algebraic groups, Math. USSR-Izvestia, **3** (1969), 1139–1147 ; Addendum to same, Math. USSR–Izvestia, **4** (1970), 784–786.

[13] Tsao, L.-C. : On Fourier coefficients of Eisenstein series, Thesis, University of Chicago, 1972 ; cf. also Bull. A. M. S., **79** (1973), 1064–1068.

[14] Weil, A. : Adèles and Algebraic Groups, Notes by M. Demazure and T. Ono, Institute for Advanced Study, Princeton, 1961.

Department of Mathematics
University of Chicago

(Received August 29, 1975)

Submanifolds and Over-determined Differential Operators

H. Goldschmidt and D. Spencer[1)]

Introduction

In this paper we study the relationship between an over-determined partial differential equation R_k of order k on a manifold Y and its restriction to a submanifold X of Y. In particular, we examine several special problems exhibiting this relationship by means of a linear differential operator D depending on X and Y and its non-linear analogue $\mathscr{D}$.

First, if the equation R_k is linear and formally integrable, we examine the Cauchy problem for R_k with initial values on X. In [7], Guillemin investigated the case of first-order equations when X is non-characteristic for R_k. We formulate the Cauchy problem for an equation R_k of arbitrary order k in terms of "naive" Spencer complexes for R_k which are supported on X with differential operator D. By means of these sequences we associate to R_k and X the cohomology groups $\{H^j_X(R_k)_m\}_{j\geq 0, m\geq k}$ which coincide with the Spencer cohomology groups $H^j(R_k)$ of R_k whenever $X=Y$ and m is sufficiently large. We define a map

$$\rho: H^j(R_k)_x \to H^j_X(R_k)_{m,x}$$

for $x\in X, j\geq 0$ and all sufficiently large m. The study of this map for $j=0$ is the Cauchy problem for R_k with initial data on X; surjectivity of the map corresponds to existence theorems and injectivity to uniqueness theorems.

If X is a non-characteristic submanifold for R_k and the equation is "involutive" then, using results of Guillemin [7], we show that the cohomology $H^j_X(R_k)_m$ is independent of m for $m\geq k+1$ and that it is isomorphic to the cohomology in degree j of a "sophisticated" version of our naive sequences on X. Guillemin's results for first-order analytic equations generalize to analytic equations of arbitrary order k; namely, existence and uniqueness hold for the Cauchy problem with non-characteristic analytic initial data (Theorem 2. 5) and $H^j_X(R_k)_m$ vanishes for all $j>0$ and $m\geq k+1$.

In § 3 a non-linear analogue of the linear complex of § 2 is defined. In particular, the non-linear operator $\mathscr{D}$ is defined in terms of the Cartan fundamental form

1) This work was supported in part by National Science Foundation Grants MPS 72–05055 A 02 and MCS 72–04357 A 04.

on Y which was constructed in § 5 of [6]. The Cartan structure equation then provides immediate justification for the introduction of the second non-linear operator $\mathscr{D}_1$ to create the desired complex. Next, it is shown that $\mathscr{D}$ is induced from the corresponding non-linear operator $\mathscr{D}_Y$ in the case $X=Y$, and various formulas are established for $\mathscr{D}$ which are analogues of formulas listed in § 2 of [6] for $\mathscr{D}_Y$. The non-linear complex is exact and to each formally integrable Lie equation R_k on Y we associate a sub-complex of the non-linear complex and the "exactness" of this sub-complex is shown to be stable if R_k is involutive and X is non-characteristic.

In § 4 formulas expressing the formal integrability of almost pseudo-complex structures of arbitrary codimension are derived. In the case of codimension 1 these formulas coincide with the integrability conditions given by Kuranishi [9] in terms of local coordinates. Our formulas are expressed in terms of brackets which are obtained from the well-known Nijenhuis bracket by means of a decomposition of the tangent space of the manifold. The decomposition seems to be unavoidable; in particular, Kuranishi's original formulas in codimension 1 depend on its choice. However, the introduction of the brackets not only rids the formulas of local coordinates but also provides a remarkably simple and natural method of computing integrability conditions. This section is self-contained and is therefore independent of the other parts of the paper.

In the final section of the paper we describe the relationship between pseudo-complex structures and the non-linear complex of § 3 associated to the Lie equation R_k for holomorphic vector fields on a complex manifold Y. From Theorem 5. 1 we conclude that sufficiently small 1-forms with values in the equation correspond to almost pseudo-complex structures, and that formally integrable almost pseudo-complex structures correspond to cocycles of the non-linear complex (i. e., to 1-forms with values in the equation which are annihilated by the operator $\mathscr{D}_1$). Moreover, a formally integrable structure is induced by local imbeddings into the ambient complex analytic manifold Y whenever the cocycle belongs (locally) to the image of the non-linear operator $\mathscr{D}$. The exactness of the non-linear complex is therefore related to the solvability of the integrability problem for almost pseudo-complex structures, but exactness cannot be expected to hold in general since additional assumptions are known to be necessary. However, in the analytic case this complex can be used to show that a formally integrable almost pseudo-complex structure is a pseudo-complex structure.

The notation and terminology employed here are the same as in the paper [6]. However, we remind the reader of one bit of notation (also used in [6]). Namely if E, F, G are finite-dimensional vector spaces, we always identify $E^*\otimes F$ with $\mathrm{Hom}(E, F)$ and, if $u \in E^*\otimes F$, $v \in F^*\otimes G$, we denote by $v \circ u$ the element of $E^*\otimes G$ defined by composition.

1. The Linear Differential Operator D

Let X be a differentiable manifold of dimension n and class C^∞ whose tangent bundle we denote by $T=T_X$. We write $\mathcal{O}_X$ for the sheaf of real-valued, differentiable functions on X. If E is a fibered manifold over X, we denote by $\mathcal{E}$ the sheaf of sections of E, and by E_x (resp. $\mathcal{E}_x$) the fiber of E (resp. the stalk of $\mathcal{E}$) at $x \in X$. We denote by $J_k(E)$ the fibered manifold of k-jets of sections of E, by $j_k : \mathcal{E} \to J_k(\mathcal{E})$ the differential operator of order k which sends a section s of E over a neighborhood of $x \in X$ into the k-jet $j_k(s)$ of this section, and by $\pi_k : J_{k+l}(E) \to J_k(E)$ and $\pi : J_k(E) \to X$ the natural projections sending $j_{k+l}(s)(x)$ into $j_k(s)(x)$ and $j_k(s)(x)$ into its source x respectively.

We shall always suppose that the fibers of a vector bundle are of the same dimension. If E is a vector bundle over X, we have an exact sequence of vector bundles

$$0 \to S^k T^* \otimes E \xrightarrow{\varepsilon} J_k(E) \xrightarrow{\pi_{k-1}} J_{k-1}(E) \to 0.$$

The Spencer operator

$$D_X : \wedge \mathcal{T}^* \otimes J_k(\mathcal{E}) \to \wedge \mathcal{T}^* \otimes J_{k-1}(\mathcal{E})$$

is the first-order differential operator characterized by the relations

$$D_X j_k(s) = 0, \qquad \text{for} \quad s \in \mathcal{E},$$

and

$$D_X(\omega \wedge u) = d\omega \wedge \pi_{k-1} u + (-1)^j \omega \wedge D_X u,$$

for $\omega \in \wedge^j \mathcal{T}^*$, $u \in \wedge \mathcal{T}^* \otimes J_k(\mathcal{E})$. The restriction of $-D_X$ to $\wedge \mathcal{T}^* \otimes \varepsilon(S^k \mathcal{T}^* \otimes \mathcal{E})$ is $\mathcal{O}_X$-linear and comes from a morphism of vector bundles

$$\delta_X : \wedge T^* \otimes S^k T^* \otimes E \to \wedge T^* \otimes S^{k-1} T^* \otimes E.$$

Let Y be a differentiable manifold of dimension m whose tangent bundle we denote by T_Y and let $\rho : X \to Y$ be a differentiable mapping. Let F be a fibered manifold over Y; we denote by $\mathcal{F}$, $\mathcal{F}_X$ the sheaves of sections of F over Y and of $\rho^{-1}F$ over X respectively, and by $J_k(F ; Y)$ the bundle of k-jets of sections of F over Y. If F is a vector bundle over Y and E is a vector bundle over X, then $E \otimes_X F$ denotes the vector bundle $E \otimes \rho^{-1} F$ over X, and $\mathcal{E} \otimes \mathcal{F}_X$ is the sheaf of sections of $E \otimes \rho^{-1} F$ over X.

The proof of the following proposition is similar to that of Proposition 3. 1 of [6] in which ρ is a submersion and will therefore be omitted. Let F be a vector bundle over Y.

Proposition 1. 1. *There exists a unique linear, first-order differential operator*

(1. 1) $$D : J_k(\mathcal{F} ; Y)_X \to \mathcal{T}^* \otimes J_{k-1}(\mathcal{F} ; Y)_X$$

satisfying one of the following equivalent conditions :

(i) *For all sections s of F over Y,*

(1. 2) $$D(j_k(s) \circ \rho) = 0$$

and

(1. 3) $$D(fu) = df \otimes \pi_{k-1} u + f D u,$$

for $f\in\mathcal{O}_X$, $u\in J_k(\mathcal{F}\,;\,Y)_X$.

(ii) If $u\in J_k(\mathcal{F}\,;\,Y)_{X,x}$ *and* $u=j_k(s)(\rho(x))$ *for* $s\in\mathcal{F}_{\rho(x)}$,

(1. 4) $$(\varepsilon Du)(x)=j_1(\pi_{k-1}u)(x)-j_1(j_{k-1}(s)\circ\rho)(x)$$

as elements of $J_1(\rho^{-1}J_{k-1}(F\,;\,Y))$.

We now define

$$D:\wedge^i\mathcal{T}^*\otimes J_k(\mathcal{F}\,;\,Y)_X\to\wedge^{i+1}\mathcal{T}^*\otimes J_{k-1}(\mathcal{F}\,;\,Y)_X$$

by setting

$$D(\alpha\otimes u)=d\alpha\otimes\pi_{k-1}u+(-1)^i\alpha\wedge Du$$

for $\alpha\in\wedge^i\mathcal{T}^*$, $u\in J_k(\mathcal{F}\,;\,Y)_X$; this is a well-defined operator because of (1. 3). The operator

(1. 5) $$D:\wedge\mathcal{T}^*\otimes J_k(\mathcal{F}\,;\,Y)_X\to\wedge\mathcal{T}^*\otimes J_{k-1}(\mathcal{F}\,;\,Y)_X$$

satisfies

(1. 6) $$D(\alpha\wedge u)=d\alpha\wedge\pi_{k-1}u+(-1)^i\alpha\wedge Du,$$

for $\alpha\in\wedge^i\mathcal{T}^*$, $u\in\wedge\mathcal{T}^*\otimes J_k(\mathcal{F}\,;\,Y)_X$, and

(1. 7) $$\langle\xi\wedge\eta,Du\rangle=\xi\,\overline{\wedge}\,D\langle\eta,u\rangle-\eta\,\overline{\wedge}\,D\langle\xi,u\rangle-\pi_{k-1}\langle[\xi,\eta],u\rangle,$$

for $\xi,\eta\in\mathcal{T}$, $u\in\mathcal{T}^*\otimes J_k(\mathcal{F}\,;\,Y)_X$. Since $D^2=0$, as is easily seen, we obtain a complex

(1. 8)

$$0\to\rho^{-1}\mathcal{F}\xrightarrow{j_k}J_k(\mathcal{F}\,;\,Y)_X\xrightarrow{D}\mathcal{T}^*\otimes J_{k-1}(\mathcal{F}\,;\,Y)_X\xrightarrow{D}\cdots\to\wedge^n\mathcal{T}^*\otimes J_{k-n}(\mathcal{F}\,;\,Y)_X\to0$$

where the map j_k is induced from $j_k:\mathcal{F}\to J_k(\mathcal{F}\,;\,Y)$ by ρ. This complex is not in general exact at $\wedge^i\mathcal{T}^*\otimes J_{k-i}(\mathcal{F}\,;\,Y)_X$ for $i\geq0$. If ρ is an immersion, it is exact at $\wedge^i\mathcal{T}^*\otimes J_{k-i}(\mathcal{F}\,;\,Y)_X$ for $i>0$.

If $X=Y$ and ρ is the identity mapping of X, this operator D is the Spencer operator D_X on X; when ρ is a submersion, this operator D was defined in § 3 of [6] and further properties of D are given there.

Let Z be a differentiable manifold whose tangent bundle we denote by T_Z. Let $\tau:X\to Z$, $\sigma:Z\to Y$ be mappings such that the diagram

(1. 9)

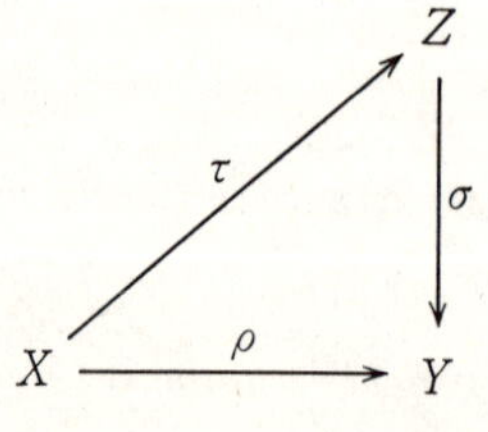

commutes. If u is a section of $\wedge^iT_Z^*\otimes_ZJ_k(F\,;\,Y)$ over Z, we define the section τ^*u of $\wedge^iT^*\otimes_XJ_k(F\,;\,Y)$ over X by

$$(\tau^*u)(x)=(\tau^*\otimes\mathrm{id})u(\tau(x)),\qquad\text{for}\quad x\in X,$$

where

$$\tau^* : T^*_{Z,\tau(x)} \to T^*_x.$$

Proposition 1. 2. *If u is a section of $\wedge^i T^*_Z \otimes_Z J_k(F\,;\,Y)$ over Z, then*

$$D(\tau^* u) = \tau^*(Du) \tag{1. 10}$$

as sections of $\wedge^{i+1} T^* \otimes_X J_{k-1}(F\,;\,Y)$ over X, where the operator D on the left-hand side is the operator (1. 5) *and the operator D on the right-hand side is the operator*

$$D : \wedge^i \mathcal{T}^*_Z \otimes J_k(\mathcal{F}\,;\,Y)_Z \to \wedge^{i+1} \mathcal{T}^*_Z \otimes J_{k-1}(\mathcal{F}\,;\,Y)_Z.$$

Proof. Because the operators D in (1. 10) satisfy relations of the form (1. 6), it suffices to verify (1. 10) for $i=0$. In fact, if $u=j_k(s)\circ\sigma$, where s is a section of F over Y, then both sides of (1. 10) vanish according to condition (i) of Proposition 1. 1. If f is a real-valued function on Z and u is a section of $\sigma^{-1}J_k(F;\,Y)$ over Z, then, for $\xi \in T_x$, $x \in X$, we have according to (1. 3)

$$\begin{aligned}\langle \xi, D(\tau^*(fu))\rangle - (f\circ\tau)\langle \xi, D(\tau^* u)\rangle &= \langle \xi, d(f\circ\tau)\otimes \pi_{k-1}u\rangle = \langle \tau_*\xi, df\otimes \pi_{k-1}u\rangle \\ &= \langle \tau_*\xi, D(fu)\rangle - (f\circ\tau)\langle \tau_*\xi, Du\rangle,\end{aligned}$$

which implies the identity (1. 10).

If we take $Z=Y$, $\tau=\rho$ and $\sigma=\mathrm{id}$ in Proposition 1.2, we obtain the formula

$$D(\rho^* u) = \rho^*(D_Y u) \tag{1. 11}$$

for all sections u of $\wedge^i T^*_Y \otimes J_k(F\,;\,Y)$ over Y, where the operator D_Y on the right-hand side is the Spencer operator on Y.

Let $E=\rho^{-1}F$; we have a natural mapping

$$\rho : \rho^{-1} J_k(F\,;\,Y) \to J_k(E)$$

sending $(x, j_k(s)(\rho(x)))$ into $j_k(s\circ\rho)(x)$, where s is a section of F over a neighborhood of $\rho(x)$ in Y. The diagram

$$\begin{array}{ccc} \rho^{-1}(S^k T^*_Y \otimes F) & \xrightarrow{\rho^*\otimes \mathrm{id}} & S^k T^* \otimes E \\ \Big\downarrow \varepsilon & & \Big\downarrow \varepsilon \\ \rho^{-1} J_k(F\,;\,Y) & \xrightarrow{\ \rho\ } & J_k(E) \end{array}$$

is easily seen to commute.

Proposition 1. 3. *Let $E=\rho^{-1}F$; the diagram*

$$\begin{array}{ccc} \wedge^j \mathcal{T}^* \otimes J_k(\mathcal{F}\,;\,Y)_X & \xrightarrow{\ D\ } & \wedge^{j+1}\mathcal{T}^* \otimes J_{k-1}(\mathcal{F}\,;\,Y)_X \\ \Big\downarrow \mathrm{id}\otimes\rho & & \Big\downarrow \mathrm{id}\otimes\rho \\ \wedge^j \mathcal{T}^* \otimes J_k(\mathcal{E}) & \xrightarrow{\ D_X\ } & \wedge^{j+1}\mathcal{T}^* \otimes J_{k-1}(\mathcal{E}) \end{array}$$

commutes.

Proof. Because the operators D and D_X satisfy relations of the form (1. 6), it is sufficient to verify that this diagram commutes when $j=0$. If $x \in X$ and s is a section of F over a neighborhood of $\rho(x)$, then

$$D(j_k(s)\circ\rho) = 0 \quad \text{and} \quad D_X(j_k(s\circ\rho)) = 0.$$

Furthermore if $u \in J_k(\mathcal{F}; Y)_X$ and $f \in \mathcal{O}_X$, then

$$(\mathrm{id} \otimes \rho)(D(fu) - fDu) = df \otimes \rho\pi_{k-1}u = df \otimes \pi_{k-1}\rho u = D_X(f\rho u) - fD_X(\rho u),$$

from which we deduce the commutativity of the diagram.

Let G be a vector bundle over Y and $\varphi : J_k(F; Y) \to G$ a morphism of vector bundles over Y. Let $p_l(\varphi) : J_{k+l}(F; Y) \to J_l(G; Y)$ be the morphism of vector bundles induced by φ sending $j_{k+l}(s)(y)$ into $j_l(\varphi j_k(s))(y)$, where s is a section of F over a neighborhood of $y \in Y$.

The proof of the following proposition is similar to that of Proposition 4.6 of [11] (cf. Lemma 1. 1 of [6]) and will therefore be omitted.

Proposition 1. 4. *The diagram*

$$\begin{array}{ccc} \wedge^j \mathcal{T}^* \otimes J_{k+l}(\mathcal{F}; Y)_X & \xrightarrow{D} & \wedge^{j+1}\mathcal{T}^* \otimes J_{k+l-1}(\mathcal{F}; Y)_X \\ \downarrow \mathrm{id} \otimes p_l(\varphi) & & \downarrow \mathrm{id} \otimes p_{l-1}(\varphi) \\ \wedge^j \mathcal{T}^* \otimes J_l(\mathcal{G}; Y)_X & \xrightarrow{D} & \wedge^{j+1}\mathcal{T}^* \otimes J_{l-1}(\mathcal{G}; Y)_X \end{array}$$

is commutative.

By (1. 3), the restriction of $-D$ to $\wedge^i \mathcal{T}^* \otimes \varepsilon(S^k \mathcal{T}^*_Y \otimes \mathcal{F})_X$ is $\mathcal{O}_X$-linear and therefore comes from a morphism

$$\delta : \wedge^i T^* \otimes_X (S^k T^*_Y \otimes F) \to \wedge^{i+1} T^* \otimes_X (S^{k-1} T^*_Y \otimes F)$$

of vector bundles, and thus we obtain a complex of vector bundles

$$\begin{aligned} 0 \to \rho^{-1}(S^k T^*_Y \otimes F) &\xrightarrow{\delta} T^* \otimes_X (S^{k-1} T^*_Y \otimes F) \\ \xrightarrow{\delta} \wedge^2 T^* \otimes_X (S^{k-2} T^*_Y \otimes F) &\xrightarrow{\delta} \cdots \to \wedge^n T^* \otimes_X (S^{k-n} T^*_Y \otimes F) \to 0, \end{aligned} \tag{1. 12}$$

for $k>0$, where

$$\delta(\omega \wedge u) = (-1)^j \omega \wedge \delta u, \tag{1. 13}$$

for $\omega \in \wedge^j T^*$, $u \in \wedge T^* \otimes_X (S^k T^*_Y \otimes F)$. If ρ is an immersion, the complex (1. 12) is exact at $\wedge^i T^* \otimes_X (S^{k-i} T^*_Y \otimes F)$ for $i>0$. Moreover

$$\delta(\omega \otimes u) = (-1)^j \omega \wedge (\rho^* \otimes \mathrm{id})\delta_Y u,$$

for $\omega \in \wedge^j T^*_x$, $u \in (S^k T^*_Y \otimes F)_{\rho(x)}$, $x \in X$, where $\rho^* : T^*_{Y,\rho(x)} \to T^*_x$, and the diagrams

$$\begin{array}{ccc} \wedge^i T^*_{Y,\rho(x)} \otimes (S^k T^*_Y \otimes F)_{\rho(x)} & \xrightarrow{\delta_Y} & \wedge^{i+1} T^*_{Y,\rho(x)} \otimes (S^{k-1} T^*_Y \otimes F)_{\rho(x)} \\ \downarrow \rho^* \otimes \mathrm{id} & & \downarrow \rho^* \otimes \mathrm{id} \\ \wedge^i T^*_x \otimes (S^k T^*_Y \otimes F)_{\rho(x)} & \xrightarrow{\delta} & \wedge^{i+1} T^*_x \otimes (S^{k-1} T^*_Y \otimes F)_{\rho(x)} \end{array} \tag{1. 14}$$

and

$$\begin{array}{ccc} \wedge^i T^*_x \otimes (S^k T^*_Y \otimes F)_{\rho(x)} & \xrightarrow{\delta} & \wedge^{i+1} T^*_x \otimes (S^{k-1} T^*_Y \otimes F)_{\rho(x)} \\ \downarrow \mathrm{id} \otimes \rho^* \otimes \mathrm{id} & & \downarrow \mathrm{id} \otimes \rho^* \otimes \mathrm{id} \\ \wedge^i T^*_x \otimes S^k T^*_x \otimes F_{\rho(x)} & \xrightarrow{\delta_X} & \wedge^{i+1} T^*_x \otimes S^{k-1} T^*_x \otimes F_{\rho(x)} \end{array} \tag{1. 15}$$

are commutative. Let $\Delta_{l,k}: S^{k+l}T_Y^* \to S^l T_Y^* \otimes S^k T_Y^*$ be the natural inclusion; then $\Delta_{1,k} = \delta_Y$ (see [3], § 5). The commutativity of the diagram

$$\begin{array}{ccc} \bigwedge^i T^* \otimes_X (S^{k+l+1} T_Y^* \otimes F)_{\rho(x)} & \xrightarrow{\delta} & \bigwedge^{i+1} T^* \otimes_X (S^{k+l} T_Y^* \otimes F) \\ \downarrow \mathrm{id} \otimes \Delta_{l+1,k} & & \downarrow \mathrm{id} \otimes \Delta_{l,k} \\ \bigwedge^i T^* \otimes_X (S^{l+1} T_Y^* \otimes S^k T_Y^* \otimes F) & \xrightarrow{\delta} & \bigwedge^{i+1} T^* \otimes_X (S^l T_Y^* \otimes S^k T_Y^* \otimes F) \end{array} \tag{1.16}$$

follows from that of diagram (6. 2) of [3].

We give a construction of the operator (1. 5) similar to the one given by Malgrange [10] for the Spencer operator D. Let Γ be the subset of $X \times Y$ which is the graph of ρ. Let $pr_1: X \times Y \to X$ be the projection onto the first factor. We shall identify a sheaf on X (resp. on Γ) with its inverse image by $pr_1: \Gamma \to X$ (resp. with its direct image by the inclusion $\Gamma \to X \times Y$). Let $\mathcal{I}_Y^{k+1}$ be the subsheaf of $\mathcal{O}_{Y\times Y}$ of functions which vanish to order k on the diagonal Δ_Y of $Y \times Y$. Let $\mathcal{I}_\rho^{k+1}$ be the inverse image of this sheaf by $\rho \times \mathrm{id}: X \times Y \to Y \times Y$. If $\mathbf{1}_Y$ is the trivial line bundle over Y, we see that $\mathcal{O}_{X\times Y}/\mathcal{I}_\rho^{k+1}$ is the sheaf of sections of $\rho^{-1} J_k(\mathbf{1}_Y; Y)$ over X. Furthermore

$$J_k(\mathcal{F}; Y)_X = (\mathcal{O}_{X\times Y}/\mathcal{I}_\rho^{k+1}) \otimes_{pr_2^{-1}\mathcal{O}_Y} pr_2^{-1}\mathcal{F}, \tag{1.17}$$

where $pr_2: X \times Y \to Y$ is the projection onto the second factor. Lifting differential forms on X to $X \times Y$ by pr_1^*, we may regard elements of

$$\bigwedge \mathcal{T}^* \otimes_{\mathcal{O}_X} (\mathcal{O}_{X\times Y}/\mathcal{I}_\rho^{k+1})$$

as germs of differential forms on $X \times Y$ modulo $\mathcal{I}_\rho^{k+1}$. The exterior differential operator on $X \times Y$ with respect to the first factor X gives by passage to the quotient a map

$$D: \bigwedge \mathcal{T}^* \otimes_{\mathcal{O}_X} (\mathcal{O}_{X\times Y}/\mathcal{I}_\rho^{k+1}) \to \bigwedge \mathcal{T}^* \otimes_{\mathcal{O}_X} (\mathcal{O}_{X\times Y}/\mathcal{I}_\rho^{k}). \tag{1.18}$$

Since D is $pr_2^{-1}\mathcal{O}_Y$-linear, by applying the functor

$$\otimes_{pr_2^{-1}\mathcal{O}_Y} pr_2^{-1}\mathcal{F}$$

to (1. 18) and using (1. 17), we obtain an operator

$$D: \bigwedge \mathcal{T}^* \otimes J_k(\mathcal{F}; Y)_X \to \bigwedge \mathcal{T}^* \otimes J_{k-1}(\mathcal{F}; Y)_X,$$

which is none other than our operator (1. 5), as it is easily seen that it satisfies conditions (i) of Proposition 1. 1 and (1. 6).

Finally, the operator (1. 18), or more generally (1. 5), is easily written in terms of local coordinates. For simplicity of notation, we shall consider only the case

$$D: \mathcal{O}_{X\times Y}/\mathcal{I}_\rho^{k+1} \to \mathcal{T}^* \otimes_{\mathcal{O}_X} (\mathcal{O}_{X\times Y}/\mathcal{I}_\rho^{k}).$$

Let $x = (x^1, \cdots, x^n)$ be a local coordinate for X and $y = (y^1, \cdots, y^m)$ a local coordinate for Y; assume that the mapping ρ is given by $\rho(x) = (\rho^1(x), \cdots, \rho^m(x))$. If u represents a germ of $\mathcal{O}_{X\times Y}/\mathcal{I}_\rho^{k+1}$, we have

$$u = \sum_{|\alpha| \le k} a_\alpha(x) \frac{(y - \rho(x))^\alpha}{\alpha\,!} \quad (\mathrm{mod}\ \mathcal{I}_\rho^{k+1}),$$

where $\alpha = (\alpha_1, \cdots, \alpha_m)$, $y'^\alpha = (y'^1)^{\alpha_1} \cdots (y'^m)^{\alpha_m}$ and $y'^i = y^i - \rho^i(x)$, $\alpha! = (\alpha_1!) \cdots (\alpha_m!)$, $|\alpha| = \alpha_1 + \cdots + \alpha_m$. Then we have

$$Du = \sum_{\substack{|\alpha|\leq k-1 \\ 1\leq i\leq n}} \left\{ dx^i \otimes \left(\frac{\partial a_\alpha}{\partial x^i}(x) - \sum_{j=1}^{m} a_{\alpha+1_j}(x) \frac{\partial \rho^j}{\partial x^i}(x) \right) \frac{(y-\rho(x))^\alpha}{\alpha!} \right\} \quad (\mathrm{mod}\ \mathcal{J}^k_\rho).$$

2. Submanifolds and Linear Complexes

Let F be a vector bundle over Y and let $R_k \subset J_k(F;Y)$ be a linear differential equation of order $k \geq 1$ on F. We denote by $R_{k+l} \subset J_{k+l}(F;Y)$ the l-th prolongation of R_k and by $\lambda_l : R_{k+l} \to J_l(R_k;Y)$ the natural injection. We shall assume throughout this paper that R_k is *formally integrable*. The operator D_Y restricts to give us the Spencer complexes

$$(2.1) \qquad 0 \to \mathcal{R}_p \xrightarrow{D_Y} \mathcal{T}^*_Y \otimes \mathcal{R}_{p-1} \xrightarrow{D_Y} \cdots \to \wedge^m \mathcal{T}^*_Y \otimes \mathcal{R}_{p-m} \to 0$$

of R_k, where $R_q = J_q(F;Y)$ for $q<k$. Let g_{k+l} be the sub-bundle of $S^{k+l}T^*_Y \otimes F$, for $l \geq 1$, such that the sequence

$$0 \to g_{k+l} \xrightarrow{\varepsilon} R_{k+l} \xrightarrow{\pi_{k+l-1}} R_{k+l-1} \to 0$$

is exact. From (2. 1), we obtain the complexes

$$(2.2) \qquad 0 \to g_p \xrightarrow{\delta_Y} T^*_Y \otimes g_{p-1} \xrightarrow{\delta_Y} \cdots \to \wedge^m T^*_Y \otimes g_{p-m} \to 0,$$

where $g_q = S^q T^*_Y \otimes F$ for $q<k$. We say that g_{k+1} is *involutive* if the complexes (2. 2) are exact at $\wedge^j T^*_Y \otimes g_q$ for all $q \geq k+1$, $j \geq 0$. If g_{k+1} is involutive, the cohomology $H^j(R_k)_q$ of the complexes (2. 1) at $\wedge^j \mathcal{T}^*_Y \otimes \mathcal{R}_q$ for $q \geq k+1$ is independent of q and is called the j-th Spencer cohomology group $H^j(R_k)$ of R_k (see [11], [4] and [5]).

From Proposition 1. 2, applied to sections of R_p, we deduce that the differential operator (1. 1) induces an operator

$$D : \mathcal{R}_{p,X} \to \mathcal{T}^* \otimes \mathcal{R}_{p-1,X};$$

hence by (1. 6), we obtain a sub-complex of (1. 8) (with k replaced by p)

$$(2.3) \qquad 0 \to \mathcal{R}_{p,X} \xrightarrow{D} \mathcal{T}^* \otimes \mathcal{R}_{p-1,X} \xrightarrow{D} \cdots \to \wedge^n \mathcal{T}^* \otimes \mathcal{R}_{p-n,X} \to 0.$$

If $H^j_\rho(R_k)_q$ denotes the cohomology of the complexes (2. 3) at $\wedge^j \mathcal{T}^* \otimes \mathcal{R}_{q,X}$, the mapping ρ induces, according to Proposition 1. 2, a map

$$(2.4) \qquad \rho : H^j(R_k)_{p,\rho(x)} \to H^j_\rho(R_k)_{p,x}$$

for all $p \geq k$, $j \geq 0$ and $x \in X$.

If X is a submanifold of Y and ρ is the inclusion mapping, the Cauchy problem for R_k is the study of the maps (2. 4) with $j=0$. Surjectivity of these maps corresponds to existence theorems and injectivity to uniqueness theorems.

From (2. 3), we obtain the sub-complex of (1. 12)

(2. 5)

$$0 \to \rho^{-1} g_p \xrightarrow{\delta} T^* \otimes_X g_{p-1} \xrightarrow{\delta} \wedge^2 T^* \otimes_X g_{p-2} \xrightarrow{\delta} \cdots \to \wedge^n T^* \otimes_X g_{p-n} \to 0.$$

We denote by

$$\delta : \wedge^{j-1} T^* \otimes_X g_{k+1} \to \wedge^j T^* \otimes_X R_k$$

the composition

$$\bigwedge^{j-1}T^*\otimes_X g_{k+1}\xrightarrow{\delta}\bigwedge^{j}T^*\otimes_X g_k\xrightarrow{\mathrm{id}\otimes\varepsilon}\bigwedge^{j}T^*\otimes_X R_k;$$

we set

$$\tilde{C}^j=\bigwedge^{j}T^*\otimes_X R_k/\delta(\bigwedge^{j-1}T^*\otimes_X g_{k+1})$$

and let $\tilde{\theta}:\bigwedge^{j}T^*\otimes_X R_k\to\tilde{C}^j$ be the natural projection. In particular, $\tilde{C}^0=\rho^{-1}R_k$.

From (1. 13), it follows that $\delta(\bigwedge T^*\otimes_X g_{k+1})$ is a graded $\bigwedge T^*$-submodule of $\bigwedge T^*\otimes_X R_k$. Therefore $\tilde{C}=\bigoplus_{j\geq 0}\tilde{C}^j$ has the structure of a graded $\bigwedge T^*$-module; if $\omega\in\bigwedge T^*$, $u\in\tilde{C}$, we denote by $\omega\wedge u$ the image of u under the action of ω on $\tilde{C}$.

Proposition 2. 1. *If $\delta:\bigwedge^{j-1}T^*\otimes_X g_{k+1}\to\bigwedge^{j}T^*\otimes_X g_k$ is of constant rank for $j\geq 1$, then there are unique first-order differential operators*

$$\hat{D}:\tilde{\mathscr{C}}^j\to\tilde{\mathscr{C}}^{j+1}$$

such that the diagrams

(2. 6)
$$\begin{array}{ccc}
\bigwedge^{j}\mathscr{T}^*\otimes\mathscr{R}_{k+1,X} & \xrightarrow{D} & \bigwedge^{j+1}\mathscr{T}^*\otimes\mathscr{R}_{k,X}\\
\downarrow{\scriptstyle \tilde{\theta}\cdot(\mathrm{id}\otimes\pi_k)} & & \downarrow{\scriptstyle\tilde{\theta}}\\
\tilde{\mathscr{C}}^j & \xrightarrow{\hat{D}} & \tilde{\mathscr{C}}^{j+1}
\end{array}$$

are commutative for $j\geq 0$. Furthermore,

(2. 7)
$$0\to\tilde{\mathscr{C}}^0\xrightarrow{\hat{D}}\tilde{\mathscr{C}}^1\xrightarrow{\hat{D}}\tilde{\mathscr{C}}^2\xrightarrow{\hat{D}}\cdots\to\tilde{\mathscr{C}}^n\to 0$$

is a complex and

(2. 8)
$$\hat{D}(\omega\wedge u)=d\omega\wedge u+(-1)^j\omega\wedge\hat{D}u$$

for $\omega\in\bigwedge^{j}\mathscr{T}^*$, $u\in\tilde{\mathscr{C}}$.

Proof. Our hypothesis implies that the $\tilde{C}^j$ are vector bundles. As the upper part of the diagram

(2. 9)
$$\begin{array}{ccc}
0 & & \\
\downarrow & & \\
\bigwedge^{j}\mathscr{T}^*\otimes g_{k+1,X} & \overset{\delta}{\searrow} & \\
\downarrow & & \\
\bigwedge^{j}\mathscr{T}^*\otimes\mathscr{R}_{k+1,X} & \xrightarrow{D} & \bigwedge^{j+1}\mathscr{T}^*\otimes\mathscr{R}_{k,X}\\
\downarrow{\scriptstyle\mathrm{id}\otimes\pi_k} & & \downarrow{\scriptstyle\tilde{\theta}}\\
\bigwedge^{j}\mathscr{T}^*\otimes\mathscr{R}_{k,X} & \xrightarrow{\bar{D}} & \tilde{\mathscr{C}}^{j+1}\\
\downarrow & & \\
0 & &
\end{array}$$

is commutative and since its left-hand column is exact, there is a unique first-order differential operator $\bar{D}:\bigwedge^{j}\mathscr{T}^*\otimes\mathscr{R}_{k,X}\to\tilde{\mathscr{C}}^{j+1}$ which makes diagram (2. 9) commutative. If $u\in\bigwedge^{j-1}\mathscr{T}^*\otimes g_{k+1,X}$, then, considering g_{k+1} as a sub-bundle of R_{k+1}, we see that there exists $u' \in \bigwedge^{j-1}\mathcal{T}^*\otimes\mathcal{R}_{k+2,X}$ such that $(\mathrm{id}\otimes\pi_{k+1})u'=u$. Then

$$\bar{D}\delta u = \bar{D}\delta(\mathrm{id}\otimes\pi_{k+1})u' = \bar{D}(\mathrm{id}\otimes\pi_k)Du' = \tilde{\theta}D\cdot Du' = 0.$$

Thus the operator $\bar{D}$ induces a unique first-order differential operator $\hat{D}:\tilde{\mathcal{C}}^j\to\tilde{\mathcal{C}}^{j+1}$ such that the diagram

$$(2.10)\qquad \begin{array}{ccc} \bigwedge^j\mathcal{T}^*\otimes\mathcal{R}_{k,X} & \xrightarrow{\bar{D}} & \tilde{\mathcal{C}}^{j+1} \\ \downarrow{\scriptstyle\tilde{\theta}} & \nearrow{\scriptstyle\hat{D}} & \\ \tilde{\mathcal{C}}^j & & \end{array}$$

commutes. The commutativity of (2. 6) is a consequence of that of (2. 9) and (2. 10). It implies the commutativity of the diagrams

$$(2.11)\qquad \begin{array}{ccccc} \bigwedge^j\mathcal{T}^*\otimes\mathcal{R}_{k+2,X} & \xrightarrow{D} & \bigwedge^{j+1}\mathcal{T}^*\otimes\mathcal{R}_{k+1,X} & \xrightarrow{D} & \bigwedge^{j+2}\mathcal{T}^*\otimes\mathcal{R}_{k,X} \\ \downarrow{\scriptstyle\tilde{\theta}\cdot(\mathrm{id}\otimes\pi_k)} & & \downarrow{\scriptstyle\tilde{\theta}\cdot(\mathrm{id}\otimes\pi_k)} & & \downarrow{\scriptstyle\tilde{\theta}} \\ \tilde{\mathcal{C}}^j & \xrightarrow{\hat{D}} & \tilde{\mathcal{C}}^{j+1} & \xrightarrow{\hat{D}} & \tilde{\mathcal{C}}^{j+2} \end{array}$$

whose vertical arrows are surjective. Since (2. 3) is a complex, we have $\hat{D}\cdot\hat{D}=0$. Relation (2. 8) follows directly from (1. 6).

Assuming that the maps $\delta:\bigwedge^{j-1}T^*\otimes_X g_{k+1}\to\bigwedge^j T^*\otimes_X g_k$ are of constant rank for $j\geq 1$, we denote by $H^j(\tilde{C})$ the cohomology of the complex (2. 7) at $\tilde{\mathcal{C}}^j$. Diagram (2. 11) gives rise to a map

$$(2.12)\qquad H^j_\rho(R_k)_{k+1}\to H^j(\tilde{C}).$$

We denote by $\delta_Y:\bigwedge^{j-1}T^*_Y\otimes g_{k+1}\to\bigwedge^j T^*_Y\otimes R_k$ the composition

$$\bigwedge^{j-1}T^*_Y\otimes g_{k+1}\xrightarrow{\delta_Y}\bigwedge^j T^*_Y\otimes g_k\xrightarrow{\mathrm{id}\otimes\varepsilon}\bigwedge^j T^*_Y\otimes R_k\,;$$

we set

$$C^j=\bigwedge^j T^*_Y\otimes R_k/\delta_Y(\bigwedge^{j-1}T^*_Y\otimes g_{k+1})$$

and let $\theta:\bigwedge^j T^*_Y\otimes R_k\to C^j$ be the natural projection. If g_{k+1} is involutive, then C^j is a vector bundle for all $j\geq 0$ (see [2]). Applying Proposition 2. 1 to the case where $Y=X$ and ρ is the identity map of Y, we see that, if g_{k+1} is involutive, there is a unique first-order differential operator $\hat{D}:\mathcal{C}^j\to\mathcal{C}^{j+1}$ such that the diagram

$$\begin{array}{ccc} \bigwedge^j\mathcal{T}^*_Y\otimes\mathcal{R}_{k+1} & \xrightarrow{D} & \bigwedge^{j+1}\mathcal{T}^*_Y\otimes\mathcal{R}_k \\ \downarrow{\scriptstyle\theta\cdot(\mathrm{id}\otimes\pi_k)} & & \downarrow{\scriptstyle\theta} \\ \mathcal{C}^j & \xrightarrow{\hat{D}} & \mathcal{C}^{j+1} \end{array}$$

is commutative and thus obtain a complex

$$(2.13)\qquad 0\to\mathcal{C}^0\xrightarrow{\hat{D}}\mathcal{C}^1\xrightarrow{\hat{D}}\mathcal{C}^2\xrightarrow{\hat{D}}\cdots\to\mathcal{C}^m\to 0$$

(see [2]). Furthermore, $C=\bigoplus_{j\geq 0}C^j$ is a graded $\bigwedge T^*_Y$-module and the operator

$\hat{D}: \mathcal{C} \to \mathcal{C}$ satisfies (2. 8) for all $\omega \in \bigwedge^j \mathcal{T}_Y^*$, $u \in \mathcal{C}$. If $H^j(C)$ is the cohomology of the complex (2. 13) at $\mathcal{C}^j$, then we have a map

(2. 14) $$H^j(R_k)_{k+1} \to H^j(C).$$

The commutativity of diagram (1. 14) implies that the diagram

$$\begin{array}{ccc} \bigwedge^j T^*_{Y,\rho(x)} \otimes R_{k,\rho(x)} & \xrightarrow{\rho^* \otimes \mathrm{id}} & \bigwedge^j T^*_x \otimes R_{k,\rho(x)} \\ \uparrow \delta_Y & & \uparrow \delta \\ \bigwedge^{j-1} T^*_{Y,\rho(x)} \otimes g_{k+1,\rho(x)} & \xrightarrow{\rho^* \otimes \mathrm{id}} & \bigwedge^{j-1} T^*_x \otimes g_{k+1,\rho(x)} \end{array}$$

is commutative. Thus we obtain a map

$$\rho^*: \rho^{-1}C \to \tilde{C}$$

satisfying

$$\rho^*(\omega \wedge u) = \rho^*\omega \wedge \rho^* u,$$

for $\omega \in \bigwedge T^*_{Y,\rho(x)}$, $u \in C_{\rho(x)}$, where $\rho^*: \bigwedge T^*_{Y,\rho(x)} \to \bigwedge T^*_x$ and $\rho^*: C_{\rho(x)} \to \tilde{C}_x$.

If the maps $\delta: \bigwedge^{j-1} T^* \otimes_X g_{k+1} \to \bigwedge^j T^* \otimes_X g_k$ are of constant rank for $j \geq 1$ and g_{k+1} is involutive, if u is a section of C over Y, we define the section $\rho^* u$ of $\tilde{C}$ over X by

$$(\rho^* u)(x) = \rho^* u(\rho(x)), \qquad \text{for} \quad x \in X,$$

where $\rho^*: C_{\rho(x)} \to \tilde{C}_x$. Under these assumptions, from Proposition 1. 2 and the construction of the operators $\hat{D}$, it follows that, if u is a section of C over Y,

$$\hat{D}(\rho^* u) = \rho^*(\hat{D} u)$$

as sections of $\tilde{C}$ over X. Therefore the mapping ρ induces a map

$$\rho: H^j(C)_{\rho(x)} \to H^j(\tilde{C})_x$$

for all $x \in X$. Clearly, for $x \in X$, the diagram

$$\begin{array}{ccc} H^j(R_k)_{k+1,\rho(x)} & \xrightarrow{\rho} & H^j_\rho(R_k)_{k+1,x} \\ \downarrow & & \downarrow \\ H^j(C)_{\rho(x)} & \xrightarrow{\rho} & H^j(\tilde{C})_x \end{array}$$

is commutative, where the vertical arrows are the mappings (2. 12) and (2. 14).

We now introduce the families of vector spaces $\tilde{g}_{k+l} \subset S^l T^* \otimes_X R_k$ which will lead to another interpretation of the bundles $\tilde{C}^j$. The map $\Delta_{l,k}: S^{k+l} T^*_Y \otimes F \to S^l T^*_Y \otimes S^k T^*_Y \otimes F$ restricts to give us a mapping

$$\Delta_{l,k}: g_{k+l} \to S^l T^*_Y \otimes g_k$$

(see [3]) ; let $\tilde{g}_{k+l} \subset S^l T^* \otimes_X R_k$ be the family of vector spaces whose fiber $\tilde{g}_{k+l,x}$ over $x \in X$ is the image of the composition

$$g_{k+l,\rho(x)} \xrightarrow{\Delta_{l,k}} (S^l T^*_Y \otimes g_k)_{\rho(x)} \xrightarrow{\rho^* \otimes \mathrm{id}} S^l T^*_x \otimes g_{k,\rho(x)} \xrightarrow{\mathrm{id} \otimes \varepsilon} S^l T^*_x \otimes R_{k,\rho(x)}.$$

We denote this mapping by $\rho^*: \rho^{-1} g_{k+l} \to \tilde{g}_{k+l}$. In particular, taking $l=0$, we see that $\tilde{g}_k$ is the family of subspaces of $\rho^{-1} R_k$ equal to $\rho^{-1}\varepsilon(g_k)$. From the commutativity of diagrams (1. 16) and (1. 15), we deduce that the diagram

$$\begin{array}{ccc}
\wedge^j T^*\otimes_X g_{k+l+1} & \xrightarrow{\delta} & \wedge^{j+1}T^*\otimes_X g_{k+l} \\
\big\downarrow \mathrm{id}\otimes\Delta_{l+1,k} & & \big\downarrow \mathrm{id}\otimes\Delta_{l,k} \\
\wedge^j T^*\otimes_X (S^{l+1}T^*_Y\otimes g_k) & \xrightarrow{\delta} & \wedge^{j+1}T^*\otimes_X(S^l T^*_Y\otimes g_k) \\
\big\downarrow \mathrm{id}\otimes\rho^*\otimes\mathrm{id} & & \big\downarrow \mathrm{id}\otimes\rho^*\otimes\mathrm{id} \\
\wedge^j T^*\otimes S^{l+1}T^*\otimes_X g_k & \xrightarrow{\delta_X} & \wedge^{j+1}T^*\otimes S^l T^*\otimes_X g_k
\end{array}$$

is commutative, where $\rho^* : \rho^{-1}T^*_Y \to T^*$. It follows that the diagram

$$\text{(2. 15)}\qquad \begin{array}{ccc}
\wedge^j T^*\otimes_X g_{k+l+1} & \xrightarrow{\delta} & \wedge^{j+1}T^*\otimes_X g_{k+l} \\
\big\downarrow \mathrm{id}\otimes\rho^* & & \big\downarrow \mathrm{id}\otimes\rho^* \\
\wedge^j T^*\otimes \tilde{g}_{k+l+1} & \xrightarrow{\delta_X} & \wedge^{j+1}T^*\otimes \tilde{g}_{k+l}
\end{array}$$

commutes, where δ_X is the restriction of

$$\delta_X : \wedge^j T^*\otimes S^{l+1}T^*\otimes_X R_k \to \wedge^{j+1}T^*\otimes S^l T^*\otimes_X R_k.$$

The commutativity of (2. 15) with $l=0$ implies that

$$\delta(\wedge^i T^*\otimes_X g_{k+1}) = \delta_X(\wedge^i T^*\otimes \tilde{g}_{k+1})$$

and hence that

$$\tilde{C}^j = \wedge^j T^*\otimes_X R_k/\delta_X(\wedge^{j-1}T^*\otimes \tilde{g}_{k+1}).$$

We obtain a complex

(2. 16)

$$0\to \tilde{g}_{k+l+1} \xrightarrow{\delta_X} T^*\otimes \tilde{g}_{k+l} \xrightarrow{\delta_X} \wedge^2 T^*\otimes \tilde{g}_{k+l-1} \xrightarrow{\delta_X} \cdots \to \wedge^n T^*\otimes \tilde{g}_{k+l+1-n} \to 0,$$

where $\tilde{g}_p=0$ for $p<k$. Let $(\tilde{g}_{k+1})_{+l}\subset S^{l+1}T^*\otimes_X R_k$ be the family of vector spaces $\Delta^{-1}_{l,1}(S^l T^*\otimes \tilde{g}_{k+1})$. Then according to [3], we have $\tilde{g}_{k+l+1}\subset(\tilde{g}_{k+1})_{+l}$. We say that $\tilde{g}_{k+1}$ is involutive if the complexes (2. 16) are exact at $\wedge^j T^*\otimes \tilde{g}_{k+l}$ for all $l\geq 1, j\geq 0$. If $\tilde{g}_{k+1}$ is involutive, then $\tilde{g}_{k+l+1}=(\tilde{g}_{k+1})_{+l}$, for $l\geq 0$.

Throughout the remainder of this section, we assume that X is a submanifold of Y and that $\rho : X\to Y$ is the inclusion map. We write

$$H^j_X(R_k)_{p,x} = H^j_\rho(R_k)_{p,x}, \qquad \text{for } p\geq k,\ j\geq 0,\ x\in X.$$

Let $N\subset T^*_{Y|X}$ be the conormal bundle of X in Y and $S^l N$ denote its l-th symmetric product.

Definition 2. 1. We say that X is non-characteristic for g_{k+1} if

$$\text{(2. 17)}\qquad (S^{k+1}N\otimes_X F)\cap g_{k+1|X} = 0.$$

If X is non-characteristic for g_{k+1}, it follows that

$$(S^{k+l}N\otimes_X F)\cap g_{k+l|X} = 0$$

for all $l\geq 1$. As it is easily seen that

$$\delta_Y(S^{p+1}N\otimes_X F) = (N\otimes_X(S^p T^*_Y\otimes F))\cap\delta_Y((S^{p+1}T^*_Y\otimes F)_{|X})$$

for $p\geq 0$, we conclude that

$$\text{(2. 18)}\qquad (N\otimes_X g_{k+l})\cap\delta_Y(g_{k+l+1|X}) = 0$$

for all $l \geq 0$. Moreover (2. 17) is equivalent to (2. 18) with $l=0$, and thus also to the fact that $\rho : g_{k+1|X} \to \tilde{g}_{k+1}$ is an isomorphism. If R_k is elliptic (see [2]), then every hypersurface of Y is non-characteristic.

Using Proposition 5. 1 of [2], Theorem A of [7], we deduce statements (i) and (ii) of :

Theorem 2.1. *If g_{k+1} is involutive and X is non-characteristic for g_{k+1}, then :*
(i) *$\tilde{g}_{k+1}$ is involutive and $\tilde{g}_{k+l+1} = (\tilde{g}_{k+1})_{+l}$ for $l \geq 0$;*
(ii) *$\rho^* : g_{k+l|X} \to \tilde{g}_{k+l}$ is an isomorphism for $l \geq 1$;*
(iii) *the sequences*
(2. 19)

$$0 \to g_{k+l+1|X} \xrightarrow{\delta} T^* \otimes_X g_{k+l} \xrightarrow{\delta} \wedge^2 T^* \otimes_X g_{k+l-1} \xrightarrow{\delta} \cdots \to \wedge^n T^* \otimes_X g_{k+l+1-n} \to 0$$

are exact at $\wedge^j T^* \otimes_X g_{k+l}$ for all $l \geq 1, j \geq 0$.

By the commutativity of diagram (2. 15), (iii) follows from (i) and (ii).

The diagram

$$\begin{array}{ccccccccc}
 & & 0 & & 0 & & & 0 & \\
 & & \downarrow & & \downarrow & & & \downarrow & \\
 & 0 \to & g_{k+l+1,X} & \xrightarrow{-\delta} & \mathcal{T}^* \otimes g_{k+l,X} & \xrightarrow{-\delta} & \cdots \to & \wedge^n \mathcal{T}^* \otimes g_{k+l+1-n,X} & \to 0 \\
 & & \downarrow \varepsilon & & \downarrow \mathrm{id} \otimes \varepsilon & & & \downarrow \mathrm{id} \otimes \varepsilon & \\
(2.\ 20) & 0 \to & \mathcal{R}_{k+l+1,X} & \xrightarrow{D} & \mathcal{T}^* \otimes \mathcal{R}_{k+l,X} & \xrightarrow{D} & \cdots \to & \wedge^n \mathcal{T}^* \otimes \mathcal{R}_{k+l+1-n,X} & \to 0 \\
 & & \downarrow \pi_{k+l} & & \downarrow \mathrm{id} \otimes \pi_{k+l-1} & & & \downarrow \mathrm{id} \otimes \pi_{k+l-n} & \\
 & 0 \to & \mathcal{R}_{k+l,X} & \xrightarrow{D} & \mathcal{T}^* \otimes \mathcal{R}_{k+l-1,X} & \xrightarrow{D} & \cdots \to & \wedge^n \mathcal{T}^* \otimes \mathcal{R}_{k+l-n,X} & \to 0 \\
 & & \downarrow & & \downarrow & & & \downarrow & \\
 & & 0 & & 0 & & & 0 &
\end{array}$$

is commutative and so induces a map $\pi_{k+l-j} : H^j_X(R_k)_{k+l+1-j} \to H^j_X(R_k)_{k+l-j}$. Its j-th column is exact if $l \geq j$ and its top row is obtained from (2. 19). Thus from Theorem 2. 1 it follows that :

Theorem 2. 2. *If g_{k+1} is involutive and X is non-characteristic for g_{k+1}, the mappings*

$$\pi_{k+l} : H^j_X(R_k)_{k+l+1} \to H^j_X(R_k)_{k+l}$$

are isomorphisms for $l \geq 1$ and are injective for $l \geq 0$.

For $j \geq 1$, the diagram

$$\begin{array}{ccccc}
 & & (\wedge^{j-1} T^*_Y \otimes g_{k+1})_{|X} & \xrightarrow{\rho^* \otimes \mathrm{id}} & \wedge^j T^* \otimes_X g_{k+1} \to 0 \\
 & & \downarrow \delta_Y & & \downarrow \delta \\
N \otimes_X (\wedge^{j-1} T^*_Y \otimes R_k) & \longrightarrow & (\wedge^j T^*_Y \otimes R_k)_{|X} & \xrightarrow{\rho^* \otimes \mathrm{id}} & \wedge^j T^* \otimes_X R_k \to 0 \\
\downarrow \mathrm{id} \otimes \theta & & \downarrow \theta & & \downarrow \tilde{\theta} \\
N \otimes_X C^{j-1} & \longrightarrow & C^j_{|X} & \longrightarrow & \tilde{C}^j \to 0 \\
\downarrow & & \downarrow & & \downarrow \\
0 & & 0 & & 0
\end{array}$$

is commutative, where the two horizontal arrows on the left-hand side are given by the $\wedge T^*_Y$-module structures of $\wedge T^*_Y \otimes R_k$ and C; its columns and two upper rows are exact. Therefore, so is the bottom row and we obtain the canonical isomorphism

$$\tilde{C}^j \simeq C^j_{|X}/(N \wedge C^{j-1}_{|X}), \qquad \text{for} \quad j \geq 1.$$

If g_{k+1} is involutive and X is non-characteristic for g_{k+1}, then by Theorem 2. 1, (iii) and the exactness of the sequences

$$0 \to g_{k+1|X} \xrightarrow{\delta} T^* \otimes_X R_k \xrightarrow{\tilde{\theta}} \tilde{C}^1 \to 0,$$

$$0 \to g_{k+j|X} \xrightarrow{\delta} T^* \otimes_X g_{k+j-1} \xrightarrow{\delta} \cdots \xrightarrow{\delta} \wedge^{j-1} T^* \otimes_X g_{k+1} \xrightarrow{\delta} \wedge^j T^* \otimes_X R_k \xrightarrow{\tilde{\theta}} \tilde{C}^j \to 0$$

for $j \geq 2$, we see that $\tilde{C}^j$ is a vector bundle for $j \geq 0$. Under these assumptions, let $p(\hat{D}) : J_l(\tilde{C}^j) \to J_{l-1}(\tilde{C}^{j+1})$ and $\sigma(\hat{D}) : S^l T^* \otimes \tilde{C}^j \to S^{l-1} T^* \otimes \tilde{C}^{j+1}$ be the morphisms of vector bundles induced by $\hat{D} : \tilde{\mathscr{C}}^j \to \tilde{\mathscr{C}}^{j+1}$. From (2. 8), it follows that $\sigma(\hat{D}) : T^* \otimes \tilde{C}^j \to \tilde{C}^{j+1}$ is the mapping given by the $\wedge T^*$-module structure of $\tilde{C}$. By the methods of § 4 of [2] or of [11], we deduce from Theorem 2. 1 :

Lemma 2. 1. *If g_{k+1} is involutive and X is non-characteristic for g_{k+1}, the sequences*

$$0 \to g_{k+l+1|X} \xrightarrow{\rho^*} S^{l+1} T^* \otimes \tilde{C}^0 \xrightarrow{\sigma(\hat{D})} S^l T^* \otimes \tilde{C}^1 \xrightarrow{\sigma(\hat{D})} \cdots \to T^* \otimes \tilde{C}^l \xrightarrow{\sigma(\hat{D})} \tilde{C}^{l+1} \to 0$$

are exact for $l \geq 0$.

We have a natural mapping

$$\rho : J_l(R_k\,;\,Y)_{|X} \to J_l(R_{k|X})$$

sending $j_l(s)(x)$ into $j_l(s_{|X})(x)$, where s is a section of R_k over a neighborhood of $x \in X$ in Y. Denote the composition

$$R_{k+l+1|X} \xrightarrow{\lambda_{l+1}} J_{l+1}(R_k\,;\,Y)_{|X} \xrightarrow{\rho} J_{l+1}(R_{k|X})$$

by ρ and its image by $\tilde{R}_{k+l+1} \subset J_{l+1}(\tilde{C}^0)$. Under the hypotheses of Theorem 2. 1, we have the commutative diagrams

$$\begin{array}{ccccccc}
 & 0 & & 0 & & 0 & \\
 & \downarrow & & \downarrow & & \downarrow & \\
0 \to & g_{k+1|X} & \xrightarrow{\rho^*} & T^* \otimes \tilde{C}^0 & \xrightarrow{\sigma(\hat{D})} & \tilde{C}^1 & \to 0 \\
 & \downarrow \varepsilon & & \downarrow \varepsilon & & \downarrow & \\
0 \to & R_{k+1|X} & \xrightarrow{\rho} & J_1(\tilde{C}^0) & \xrightarrow{p(\hat{D})} & \tilde{C}^1 & \to 0 \\
 & \downarrow & & \downarrow \pi_0 & & \downarrow & \\
0 \to & \tilde{C}^0 & \xrightarrow{\text{id}} & \tilde{C}^0 & \longrightarrow & 0 & \\
 & \downarrow & & \downarrow & & & \\
 & 0 & & 0 & & &
\end{array}$$

and

$$\begin{array}{ccccccccccc}
 & 0 & & 0 & & 0 & & 0 & & 0 & \\
 & \downarrow & & \downarrow & & \downarrow & & \downarrow & & \downarrow & \\
0 \to & g_{k+l+1|X} & \xrightarrow{\rho^*} & S^{l+1}T^*\otimes \tilde{C}^0 & \xrightarrow{\sigma(\hat{D})} & S^{l}T^*\otimes \tilde{C}^1 & \longrightarrow \cdots \longrightarrow & T^*\otimes \tilde{C}^l & \xrightarrow{\sigma(\hat{D})} & \tilde{C}^{l+1} & \to 0 \\
 & \downarrow{\scriptstyle \varepsilon} & & \downarrow{\scriptstyle \varepsilon} & & \downarrow{\scriptstyle \varepsilon} & & \downarrow{\scriptstyle \varepsilon} & & \downarrow{\scriptstyle \mathrm{id}} & \\
0 \to & R_{k+l+1|X} & \xrightarrow{\rho} & J_{l+1}(\tilde{C}^0) & \xrightarrow{p(\hat{D})} & J_{l}(\tilde{C}^1) & \xrightarrow{p(\hat{D})} \cdots \longrightarrow & J_1(\tilde{C}^l) & \xrightarrow{p(\hat{D})} & \tilde{C}^{l+1} & \to 0 \\
 & \downarrow{\scriptstyle \pi_{k+l}} & & \downarrow{\scriptstyle \pi_{l}} & & \downarrow{\scriptstyle \pi_{l-1}} & & \downarrow{\scriptstyle \pi_{0}} & & \downarrow & \\
0 \to & R_{k+l|X} & \xrightarrow{\rho} & J_{l}(\tilde{C}^0) & \xrightarrow{p(\hat{D})} & J_{l-1}(\tilde{C}^1) & \xrightarrow{p(\hat{D})} \cdots \longrightarrow & \tilde{C}^l & \longrightarrow & 0 & \\
 & \downarrow & & \downarrow & & \downarrow & & \downarrow & & & \\
 & 0 & & 0 & & 0 & & 0 & & &
\end{array}$$

for $l \geq 1$. Using Lemma 2. 1, by induction on l, we conclude that these diagrams are exact. Thus:

Theorem 2. 3. *If g_{k+1} is involutive and X is non-characteristic for g_{k+1}, then the complex of vector bundles* (2. 7) *is formally exact in the sense that the sequences*

$$0 \to R_{k+l+1|X} \xrightarrow{\rho} J_{l+1}(\tilde{C}^0) \xrightarrow{p(\hat{D})} J_{l}(\tilde{C}^1) \xrightarrow{p(\hat{D})} \cdots \to J_1(\tilde{C}^l) \xrightarrow{p(\hat{D})} \tilde{C}^{l+1} \to 0$$

are exact for $l \geq 0$. Moreover, the equation $\tilde{R}_{k+1} \subset J_1(\tilde{C}^0)$ corresponding to the differential operator $\hat{D} : \tilde{\mathcal{C}}^0 \to \tilde{\mathcal{C}}^1$ is formally integrable and its l-th prolongation is isomorphic to $R_{k+l+1|X}$.

From the commutativity of the diagram

$$\begin{array}{ccc}
\bigwedge^j \mathcal{T}^* \otimes \mathcal{R}_{k+l+1,X} & \xrightarrow{D} & \bigwedge^{j+1} \mathcal{T}^* \otimes \mathcal{R}_{k+l,X} \\
\downarrow{\scriptstyle \mathrm{id}\otimes\lambda_{l+1}} & & \downarrow{\scriptstyle \mathrm{id}\otimes\lambda_{l}} \\
\bigwedge^j \mathcal{T}^* \otimes J_{l+1}(\mathcal{R}_k;\, Y)_X & \xrightarrow{D} & \bigwedge^{j+1} \mathcal{T}^* \otimes J_{l}(\mathcal{R}_k;\, Y)_X
\end{array}$$

given by Proposition 1. 4, and from Proposition 1. 3 with F replaced by R_k, we deduce the commutativity of the diagram

$$\begin{array}{ccc}
\bigwedge^j \mathcal{T}^* \otimes \mathcal{R}_{k+l+1,X} & \xrightarrow{D} & \bigwedge^{j+1} \mathcal{T}^* \otimes \mathcal{R}_{k+l,X} \\
\downarrow{\scriptstyle \mathrm{id}\otimes\rho} & & \downarrow{\scriptstyle \mathrm{id}\otimes\rho} \\
\bigwedge^j \mathcal{T}^* \otimes J_{l+1}(\tilde{\mathcal{C}}^0) & \xrightarrow{D_X} & \bigwedge^{j+1} \mathcal{T}^* \otimes J_{l}(\tilde{\mathcal{C}}^0).
\end{array}$$

Thus under the assumptions of Theorem 2. 1, we have the commutative diagram

$$\begin{array}{ccccccccc}
 & & & 0 & & 0 & & 0 & \\
 & & & \downarrow & & \downarrow & & \downarrow & \\
 & 0 & \longrightarrow & \mathcal{R}_{k+l+1,X} & \xrightarrow{D} & \mathcal{T}^*\otimes \mathcal{R}_{k+l,X} & \xrightarrow{D} \cdots \longrightarrow & \bigwedge^n \mathcal{T}^*\otimes \mathcal{R}_{k+l-n+1,X} & \to 0 \\
 & \downarrow & & \downarrow{\scriptstyle \rho} & & \downarrow{\scriptstyle \mathrm{id}\otimes\rho} & & \downarrow{\scriptstyle \mathrm{id}\otimes\rho} & \\
0 \to & \tilde{\mathcal{C}}^0 & \xrightarrow{j_{l+1}} & J_{l+1}(\tilde{\mathcal{C}}^0) & \xrightarrow{D_X} & \mathcal{T}^*\otimes J_{l}(\tilde{\mathcal{C}}^0) & \xrightarrow{D_X} \cdots \longrightarrow & \bigwedge^n \mathcal{T}^*\otimes J_{l-n+1}(\tilde{\mathcal{C}}^0) & \to 0 \\
 & \downarrow{\scriptstyle \hat{D}} & & \downarrow{\scriptstyle p(\hat{D})} & & \downarrow{\scriptstyle \mathrm{id}\otimes p(\hat{D})} & & \downarrow{\scriptstyle \mathrm{id}\otimes p(\hat{D})} & \\
0 \to & \tilde{\mathcal{C}}^1 & \xrightarrow{j_{l}} & J_{l}(\tilde{\mathcal{C}}^1) & \xrightarrow{D_X} & \mathcal{T}^*\otimes J_{l-1}(\tilde{\mathcal{C}}^1) & \xrightarrow{D_X} \cdots \longrightarrow & \bigwedge^n \mathcal{T}^*\otimes J_{l-n}(\tilde{\mathcal{C}}^1) & \to 0 \\
 & \downarrow{\scriptstyle \hat{D}} & & \downarrow{\scriptstyle p(\hat{D})} & & \downarrow{\scriptstyle \mathrm{id}\otimes p(\hat{D})} & & \downarrow{\scriptstyle \mathrm{id}\otimes p(\hat{D})} & \\
 & \vdots & & \vdots & & \vdots & & \vdots &
\end{array}$$

for $l \geq n$, where $J_p(\tilde{\mathcal{C}}^j) = 0$ for $p < 0$; its rows other than the top one are exact (see [4]) and its columns other than the first one are exact by Theorem 2. 3. Therefore it induces an isomorphism

$$H^j_X(R_k)_{k+l+1-j} \to H^j(\tilde{C}).$$

Combining this isomorphism with the isomorphisms of Theorem 2. 2, we obtain :

Theorem 2. 4. *If g_{k+1} is involutive and X is non-characteristic for g_{k+1}, then the mapping* (2. 21)

$$H^j_X(R_k)_{k+1} \to H^j(\tilde{C})$$

is an isomorphism for $j \geq 0$.

If all objects under consideration are real-analytic and we consider only real-analytic sections of our vector bundles, under the assumptions of Theorem 2. 1, we may apply Theorem 7. 2 of [7] to conclude that the mapping

$$H^0(C)_{|X} \to H^0(\tilde{C})$$

is an isomorphism and Theorem 7. 2 of [2] to infer from Theorem 2. 3 that

$$H^j(\tilde{C}) = 0, \qquad \text{for } j>0.$$

Thus by Theorem 2. 4, we have :

Theorem 2. 5. *If all objects are real-analytic and we consider only real-analytic sections of our vector bundles, and if g_{k+1} is involutive and X is non-characteristic for g_{k+1}, then the mappings* (2. 4)

$$H^0(R_k)_{|X} \to H^0_X(R_k)_{k+l}$$

are isomorphisms and

$$H^j_X(R_k)_{k+l} = 0$$

for $l \geq 1, j>0$.

3. The Non-linear Operator $\mathscr{D}$

Let $Q_k(Y)$ be the bundle of k-jets of local diffeomorphisms $Y \to Y$, which we consider as a bundle over Y via the projection source. We identify $Q_0(Y)$ with $Y \times Y$ and denote by $\tilde{\mathcal{Q}}_k(Y)$ the sub-sheaf of $\mathcal{Q}_k(Y)$ whose sections are local mappings $Y \to Q_k(Y)$ such that $\pi_0\phi$ is the graph of an immersion. Let

$$\omega_Y : T(Q_{k+1}(Y)) \to J_k(T_Y ; Y)$$

be the Cartan fundamental form on $Q_{k+1}(Y)$ defined in § 5 of [6].

We now define an operator

$$(3.1) \qquad \mathscr{D} : \mathcal{Q}_{k+1}(Y)_X \to \mathcal{T}^* \otimes J_k(\mathcal{T}_Y ; Y)_X$$

sending a section ϕ of $\rho^{-1}Q_{k+1}(Y)$ into the section $\phi^*\omega_Y$ of $T^* \otimes_X J_k(T_Y ; Y)$ which is given by

$$\langle \xi, \phi^*\omega_Y \rangle = \langle \phi_*\xi, \omega_Y \rangle,$$

for $\xi \in T_x$, $x \in X$, where $\phi_*\xi \in T_{\phi(x)}(Q_{k+1}(Y))$.

We define a bracket

(3. 2)
$$(\wedge^i T^* \otimes_X J_k(T_Y; Y)) \otimes (\wedge^j T^* \otimes_X J_k(T_Y; Y)) \to \wedge^{i+j} T^* \otimes_X J_{k-1}(T_Y; Y)$$
by the formula
$$[\alpha \otimes \xi, \beta \otimes \eta] = (\alpha \wedge \beta) \otimes [\xi, \eta],$$
for $\alpha \in \wedge^i T^*_x$, $\beta \in \wedge^j T^*_x$, ξ, $\eta \in J_k(T_Y; Y)_{\rho(x)}$, $x \in X$, where $[\xi, \eta]$ is given by the bracket
$$J_k(T_Y; Y) \times_Y J_k(T_Y; Y) \to J_{k-1}(T_Y; Y)$$
(see [6], § 1). Then Jacobi's identity holds, namely

(3. 3) $$[\pi_{k-1}u, [v, w]] = [[u, v], \pi_{k-1}w] + (-1)^{ij}[\pi_{k-1}v, [u, w]],$$
for $u \in \wedge^i T^* \otimes_X J_k(T_Y; Y)$, $v \in \wedge^j T^* \otimes_X J_k(T_Y; Y)$, $w \in \wedge T^* \otimes_X J_k(T_Y; Y)$. Moreover,

(3. 4) $$D[u, v] = [Du, \pi_{k-1}v] + (-1)^i[\pi_{k-1}u, Dv]$$
for $u \in \wedge^i \mathcal{T}^* \otimes J_k(\mathcal{T}_Y; Y)_X$, $v \in \wedge \mathcal{T}^* \otimes J_k(\mathcal{T}_Y; Y)_X$. Let
$$\mathcal{D}_1 : \mathcal{T}^* \otimes J_k(\mathcal{T}_Y; Y)_X \to \wedge^2 \mathcal{T}^* \otimes J_{k-1}(\mathcal{T}_Y; Y)_X$$
be the operator defined by
$$\mathcal{D}_1 u = Du - \frac{1}{2}[u, u], \qquad u \in \mathcal{T}^* \otimes J_k(\mathcal{T}_Y; Y).$$

According to Proposition 5. 2 of [6], the form ω_Y on $Q_{k+1}(Y)$ satisfies the Cartan structure equation
$$D\omega_Y - \frac{1}{2}[\omega_Y, \omega_Y] = 0,$$
where ω_Y is considered as a section of $T^*_{Q_{k+1}(Y)} \otimes_{Q_{k+1}(Y)} J_k(T_Y; Y)$ over Y and the bracket is given by (3. 2) (with X replaced by $Q_{k+1}(Y)$). Therefore by Proposition 1. 2, if ϕ is a section of $\rho^{-1}Q_{k+1}(Y)$ over X, then
$$\begin{aligned} D\mathcal{D}\phi - \frac{1}{2}[\mathcal{D}\phi, \mathcal{D}\phi] &= D\phi^*\omega_Y - \frac{1}{2}[\phi^*\omega_Y, \phi^*\omega_Y] \\ &= \phi^* D\omega_Y - \frac{1}{2}\phi^*[\omega_Y, \omega_Y] \\ &= \phi^*\Big(D\omega_Y - \frac{1}{2}[\omega_Y, \omega_Y]\Big) = 0. \end{aligned}$$

Thus we obtain a complex

(3. 5) $$Q_{k+1}(Y)_X \xrightarrow{\mathcal{D}} \mathcal{T}^* \otimes J_k(\mathcal{T}_Y; Y)_X \xrightarrow{\mathcal{D}_1} \wedge^2 \mathcal{T}^* \otimes J_{k-1}(\mathcal{T}_Y; Y)_X.$$

Consider the commutative diagram (1. 9) ; if ϕ is a section of $Q_{k+1}(Y)_Z$ over Z, then $\phi \circ \tau$ is a section of $Q_{k+1}(Y)_X$ over X.

Proposition 3. 1. *If ϕ is a section of $Q_{k+1}(Y)_Z$ over Z, then*

(3. 6) $$\mathcal{D}(\phi\circ\tau) = \tau^*(\mathcal{D}\phi)$$

as sections of $T^*\otimes_X J_k(T_Y; Y)$ *over* X, *where the operator* $\mathcal{D}$ *on the left-hand side is the operator* (3. 1) *and the operator* $\mathcal{D}$ *on the right-hand side is the operator*

$$\mathcal{D}: Q_{k+1}(Y)_Z \to \mathcal{T}_Z^*\otimes J_k(\mathcal{T}_Y; Y)_Z.$$

Proof. If ϕ is a section of $Q_{k+1}(Y)_Z$ over Z, then

$$\mathcal{D}(\phi\circ\tau) = (\phi\circ\tau)^*\omega_Y = \tau^*\phi^*\omega_Y = \tau^*(\mathcal{D}\phi).$$

If we take $Z=Y$, $\tau=\rho$ and $\sigma=\mathrm{id}$ in Proposition 3. 1, we obtain the formula

(3. 7) $$\mathcal{D}(\phi\circ\rho) = \rho^*(\mathcal{D}_Y\phi)$$

for all sections ϕ of $Q_{k+1}(Y)$ over Y, where the operator $\mathcal{D}=\mathcal{D}_Y: Q_{k+1}(Y)\to\mathcal{T}_Y^*\otimes J_k(\mathcal{T}_Y; Y)$ is an extension of the operator $\mathcal{D}$ of [10], [8] and [6].

Let $\tilde{J}_k(T_Y; Y)$ be the vector bundle over Y whose sections are the "diagonal" vector fields on $Y\times Y$ modulo those which vanish to order k on the diagonal of $Y\times Y$. We identify $\tilde{J}_0(T_Y; Y)$ with T_Y and let $\nu: \tilde{J}_k(T_Y; Y)\to J_k(T_Y; Y)$ be the canonical isomorphism (see [6], § 1). We write $J_0(T_Y)=J_0(T_Y; Y)$. For $k\geq 1$, we shall identify $S^kJ_0(T_Y)^*\otimes J_0(T_Y)$ with the kernels of the projections $\pi_{k-1}: J_k(T_Y; Y)\to J_{k-1}(T_Y; Y)$ and $\pi_{k-1}: \tilde{J}_k(T_Y; Y)\to\tilde{J}_{k-1}(T_Y; Y)$. We denote by

$$\delta_Y: S^kJ_0(T_Y)^*\otimes J_0(T_Y)\to T_Y^*\otimes S^{k-1}J_0(T_Y)^*\otimes J_0(T_Y)$$

the morphism of vector bundles sending u into $(\mathrm{id}\otimes\nu^*\otimes\nu)\cdot\delta_Y\cdot(\nu^*\otimes\nu^{-1})u$, where δ_Y is the morphism of § 1 (with $F=T_Y$). The restriction of the operator (1. 5), with $F=T_Y$, to $\wedge^j\mathcal{T}^*\otimes(S^kJ_0(\mathcal{T}_Y)^*\otimes J_0(\mathcal{T}_Y))_X$ is $\mathcal{O}_X$-linear and comes from a morphism of vector bundles

$$\delta: \wedge^jT^*\otimes_X(S^kJ_0(T_Y)^*\otimes J_0(T_Y))\to\wedge^{j+1}T^*\otimes_X(S^{k-1}J_0(T_Y)^*\otimes J_0(T_Y)),$$

which by (1. 13) is given by

$$\delta(\omega\otimes u) = (-1)^j\omega\wedge(\rho^*\otimes\mathrm{id})\delta_Yu,$$

for $\omega\in\wedge^jT_x^*$, $u\in(S^kJ_0(T_Y)^*\otimes J_0(T_Y))_{\rho(x)}$, $x\in X$.

Recall that we have a bracket

$$\tilde{J}_{k+1}(\mathcal{T}_Y; Y)\times_Y J_k(\mathcal{T}_Y; Y)\to J_k(\mathcal{T}_Y; Y)$$

sending $(\tilde{\xi}, \eta)$ into $[\tilde{\xi}, \eta]=\mathcal{L}(\tilde{\xi})\eta$, and that

(3. 8) $$\mathcal{L}(\tilde{\xi})\pi_0\eta = \nu[\pi_0\tilde{\xi}, \pi_0\tilde{\eta}]+\pi_0\tilde{\eta}\;\overline{\wedge}\;D\xi,$$

(3. 9) $$[\xi, \eta] = \mathcal{L}(\tilde{\xi})\pi_0\eta-\mathcal{L}(\tilde{\eta})\pi_0\xi-\nu[\pi_0\tilde{\xi}, \pi_0\tilde{\eta}],$$

for $\tilde{\xi}, \tilde{\eta}\in\tilde{J}_1(\mathcal{T}_Y; Y)$, with $\xi=\nu\tilde{\xi}$, $\eta=\nu\tilde{\eta}$ (see [6], formulas (1. 15) and (1. 16)).

If $X=Y$ and ρ is the identity map of X, then

$$\mathcal{D}_1 = \mathcal{D}_{1,Y}: \mathcal{T}_Y^*\otimes J_k(\mathcal{T}_Y; Y)\to\wedge^2\mathcal{T}_Y^*\otimes J_{k-1}(\mathcal{T}_Y; Y)$$

is the operator $\mathcal{D}_1$ of [10], [8] and [6] given by

$$\mathcal{D}_{1,Y}u = D_Yu-\frac{1}{2}[u, u],$$

for $u \in \mathcal{T}_Y^* \otimes J_k(\mathcal{T}_Y ; Y)$, where

$$\left\langle \xi \wedge \eta, \frac{1}{2}[u, u] \right\rangle = [u(\xi), u(\eta)], \qquad \text{if} \quad \xi, \eta \in \mathcal{T}_Y.$$

Moreover, by (1. 11), we obtain the formula

$$\mathcal{D}_1(\rho^* u) = \rho^*(\mathcal{D}_{1,Y} u), \tag{3. 10}$$

for $u \in (\mathcal{T}_Y^* \otimes J_k(\mathcal{T}_Y ; Y))_{\rho(x)}$, $x \in X$.

Lemma 3. 1. *Let $u \in \mathcal{T}_Y^* \otimes J_1(\mathcal{T}_Y ; Y)$ and $u_0 = \pi_0 u$, $\tilde{u}_0 = \nu^{-1} \circ u_0$. Then for ξ, $\eta \in \mathcal{T}_Y$, we have*

$$\begin{aligned} \langle \xi \wedge \eta, \mathcal{D}_{1,Y} u \rangle &= \nu[(\mathrm{id} + \tilde{u}_0)\xi, (\mathrm{id} + \tilde{u}_0)\eta] - (\nu + u_0)[\xi, \eta] \\ &\quad - \mathcal{L}(\nu^{-1} u(\xi))((\nu + u_0)\eta) + \mathcal{L}(\nu^{-1} u(\eta))((\nu + u_0)\xi). \end{aligned}$$

Proof. By (1. 7), (3. 8) and (3. 9), we have

$$\begin{aligned} \langle \xi \wedge \eta, \mathcal{D}_{1,Y} u \rangle &= \xi \,\overline{\wedge}\, Du(\eta) - \eta \,\overline{\wedge}\, Du(\xi) - u_0([\xi, \eta]) - [u(\xi), u(\eta)] \\ &= \mathcal{L}(\nu^{-1} u(\eta))(\nu \xi) - \nu[\tilde{u}_0(\eta), \xi] \\ &\quad - \mathcal{L}(\nu^{-1} u(\xi))(\nu \eta) + \nu[\tilde{u}_0(\xi), \eta] - u_0([\xi, \eta]) \\ &\quad - \mathcal{L}(\nu^{-1} u(\xi))(u_0(\eta)) + \mathcal{L}(\nu^{-1} u(\eta))(u_0(\xi)) + \nu[\tilde{u}_0(\xi), \tilde{u}_0(\eta)], \end{aligned}$$

from which the desired identity follows immediately.

An element $F \in Q_{k+1}(Y)$, with source $F = a$, target $F = b$, determines, according to [6], § 2, a mapping

$$J_k(T_Y ; Y)_a \rightarrow J_k(T_Y ; Y)_b.$$

It is easily seen that if ϕ is a section of $\rho^{-1} Q_1(Y)$ over a neighborhood of $x \in X$ and $f = \mathrm{target} \circ \phi$, then

$$\langle \xi, \mathcal{D}\phi \rangle = \phi(x)^{-1}(\nu f_* \xi) - \nu \rho_* \xi, \tag{3. 11}$$

for $\xi \in T_x$ (cf. [6], formula (2. 25)).

An alternative definition of the operator $\mathcal{D}$ can be given generalizing formula (2. 27) of [6] as follows. Set $E = \rho^{-1} Q_k(Y)$ and recall that $J_1(E)$ is an affine bundle over E whose associated vector bundle is $T^* \otimes_E V(Q_k(Y))$, where $V(Q_k(Y))$ is the bundle of vertical tangent vectors of $Q_k(Y)$. If $I_{Y,k}$ denotes the section of $Q_k(Y)$ over Y which is the k-jet of the identity map of Y, then $j_1(I_{Y,k} \circ \rho)$ is a section of $J_1(E)$. According to [10] or [6], § 2, we identify $V_{I_{Y,k}(y)}(Q_k(Y))$ with $\tilde{J}_k(T_Y ; Y)_y$, for $y \in Y$. Thus we obtain an injective map

$$T^* \otimes_X \tilde{J}_k(T_Y ; Y) \rightarrow J_1(E) \tag{3. 12}$$

sending $u \in T_x^* \otimes \tilde{J}_k(T_Y ; Y)_{\rho(x)}$ into $j_1(I_{Y,k} \circ \rho)(x) + u$, where $x \in X$. We denote by ∂F the inverse image by the map (3. 12) of an element $F \in J_1(E)$ belonging to the image of (3. 12). If $F \in Q_{k+1}(Y)$ is equal to $j_{k+1}(f)(y)$, where f is a local diffeomorphism $Y \rightarrow Y$ defined on a neighborhood of $y \in Y$, then the 1-jet at $\pi_k F^{-1}$ of the local diffeomorphism $Q_k(Y) \rightarrow Q_k(Y)$ sending G into $j_k(f)(\mathrm{target}\ G) \cdot G$ depends only on F and will be denoted by $\mu_1 F$. If ϕ is a section of $Q_{k+1}(Y)_X$, then the composition of 1-jets

$\mu_1\phi(x)^{-1}\cdot j_1(\pi_k\phi)(x)$ is a well-defined section of $J_1(E)$, where $x\in X$. Since $\pi_0(\mu_1\phi(x)^{-1}\cdot j_1(\pi_k\phi)(x))=I_{Y,k}(\rho(x))$, it follows that this section of $J_1(E)$ is in fact a section of the image of the map (3. 12). The proof of the formula

$$\mathscr{D}\phi=(\mathrm{id}\otimes\nu)\cdot\partial(\mu_1\phi^{-1}\cdot j_1(\pi_k\phi)),\qquad \text{for}\quad \phi\in Q_{k+1}(Y)_X \tag{3. 13}$$

follows the same lines as that of Proposition 5. 1, (ii) of [6].

Consider the mapping

$$\rho:\rho^{-1}J_1(Q_k(Y);Y)\to J_1(E)$$

sending $(x, j_1(F)(\rho(x)))$ into $j_1(F\circ\rho)(x)$, where $x\in X$ and F is a section of $Q_k(Y)$ over a neighborhood of $\rho(x)$. Recall that $J_1(Q_k(Y);Y)$ is an affine bundle over $Q_k(Y)$ whose associated vector bundle is $T_Y^*\otimes_{Q_k(Y)}V(Q_k(Y))$. Then, for $x\in X$, $H\in J_1(Q_k(Y);Y)_{\rho(x)}$ and $u\in T^*_{Y,\rho(x)}\otimes V_{\pi_0 H}(Q_k(Y))$, we have

$$\rho(x, H+u)=\rho(x, H)+(\rho^*\otimes\mathrm{id})u, \tag{3. 14}$$

where $\rho^*: T^*_{Y,\rho(x)}\to T^*_x$. The proof of this formula is essentially the same as that of Proposition 2. 1, (ii) of [6] and will be omitted.

For $k\geq 0$, let $Q^0_{(1,k)}(Y)$ be the bundle of 1-jets of invertible (local) sections of $Q_k(Y)$ over Y which project onto $I_{Y,k}$. Let

$$\partial: Q^0_{(1,k)}(Y)\to T_Y^*\otimes\tilde{J}_k(T_Y;Y)$$

be the mapping sending H into $H-j_1(I_{Y,k})(y)$, where $\pi_0 H=I_{Y,k}(y)$. If $x\in X$, $H\in Q^0_{(1,k)}$, with $\pi_0 H=I_{Y,k}(\rho(x))$, then $\rho(x, H)$ belongs to the image of (3. 12), and by (3. 14) we have

$$\partial\rho(x, H)=(\rho^*\otimes\mathrm{id})\partial H, \tag{3. 15}$$

where $\rho^*: T^*_{Y,\rho(x)}\to T^*_x$.

Let $Q^k_{k+1}(Y)$ be the bundle of the $G\in Q_{k+1}(Y)$ satisfying $\pi_k G=I_{Y,k}(y)$, where $y=$ source G. We have the mapping

$$\partial: Q^k_{k+1}(Y)\to S^{k+1}J_0(T_Y)^*\otimes J_0(T_Y)$$

given in § 2 of [6]. For $k\geq 1$, it is bijective; if $\phi\in Q^0_1(Y)$, then $\mathrm{id}+\partial\phi: J_0(T_Y)\to J_0(T_Y)$ is bijective. If

$$\lambda_1: Q^k_{k+1}(Y)\to Q^0_{(1,k)}(Y)$$

is the mapping sending $j_{k+1}(f)(y)$ into $j_1(j_k(f))(y)$, where f is a local diffeomorphism of Y defined on a neighborhood of $y\in Y$, then clearly

$$\lambda_1 F=\mu_1 F\cdot j_1(I_{Y,k})(y) \tag{3. 16}$$

for $F\in Q^k_{k+1}(Y)$, with source $F=y$. Thus, if $\phi\in Q^k_{k+1}(Y)_X$, by (3. 13), (3. 16) and (3. 15), we have for $x\in X$

$$\begin{aligned}(\mathscr{D}\phi)(x)&=(\mathrm{id}\otimes\nu)\partial(\mu_1\phi^{-1}(x)\cdot j_1(I_{Y,k}\circ\rho)(x))\\&=(\mathrm{id}\otimes\nu)\partial(\rho(x, \mu_1\phi^{-1}(x)\cdot j_1(I_{Y,k})(\rho(x))))\\&=(\mathrm{id}\otimes\nu)\partial(\rho(x, \lambda_1\phi^{-1}(x)))\\&=(\mathrm{id}\otimes\nu)(\rho^*\otimes\mathrm{id})\partial(\lambda_1\phi^{-1}(x)).\end{aligned}$$

From the commutativity of diagram (2. 22) of [6] and Proposition 2. 1, (iii) of [6], we deduce that for $k\geq 1$

$$\begin{aligned}(\mathcal{D}\phi)(x) &= (\rho^* \otimes \nu)\delta_Y\partial\phi^{-1}(x)\\ &= -(\rho^* \otimes \nu)\delta_Y\partial\phi(x)\\ &= -\delta\partial\phi(x),\end{aligned}$$

and that for $k=0$

$$\begin{aligned}(\mathcal{D}\phi)(x) &= (\rho^* \circ \nu^* \otimes \mathrm{id})\partial\phi^{-1}(x)\\ &= [(\mathrm{id}+\partial\phi(x))^{-1}-\mathrm{id}]\circ\nu\circ\rho_*,\end{aligned}$$

where $\rho_* : T_x \to T_{Y,\rho(x)}$. Thus if $\phi \in Q^k_{k+1}(Y)_X$, we have

(3. 17) $$\mathcal{D}\phi = -\delta\partial\phi, \qquad \text{if } k\geq 1,$$

(3. 18) $$\mathcal{D}\phi = [(\mathrm{id}+\partial\phi)^{-1}-\mathrm{id}]\circ\nu\circ\rho_*, \qquad \text{if } k\geq 0,$$

generalizing formulas (2. 29) and (2. 30) of [6].

Assume now that $\rho : X\to Y$ is everywhere of maximal rank. Let $\tilde{Q}_k(Y)_X$ be the sub-sheaf of $Q_k(Y)_X$ whose sections are local mappings $\phi : X\to Q_k(Y)$ such that source$\circ\phi=\rho$ and such that target$\circ\phi : X\to Y$ is everywhere of maximal rank.

If ρ is an immersion, then for $x \in X$ and $\psi \in \tilde{Q}_k(Y)_{\rho(x)}$, it is clear that $\psi\circ\rho$ belongs to $\tilde{Q}_k(Y)_{X,x}$; conversely, if $\phi \in \tilde{Q}_k(Y)_{X,x}$, it is easily seen that there is an element F of $\tilde{Q}_k(Y)_{\rho(x)}$ such that $\phi=F\circ\rho$.

If ρ is a surjective submersion, let $Q_k(\rho)$ denote the bundle of invertible jets of order k of ρ-projectable mappings $X\to X$ (i.e., which induce mappings $Y\to Y$) and let $\tilde{Q}_k(\rho)$ be the sub-sheaf of invertible sections of $Q_k(\rho)$. Let $J_k(T;\rho)$ denote the sub-bundle of $J_k(T)$ of k-jets of ρ-projectable vector fields on X. We have natural projections over $\rho : X\to Y$,

$$\begin{aligned}\rho &: Q_k(\rho) \to Q_k(Y),\\ \rho &: J_k(T;\rho) \to J_k(T_Y;Y).\end{aligned}$$

The map

$$\rho : Q_k(\rho) \to Q_k(Y)_X$$

is surjective and induces by restriction a surjective mapping

(3. 19) $$\rho : \tilde{Q}_k(\rho) \to \tilde{Q}_k(Y)_X.$$

Furthermore, the operator (3. 1) (with $X=Y$ and $\rho=\mathrm{id}$) gives us according to [6], § 6 an operator

$$\mathcal{D}_X : Q_{k+1}(\rho) \to \mathcal{T}^* \otimes J_k(\mathcal{T};\rho)$$

sending F into $F^*\omega_X$, where

$$\omega_X : T(Q_{k+1}(\rho)) \to J_k(T;\rho)$$

is the Cartan fundamental form on $Q_{k+1}(\rho)$. The diagram

(3. 20) $$\begin{array}{ccc} Q_{k+1}(\rho) & \xrightarrow{\mathcal{D}_X} & \mathcal{T}^*\otimes J_k(\mathcal{T};\rho)\\ \downarrow{\scriptstyle\rho} & & \downarrow{\scriptstyle \mathrm{id}\otimes\rho}\\ Q_{k+1}(Y)_X & \xrightarrow{\mathcal{D}} & \mathcal{T}^*\otimes J_k(\mathcal{T}_Y;Y)_X \end{array}$$

is commutative; indeed, if $F \in Q_{k+1}(\rho)$ and $\xi \in T$, by the commutativity of diagram

(6. 29) of [6], we have

$$\begin{aligned}\langle \xi, (\mathrm{id}\otimes\rho)\mathcal{D}_X F\rangle &= \rho\langle \xi, \mathcal{D}_X F\rangle = \rho\langle F_*\xi, \omega_X\rangle \\ &= \langle \rho_* F_*\xi, \omega_Y\rangle = \langle(\rho F)_*\xi, \omega_Y\rangle \\ &= \langle \xi, \mathcal{D}(\rho F)\rangle,\end{aligned}$$

where $\rho F\in Q_{k+1}(Y)_X$. The commutativity of the left-hand square of diagram (6. 39) of [6] is a consequence of that of diagram (3. 20).

Let $(T^*\otimes_X J_k(T_Y\,;\,Y))^\wedge$ be the sub-bundle of $T^*\otimes_X J_k(T_Y\,;\,Y)$ whose fiber at $x\in X$ consists of those elements u of $T^*_x\otimes J_k(T_Y\,;\,Y)_{\rho(x)}$ such that $\nu\circ\rho_*+\pi_0 u : T_x\to J_0(T_Y)_{\rho(x)}$ is of maximal rank, where $\nu : T_{Y,\rho(x)}\to J_0(T_Y)_{\rho(x)}$ is the canonical isomorphism. We denote by $(\mathcal{T}^*\otimes J_k(\mathcal{T}_Y\,;\,Y)_X)^\wedge$ its sheaf of sections. If ρ is an immersion or a surjective submersion, we have the sub-complex

$$(3.\ 21)\qquad \tilde{Q}_{k+1}(Y)_X\xrightarrow{\mathcal{D}}(\mathcal{T}^*\otimes J_k(\mathcal{T}_Y\,;\,Y)_X)^\wedge\xrightarrow{\mathcal{D}_1}\wedge^2\mathcal{T}^*\otimes J_{k-1}(\mathcal{T}_Y\,;\,Y)_X$$

of (3. 5). First, if ρ is an immersion and $\phi\in\tilde{Q}_{k+1}(Y)_{X,x}$, $x\in X$, we may write $\phi=F\circ\rho$, for some $F\in\tilde{Q}_{k+1}(Y)_{\rho(x)}$. By (3. 7), we have

$$\begin{aligned}\nu\circ\rho_*+\pi_0\mathcal{D}\phi &= \nu\circ\rho_*+\pi_0\mathcal{D}(F\circ\rho) = \nu\circ\rho_*+\pi_0\rho^*(\mathcal{D}_Y F) \\ &= \rho^*(\nu+\pi_0\mathcal{D}_Y F)\,;\end{aligned}$$

according to Proposition 5. 1, (ii) of [6] and [10], $(\nu+\pi_0\mathcal{D}_Y F)(\rho(x)) : T_{Y,\rho(x)}\to J_0(T_Y)_{\rho(x)}$ is invertible, and so $\mathcal{D}\phi\in(\mathcal{T}^*\otimes J_k(\mathcal{T}_Y\,;\,Y)_X)^\wedge$. Next, if ρ is a surjective submersion, let $(T^*\otimes J_k(T\,;\,\rho))^\wedge$ be the sub-bundle of $T^*\otimes J_k(T\,;\,\rho)$ whose fiber at $x\in X$ consists of those elements u of $(T^*\otimes J_k(T\,;\,\rho))_x$ such that $\nu+\pi_0 u$: $T_x\to J_0(T)_x$ is an isomorphism, where $\nu : T\to J_0(T)$ is the canonical isomorphism. As the map (3. 19) is surjective and

$$\mathcal{D}_X(\tilde{Q}_{k+1}(\rho))\subset(\mathcal{T}^*\otimes J_k(\mathcal{T}\,;\,\rho))^\wedge,$$

the diagram (3. 20) gives us by restriction the commutative diagram

$$(3.\ 22)\qquad \begin{array}{ccc}\tilde{Q}_{k+1}(\rho) & \xrightarrow{\mathcal{D}_X} & (\mathcal{T}^*\otimes J_k(\mathcal{T}\,;\,\rho))^\wedge \\ \downarrow{\rho} & & \downarrow{\mathrm{id}\otimes\rho} \\ \tilde{Q}_{k+1}(Y)_X & \xrightarrow{\mathcal{D}} & (\mathcal{T}^*\otimes J_k(\mathcal{T}_Y\,;\,Y)_X)^\wedge.\end{array}$$

The complexes (3. 5) and (3. 21) are finite forms of the linear complex

$$J_{k+1}(\mathcal{T}_Y\,;\,Y)_X\xrightarrow{D}\mathcal{T}^*\otimes J_k(\mathcal{T}_Y\,;\,Y)_X\xrightarrow{D}\wedge^2\mathcal{T}^*\otimes J_{k-1}(\mathcal{T}_Y\,;\,Y)_X$$

given by (1. 8). If $X=Y$ and ρ is the identity map of X, then (3. 21) is the complex (5. 5) of [10] or the complex (2. 23) of [6]. When ρ is a surjective submersion, further properties of $\mathcal{D}$ are given in § 6 of [6].

We assume throughout the remainder of this section that $\rho : X\to Y$ is an immersion. Let ψ be a section of $\tilde{Q}_{k+1}(Y)_X$ over X satisfying $\pi_0\psi=I_{Y,0}\circ\rho$ and let u be a section of $\wedge^i T^*\otimes_X J_k(T_Y\,;\,Y)$ over X. We define the section $\psi^{-1}(u)$ of $\wedge^i T^*\otimes_X J_k(T_Y\,;\,Y)$ over X by means of the formula

$$\langle\xi, (\psi^{-1}(u))(x)\rangle = \langle\xi, \psi(x)^{-1}\cdot u(x)\rangle\in J_k(T_Y\,;\,Y)_{\rho(x)},$$

where $x\in X$, $\xi\in\wedge^i T_x$ and $\psi(x)^{-1}$ is the automorphism of $J_k(T_Y\,;\,Y)_{\rho(x)}$ determined by the action of $Q_{k+1}(Y)$ on $J_k(T_Y\,;\,Y)$ as described in § 2 of [6]. If $i=1$, we set

(3. 23) $$u^{\psi} = \psi^{-1}(u) + \mathcal{D}\psi.$$

Next, let ϕ be a section of $\tilde{Q}_{k+1}(Y)_X$; we let $\phi\cdot\psi$ be the section of $\tilde{Q}_{k+1}(Y)_X$ defined by

$$(\phi\cdot\psi)(x) = \phi(x)\cdot\psi(x), \qquad \text{for} \quad x\in X.$$

This formula is meaningful because target $\psi(x)=$source $\psi(x)=\rho(x)$. Clearly $\pi_0(\phi\cdot\psi)=\pi_0\phi$.

We now list analogues of the formulas (21. 9)–(21. 13) of [8] for $\mathcal{D}_Y$. If $\phi, \psi\in \tilde{Q}_{k+1}(Y)_X$ and $\pi_0\psi=I_{Y,0}\circ\rho$, and if $u\in\bigwedge^i\mathcal{T}^*\otimes J_k(\mathcal{T}_Y\,;\,Y)_X$, then

(3. 24) $$\mathcal{D}(\phi\cdot\psi) = \psi^{-1}(\mathcal{D}\phi) + \mathcal{D}\psi,$$

(3. 25) $$D\psi^{-1}(u) - \psi^{-1}(Du) = -[\mathcal{D}\psi, u],$$

and, if $i=1$,

(3. 26) $$\mathcal{D}_1(u^{\psi}) = \psi^{-1}(\mathcal{D}_1 u);$$

moreover if $i=1$ and we also have $\pi_0\phi=I_{Y,0}\circ\rho$, then

(3. 27) $$u^{\phi\cdot\psi} = (u^{\phi})^{\psi}.$$

Since the analogous formulas hold for $\mathcal{D}_Y$ and we can write $\phi=F\circ\rho$, $\psi=G\circ\rho$, where $F, G\in\tilde{Q}_{k+1}(Y)_{\rho(x)}$ with $\pi_0 G=I_{Y,0}$, it follows from Proposition 3. 1 with $Z=Y$, $\tau=\rho$ and $\sigma=\mathrm{id}$ that the formulas for $\mathcal{D}_Y$ involving F, G are carried over into the corresponding formulas for $\mathcal{D}$ by means of ρ^*. The same method can be used to deduce formulas (3. 17) and (3. 18) from formulas (2. 29) and (2. 30) of [6] when ρ is an immersion.

Next, we define a twisted δ-operator analogous to one mentioned in § 7 of [6] (see the remark following Proposition 7. 4). Let v be a section of $T^*\otimes_X J_0(T_Y)$ over X; we then have the operator

$$\delta_v : \bigwedge^j T^*\otimes_X(S^k J_0(T_Y)^*\otimes J_0(T_Y)) \to \bigwedge^{j+1}T^*\otimes_X(S^{k-1}J_0(T_Y)^*\otimes J_0(T_Y))$$

defined by

$$\delta_v w = [v, w] = [v_1, w],$$

where $w\in\bigwedge^j T^*\otimes_X(S^k J_0(T_Y)^*\otimes J_0(T_Y))$ and v_1 is any section of $T^*\otimes_X J_k(T_Y\,;\,Y)$ over X such that $\pi_0 v_1=v$. Let $x\in X$ and $v^* : J_0(T_Y)^*_{\rho(x)}\to T^*_x$ be the mapping dual to $v : T_x\to J_0(T_Y)_{\rho(x)}$. Then

(3. 28) $$\delta_v(\omega\otimes u) = (-1)^j\omega\wedge(v^*\circ\nu^{*-1}\otimes\mathrm{id})\delta_Y u,$$

for $\omega\in\bigwedge^j T^*_x$, $u\in(S^k J_0(T_Y)^*\otimes J_0(T_Y))_{\rho(x)}$. Therefore if v is the section of $T^*\otimes_X J_0(T_Y)$ corresponding to $\nu\circ\rho_* : T\to J_0(T_Y)$, then $\delta_v=\delta$. It is easily seen that we obtain a complex

(3. 29) $$\begin{aligned}0\to\rho^{-1}(S^k J_0(T_Y)^*\otimes J_0(T_Y)) &\xrightarrow{\delta_v} T^*\otimes_X(S^{k-1}J_0(T_Y)^*\otimes J_0(T_Y))\xrightarrow{\delta_v}\cdots\\ &\to\bigwedge^n T^*\otimes_X(S^{k-n}J_0(T_Y)^*\otimes J_0(T_Y))\to 0,\end{aligned}$$

for $k>0$; if $v : T\to J_0(T_Y)$ is injective, it is exact at $\bigwedge^i T^*\otimes_X(S^{k-i}J_0(T_Y)^*\otimes J_0(T_Y))$ for $i>0$. Indeed, let $V\subset T_{Y|X}$ be the image of $\nu^{-1}\circ v$ and $i : V\to T_{Y|X}$ the inclusion map. If

$$\delta : \wedge^j V^* \otimes_X (S^k J_0(T_Y)^* \otimes J_0(T_Y)) \to \wedge^{j+1} V^* \otimes_X (S^{k-1} J_0(T_Y)^* \otimes J_0(T_Y))$$

is the mapping defined by

$$\delta(\omega \otimes u) = (-1)^j \omega \wedge (i^* \otimes \mathrm{id}) \delta_Y u,$$

and if $\nu^{-1} \circ v$ is the mapping $T \to V$, the diagram

$$\begin{array}{ccc} \wedge^j V^* \otimes_X (S^k J_0(T_Y)^* \otimes J_0(T_Y)) & \xrightarrow{\delta} & \wedge^{j+1} V^* \otimes_X (S^{k-1} J_0(T_Y)^* \otimes J_0(T_Y)) \\ \downarrow (\nu^{-1} \circ v)^* \otimes \mathrm{id} & & \downarrow (\nu^{-1} \circ v)^* \otimes \mathrm{id} \\ \wedge^j T^* \otimes_X (S^k J_0(T_Y)^* \otimes J_0(T_Y)) & \xrightarrow{\delta_v} & \wedge^{j+1} T^* \otimes_X (S^{k-1} J_0(T_Y)^* \otimes J_0(T_Y)) \end{array} \tag{3. 30}$$

is commutative. Thus if v is injective, the exactness of (3. 29) is equivalent to that of a sequence of the type of (1. 12).

Proposition 3. 2. *If ρ is an immersion, the sequence* (3. 21) *is exact, that is,*

$$\mathcal{D}(\tilde{\hat{Q}}_{k+1}(Y)_X) = \{u \in (\mathcal{T}^* \otimes J_k(\mathcal{T}_Y\,;\,Y)_X)^\wedge | \mathcal{D}_1 u = 0\}.$$

In fact, if $u \in (\mathcal{T}^* \otimes J_k(\mathcal{T}_Y\,;\,Y)_X)^\wedge$ satisfies $\mathcal{D}_1 u = 0$, then there is an element $\phi \in \tilde{\hat{Q}}_{k+1}(Y)_X$ such that $\pi_0 \phi = I_{Y,0} \circ \rho$ and $\mathcal{D}\phi = u$.

Proof. We proceed by induction on k. If $x \in X$, $u \in (\mathcal{T}^* \otimes J_0(\mathcal{T}_Y)_X)^\wedge_x$, we may write

$$u + \nu \circ \rho_* = v \circ \rho_*$$

for some $v \in (\mathcal{T}_Y^* \otimes J_0(\mathcal{T}_Y))_{\rho(x)}$ such that $v(x) : T_{Y,\rho(x)} \to J_0(T_Y)_{\rho(x)}$ is invertible. Let $w \in (J_0(\mathcal{T}_Y)^* \otimes J_0(\mathcal{T}_Y))_{\rho(x)}$ be the element such that

$$w^{-1} \circ \nu = v.$$

Let F be the unique element of $Q_1^0(Y)_{\rho(x)}$ satisfying

$$\partial F = w - \mathrm{id}.$$

If $\phi = F \circ \rho$, then according to (3. 18),

$$\mathcal{D}\phi = (v \circ \nu^{-1} - \mathrm{id}) \circ \nu \circ \rho_* = v \circ \rho_* - \nu \circ \rho_* = u.$$

Now suppose that $u \in (\mathcal{T}^* \otimes J_k(\mathcal{T}_Y\,;\,Y)_X)^\wedge$, with $k \geq 1$ and $\mathcal{D}_1 u = 0$, and assume that there exists $\phi_1 \in \tilde{\hat{Q}}_k(Y)_X$ such that $\mathcal{D}\phi_1 = \pi_{k-1} u$ and $\pi_0 \phi_1 = I_{Y,0} \circ \rho$. Let $\phi \in \tilde{\hat{Q}}_{k+1}(Y)_X$ with $\pi_k \phi = \phi_1$; then, since $\pi_{k-1} \mathcal{D}\phi = \pi_{k-1} u$, we see that $\mathcal{D}\phi - u$ belongs to $\mathcal{T}^* \otimes (S^k J_0(\mathcal{T}_Y)^* \otimes J_0(\mathcal{T}_Y))_X$. Writing $v = \nu \circ \rho_* + \pi_0 u$, we have

$$\begin{aligned} \delta_v(\mathcal{D}\phi - u) &= -\delta(u - \mathcal{D}\phi) - [u, u - \mathcal{D}\phi] \\ &= D(u - \mathcal{D}\phi) - [u, u - \mathcal{D}\phi] \\ &= \mathcal{D}_1(u - \mathcal{D}\phi) - [u, u - \mathcal{D}\phi], \end{aligned}$$

since $[u - \mathcal{D}\phi, u - \mathcal{D}\phi] = 0$, because $\pi_{k-1}(u - \mathcal{D}\phi) = 0$ with $k \geq 1$. Thus

$$\begin{aligned} \delta_v(\mathcal{D}\phi - u) &= Du - \frac{1}{2}[u, u] - D\mathcal{D}\phi - \frac{1}{2}[\mathcal{D}\phi, \mathcal{D}\phi] + [u, \mathcal{D}\phi] - [u, u - \mathcal{D}\phi] \\ &= -[\mathcal{D}\phi, \mathcal{D}\phi] + [u, \mathcal{D}\phi] - [u, u - \mathcal{D}\phi], \end{aligned}$$

since $\mathcal{D}_1 u = 0$ and $\mathcal{D}_1 \mathcal{D}\phi = 0$. We obtain

$$\delta_v(\mathcal{D}\phi - u) = -[u - \mathcal{D}\phi, u - \mathcal{D}\phi] = 0.$$

As u belongs to $(\mathcal{T}^* \otimes J_k(\mathcal{T}_Y\,;\,Y)_X)^\wedge$, v comes from an injective section of

$T^*\otimes_X J_0(T_Y)$. By the exactness of (3. 29) in positive degrees, there exists $h\in(S^{k+1}J_0(\mathcal{T}_Y)^*\otimes J_0(\mathcal{T}_Y))_X$ such that $\mathcal{D}\phi-u=\delta_v h$. Let $\psi=\partial^{-1}h\in\mathcal{Q}^k_{k+1}(Y)_X$; we have by (3. 24) and (3. 17)

$$\begin{aligned}\mathcal{D}(\phi\circ\psi) &= \psi^{-1}(\mathcal{D}\phi)+\mathcal{D}\psi = \psi^{-1}(u+\delta_v h)-\delta h \\ &= \psi^{-1}(u+\delta h+[u_1, h])-\delta h,\end{aligned} \tag{3. 31}$$

where u_1 is an element of $\mathcal{T}^*\otimes J_{k+1}(\mathcal{T}_Y\,;\,Y)_X$ satisfying $\pi_k u_1=u$. Since $\psi^{-1}\in\mathcal{Q}^k_{k+1}(Y)_X$, with $k\geq 1$, and $\psi^{-1}=-\partial^{-1}h$, by formula (2. 20) of [6] and (3. 28), we see that

$$\psi^{-1}(w) = w-(\delta_Y h)\circ\nu^{-1}\circ\pi_0 w = w-[w_1, h]$$

for $w\in\mathcal{T}^*\otimes J_k(\mathcal{T}_Y\,;\,Y)_X$ and $w_1\in\mathcal{T}^*\otimes J_{k+1}(\mathcal{T}_Y\,;\,Y)_X$ with $\pi_k w_1=w$. Thus $\psi^{-1}(\delta h)=\delta h$, $\psi^{-1}([u_1, h])=[u_1, h]$ and

$$\psi^{-1}(u) = u-[u_1, h].$$

Substituting these expressions into (3. 31), we conclude that $\mathcal{D}(\phi\cdot\psi)=u$.

If $R_k\subset J_k(T_Y\,;\,Y)$ is a formally integrable Lie equation (with $k\geq 1$) and $P_k\subset Q_k(Y)$ is a formally integrable finite form of R_k whose l-th prolongation we denote by P_{k+l}, the restriction of the fundamental form ω_Y to P_{k+l+1} is a mapping

$$\omega_Y : T(P_{k+l+1})\rightarrow R_{k+l}.$$

Thus we obtain sub-complexes

$$\mathcal{P}_{k+l+1,X}\xrightarrow{\mathcal{D}}\mathcal{T}^*\otimes\mathcal{R}_{k+l,X}\xrightarrow{\mathcal{D}_1}\wedge^2\mathcal{T}^*\otimes J_{k+l-1}(\mathcal{T}_Y\,;\,Y)_X \tag{3. 32}$$

of (3. 5) and

$$\tilde{\mathcal{P}}_{+kl+1,X}\xrightarrow{\mathcal{D}}(\mathcal{T}^*\otimes\mathcal{R}_{k+l,X})^\wedge\xrightarrow{\mathcal{D}_1}\wedge^2\mathcal{T}^*\otimes J_{k+l-1}(\mathcal{T}_Y\,;\,Y)_X \tag{3. 33}$$

of (3. 21), where

$$\begin{aligned}\tilde{\mathcal{P}}_{k+l+1,X} &= \mathcal{P}_{k+l+1,X}\cap\tilde{\mathcal{Q}}_{k+l+1}(Y)_X, \\ (\mathcal{T}^*\otimes\mathcal{R}_{k+l,X})^\wedge &= (\mathcal{T}^*\otimes\mathcal{R}_{k+l,X})\cap(\mathcal{T}^*\otimes J_{k+l}(\mathcal{T}_Y\,;\,Y)_X)^\wedge.\end{aligned}$$

The complexes (3. 32) and (3. 33) are finite forms of the linear complex

$$\mathcal{R}_{k+l+1,X}\xrightarrow{D}\mathcal{T}^*\otimes\mathcal{R}_{k+l,X}\xrightarrow{D}\wedge^2\mathcal{T}^*\otimes\mathcal{R}_{k+l-1,X}$$

given by (2. 1). If $X=Y$ and ρ is the identity map of X, then (3. 33) is the complex considered in § 7 of [6].

If R_k is the Lie equation corresponding to a Lie pseudogroup acting on Y and u is a section of $T^*\otimes_X R_k$ over X the equation

$$\mathcal{D}\phi = u, \qquad \text{for}\quad \phi\in\mathcal{P}_{k+1,X}$$

was discussed by É. Cartan [1] in connection with the deformation problem of submanifolds.

We shall consider g_{k+l} as a sub-bundle of $S^{k+l}J_0(T_Y)^*\otimes J_0(T_Y)$. Let $v\in T^*_x\otimes J_0(T_Y)_{\rho(x)}$, $x\in X$; we say that v is non-characteristic for g_{k+1} if

$$(S^{k+1}N'_x\otimes J_0(T_Y)_{\rho(x)})\cap g_{k+1,\rho(x)} = 0,$$

where $N'_x\subset J_0(T_Y)^*_{\rho(x)}$ is the kernel of the dual mapping of $\nu\circ\rho_*+v : T_x\rightarrow J_0(T_Y)_{\rho(x)}$.

If X is non-characteristic for g_{k+1}, then all $v \in T^*_x \otimes J_0(T_Y)_{\rho(x)}$ belonging to a neighborhood of 0 are non-characteristic for g_{k+1}. If v is a section of $(T^* \otimes_X J_0(T_Y))^\wedge$ over a neighborhood of $x_0 \in X$ and $v(x_0)$ is non-characteristic for g_{k+1}, it follows that $v(x)$ is non-characteristic for g_{k+1} for all x belonging to a neighborhood of x. If R_k is elliptic, X is a hypersurface of Y, then every element of $(T^* \otimes_X J_0(T_Y))^\wedge$ is non-characteristic for g_{k+1}.

Let w be a section of $T^* \otimes_X J_0(T_Y)$ over X; then by (3. 28)

$$\delta_w(g_{k+l}) \subset T^* \otimes_X g_{k+l-1}$$

for all $l \geq 1$ and thus we obtain a complex

(3. 34)

$$0 \to g_{k+l+1|X} \xrightarrow{\delta_w} T^* \otimes_X g_{k+l} \xrightarrow{\delta_w} \wedge^2 T^* \otimes_X g_{k+l-1} \xrightarrow{\delta_w} \cdots \to \wedge^n T^* \otimes_X g_{k+l+1-n} \to 0.$$

By Theorem 2. 1, (iii), if g_{k+1} is involutive and $w(x) - \nu \circ \rho_*$ is non-characteristic for g_{k+1}, then the sequence (3. 34) is exact at $\wedge^j T^*_x \otimes g_{k+l,\rho(x)}$ for all $j \geq 0$, $l \geq 1$, since diagram (3. 30) commutes (with v replaced by w).

Proposition 3. 3. *Let $R_k \subset J_k(T_Y ; Y)$ be a formally integrable Lie equation and $x \in X$. Suppose that g_{k+1} is involutive and that $u \in (\mathcal{T}^* \otimes \mathcal{R}_{k+l,X})^\wedge_x$ satisfies $\mathcal{D}_1 u = 0$, with $l \geq 1$. If $\pi_0 u(x)$ is non-characteristic for g_{k+1}, then there exists $u_1 \in (\mathcal{T}^* \otimes \mathcal{R}_{k+l+1,X})^\wedge_x$ such that $\pi_{k+l} u_1 = u$ and $\mathcal{D}_1 u_1 = 0$.*

Proof. Choose $u' \in \mathcal{T}^* \otimes \mathcal{R}_{k+l+1,X}$ such that $\pi_{k+l} u' = u$. Then $\mathcal{D}_1 u' \in \wedge^2 \mathcal{T}^* \otimes g_{k+l,X}$ and, writing $w = \nu \circ \rho_* + \pi_0 u$, we have

$$\begin{aligned}
\delta_w(\mathcal{D}_1 u') &= -D\Big(Du' - \frac{1}{2}[u', u']\Big) + [u, \mathcal{D}_1 u'] \\
&= \frac{1}{2} D[u', u'] + [u, Du'] - \frac{1}{2}[u, [u', u']] \\
&= [Du', u] + [u, Du'] = 0
\end{aligned}$$

by (3. 3) and (3. 4). Since g_{k+1} is involutive and $\pi_0 u(x)$ is non-characteristic for g_{k+1}, the sequence (3. 34) is exact at $\wedge^2 T^* \otimes_X g_{k+l}$ over a neighborhood of x. Therefore, there is an element $u'' \in \mathcal{T}^* \otimes g_{k+l+1,X}$ satisfying $\delta_w u'' = \mathcal{D}_1 u'$. Then

$$\begin{aligned}
\mathcal{D}_1(u' + u'') &= Du' - \delta u'' - \frac{1}{2}[u', u'] - [u', u''] \\
&= \mathcal{D}_1 u' - \delta_w u'' = 0
\end{aligned}$$

and $\pi_{k+l}(u' + u'') = u$.

Proposition 3. .4 *Let $R_k \subset J_k(T_Y ; Y)$ be a formally integrable Lie equation and $P_k \subset Q_k(Y)$ a formally integrable finite form of R_k; let $x \in X$. Suppose that g_{k+1} is involutive and that $u \in (\mathcal{T}^* \otimes \mathcal{R}_{k+l+1,X})^\wedge_x$ satisfies $\mathcal{D}_1 u = 0$, with $l \geq 0$. If $\pi_0 u(x)$ is non-characteristic for g_{k+1} and there exists $\phi \in \tilde{\mathcal{P}}_{k+l+1,X,x}$ satisfying $\mathcal{D}\phi = \pi_{k+l} u$, then there is an element ϕ_1 of $\tilde{\mathcal{P}}_{k+l+2,X,x}$ such that $\pi_{k+l+1}\phi_1 = \phi$ and $\mathcal{D}\phi_1 = u$.*

Proof. Choose $\phi' \in \tilde{\mathcal{P}}_{k+l+2,X,x}$ with $\pi_{k+l+1}\phi' = \phi$. Then $\mathcal{D}\phi' - u \in \mathcal{T}^* \otimes g_{k+l+1,X,x}$ and by the proof of Proposition 3. 2, $\delta_w(\mathcal{D}\phi' - u) = 0$, where $w = \nu \circ \rho_* + \pi_0 u$. As (3. 34) is exact at $T^*\otimes_X g_{k+l+1}$ over a neighborhood of x, there exists $v\in g_{k+l+2,X}$ such that $\delta_w v=\mathcal{D}\phi'-u$. Then the element $\psi=\partial^{-1}v$ of $Q_{k+l+2}^{k+l+1}(Y)_X$ belongs to $\tilde{\mathcal{P}}_{k+l+2}$ (see [6], § 7). By the proof of Proposition 3. 2, $\mathcal{D}(\phi'\cdot\psi)=u$ and $\pi_{k+l+1}(\phi'\cdot\psi)=\phi$.

If $x\in X$ and U is an open subset of $T_x^*\otimes J_0(T_Y)_{\rho(x)}$, set

$$Z_U^1(R_{k+l})_x=\{u\in(\mathcal{T}^*\otimes\mathcal{R}_{k+l,X})_x^\wedge|\mathcal{D}_1u=0,\ \pi_0u(x)\in U\},$$

for $l\geq 0$.

From the remarks concerning non-characteristic elements of $T^*\otimes_X J_0(T_Y)$ for g_{k+1} and Propositions 3. 3 and 3. 4, we deduce :

Corollary 3. 1. *Let $R_k\subset J_k(T_Y\,;Y)$ be a formally integrable Lie equation and $P_k\subset Q_k(Y)$ a formally integrable finite form of R_k. Suppose that g_{k+1} is involutive and X is non-characteristic. Let $x\in X$; then there exists a neighborhood U of $0\in T_x^*\otimes J_0(T_Y)_{\rho(x)}$ such that the mappings*

$$\pi_{k+l}: Z_U^1(R_{k+l+1})_x\rightarrow Z_U^1(R_{k+l})_x$$

are surjective for all $l\geq 1$, and such that the following conditions are equivalent :

(i) *for all $l\geq 0$, we have*

$$Z_U^1(R_{k+l})_x\subset\mathcal{D}(\tilde{\mathcal{P}}_{k+l+1,X})\,;$$

(ii) *for some $l\geq 0$, we have*

$$Z_U^1(R_{k+l})_x\subset\mathcal{D}(\tilde{\mathcal{P}}_{k+l+1,X}).$$

If R_k is elliptic, X is a hypersurface of Y, then we may take $U=(T^*\otimes_X J_0(T_Y))_x^\wedge$.

4. Pseudo-complex Structures ; Formulas of Kuranishi

The purpose of this section is to redefine and reformulate the formulas of Kuranishi [9] in such a way that they can be extended to pseudo-complex structures of arbitrary codimension. We begin by recalling the definitions of an almost pseudo-complex structure and its formal integrability.

Definition 4. 1. Let Z be a manifold of dimension $2n-k$, $0\leq k\leq n-1$. An almost pseudo-complex structure of type k on Z is defined by a complex sub-bundle E'' of rank $n-k$ (over **C**) of the complexified tangent bundle $\mathbf{C}T_Z$ of Z such that E'' and its complex conjugate E' have a zero intersection. An almost pseudo-complex structure of type k on Z defined by E'' is said to be formally integrable if the space of sections of E'' is stable under the Lie bracket.

We shall sometimes call E'' the almost pseudo-complex structure (of type k), omitting for simplicity the phrase "defined by".

Definition 4. 2. An almost pseudo-complex structure E'' of type k on Z is a pseudo-complex structure of type k if and only if it is induced by an imbedding of

Z into a complex analytic manifold W with $\dim_{\mathbf{C}} W=n$, in which case E'' is the intersection of $\mathbf{C}T_Z$ with the restriction to Z of the bundle T''_W of complex vectors of type (0, 1) tangent to W.

Formal integrability is an obvious necessary condition in order that an almost pseudo-complex structure be pseudo-complex. We remark that a pseudo-complex structure of type 0 is a complex analytic structure. The restriction on an almost pseudo-complex structure that $E' \cap E''=0$ will not be used throughout the remainder of this section.

The formal integrability of an almost pseudo-complex structure E'' (of type k) on Z is equivalent to the existence of a canonically defined differential operator

(4. 1) $$\bar{\partial}_{E''} : \wedge^p \mathcal{E}''^* \to \wedge^{p+1} \mathcal{E}''^*$$

for all p, $0 \leq p \leq n-k$, satisfying $\bar{\partial}^2_{E''}=0$. In fact, let I be the graded ideal of $\wedge \mathbf{C} T^*_Z$ generated by the annihilator of E''. Formal integrability of E'' is equivalent to the stability of $\mathcal{I}$ under the usual exterior differential $d : \wedge \mathbf{C}\mathcal{T}^*_Z \to \wedge \mathbf{C}\mathcal{T}^*_Z$; if E'' is formally integrable, d yields, by passage to the quotient by $\mathcal{I}$, the operator (4. 1). We observe that, for $p=0$, the operator (4. 1) is just d followed by restriction from $\mathbf{C}\mathcal{T}^*_Z$ to $\mathcal{E}''$, and therefore exists in this case even if the structure is not formally integrable.

Assume now that X is a submanifold of Y of codimension k, that Y is a complex analytic manifold and that $\rho : X \to Y$ is the inclusion map. We denote by $\mathcal{O}_X$ the sheaf of complex-valued functions on X. We write

$$T' = T'_{Y|X} \cap \mathbf{C}T, \qquad T'' = T''_{Y|X} \cap \mathbf{C}T,$$

where T'_Y and T''_Y are the sub-bundles of $\mathbf{C}T_Y$ of tangent vectors of types (1, 0) and (0, 1) respectively, and $\mathbf{C}T$ is the complexified tangent bundle of X. Suppose that T' and T'' are sub-bundles of $\mathbf{C}T$; if X is a hypersurface of Y, this is always true. Then T'' is a pseudo-complex structure of type k on X. Set

$$\bar{\partial}_b = \bar{\partial}_{T''} : \wedge \mathcal{T}''^* \to \wedge \mathcal{T}''^*,$$

where $\bar{\partial}_{T''}$ is the operator (4. 1) for the pseudo-complex structure on X defined by $E''=T''$. Since T'_Y is a holomorphic bundle over Y, the operator $\bar{\partial}_b$ gives rise to an operator

(4. 2) $$\bar{\partial}_b : \wedge \mathcal{T}''^* \otimes \mathcal{T}'_{Y|X} \to \wedge \mathcal{T}''^* \otimes \mathcal{T}'_{Y|X},$$

where $\mathcal{T}'_{Y|X}$ is the sheaf of sections of $T'_{Y|X}$ over X, satisfying

(4. 3) $$\bar{\partial}_b(\alpha \wedge \theta) = (\bar{\partial}_b \alpha) \wedge \theta + (-1)^i \alpha \wedge \bar{\partial}_b \theta,$$

for all $\alpha \in \wedge^i \mathcal{T}''^*$, $\theta \in \wedge \mathcal{T}''^* \otimes \mathcal{T}'_{Y|X}$. Let

$$\varpi' : \mathbf{C}T_Y \to T'_Y, \qquad \varpi'' : \mathbf{C}T_Y \to T''_Y$$

be the projections ; then if $\xi \in T''_x$, $x \in X$ and θ is a section of $T'_{Y|X}$ over a neighborhood of x,

(4. 4) $$\langle \xi, \bar{\partial}_b \theta \rangle = \varpi'[\xi, \tilde{\theta}](x),$$

where $\tilde{\xi}$, $\tilde{\theta}$ are sections of T''_Y and T'_Y respectively over a neighborhood of x in Y satisfying $\tilde{\xi}(x)=\xi$ and $\tilde{\theta}_{|X}=\theta$. Relations (4. 3) and (4. 4) determine the operator (4. 2) uniquely.

We choose a sub-bundle H of $\mathbf{C}T$ containing T' such that

$$(4.5) \qquad \mathbf{C}T = H \oplus T''.$$

Since the sub-bundle $T' \oplus T''$ of $\mathbf{C}T$ is invariant under conjugation, there exists a real sub-bundle F of T of rank k such that

$$(4.6) \qquad H = T' \oplus \mathbf{C}F,$$

where $\mathbf{C}F$ is the complexification of F, and

$$\mathbf{C}T = T' \oplus T'' \oplus \mathbf{C}F.$$

The subsequent considerations depend only on the choice of H and not on the choice of a real sub-bundle satisfying (4. 6). From (4. 5), we deduce that

$$H \cap T''_{Y|X} = 0;$$

hence $\varpi'_{|H}$ is injective and, by counting dimensions, we see that it is an isomorphism. Denote its inverse by

$$\tau : T'_{Y|X} \to H.$$

Clearly τ is the identity on T'. Let

$$\pi_H : \mathbf{C}T \to H, \qquad \pi'' : \mathbf{C}T \to T''$$

be the projections corresponding to the decomposition (4. 5). Then

$$(4.7) \qquad \varpi'_{|\mathbf{C}T} = \varpi' \circ \pi_H = \tau^{-1} \circ \pi_H.$$

We now derive other formulas for the operator (4. 2). If $\theta \in \mathcal{T}'_{Y|X}$, then

$$(4.8) \qquad \langle \xi, \bar{\partial}_b \theta \rangle = \varpi'[\xi, \tau\theta]$$

for $\xi \in \mathcal{T}''$; indeed, if ξ, θ are restrictions to X of elements $\tilde{\xi} \in \mathcal{T}''_Y$ and $\tilde{\theta} \in \mathcal{T}'_Y$ respectively, there is an extension $\widetilde{\tau\theta} \in \mathbf{C}\mathcal{T}_Y$ of $\tau\theta$ satisfying $\varpi'\widetilde{\tau\theta} = \tilde{\theta}$, since $\varpi'\tau\theta = \theta$. We have by (4. 4)

$$\begin{aligned} \varpi'[\xi, \tau\theta] &= \varpi'[\tilde{\xi}, \widetilde{\tau\theta}]_{|X} = \varpi'[\tilde{\xi}, \tilde{\theta}]_{|X} - \varpi'[\tilde{\xi}, \tilde{\theta} - \widetilde{\tau\theta}]_{|X} \\ &= \varpi'[\tilde{\xi}, \tilde{\theta}]_{|X} = \langle \xi, \bar{\partial}_b \theta \rangle, \end{aligned}$$

as $\tilde{\theta} - \widetilde{\tau\theta} \in \mathcal{T}''_Y$. From (4. 3) and (4. 8), we deduce

$$\langle \xi \wedge \eta, \bar{\partial}_b \theta \rangle = \langle \xi, \bar{\partial}_b(\theta(\eta)) \rangle - \langle \eta, \bar{\partial}_b(\theta(\xi)) \rangle - \langle [\xi, \eta], \theta \rangle$$

and

$$(4.9) \qquad \langle \xi \wedge \eta, \bar{\partial}_b \theta \rangle = \varpi'([\xi, \tau\theta(\eta)] - [\eta, \tau\theta(\xi)]) - \langle [\xi, \eta], \theta \rangle,$$

for $\xi, \eta \in \mathcal{T}''$, $\theta \in \mathcal{T}''^* \otimes \mathcal{T}'_{Y|X}$.

Definition 4. 3. A (complex) differential form θ of degree p on X will be said to be of type $(0, p)_b$ if $i(u)\theta = u \barwedge \theta = 0$ for all $u \in H$.

The restriction map $\mathbf{C}T^* \to T''^*$ induces an isomorphism of the sheaf of differential forms of type $(0, p)_b$ with $\wedge^p \mathcal{T}''^*$. We shall identify these two sheaves.

Definition 4. 4. An almost pseudo-complex structure E'' of type k on X is said to be at finite distance from T'' if

$$\pi''_{|E''} : E'' \to T''$$

is an isomorphism.

If an almost pseudo-complex structure E'' of type k is at finite distance from T'', then

$$\varphi = -\varpi' \circ (\pi''_{|E''})^{-1} = -\tau^{-1} \circ \pi_H \circ (\pi''_{|E''})^{-1} : T'' \to T'_{Y|X}$$

is the unique $T'_{Y|X}$-valued differential form of type $(0, 1)_b$ such that

$$E'' = \{\xi - \tau\varphi(\xi) | \xi \in T''\} \tag{4. 10}$$

(see [9]). The sub-bundle E'' defined by (4. 10) will frequently be denoted by T''_φ. Conversely, given any sufficiently small differential form φ of type $(0, 1)_b$ with values in $T'_{Y|X}$, the sub-bundle E'' defined by (4. 10) is an almost pseudo-complex structure of type k at finite distance from T''.

Let $E'' = T''_\varphi$ be the sub-bundle (4. 10) determined by a differential form φ of type $(0, 1)_b$ with values in $T'_{Y|X}$. We have the decomposition

$$\mathbf{C}T = H \oplus E''. \tag{4. 11}$$

Let

$$\pi_{H,\varphi} : \mathbf{C}T \to H, \qquad \pi''_\varphi : \mathbf{C}T \to E''$$

be the projections corresponding to the decomposition (4. 11). Then

$$\pi_{H,\varphi} = \pi_H + \tau \cdot \varphi \cdot \pi'', \qquad \pi''_\varphi = \pi'' - \tau \cdot \varphi \cdot \pi''$$

and $\lambda_\varphi = \pi''_{\varphi|T''} : T'' \to E''$ is an isomorphism. Following [9], define

$$\bar\partial_{E''} : \wedge \mathcal{E}''^* \to \wedge \mathcal{E}''^*$$

by the formula

$$\bar\partial_{E''} = \gamma \cdot d \cdot \pi''^*_\varphi, \tag{4. 12}$$

where $\gamma : \wedge \mathbf{C}T^* \to \wedge E''^*$ is the restriction map. Since $\gamma \cdot \pi''^*_\varphi$ is the identity map of $\wedge E''^*$, the operator (4. 12) coincides with the canonical operator (4. 1), whenever E'' is a formally integrable almost pseudo-complex structure of type k. In fact, it is not difficult to see that (4. 12) has vanishing square if and only if $\mathcal{E}''$ is stable under the Lie bracket.

Define

$$\bar\partial_{b,\varphi} : \wedge^p \mathcal{T}''^* \to \wedge^{p+1} \mathcal{T}''^*$$

by

$$\bar\partial_{b,\varphi} = \lambda^*_\varphi \cdot \bar\partial_{E''} \cdot \lambda^{-1*}_\varphi.$$

Then

$$\bar\partial_{b,\varphi} = \lambda^*_\varphi \cdot \gamma \cdot d \cdot \pi''^*.$$

If $\zeta \in \wedge^{p+1}\mathcal{T}''$, $\beta \in \wedge^{p}\mathcal{T}''^{*}$, then

$$(4.13)\qquad \langle \zeta, \bar{\partial}_{b,\varphi}\beta\rangle = \langle \lambda_{\varphi}\zeta, d\beta\rangle,$$

where λ_{φ} is induced by $\mathrm{id}-\tau\varphi: T''\to \mathbf{C}T$. In particular, we have

$$(4.14)\qquad \langle \zeta, \bar{\partial}_{b,\varphi}f\rangle = (\zeta-\tau\varphi(\zeta))\cdot f,$$

for $\zeta \in \mathcal{T}''$, $f \in \mathcal{O}_X$. We have

$$(4.15)\qquad \bar{\partial}_{b,\varphi}(\beta_1\wedge\beta_2) = (\bar{\partial}_{b,\varphi}\beta_1)\wedge\beta_2 + (-1)^{p}\beta_1\wedge(\bar{\partial}_{b,\varphi}\beta_2),$$

for $\beta_1 \in \wedge^{p}\mathcal{T}''^{*}$, $\beta_2 \in \wedge\mathcal{T}''^{*}$, and $\bar{\partial}_{b,0}=\bar{\partial}_b$. Clearly $\mathcal{E}''$ is stable under the Lie bracket if and only if

$$\bar{\partial}^{2}_{b,\varphi} = 0.$$

We now define an operator

$$(4.16)\qquad \bar{\partial}_{b,\varphi}: \wedge^{p}\mathcal{T}''^{*}\otimes\mathcal{T}'_{Y|X} \to \wedge^{p+1}\mathcal{T}''^{*}\otimes\mathcal{T}'_{Y|X}$$

satisfying

$$(4.17)\qquad \bar{\partial}_{b,\varphi}(\alpha\wedge\theta) = (\bar{\partial}_{b,\varphi}\alpha)\wedge\theta + (-1)^{i}\alpha\wedge\bar{\partial}_{b,\varphi}\theta$$

for all $\alpha \in \wedge^{i}\mathcal{T}''^{*}$, $\theta \in \wedge\mathcal{T}''^{*}\otimes\mathcal{T}'_{Y|X}$. First, if $\theta \in \mathcal{T}'_{Y|X}$, set

$$(4.18)\qquad \langle \xi, \bar{\partial}_{b,\varphi}\theta\rangle = (\varpi'+\varphi)[\lambda_{\varphi}\xi, \tau\theta],$$

for $\xi \in \mathcal{T}''$; since $(\varpi'+\varphi)\circ\lambda_{\varphi}=0$, for $f \in \mathcal{O}_X$ we have

$$\langle f\xi, \bar{\partial}_{b,\varphi}\theta\rangle = f\langle \xi, \bar{\partial}_{b,\varphi}\theta\rangle,$$

and so the operator (4. 16) with $p=0$ given by (4. 18) is well-defined; by (4. 14), it satisfies (4. 17) for all $\alpha \in \mathcal{O}_X$, $\theta \in \mathcal{T}'_{Y|X}$. Using this last relation, it is easily seen that the operator (4. 16) given by means of (4. 17) with $\alpha \in \wedge^{p}\mathcal{T}''^{*}$, $\theta \in \mathcal{T}'_{Y|X}$ is well-defined and satisfies (4. 17) for all $\alpha \in \wedge^{i}\mathcal{T}''^{*}$, $\theta \in \wedge\mathcal{T}''^{*}\otimes\mathcal{T}'_{Y|X}$. From (4. 3) and (4. 8), we see that, if $\varphi=0$, the operator (4. 16) is equal to (4. 2). In the case $k=1$, the operator (4. 16) coincides with the operator $\bar{\partial}^{\xi}_{b}$ defined by Kuranishi [9] in terms of local coordinates and a connection.

Next, we define the required brackets. We recall that we have the Nijenhuis bracket

$$(\wedge\mathbf{C}\mathcal{T}^{*}\otimes\mathbf{C}\mathcal{T})\otimes(\wedge\mathbf{C}\mathcal{T}^{*}\otimes\mathbf{C}\mathcal{T}) \to \wedge\mathbf{C}\mathcal{T}^{*}\otimes\mathbf{C}\mathcal{T}$$

which assigns to $\wedge\mathbf{C}\mathcal{T}^{*}\otimes\mathbf{C}\mathcal{T}$ a structure of sheaf of graded Lie algebras. If $u \in \wedge^{p}\mathbf{C}\mathcal{T}^{*}\otimes\mathbf{C}\mathcal{T}$, we define the Lie derivative $\mathcal{L}(u)$ on $\wedge\mathbf{C}\mathcal{T}^{*}$ by the formula

$$\mathcal{L}(u) = [i(u), d] = i(u)\cdot d - (-1)^{p-1}d\cdot i(u);$$

if $u=\alpha\otimes\xi$ and $\beta \in \wedge\mathbf{C}\mathcal{T}^{*}$, then

$$\mathcal{L}(\alpha\otimes\xi)\beta = \alpha\wedge\mathcal{L}(\xi)\beta + (-1)^{p}d\alpha\wedge i(\xi)\beta.$$

For $u=\alpha\otimes\xi \in \wedge^{p}\mathbf{C}\mathcal{T}^{*}\otimes\mathbf{C}\mathcal{T}$, $v=\beta\otimes\eta \in \wedge^{q}\mathbf{C}\mathcal{T}^{*}\otimes\mathbf{C}\mathcal{T}$, we define the Nijenhuis bracket $[u, v]$ by

$$[\alpha\otimes\xi, \beta\otimes\eta] = (\alpha\wedge\beta)\otimes[\xi, \eta] + \mathcal{L}(\alpha\otimes\xi)\beta\otimes\eta - (-1)^{pq}\mathcal{L}(\beta\otimes\eta)\alpha\otimes\xi,$$

and extend this definition to arbitrary u, v by bilinearity. If $u \in \mathbf{C}\mathcal{T}^{*}\otimes\mathbf{C}\mathcal{T}$ and $\xi, \eta \in$

$\mathbf{C}\mathcal{T}$, we have

$$\langle \xi, \mathcal{L}(u)f\rangle = u(\xi)\cdot f, \quad \text{for} \quad f \in \mathcal{O}_X, \tag{4.19}$$

and

$$\begin{aligned}\langle \xi\wedge\eta, \mathcal{L}(u)\alpha\rangle &= u(\xi)\cdot\langle\eta,\alpha\rangle - u(\eta)\cdot\langle\xi,\alpha\rangle \\ &\quad -\langle[\xi,u(\eta)],\alpha\rangle + \langle[\eta,u(\xi)],\alpha\rangle - \langle u[\xi,\eta],\alpha\rangle,\end{aligned} \tag{4.20}$$

for $\alpha\in\mathbf{C}\mathcal{T}^*$. If $u, v\in\mathbf{C}\mathcal{T}^*\otimes\mathbf{C}\mathcal{T}$, then

$$\begin{aligned}\langle \zeta_1\wedge\zeta_2, [u,v]\rangle &= [u(\zeta_1), v(\zeta_2)] - [u(\zeta_2), v(\zeta_1)] \\ &\quad - v[u(\zeta_1),\zeta_2] + v[u(\zeta_2),\zeta_1] \\ &\quad - u[v(\zeta_1),\zeta_2] + u[v(\zeta_2),\zeta_1] \\ &\quad + v(u([\zeta_1,\zeta_2])) + u(v([\zeta_1,\zeta_2])),\end{aligned} \tag{4.21}$$

for $\zeta_1, \zeta_2\in\mathbf{C}\mathcal{T}$.

By means of the isomorphism

$$\tau^{-1}\oplus \mathrm{id} : H\oplus T'' = \mathbf{C}T \to T'_{Y|X}\oplus T'',$$

the Nijenhuis bracket can be transported into a bracket on $\wedge\mathbf{C}\mathcal{T}^*\otimes(\mathcal{T}'_{Y|X}\oplus\mathcal{T}'')$. Since we identify T''^* with a sub-bundle of $\mathbf{C}T^*$, we may restrict this bracket to the sub-sheaf $\wedge\mathcal{T}''^*\otimes(\mathcal{T}'_{Y|X}\oplus\mathcal{T}'')$ and regard the values of this restricted bracket as differential forms on X with values in $T'_{Y|X}\oplus T''$; if we restrict these forms to $\wedge T''$, we obtain a bracket on $\wedge\mathcal{T}''^*\otimes(\mathcal{T}'_{Y|X}\oplus\mathcal{T}'')$. If $\psi_1\in\wedge^p\mathcal{T}''^*\otimes(\mathcal{T}'_{Y|X}\oplus\mathcal{T}'')$, $\psi_2\in\wedge^q\mathcal{T}''^*\otimes(\mathcal{T}'_{Y|X}\oplus\mathcal{T}'')$, set

$$\bar{\psi}_i = [\mathrm{id}\otimes(\tau\oplus\mathrm{id})]\psi_i, \quad i = 1, 2;$$

the bracket $⟦\psi_1,\psi_2⟧\in\wedge^{p+q}\mathcal{T}''^*\otimes(\mathcal{T}'_{Y|X}\oplus\mathcal{T}'')$ is determined by

$$\langle\xi, ⟦\psi_1,\psi_2⟧\rangle = (\tau^{-1}\oplus\mathrm{id})\langle\xi, [\bar{\psi}_1,\bar{\psi}_2]\rangle,$$

for $\xi\in\wedge^{p+q}\mathcal{T}''$. We write

$$⟦\psi_1,\psi_2⟧ = [\psi_1,\psi_2] + [\![\psi_1,\psi_2]\!],$$

where $[\psi_1,\psi_2]\in\wedge^{p+q}\mathcal{T}''^*\otimes\mathcal{T}'_{Y|X}$ and $[\![\psi_1,\psi_2]\!]\in\wedge^{p+q}\mathcal{T}''^*\otimes\mathcal{T}''$. Thus we obtain brackets

$$(\wedge\mathcal{T}''^*\otimes\mathcal{T}'_{Y|X})\times_X(\wedge\mathcal{T}''^*\otimes\mathcal{T}'_{Y|X}) \to \wedge\mathcal{T}''^*\otimes\mathcal{T}'_{Y|X}, \tag{4.22}$$

$$(\wedge\mathcal{T}''^*\otimes\mathcal{T}'_{Y|X})\times_X(\wedge\mathcal{T}''^*\otimes\mathcal{T}'_{Y|X}) \to \wedge\mathcal{T}''^*\otimes\mathcal{T}'' \tag{4.23}$$

sending (ψ_1,ψ_2) into $[\psi_1,\psi_2]$, $[\![\psi_1,\psi_2]\!]$ respectively.

As we have identified $\wedge\mathcal{T}''^*$ with a sub-sheaf of $\wedge\mathbf{C}\mathcal{T}^*$, we may define an operation of $\wedge\mathcal{T}''^*\otimes\mathcal{T}'_{Y|X}$ on $\wedge\mathcal{T}''^*$ as follows. For $\varphi\in\wedge^p\mathcal{T}''^*\otimes\mathcal{T}'_{Y|X}$, $\beta\in\wedge^q\mathcal{T}''^*$, we define $\mathcal{L}(\varphi)\cdot\beta$ to be the restriction to $\wedge^{p+q}\mathcal{T}''^*$ of $\mathcal{L}(\tau\circ\varphi)\beta\in\wedge^{p+q}\mathbf{C}\mathcal{T}^*$. If $\varphi=\alpha\otimes\xi$, then, since $i(\tau\xi)\beta=0$, we have

$$\mathcal{L}(\varphi)\cdot\beta = \alpha\wedge\mathcal{L}(\xi)\cdot\beta,$$

where $\mathcal{L}(\xi)\cdot\beta$ is the restriction of $i(\tau\xi)d\beta$ to $\wedge^q\mathcal{T}''^*$.

If $\varphi=\alpha\otimes\xi\in\wedge^p\mathcal{T}''^*\otimes\mathcal{T}'_{Y|X}$, $\psi=\beta\otimes\eta\in\wedge^q\mathcal{T}''^*\otimes\mathcal{T}'_{Y|X}$, we have by (4.7)

$$\begin{aligned}[\varphi,\psi] &= (\alpha\wedge\beta)\otimes\varpi'[\tau\xi,\tau\eta] + \alpha\wedge\mathcal{L}(\xi)\beta\otimes\eta - (-1)^{pq}\beta\wedge\mathcal{L}(\eta)\alpha\otimes\xi, \\ [\![\varphi,\psi]\!] &= (\alpha\wedge\beta)\otimes\pi''[\tau\xi,\tau\eta].\end{aligned}$$

Thus, in particular, the bracket (4. 23) is $\mathcal{O}_X$-bilinear. Moreover, if $\varphi \in \bigwedge^p \mathcal{T}''^* \otimes \mathcal{T}'_{Y|X}$, $\beta \otimes \eta \in \bigwedge^q \mathcal{T}''^* \otimes \mathcal{T}'_{Y|X}$, we have

(4. 24) $$[\varphi, \beta \otimes \eta] = (\mathcal{L}(\varphi)\beta) \otimes \eta + (-1)^{pq} \beta \wedge [\varphi, \eta],$$

and if $\psi \in \bigwedge \mathcal{T}''^* \otimes \mathcal{T}'_{Y|X}$, $\alpha \in \bigwedge \mathcal{T}''^*$,

(4. 25) $$[\![\alpha \wedge \varphi, \psi]\!] = \alpha \wedge [\![\varphi, \psi]\!].$$

From (4. 19) and (4. 20) we deduce that, if $\varphi \in \mathcal{T}''^* \otimes \mathcal{T}'_{Y|X}$,

(4. 26) $$\langle \zeta, \mathcal{L}(\varphi) f \rangle = (\tau\varphi(\zeta)) \cdot f,$$

for $\zeta \in \mathcal{T}''$, $f \in \mathcal{O}_X$ and

(4. 27) $$\begin{aligned} \langle \zeta_1 \wedge \zeta_2, \mathcal{L}(\varphi)\beta \rangle &= (\tau\varphi(\zeta_1)) \cdot \langle \zeta_2, \beta \rangle - (\tau\varphi(\zeta_2)) \cdot \langle \zeta_1, \beta \rangle \\ &\quad - \langle [\zeta_1, \tau\varphi(\zeta_2)], \beta \rangle + \langle [\zeta_2, \tau\varphi(\zeta_1)], \beta \rangle, \end{aligned}$$

for $\zeta_1, \zeta_2 \in \mathcal{T}''$, $\beta \in \mathcal{T}''^*$. Furthermore, if $\varphi \in \mathcal{T}''^* \otimes \mathcal{T}'_{Y|X}$, $\psi \in \mathcal{T}'_{Y|X}$,

(4. 28) $$\langle \zeta, [\varphi, \psi] \rangle = \varpi'[\tau\varphi(\zeta), \tau\psi] - \varphi[\zeta, \tau\psi],$$

and

(4. 29) $$\langle \zeta, [\![\varphi, \psi]\!] \rangle = \pi''[\tau\varphi(\zeta), \tau\psi],$$

for $\zeta \in \mathcal{T}''$. From (4. 21) it follows that, if $\varphi, \psi \in \mathcal{T}''^* \otimes \mathcal{T}'_{Y|X}$,

(4. 30) $$\begin{aligned} \langle \zeta_1 \wedge \zeta_2, [\varphi, \psi] \rangle &= \varpi'[\tau\varphi(\zeta_1), \tau\psi(\zeta_2)] - \varpi'[\tau\varphi(\zeta_2), \tau\psi(\zeta_1)] \\ &\quad - \psi[\tau\varphi(\zeta_1), \zeta_2] + \psi[\tau\varphi(\zeta_2), \zeta_1] \\ &\quad - \varphi[\tau\psi(\zeta_1), \zeta_2] + \varphi[\tau\psi(\zeta_2), \zeta_1], \end{aligned}$$

(4. 31) $$\langle \zeta_1 \wedge \zeta_2, [\![\varphi, \psi]\!] \rangle = \pi''[\tau\varphi(\zeta_1), \tau\psi(\zeta_2)] - \pi''[\tau\varphi(\zeta_2), \tau\psi(\zeta_1)],$$

for $\zeta_1, \zeta_2 \in \mathcal{T}''$.

If $u = \alpha \otimes \xi \in \bigwedge T''^* \otimes T''$, $\beta \in \bigwedge T''^*$, we set

$$u \,\bar{\wedge}\, \beta = \alpha \wedge i(\xi)\beta,$$

and if $v = \beta \otimes \eta \in \bigwedge T''^* \otimes T'_{Y|X}$, we set

$$u \,\bar{\wedge}\, v = (\alpha \wedge i(\xi)\beta) \otimes \eta;$$

in particular if $v \in T''^* \otimes T'_{Y|X}$, then $u \,\bar{\wedge}\, v = v \circ u$.

Proposition 4. 1. *Let φ be a differential form of type $(0, 1)_b$ with values in $T'_{Y|X}$ and let $E'' = T''_\varphi$ be the sub-bundle (4. 10). Then $\mathcal{E}''$ is stable under the Lie bracket if and only if the section $P(\varphi)$ of $\bigwedge^2 \mathcal{T}''^* \otimes \mathcal{T}'_{Y|X}$ given by*

(4. 32) $$P(\varphi) = \bar{\partial}_b \varphi - \frac{1}{2}[\varphi, \varphi] - \frac{1}{2}[\![\varphi, \varphi]\!] \,\bar{\wedge}\, \varphi$$

vanishes. For $\beta \in \bigwedge \mathcal{T}''^*$, we have

(4. 33) $$\bar{\partial}_{b,\varphi} \beta = \bar{\partial}_b \beta - \mathcal{L}(\varphi)\beta - \frac{1}{2}[\![\varphi, \varphi]\!] \,\bar{\wedge}\, \beta$$

and

(4. 34) $$\bar{\partial}^2_{b,\varphi} \beta = -\mathcal{L}(P(\varphi))\beta - [\![P(\varphi), \varphi]\!] \,\bar{\wedge}\, \beta.$$

For $\theta \in \bigwedge \mathcal{T}''^* \otimes \mathcal{T}'_{Y|X}$, we have

(4. 35) $$\bar{\partial}_{b,\varphi}\theta = \bar{\partial}_b\theta - [\varphi, \theta] - [\![\varphi, \theta]\!] \barwedge \varphi - \frac{1}{2}[\![\varphi, \varphi]\!] \barwedge \theta$$

and

(4. 36)
$$\bar{\partial}^2_{b,\varphi}\theta = -[P(\varphi), \theta] - [\![P(\varphi), \varphi]\!] \barwedge \theta - [\![\theta, \varphi]\!] \barwedge P(\varphi) + [\![\theta, P(\varphi)]\!] \barwedge \varphi.$$

Moreover, for $\theta \in \mathcal{T}''^* \otimes \mathcal{T}'_{Y|X}$,

(4. 37) $$\frac{d}{dt}P(\varphi + t\theta)_{|t=0} = \bar{\partial}_{b,\varphi}\theta,$$

(4. 38) $$\bar{\partial}_{b,\varphi}P(\varphi) = 0.$$

Proof. The first assertion of the proposition is an immediate consequence of the formula

(4. 39) $$\varpi' \cdot \pi_{H,\varphi}[\xi - \tau\varphi(\xi), \eta - \tau\varphi(\eta)] = -\langle \xi \wedge \eta, P(\varphi) \rangle,$$

where $\xi, \eta \in \mathcal{T}''$. We now verify (4. 39) ; first note that by (4. 7)

$$\varpi' \cdot \pi_{H,\varphi} = \varpi' \cdot (\pi_H + \tau\varphi \cdot \pi'') = \varpi' + \varphi.$$

Therefore, if $\xi, \eta \in \mathcal{T}''$,

$$\begin{aligned}
\varpi' \cdot \pi_{H,\varphi}&[\xi - \tau\varphi(\xi), \eta - \tau\varphi(\eta)] \\
&= (\varpi' + \varphi)([\xi, \eta] - [\xi, \tau\varphi(\eta)] - [\tau\varphi(\xi), \eta] + [\tau\varphi(\xi), \tau\varphi(\eta)]) \\
&= -\varpi'([\xi, \tau\varphi(\eta)] + [\tau\varphi(\xi), \eta]) + \varphi[\xi, \eta] + \varpi'[\tau\varphi(\xi), \tau\varphi(\eta)] \\
&\quad - \varphi[\xi, \tau\varphi(\eta)] - \varphi[\tau\varphi(\xi), \eta] + \varphi[\tau\varphi(\xi), \tau\varphi(\eta)] \\
&= -\langle \xi \wedge \eta, P(\varphi) \rangle,
\end{aligned}$$

by (4. 9), (4. 30), (4. 31) and (4. 32). Since $\bar{\partial}_{b,\varphi}$ and the operator given by (4. 33) both satisfy (4. 15), to prove (4. 33) it suffices to verify it for $\beta \in \mathcal{O}_X$ and $\beta \in \mathcal{T}''^*$. From (4. 14) and (4. 26), it follows that

$$\langle \zeta, \bar{\partial}_{b,\varphi} f \rangle = \zeta \cdot f - \tau\varphi(\zeta) \cdot f = \langle \zeta, \bar{\partial}_b f \rangle - \langle \zeta, \mathcal{L}(\varphi) \cdot f \rangle,$$

for $\zeta \in \mathcal{T}''$, $f \in \mathcal{O}_X$. Next, if $\beta \in \mathcal{T}''^*$, $\zeta_1, \zeta_2 \in \mathcal{T}''$, we have by (4. 13), (4. 27) and (4. 31)

$$\begin{aligned}
\langle \zeta_1 \wedge \zeta_2, \bar{\partial}_{b,\varphi}\beta \rangle &= \langle (\zeta_1 - \tau\varphi(\zeta_1)) \wedge (\zeta_2 - \tau\varphi(\zeta_2)), d\beta \rangle \\
&= \langle \zeta_1 \wedge \zeta_2, d\beta \rangle - \langle \zeta_1 \wedge \tau\varphi(\zeta_2), d\beta \rangle \\
&\quad - \langle \tau\varphi(\zeta_1) \wedge \zeta_2, d\beta \rangle + \langle \tau\varphi(\zeta_1) \wedge \tau\varphi(\zeta_2), d\beta \rangle \\
&= \langle \zeta_1 \wedge \zeta_2, \bar{\partial}_b\beta \rangle + \tau\varphi(\zeta_2) \cdot \langle \zeta_1, \beta \rangle + \langle [\zeta_1, \tau\varphi(\zeta_2)], \beta \rangle \\
&\quad - \tau\varphi(\zeta_1) \cdot \langle \zeta_2, \beta \rangle - \langle [\zeta_2, \tau\varphi(\zeta_1)], \beta \rangle - \langle [\tau\varphi(\zeta_1), \tau\varphi(\zeta_2)], \beta \rangle \\
&= \langle \zeta_1 \wedge \zeta_2, \bar{\partial}_b\beta \rangle - \langle \zeta_1 \wedge \zeta_2, \mathcal{L}(\varphi)\beta \rangle - \beta\left(\left\langle \zeta_1 \wedge \zeta_2, \frac{1}{2}[\![\varphi, \varphi]\!] \right\rangle\right),
\end{aligned}$$

from which we deduce (4. 33).

Now, if $\theta \in \mathcal{T}'_{Y|X}$, $\zeta \in \mathcal{T}''$, we have by (4. 18), (4. 8), (4. 28) and (4. 29),

$$\begin{aligned}
\langle \zeta, \bar{\partial}_{b,\varphi}\theta \rangle &= (\varpi' + \varphi)[\zeta - \tau\varphi(\zeta), \tau\theta] \\
&= \varpi'[\zeta, \tau\theta] + \varphi[\zeta, \tau\theta] - \varpi'[\tau\varphi(\zeta), \tau\theta] - \varphi[\tau\varphi(\zeta), \tau\theta] \\
&= \langle \zeta, \bar{\partial}_b\theta \rangle - \langle \zeta, [\varphi, \theta] \rangle - \varphi\langle \zeta, [\![\varphi, \theta]\!] \rangle,
\end{aligned}$$

which implies (4. 35) for $\theta \in \mathcal{T}'_{Y|X}$. Next, if $\alpha \otimes \theta \in \wedge^p \mathcal{T}''^* \otimes \mathcal{T}'_{Y|X}$, we obtain from (4. 17)

$$\begin{aligned}\bar\partial_{b,\varphi}(\alpha\otimes\theta) &= (\bar\partial_{b,\varphi}\alpha)\otimes\theta+(-1)^p\alpha\wedge\bar\partial_{b,\varphi}\theta\\ &= \left(\bar\partial_b\alpha-\mathcal{L}(\varphi)\alpha-\frac{1}{2}[\![\varphi,\varphi]\!]\barwedge\alpha\right)\otimes\theta\\ &\quad +(-1)^p\alpha\wedge(\bar\partial_b\theta-[\varphi,\theta]-[\![\varphi,\theta]\!]\barwedge\varphi)\\ &= (\bar\partial_b\alpha)\otimes\theta+(-1)^p\alpha\wedge\bar\partial_b\theta-\mathcal{L}(\varphi)\alpha\otimes\theta\\ &\quad -(-1)^p\alpha\wedge[\varphi,\theta]-\frac{1}{2}[\![\varphi,\varphi]\!]\barwedge(\alpha\otimes\theta)-(-1)^p\alpha\wedge[\![\varphi,\theta]\!]\barwedge\varphi\\ &= \bar\partial_b(\alpha\otimes\theta)-[\varphi,\alpha\otimes\theta]-\frac{1}{2}[\![\varphi,\varphi]\!]\barwedge(\alpha\otimes\theta)-[\![\varphi,\alpha\otimes\theta]\!]\barwedge\varphi,\end{aligned}$$

by (4. 33), (4. 3), (4. 24) and (4. 25), so (4. 35) holds for all $\theta\in\wedge\mathcal{T}''^*\otimes\mathcal{T}'_{Y|X}$.

Relation (4. 37) follows directly from (4. 32) and (4. 35). We omit the proofs of (4. 34), (4. 36) and (4. 38) ; these formulas can be verified by the methods of the proposition and are given (in local coordinates) by Propositions 4. 4, 4. 7 and 4. 5 of [9] in the case that X is of codimension 1 in Y.

If $X=Y$ and ρ is the identity map of X, then $H=T'$, τ is the identity map of T' and $\bar\partial_b=\bar\partial$. Thus the bracket (4. 22) is the Nijenhuis bracket on $\wedge\mathcal{T}''^*\otimes\mathcal{T}'$ and the bracket (4. 23) vanishes. In this case, Proposition 3. 1 gives us the formulas, for $\varphi\in\mathcal{T}''^*\otimes\mathcal{T}'$,

$$P(\varphi)=\bar\partial\varphi-\frac{1}{2}[\varphi,\varphi],$$

$$\begin{aligned}\bar\partial_\varphi\beta &= \bar\partial\beta-\mathcal{L}(\varphi)\beta, &\quad \bar\partial_\varphi^2\beta &= -\mathcal{L}(P(\varphi))\beta, &\quad \beta&\in\wedge\mathcal{T}''^*,\\ \bar\partial_\varphi\theta &= \bar\partial\theta-[\varphi,\theta], & \bar\partial_\varphi^2\theta &= -[P(\varphi),\theta], & \theta&\in\wedge\mathcal{T}''^*\otimes\mathcal{T}',\end{aligned}$$

if we set $\bar\partial_\varphi=\bar\partial_{b,\varphi}$.

5. Pseudo-complex Structures and the Non-linear Complex

In this section, we describe the relationship between the complex (3. 33) of § 3, in which R_k is the Lie equation for holomorphic vector fields on a complex manifold Y, and pseudo-complex structures.

Assume that Y is a complex analytic manifold. We identify T'^*_Y and T''^*_Y with sub-bundles of $\mathbf{C}T^*_Y$; we write

$$J_0(T'_Y)=J_0(T'_Y\,;Y),\qquad J_0(T''_Y)=J_0(T''_Y\,;Y)$$

and we thus have

$$J_0(\mathbf{C}T_Y)^*=J_0(T'_Y)^*\oplus J_0(T''_Y)^*.$$

Since T'_Y is a holomorphic vector bundle, we have a first-order differential operator

$$\bar\partial:\mathcal{T}'_Y\to\mathcal{T}''^*_Y\otimes\mathcal{T}'_Y$$

whose solutions are the holomorphic vector fields on Y. The first-order differential equation $R_1\subset J_1(T'_Y\,;Y)$ corresponding to $\bar\partial$ is a formally integrable Lie equation. If $\tilde R_1=\nu^{-1}(R_1)$, then

$$[\tilde{\mathcal{R}}_1, J_0(\mathcal{T}''_Y)]\subset J_0(\mathcal{T}''_Y)\tag{5. 1}$$

and $\pi_0 : R_1 \to J_0(T'_Y)$ is surjective ; the kernel $g_1 \subset J_0(\mathbf{C}T_Y)^* \otimes J_0(T'_Y)$ of this map is equal to $J_0(T'_Y)^* \otimes J_0(T'_Y)$. It is easily seen that

$$\delta_Y(\mathbf{C}T_Y^* \otimes g_1) = \{u \in \wedge^2 \mathbf{C}T_Y^* \otimes J_0(T'_Y) | u_{|\wedge^2 T''_Y} = 0\}.$$

Let P_k be the bundle of k-jets of local biholomorphic mappings $Y \to Y$; then P_1 is a formally integrable finite form of R_1 whose k-th prolongation is P_{k+1}. An element $F \in P_1$, with source $F = a$ and target $F = b$, determines mappings

$$J_0(\mathbf{C}T_Y)_a \to J_0(\mathbf{C}T)_b, \qquad J_0(T''_Y)_a \to J_0(T''_Y)_b.$$

According to § 3, we obtain the complex

$$\mathcal{P}_2 \xrightarrow{\mathcal{D}_Y} \mathbf{C}\mathcal{T}_Y^* \otimes \mathcal{R}_1 \xrightarrow{\mathcal{D}_{1,Y}} \wedge^2 \mathbf{C}\mathcal{T}_Y^* \otimes J_0(\mathcal{T}'_Y), \tag{5.2}$$

since $\mathcal{T}'_Y$ is stable under the Lie bracket ; if $\tilde{\mathcal{P}}_k$ is the sub-sheaf of $\mathcal{P}_k$ of sections F for which $\pi_0 F$ is a local diffeomorphism of Y, and if $(\mathbf{C}T_Y^* \otimes R_1)^\wedge$ is the sub-bundle of $\mathbf{C}T_Y^* \otimes R_1$ consisting of those elements $u \in \mathbf{C}T_Y^* \otimes R_1$ for which $\mathrm{id} + \nu^{-1} \circ \pi_0 u : \mathbf{C}T_Y \to \mathbf{C}T_Y$ is invertible, we have the sub-complex

$$\tilde{\mathcal{P}}_2 \xrightarrow{\mathcal{D}_Y} (\mathbf{C}\mathcal{T}_Y^* \otimes \mathcal{R}_1)^\wedge \xrightarrow{\mathcal{D}_{1,Y}} \wedge^2 \mathbf{C}\mathcal{T}_Y^* \otimes J_0(\mathcal{T}'_Y) \tag{5.3}$$

of (5. 2).

Assume that X is a submanifold of Y of codimension k and that $\rho : X \to Y$ is the inclusion map. We suppose, as in § 4, that $T' = T'_{Y|X} \cap \mathbf{C}T$ and $T'' = T''_{Y|X} \cap \mathbf{C}T$ are sub-bundles of $\mathbf{C}T$. It is easily seen that

$$\delta(\mathbf{C}T^* \otimes_X g_1) = \{u \in \wedge^2 \mathbf{C}T^* \otimes_X J_0(T'_Y) | u_{|\wedge^2 T''} = 0\} ;$$

more generally, if v is a section of $\mathbf{C}T^* \otimes_X J_0(\mathbf{C}T_Y)$ over X and

$$E'' = \{\xi \in \mathbf{C}T | \nu^{-1} \circ v(\xi) \in T''_Y\},$$

then

$$\delta_v(\mathbf{C}T^* \otimes_X g_1) = \{u \in \wedge^2 \mathbf{C}T^* \otimes_X J_0(T'_Y) | u_{|\wedge^2 E''} = 0\}. \tag{5.4}$$

Let $\tilde{\mathcal{P}}_{k,X}$ be the sub-sheaf of $\mathcal{P}_{k,X}$ of sections ϕ for which $\pi_0 \phi$ is a local immersion of X into Y and let $(\mathbf{C}T^* \otimes_X R_1)^\wedge$ be the sub-bundle of $\mathbf{C}T^* \otimes_X R_1$ whose fiber at $x \in X$ consists of those elements $u \in \mathbf{C}T_x^* \otimes R_{1,x}$ for which $\mathrm{id} + \nu^{-1} \circ \pi_0 u : \mathbf{C}T_x \to \mathbf{C}T_{Y,x}$ is injective ; we denote by $(\mathbf{C}\mathcal{T}^* \otimes \mathcal{R}_{1,X})^\wedge$ the sheaf of sections of $(\mathbf{C}T^* \otimes_X R_1)^\wedge$. From (3. 21) and (5. 3), we obtain the complex

$$\tilde{\mathcal{P}}_{2,X} \xrightarrow{\mathcal{D}} (\mathbf{C}\mathcal{T}^* \otimes \mathcal{R}_{1,X})^\wedge \xrightarrow{\mathcal{D}_1} \wedge^2 \mathbf{C}\mathcal{T}^* \otimes J_0(\mathcal{T}'_Y)_X. \tag{5.5}$$

As in § 4, we choose a sub-bundle H of $\mathbf{C}T$ containing T' such that (4. 5) holds and continue to identify T''^* with a sub-bundle of $\mathbf{C}T^*$.

Theorem 5. 1. (i) *Let u be a section of $(\mathbf{C}T^* \otimes_X R_1)^\wedge$ over X; set $u_0 = \pi_0 u$, $\tilde{u}_0 = \nu^{-1} \circ u_0$. Suppose that*

$$E'' = \{\xi \in \mathbf{C}T | (\mathrm{id} + \tilde{u}_0)\xi \in T''_{Y|X}\} \tag{5.6}$$

is a sub-bundle of $\mathbf{C}T$. Then $\mathcal{E}''$ is stable under the Lie bracket if and only if $\mathcal{D}_1 u_{|\wedge^2 E''} = 0$.

(ii) *If $\tilde{u}_0$ is a section of $T''^* \otimes T'_{Y|X}$ over X, then the subset E'' of $\mathbf{C}T$ given by* (5. 6) *is* *a sub-bundle and is given by* (4. 10) *with* $\varphi=\tilde{u}_0$.

(iii) *If* u *is a section of* $(\mathbf{C}T^*\otimes_X R_1)^\wedge$ *over* X *satisfying* $\mathcal{D}_1 u=0$ *and if the subset* E'' *of* $\mathbf{C}T$ *given by* (5. 6) *is a sub-bundle, then* $\mathcal{E}''$ *is stable under the Lie bracket. Moreover, if* $u=\mathcal{D}\phi$ *for some section* ϕ *of* $\tilde{\mathcal{P}}_{2,X}$ *and* $f: X\to Y$ *is the immersion* target$\circ\phi$, *then*

$$f_*(E''_x)=f_*(\mathbf{C}T_x)\cap T''_{Y,f(x)}, \quad \text{for} \quad x\in X. \tag{5. 7}$$

(iv) *Let* u_0 *be a section of* $\mathbf{C}T^*\otimes_X J_0(T'_Y)$ *over* X. *Assume that the subset* E'' *of* $\mathbf{C}T$ *given by* (5. 6), *with* $\tilde{u}_0=\nu^{-1}\circ u_0$, *is a sub-bundle and that* $\mathrm{id}+\tilde{u}_0: \mathbf{C}T\to\mathbf{C}T_{Y|X}$ *is injective. Then if* $\mathcal{E}''$ *is stable under the Lie bracket, there exists a section* u *of* $(\mathbf{C}T^*\otimes_X R_1)^\wedge$ *over* X *satisfying* $\mathcal{D}_1u=0$ *and* $\pi_0 u=u_0$.

Proof. (i) Let $x\in X$ and let v be a section of $(\mathbf{C}T^*_Y\otimes R_1)^\wedge$ over a neighborhood of x in Y such that $\rho^*v=u$. We set $v_0=\pi_0 v$ and $\tilde{v}_0=\nu^{-1}\circ v_0$. If $\xi,\eta\in\mathcal{E}''_x$, there exist elements ξ_1,η_1 of $\mathbf{C}\mathcal{T}_{Y,x}$ whose restrictions to X are equal to ξ,η respectively such that $(\mathrm{id}+\tilde{v}_0)\xi_1$, $(\mathrm{id}+\tilde{v}_0)\eta_1$ belong to $\mathcal{T}''_{Y,x}$. Then by (3. 10) and Lemma 3. 1,

$$\begin{aligned}\langle\xi\wedge\eta,\mathcal{D}_1u\rangle &= \langle\xi_1\wedge\eta_1,\mathcal{D}_{1,Y}v\rangle_{|X}\\ &= -(\nu+u_0)[\xi,\eta]+\nu[(\mathrm{id}+\tilde{v}_0)\xi_1,(\mathrm{id}+\tilde{v}_0)\eta_1]_{|X}\\ &\quad -\mathcal{L}(\nu^{-1}v(\xi_1))((\nu+v_0)\eta_1)_{|X}+\mathcal{L}(\nu^{-1}v(\eta_1))((\nu+v_0)\xi_1)_{|X}.\end{aligned}$$

Since $\mathcal{T}''_Y$ is stable under the Lie bracket and (5. 1) holds, the last three terms of the right-hand member of the above equation belong to $J_0(\mathcal{T}''_Y)_{X,x}$. As $\mathcal{D}_1u\in\wedge^2\mathbf{C}\mathcal{T}^*\otimes J_0(\mathcal{T}'_Y)_X$, for $\xi,\eta\in\mathcal{E}''$, we see therefore that $\langle\xi\wedge\eta,\mathcal{D}_1u\rangle=0$ if and only if

$$(\mathrm{id}+\tilde{u}_0)\ [\xi,\eta]\in\mathcal{T}''_{Y|X},$$

that is, if and only if $[\xi,\eta]\in\mathcal{E}''$.

(ii) For $\xi\in T''$, we have

$$(\mathrm{id}+\tilde{u}_0)(\xi-\tau\tilde{u}_0(\xi))=\xi-\tau\tilde{u}_0(\xi)+\tilde{u}_0(\xi)\ ;$$

since $\varpi'\tau\tilde{u}_0(\xi)=\tilde{u}_0(\xi)$, the right-hand member of the above equality belongs to T''_Y and therefore $\xi-\tau\tilde{u}_0(\xi)\in E''$. We thus obtain an injective mapping $\lambda_{u_0}: T''\to E''$ sending ξ into $\xi-\tau\tilde{u}_0(\xi)$. Clearly $\pi''\circ\lambda_{u_0}$ is the identity map of T'' and so λ_{u_0} is an isomorphism ; therefore E'' is given by (4. 10) with $\varphi=\tilde{u}_0$.

(iii) The first assertion follows directly from (i). If ϕ is a section of $\tilde{\mathcal{P}}_{1,X}$ over X with $f=$target$\circ\phi$ and $u_0=\mathcal{D}\phi$, then by (3. 11), for $x\in X$, we have

$$(\nu+u_0)(x)=\phi(x)^{-1}\circ\nu\circ f_* \tag{5. 8}$$

as mappings $\mathbf{C}T_x\to\mathbf{C}T_{Y,x}$. Since $\phi(x)\in P_{1,x}$, it determines an isomorphism of $J_0(T'_Y)_x$ onto $J_0(T'_Y)_{f(x)}$ and it follows from (5. 8) that $\xi\in\mathbf{C}T_x$ belongs to E''_x if and only if $f_*\xi\in T''_{Y,f(x)}$.

(iv) Let u_1 be a section of $\mathbf{C}T^*\otimes_X R_1$ such that $\pi_0u_1=u_0$. Then by our hypotheses and (i), u_1 is a section of $(\mathbf{C}T^*\otimes_X R_1)^\wedge$ and $\mathcal{D}_1u_{1|\wedge^2E''}=0$. Therefore by (5. 4), there exists a section v of $\mathbf{C}T^*\otimes_X g_1$ over X such that

$$\delta_{\nu+u_0}v=\mathcal{D}_1u_1.$$

Then

$$\mathcal{D}_1(u_1+v)=Du_1-\delta v-\frac{1}{2}[u_1,u_1]-[u_1,v]=\mathcal{D}_1u_1-\delta_{\nu+u_0}v=0$$

and $\pi_0(u_1+v)=u_0$, so u is a section of $(T^*\otimes_X R_1)^\wedge$.

If E'' is a sub-bundle of $\mathbf{C}T$, we remark that for (5. 7) to hold, we must necessarily have $E'\cap E''=0$, where E' is the complex conjugate of E''. If $\tilde{u}_0=0$ in (5. 6), then $E''=T''$. Thus if u is a section of $\mathbf{C}T^*\otimes_X R_1$ and if $u_0=\pi_0 u$ is sufficiently small, then u is a section of $(\mathbf{C}T^*\otimes_X R_1)^\wedge$ and the subset E'' given by (5. 6), with $\tilde{u}_0=\nu^{-1}\circ u_0$, is an almost pseudo-complex structure on X.

Theorem 5. 1 provides a mechanism for formulating the integrability of almost pseudo-complex structures and, in the analytic case, for proving the existence of solutions to the integrability problem. However, the linear complex

$$\mathcal{R}_{2,X}\xrightarrow{D}\mathbf{C}\mathcal{T}^*\otimes\mathcal{R}_{1,X}\xrightarrow{D}\wedge^2\mathbf{C}\mathcal{T}^*\otimes J_0(\mathcal{T}'_Y)_X$$

obtained from (2. 3) is not sub-elliptic if X is of codimension 1 in Y (or even of codimension 0), whereas, whenever $P(\varphi)=0$, the complex

$$\mathcal{T}'_{Y|X}\xrightarrow{\bar\partial_{b,\varphi}}\mathcal{T}''^*\otimes\mathcal{T}'_{Y|X}\xrightarrow{\bar\partial_{b,\varphi}}\wedge^2\mathcal{T}''^*\otimes\mathcal{T}'_{Y|X}$$

is sub-elliptic if X is of codimension 1 (or 0) (see [9]).

References

[1] Cartan, É. : Sur le problème général de la déformation, C. R. Congrès Strasbourg, 1920, pp. 397–406 ; Oeuvres complètes, Partie III, Vol. 1, pp. 539–548.

[2] Goldschmidt, H. : Existence theorems for analytic linear partial differential equations, Ann. of Math., **86** (1967), 246–270.

[3] ——: Integrability criteria for systems of non-linear partial differential equations, J. Differential Geometry, **1** (1967), 269–307.

[4] ——: Prolongations of linear partial differential equations. I. A conjecture of Élie Cartan, Ann. scient. Éc. Norm. Sup., (4) **1** (1968), 417–444.

[5] ——: Prolongements d'équations différentielles linéaires. III. La suite exacte de cohomologie de Spencer, Ann. scient. Éc. Norm. Sup., (4) **7** (1974), 5–27.

[6] Goldschmidt, H. and Spencer, D. : On the non-linear cohomology of Lie equations. I, II, Acta Math., **136** (1976), 103–239.

[7] Guillemin, V. W. : Some algebraic results concerning the characteristics of overdetermined partial differential equations, Amer. J. Math., **90** (1968), 270–284.

[8] Kumpera, A. and Spencer, D. : Lie equations. Volume I : general theory, Ann. of Math. Studies No. **73**, Princeton University Press and University of Tokyo Press, 1972.

[9] Kuranishi, M. : Deformations of isolated singularities and $\bar\partial_b$ (to appear).

[10] Malgrange, B. : Équations de Lie. I, II, J. Differential Geometry, **6** (1972), 503–522 ; **7** (1972), 117–141.

[11] Quillen, D. G. : Formal properties of over-determined systems of linear partial differential equations, Ph. D. thesis, Harvard University, 1964.

Institute for Advanced Study
Princeton University

(Received December 15, 1975)

On the First Terms of Certain Asymptotic Expansions[1)]

J. Igusa

Introduction

We shall recall the main feature of our theory of asymptotic expansions: let K denote a local field of characteristic 0, $|\ |_K$ the usual absolute value on K, and ψ a non-trivial character of K; let $f(x)$ denote a polynomial in n variables $x_1, x_2, \cdots, x_n$ with coefficients in K and $|dx|_K$ the usual Haar measure on K^n. We shall assume that $df=0$ at a point of K^n only if $f=0$ at that point; in view of Bertini's theorem this is not a restriction for our purpose. Finally let Φ denote a Schwartz-Bruhat function on K^n and for every i^* in K consider the following integral:

$$F_\Phi^*(i^*) = \int_{K^n} \Phi(x)\psi(i^*f(x))|dx|_K.$$

Then we get an L^∞-function F_Φ^* on K with an asymptotic expansion of the following form:

$$F_\Phi^*(i^*) \approx \sum_\lambda \sum_{m=1}^{m_\lambda} a_{\lambda,m}^* \cdot |i^*|_K^{-\lambda} (\log|i^*|_K)^{m-1}$$

as $|i^*|_K \to \infty$. We recall that $a_{\lambda,m}^*$ depends not only on λ, m but also on (the angular component of) i^* and Φ and that for a fixed i^* it defines a tempered distribution on K^n. Furthermore we can write down such an asymptotic expansion once we explicitly construct a desingularization (relative to K) of the embedded hypersurface $f(x)=0$; we refer to [1] for the details.

We shall explain the content of this paper: there are several cases where we can explicitly construct a desingularization of $f(x)=0$. For instance the desingularization in the case where $n=2$ and $K=\boldsymbol{C}$ is classical; cf. [3]. Suppose that the origin 0 of $\boldsymbol{C}^2$ is a singular point of $f(x)=0$ and, for the sake of simplicity, assume that $f(x)$ is irreducible in $\boldsymbol{C}[[x_1, x_2]]$. Let $f_m(x)$ denote the leading form of $f(x)$ and, by replacing x_1 by $x_1+\text{const.}\ x_2$ if necessary, assume that $f_m(0, 1)\neq 0$; then we can solve $f(x)=0$ in some open neighborhood of 0 by a power series $x_2=x_2(x_1)$ in $x_1^{1/m}$. Let x_1^r denote the first non-integral power of x_1 which does appear in $x_2(x_1)$; then $r>1$ and, just as m, it has an invariant meaning. We shall show that if we put

1) This work was partially supported by the National Science Foundation.

$$\lambda = (1+1/r)/m,$$

then the asymptotic expansion of $F_{\Phi}^{*}(i^{*})$ starts as

$$F_{\Phi}^{*}(i^{*}) \approx c \cdot \Phi(0)|i^{*}|_{K}^{-\lambda} + \cdots$$

provided that the support of Φ does not contain any other singular points of $f(x)=0$. The numerical factor c is essentially, i.e., except for an elementary factor, the total measure of a particular exceptional curve; it is also the coefficient of δ_0 in the residue of the complex power $|f|_C^s$ at $s=-\lambda$.

A similar asymptotic formula holds even if K is an arbitrary local field (of characteristic 0) provided that $f(x)$ is irreducible in $\bar{K}[[x_1, x_2]]$, where $\bar{K}$ is the algebraic closure of K, and r is defined relative to $\bar{K}$; we refer to § 5, Th. 3 for the details. We might add that a major part of the paper is devoted to the settling of an extremum problem for a certain function defined on the set of exceptional curves.

1. Characteristic Exponents

We shall review the definition and basic properties of "characteristic exponents"; we refer to [4], pp. 993–997 for the details. Let $\mathfrak{o}$ denote the local ring of an irreducible plane algebroid curve $f=0$ over an algebraically closed field K of characteristic 0 and $\mathfrak{m}$ the maximal ideal of $\mathfrak{o}$; then we have

$$\mathfrak{m} = \mathfrak{o}x + \mathfrak{o}y$$

for some x, y in $\mathfrak{m}$. Let "ord" denote the normalized discrete valuation on the field of quotients of $\mathfrak{o}$; then the integral closure of $\mathfrak{o}$ becomes the ring of formal power series in any element of order 1 with coefficients in K. We shall assume that $\mathfrak{o}$ is not regular, i.e., that $\mathrm{ord}(x)$, $\mathrm{ord}(y) \geqq 2$.

If $\mathrm{ord}(x)=m$, then $x^{1/m}$ is an element of the field of quotients of $\mathfrak{o}$ and it is of order 1; hence we get

$$y = y(x) = \sum_{i=1}^{\infty} a_i x^{i/m}$$

with a_i in K for $i=1, 2, \cdots$. We can rewrite this "Puiseux series" as

$$y(x) = \sum_{i=1}^{k_0} a_{0,i} x^{i} + \sum_{i=0}^{k_1} a_{1,i} x^{(\mu_1+i)/\nu_1}$$
$$+ \cdots + \sum_{i=0}^{\infty} a_{g,i} x^{(\mu_g+i)/\nu_1\nu_2\cdots\nu_g},$$

in which the exponents are strictly increasing, $a_{1,0}a_{2,0}\cdots a_{g,0} \neq 0$, μ_i, ν_i are relatively prime integers for $1 \leqq i \leqq g$, and $\nu_1, \nu_2, \cdots, \nu_g \geqq 2$. We then have

$$\mathrm{ord}(x) = m = \nu_1\nu_2 \cdots \nu_g;$$

the g exponents

$$\mu_1/\nu_1,\ \mu_2/\nu_1\nu_2,\ \cdots,\ \mu_g/\nu_1\nu_2\cdots\nu_g$$

are called *characteristic exponents of the series* $y(x)$.

Since x and y are "symmetric," we can expand x into a power series in $y^{1/n}$ if

$\mathrm{ord}(y)=n$. Let

$$\mu_1'/\nu_1',\ \mu_2'/\nu_1'\nu_2',\ \cdots,\ \mu_{g'}'/\nu_1'\nu_2'\cdots\nu_{g'}'$$

denote the characteristic exponents of the series $x(y)$; then we have

$$\mathrm{ord}(y) = n = \nu_1'\nu_2'\cdots\nu_{g'}'.$$

There is a relation between these two sets of characteristic exponents which Zariski called an *inversion formula* ; it is as follows :

Abandoning the symmetry of x and y, assume that $\mathrm{ord}(x)\leqq\mathrm{ord}(y)$; then we have the following alternatives :

(case 1) $\mathrm{ord}(x) < \mathrm{ord}(y) < \mu_1/\nu_1\cdot\mathrm{ord}(x)$.

In this case we have $g'=g+1$ and

$$\begin{cases}\mu_1'/\nu_1'\cdot\mathrm{ord}(y) = \mathrm{ord}(x)\\ (\mu_i'/\nu_1'\cdots\nu_i'+1)\mathrm{ord}(y) = (\mu_{i-1}/\nu_1\cdots\nu_{i-1}+1)\mathrm{ord}(x)\end{cases}$$

for $1<i\leqq g+1$.

(case 2) $\mathrm{ord}(y) = \mathrm{ord}(x)$ or $\mathrm{ord}(y) = \mu_1/\nu_1\cdot\mathrm{ord}(x)$.

In this case we have $g'=g$ and

$$(\mu_i'/\nu_1'\cdots\nu_i'+1)\mathrm{ord}(y) = (\mu_i/\nu_1\cdots\nu_i+1)\mathrm{ord}(x)$$

for $1\leqq i\leqq g$.

In particular if $\mathrm{ord}(x)=\mathrm{ord}(y)$, the two series $y(x)$ and $x(y)$ have the same characteristic exponents. Actually these characteristic exponents depend only on $\mathfrak{o}$ and they are called the *characteristic exponents of the algebroid curve* $f=0$. We observe that the characteristic exponents of $y(x)$ become those of $f=0$ if and only if $\mathrm{ord}(y)\geqq\mathrm{ord}(x)$, i.e., if and only if $\mu_1/\nu_1>1$.

2. A Geometric Theorem

Let X denote a non-singular algebraic surface over an algebraically closed field K (of characteristic 0) and C a finite linear combination with positive integer coefficients of irreducible curves on X ; by replacing each coefficient by 1 we get C_{red}. We shall regard C as "desingularized" if C_{red} has only normal crossings. It is well known that there exists a sequence of successive quadratic transformations of X such that the total transform C^* of C under the product morphism $X^*\rightarrow X$ is desingularized. Moreover if none of the quadratic transformations is redundant, then the morphism is unique. Since the above desingularization process is local in X, we may assume that C has only one singular point and, for the sake of simplicity, we shall assume that C is analytically irreducible at that point. Then the sequence of quadratic transformations becomes unique and it can be described by the characteristic exponents of the corresponding irreducible algebroid curve ; cf. [3], pp. 5–10. We shall formulate a *quantitative theorem* concerning this process ; first we shall recall some of its details :

Let

$$\mu_1/\nu_1,\ \mu_2/\nu_1\nu_2,\ \cdots,\ \mu_g/\nu_1\nu_2\cdots\nu_g$$

denote the characteristic exponents in question and expand each $\mu_i/\nu_i-\mu_{i-1}$, where $\mu_0=0$, into a continued fraction

$$\mu_i/\nu_i-\mu_{i-1} = [k_{i0}, k_{i1}, \cdots, k_{i,t_i}]$$

for $1\leqq i\leqq g$; the k_{ij} are non-negative integers and

$$k_{10}, k_{i1}, \cdots, k_{i,t_{i-1}} \geqq 1, \qquad k_{i,t_i} \geqq 2, \qquad t_i \geqq 1$$

for $1\leqq i\leqq g$. We note that unlike $k_{10}\geqq 1$ we may have $k_{i0}=0$ for some i. With this notation the number of quadratic transformations becomes the sum of all k_{ij}. Moreover if C' denotes the strict transform of C under the morphism $X^*\to X$ and if E_I denotes the exceptional curve of X^* introduced by the I-th quadratic transformation, then the total transform C^* of C is of the form

$$C^* = \sum_I N_I E_I + C',$$

in which $N_I\geqq 1$ for every I.

On the other hand, by making X smaller if necessary, we may assume that there exists a gauge-form ϖ on X, i.e., a 2-form on X without zeros and poles. Let ϖ^* denote the preimage of ϖ under $X^*\to X$; then its divisor (ϖ^*) is of the form

$$(\varpi^*) = \sum_I (n_I-1)E_I,$$

in which $n_I\geqq 2$ for every I and it is independent of the choice of ϖ.

We shall later express N_I and n_I in terms of k_{ij} and examine the relation of all quotients such as n_I/N_I; in this way we shall prove the following theorem:

Theorem 1. *We put*

$$I_i = k_{10}+k_{11}+\cdots+k_{i,t_i}$$

for $1\leqq i\leqq g$; *then we have*

$$n_{I_1}/N_{I_1} = (1+\nu_1/\mu_1)/\nu_1\nu_2\cdots\nu_g$$

$$n_I/N_I > n_{I_1}/N_{I_1} \quad (I<I_1), \qquad n_I/N_I > n_{I_i}/N_{I_i} \quad (I>I_i)$$

for $1\leqq i<g$. *In particular* n_{I_1}/N_{I_1} *is smaller than any other* n_I/N_I.

Actually we shall prove a finer property of the function $I\to n_I/N_I$: in the first interval $0<I\leqq I_1$ it is so-to-speak stepwise strictly decreasing, i.e., it is strictly decreasing in the subinterval

$$k_{10}+\cdots+k_{1,s-1} < I \leqq k_{10}+\cdots+k_{1,s}$$

for $0\leqq s\leqq t_1$. In every other interval $I_{i-1}<I\leqq I_i$ it is oscillating, i.e., it is strictly increasing or decreasing in the subinterval

$$I_{i-1}+k_{i0}+\cdots+k_{i,s-1} < I \leqq I_{i-1}+k_{i0}+\cdots+k_{i,s}$$

according as s is even or odd for $0\leqq s\leqq t_i$, $1<i\leqq g$.

For our later purpose we shall add the following remark: suppose that K_0 is a subfield of K over which X, C are defined and the singular point is rational. Then all successive quadratic transformations are defined over K_0. This follows, e.g., from the fact that the center of each quadratic transformation is unique and hence rational over K_0; we have tacitly used the assumptions that C is analytically irreducible at the singular point and that K_0 is a perfect field.

3. Algebraic Preliminaries

Let $p_n = p_n(k_0, k_1, \cdots, k_n)$ denote a polynomial in $n+1$ variables $k_0, k_1, \cdots, k_n$ with integer coefficients defined inductively as follows : it represents 0, 1, respectively, for $n=-2, -1$ and

$$p_n = k_0 p_{n-1}(k_1, \cdots, k_n) + p_{n-2}(k_2, \cdots, k_n)$$

for every $n \geqq 0$. Since there will be no confusion, we shall drop n from p_n. It is well known (and can easily be proved) that $p(k_0, k_1, \cdots, k_n)$ is unchanged even if we replace $k_0, k_1, \cdots, k_n$ by $k_n, \cdots, k_1, k_0$. In the following we shall fix a positive integer t and limit ourselves to the $t+1$ variables $k_0, k_1, \cdots, k_t$.

For any integer pair (r, s) satisfying $0 \leqq r \leqq s \leqq t$ we put

$$a(r, s) = \sum_{i=s-r}^{s} k_i p(k_{i+1}, \cdots, k_s) p(k_{i+1}, \cdots, k_t)\ ;$$

then we get

$$a(r, s) = \begin{cases} p(k_{s-r}, \cdots, k_s) p(k_{s-r+1}, \cdots, k_t) & r \text{ even} \\ p(k_{s-r+1}, \cdots, k_s) p(k_{s-r}, \cdots, k_t) & r \text{ odd.} \end{cases}$$

This can be proved by an induction on r. In particular if we put $a(s, s) = a(s)$, we get

$$a(s) = \begin{cases} p(k_0, \cdots, k_s) p(k_1, \cdots, k_t) & s \quad \text{even} \\ p(k_1, \cdots, k_s) p(k_0, \cdots, k_t) & s \quad \text{odd}\ ; \end{cases}$$

this remains valid for $s=-2, -1$ if we put $a(-2) = a(-1) = 0$. We define $b(s)$ for $-2 \leqq s \leqq t$ as

$$b(s) = a(s) + p(k_{s+2}, \cdots, k_t)\ ;$$

we obviously have $b(t) = a(t)$. Finally we define $c(s)$ for $s \geqq 0$ as

$$c(s) = \sum_{i=0}^{s} k_i p(k_{i+1}, \cdots, k_s) + 1$$

and we put $c(-2) = c(-1) = 1$; then we get

$$c(s) = p(k_0, \cdots, k_s) + p(k_1, \cdots, k_s)$$

for $s \geqq -1$.

We shall show that $a(s)$, $b(s)$, $c(s)$ for $0 \leqq s \leqq t$ satisfy the following identities:

$$a(s) = k_s b(s-1) + a(s-2), \qquad b(s) = k_s a(s-1) + b(s-2)$$
$$c(s) = k_s c(s-1) + c(s-2).$$

The first identity can easily be verified for $s=0$, 1 and if $s \geqq 2$, then

$$\begin{aligned} a(s)-a(s-2) &= \sum_{i=0}^{s} k_i p(k_{i+1}, \cdots, k_s) p(k_{i+1}, \cdots, k_t) \\ &\quad - \sum_{i=0}^{s-2} k_i p(k_{i+1}, \cdots, k_{s-2}) p(k_{i+1}, \cdots, k_t) \\ &= \sum_{i=0}^{s-1} k_i k_s p(k_{i+1}, \cdots, k_{s-1}) p(k_{i+1}, \cdots, k_t) \\ &\quad + k_s p(k_{s+1}, \cdots, k_t) \\ &= k_s b(s-1). \end{aligned}$$

The second identity follows from the first ; the proof of the third identity is similar.

We can express $p(k_0, \cdots, k_s)$ and $p(k_1, \cdots, k_s)$ in terms of $a(s)$, $b(s)$, $c(s)$; for instance we have

$$p(k_1, \cdots, k_s) = \begin{cases} c(s)-a(s)/b(-1) & s \text{ even} \\ c(s)-b(s)/b(-1) & s \text{ odd} \end{cases}$$

for $-1 \leqq s \leqq t$. In fact if we denote the right hand side by $P(s)$, we get $P(-1)=0$, $P(0)=1$, and $P(s)=k_s P(s-1)+P(s-2)$ for $1 \leqq s \leqq t$. Since $p(k_1, \cdots, k_s)$ has the same property, we get

$$P(s) = p(k_1, \cdots, k_s)$$

for $-1 \leqq s \leqq t$. We observe that $P(-2)=1$ and that the above recursion formula for $P(s)$ is valid for $0 \leqq s \leqq t$.

We shall calculate two determinants both depending on s between 0 and $t+1$; one is as follows:

$$\det \begin{bmatrix} a(s-2) & b(s-1) \\ c(s-2) & c(s-1) \end{bmatrix}.$$

We apply the identities such as $b(s-1)=k_{s-1}a(s-2)+b(s-3)$ to the second column, then to the first column, etc. ; in this way we see that the above determinant is equal to $-b(-1)$ or $-b(-2)$ according as s is even or odd. Another determinant is a generalization of the above and it involves additional variables M, m:

$$\det \begin{bmatrix} (m-1)P(s-2)+c(s-2) & (m-1)P(s-1)+c(s-1) \\ MP(s-2)+a(s-2)/b(-1) & MP(s-1)+b(s-1)/b(-1) \end{bmatrix}.$$

By applying the same argument we see that this determinant is equal to $m-M$ or $M+b(-2)/b(-1)$ according as s is even or odd.

We recall that if a, b, c, d are real numbers satisfying $ad-bc<0$ or >0, then $(a+bx)/(c+dx)$, as a function of x, is strictly increasing or decreasing. In particular $(a+x)/(c+x)$ is strictly increasing if $c>a$.

We have so far assumed that $k_0, k_1, \cdots, k_t, M, m$ are variables. Suppose that they are real numbers satisfying $k_0 \geqq 0$ and $k_1, \cdots, k_t > 0$; then we have $a(s) \geqq 0$, $b(s)>0$, and $c(s) \geqq 1$ for $-2 \leqq s \leqq t$. In particular $b(-1) \neq 0$, and hence all identities remain valid under the above specialization.

We divide

$$a(s-2)c(s-1)-b(s-1)c(s-2) = -b(-1) \quad \text{or} \quad -b(-2)$$

by $c(s-1)c(s-2)$; then we get

$$a(s-2)/c(s-2) < b(s-1)/c(s-1)$$

for $0 \leqq s \leqq t+1$. Moreover

$$(*) \qquad (a(s-2)+jb(s-1))/(c(s-2)+jc(s-1))$$

is a strictly increasing function of j for $0 \leqq s \leqq t+1$; hence

$$\frac{a(s-2)}{c(s-2)} < \frac{a(s-2)+k_s b(s-1)}{c(s-2)+k_s c(s-1)} = \frac{a(s)}{c(s)}$$

for $1 \leqq s \leqq t$. By putting these together we see that $a(t)/c(t) = b(t)/c(t)$ is larger than $a(s)/c(s)$ for $0 \leqq s < t$, hence larger than (*) for $j \leqq k_s$, $0 \leqq s \leqq t$, where $(j, s) \neq (k_t, t)$.

We shall examine similar quotients by assuming that $M > m \geqq 0$: if s is even and $0 \leqq s \leqq t+1$, then

$$(**) \qquad \frac{(m-1)(P(s-2)+jP(s-1))+c(s-2)+jc(s-1)}{M(P(s-2)+jP(s-1))+(a(s-2)+jb(s-1))/b(-1)}$$

is a strictly increasing function of j. Moreover for $j=0$ it takes the form $(m+\alpha)/(M+\alpha)$ with

$$\alpha = a(s-2)/b(-1)P(s-2) \geqq 0;$$

hence it is at least equal to m/M. Since the limit of (**) as $j \to \infty$ is either 1 or of the same form, it is at most equal to 1. Therefore we get

$$m/M < (**) < 1$$

for $j > 0$. On the other hand, if s is odd and $0 \leqq s \leqq t+1$, then (**) is a strictly decreasing function of j. Moreover, for $j = k_s$, where $s \leqq t$, it takes the form $(m+\beta)/(M+\alpha)$ with

$$\alpha = a(s)/b(-1)P(s) = b(-2)/b(-1), \qquad \beta = b(s)/b(-1)P(s).$$

Since $\beta \geqq \alpha > 0$, it is at least equal to $(m+\alpha)/(M+\alpha)$ and this is larger than m/M. In the special case where $s=t$, it is equal to $(m+\alpha)/(M+\alpha)$ and this is smaller than 1. Therefore we get $(**) > m/M$ for $j \leqq k_s$ and $(**) < 1$ for $(j, s) = (k_t, t)$.

4. Proof of the Theorem

We may assume that X is an affine plane minus a finite number of points, C is an irreducible curve on X with the origin 0 of X as its only singular point, and C is analytically irreducible at 0 ; then C is defined by $f=0$ and X has a gauge-form ϖ. We shall denote by X_I the I-th quadratic transform of X and by 0_I the unique point of X_I where the strict transform of C and the I-th exceptional curve intersect; in calculating N_I and n_I we have only to examine X_I around 0_I. We shall denote by $f_I=0$ and ϖ_I a local equation for the strict transform of C and a local gauge-form, respectively, both on X_I around 0_I.

We choose affine coordinates (x, y) on X such that $x=0$ is not tangent to C at 0 ; then we get a Puiseux series

$$y = y(x) = \sum_{i=1}^{k_0} a_{0,i}x^i + \sum_{i=0}^{k_1} a_{1,i}x^{(\mu_1+i)/\nu_1} + \cdots$$

such that its characteristic exponents

$$\mu_1/\nu_1,\ \mu_2/\nu_1\nu_2,\ \cdots,\ \mu_g/\nu_1\nu_2\cdots\nu_g$$

are also those for the algebroid curve $f=0$ at 0. If we denote by $\varepsilon(x, y)$ a unit in the ring of formal power series in x, y, then we can write

$$f(x,y) = \varepsilon(x,y)\cdot\prod(y-\mathrm{conj}\cdot y(x)),$$

in which the product extends over $\nu_1\nu_2\cdots\nu_g$ conjugates of $y(x)$; we shall use $dx\wedge dy$ as $\varpi(x,y)$. And, as before, we put

$$\mu_i/\nu_i-\mu_{i-1} = [k_{i0}, k_{i1}, \cdots, k_{i,t_i}]$$
$$I_i = k_{10}+k_{11}+\cdots+k_{i,t_i}$$

for $1\leqq i\leqq g$, in which $\mu_0=0$; we shall denote by $a_i(s)$, $b_i(s)$, $c_i(s)$, $P_i(s)$ the $a(s)$, $b(s)$, $c(s)$, $P(s)$ for the sequence $k_{i0}, k_{i1}, \cdots, k_{i,t_i}$. Then we have the following lemma :

Lemma 1. *If we take*

$$I = k_{10}+k_{11}+\cdots+k_{1,s-1}+j,$$

where $1\leqq j\leqq k_{1,s}$, $0\leqq s\leqq t_1$, *then we get*

$$N_I = (a_1(s-2)+jb_1(s-1))\nu_2\cdots\nu_g$$
$$n_I = c_1(s-2)+jc_1(s-1).$$

Proof. We shall construct local coordinates (u, v) on X_I valid in an open subset containing 0_I. For the sake of simplicity we shall omit "1" from $k_{10}, k_{11}, \cdots, k_{1,t_1}$, $a_1(s)$, $b_1(s)$, $c_1(s)$, $P_1(s)$ and we put

$$q_i = p(k_{i+1}, \cdots, k_t)$$

for $-1\leqq i\leqq t+1$. By passing from (x,y) to (x',y') defined as

$$x' = x, \qquad y' = y-\sum_{i=1}^{k_0} a_{0,i}x^i,$$

we may assume that $a_{0,1}=\cdots=a_{0,k_0}=0$. We put $x=y_{-1}$, $y=x_{-1}$ and introduce (x_0, y_0), $\cdots$, (x_{s-1}, y_{s-1}) as

$$x_{i-1} = x_i^{k_i}y_i, \qquad y_{i-1} = x_i$$

for $0\leqq i<s$; finally we put

$$x_{s-1} = u^jv, \qquad y_{s-1} = u.$$

Then by repeatedly applying the inversion formula we get

$$\begin{aligned}
x = y_{-1} = x_0 &= x_{s-1}^{P(s-1)}y_{s-1}^{P(s-2)}\\
&= u^{P(s-2)+jP(s-1)}v^{P(s-1)}\\
f(x,y) &= (x_0^{k_0q_0}\cdots x_{s-1}^{k_{s-1}q_{s-1}})^{\nu_2\cdots\nu_g}f_{I-j}(x_{s-1}, y_{s-1})\\
&= (x_{s-1}^{a(s-1)}y_{s-1}^{a(s-2)})^{\nu_2\cdots\nu_g}f_{I-j}(x_{s-1}, y_{s-1})\\
&= (u^{a(s-2)+jb(s-1)}v^{a(s-1)})^{\nu_2\cdots\nu_g}f_I(u, v)\\
\varpi(x,y) &= \pm x_0^{k_0}\cdots x_{s-1}^{k_{s-1}}dx_{s-1}\wedge dy_{s-1}\\
&= \pm x_{s-1}^{c(s-1)-1}y_{s-1}^{c(s-2)-1}dx_{s-1}\wedge dy_{s-1}\\
&= \pm u^{c(s-2)+jc(s-1)-1}v^{c(s-1)-1}du\wedge dv.
\end{aligned}$$

We have only to observe that $u=0$ is a local equation for the I-th exceptional curve. q. e. d.

Although we did not mention this in the above proof, we get a Puiseux series

$$x_{s-1}=x_{s-1}(y_{s-1})=(a_{1,0})^{(-1)^s q_0/q_s} y_{s-1}^{q_{s-1}/q_s}+\cdots;$$

and this series has

$$\frac{\mu_i-(k_0q_0+\cdots+k_{s-1}q_{s-1}-q_0+q_s)\nu_2\cdots\nu_i}{q_s\nu_2\cdots\nu_i}$$

for $1\leqq i\leqq g$ as its characteristic exponents. There is an exception, however, in the case where $s=t$; the above g exponents then become

$$\mu_i/\nu_2\cdots\nu_i-\mu_1+k_t$$

for $1\leqq i\leqq g$ and we simply omit the first exponent k_t which is an integer.

It follows from this that in the extreme case where $I=I_1$ we get

$$v=v(u)=\sum_{i=0}^{\kappa_0}\alpha_{0,i}u^i+\alpha_{1,0}u^{\mu_2/\nu_2-\mu_1}+\cdots,$$

in which $\kappa_0=k_{2,0}$ and

$$\alpha_{0,0}=(a_{1,0})^{\pm\nu_1}\neq 0,\qquad \alpha_{1,0}\neq 0,\cdots.$$

Therefore if we pass from (u, v) to (ξ, η) defined as

$$\xi=u,\qquad \eta=v-\sum_{i=0}^{\kappa_0}\alpha_{0,i}u^i,$$

we get

$$f(x,y)=\xi^{\mu_1\nu_1\cdots\nu_g}f_{I_1}(\xi,\eta)$$
$$\varpi(x,y)=\xi^{\mu_1+\nu_1-1}\varpi_{I_1}(\xi,\eta);$$

and the Puiseux series $\eta=\eta(\xi)$ has

$$\mu_i/\nu_2\cdots\nu_i-\mu_1$$

for $1<i\leqq g$ as its characteristic exponents. The point is that we can apply Lemma 1 to $f_{I_1}(\xi,\eta)$ to determine N_I, n_I for $I_1<I\leqq I_2$ and repeat the same argument. In this way we get the following lemma:

Lemma 2. *For $1<i\leqq g$ we take*

$$I=I_{i-1}+k_{i0}+k_{i1}+\cdots+k_{i,s-1}+j,$$

in which $1\leqq j\leqq k_{i,s}$, $0\leqq s\leqq t_i$; then we get

$$N_I=(P_i(s-2)+jP_i(s-1))N_{I_{i-1}}$$
$$+(a_i(s-2)+jb_i(s-1))\nu_{i+1}\cdots\nu_g$$
$$n_I=(P_i(s-2)+jP_i(s-1))(n_{I_{i-1}}-1)$$
$$+c_i(s-2)+jc_i(s-1).$$

We are ready to prove Th. 1: we apply our observation on (*) in the previous section to the expression for n_I/N_I obtained from Lemma 1; then we see that

$n_I/N_I > n_{I_1}/N_{I_1}$ for $I < I_1$. Moreover if we put

$$M_i = N_{I_i}/\nu_{i+1}\cdots\nu_g, \qquad m_i = n_{I_i}$$

for $1 \leqq i \leqq g$, then we get

$$M_1 = \mu_1\nu_1, \qquad m_1 = \mu_1 + \nu_1;$$

in particular we have $m_1 < M_1$. We take $i > 1$ and assume that $m_{i-1} < M_{i-1}$; then we can apply our observation on (**) to the expression for n_I/N_I obtained from Lemma 2. In this way we see that $n_I/N_I > n_{I_{i-1}}/N_{I_{i-1}}$ for $I_{i-1} < I \leqq I_i$. Moreover we get

$$M_i = (M_{i-1} + \mu_i/\nu_i - \mu_{i-1})\nu_i^2$$
$$m_i = (m_{i-1} + \mu_i/\nu_i - \mu_{i-1})\nu_i$$

and $m_i/M_i < 1/\nu_i$; in particular we have $m_i < M_i$. Therefore our induction is complete and the theorem is proved.

5. Asymptotic Formulas

We shall first recall some definitions so that we can later state our theorems without ambiguity: let K denote a local field; K is $\boldsymbol{R}$, $\boldsymbol{C}$, or a p-field. The Haar measure on K such that the unit disc has measure 2, 2π, or 1, respectively, is called the usual Haar measure; the usual Haar measure $|dx|_K$ on K^n is its product measure. We define the usual absolute value $|\ |_K$ on K as $|0|_K = 0$ and $|d(tx)|_K = |t|_K |dx|_K$, where $n = 1$, for every t in $K^\times = K - \{0\}$. We shall denote by $K_1^\times$ the compact subgroup of $K^\times$ defined by $|t|_K = 1$. For every t in $K^\times$ we define its angular component, denoted by $ac(t)$, as follows:

$$ac(t) = \begin{cases} t/|t| & K = \boldsymbol{R}, \boldsymbol{C} \\ t/\pi^{\mathrm{ord}(t)} & K = p\text{-field}, \end{cases}$$

in which "π" is a fixed element of K of order 1. Let ω denote a quasicharacter of $K^\times$, i.e., a continuous homomorphism of $K^\times$ to $\boldsymbol{C}^\times$, and χ its restriction to $K_1^\times$; then for every t in $K^\times$ we have

$$\omega(t) = |t|_K^s \chi(ac(t)),$$

in which s is a complex number independent of t. We shall denote by ψ a fixed nontrivial character of K; we have $\psi(t) = \mathbf{e}(\gamma t)$ or $\mathbf{e}(2\mathrm{Re}(\gamma t))$ for some γ in $K^\times$ if K is $\boldsymbol{R}$ or $\boldsymbol{C}$. If K is a p-field, we shall denote by R the maximal compact subring of K, by P its maximal ideal, and by q the cardinality of the residue class field R/P; we have $\psi = 1$ on P^{-d} but not on P^{-d-1} for some integer d. Still in the p-field case we shall denote by e_χ the smallest positive integer such that $\chi = 1$ on $1 + P^{e_\chi}$; and we put $g_1 = -q^{-1}$ and

$$g_\chi = \int_{K_1^\times} \chi(x)\psi(\pi^{-d-e_\chi}x)|dx|_K$$

for $\chi \neq 1$; this is a complex number of absolute value $q^{-\frac{1}{2}e_\chi}$. Finally we put

$$m_K = 2, \quad 2\pi, \quad \text{or} \quad 1 - q^{-1}$$

according as K is $\boldsymbol{R}$, $\boldsymbol{C}$, or a p-field.

Let $f(x)$ denote a polynomial in n variables $x_1, x_2, \cdots, x_n$ with coefficients in K; for any Schwartz-Bruhat function Φ on K^n put

$$Z(\omega, \Phi) = \int_{K^n} \omega(f(x))\Phi(x)|dx|_K$$
$$F_\Phi^*(i^*) = \int_{K^n} \psi(i^* f(x))\Phi(x)|dx|_K,$$

in which $\mathrm{Re}(s)>0$ and i^* is in K. Then, at least in the case where the characteristic of K is 0, we can examine general properties of $Z(\omega, \Phi)$ via Hironaka's fundamental theorem and from them we can derive an asymptotic expansion of $F_\Phi^*(i^*)$ as $|i^*|_K \to \infty$. This has been worked out in [1]; also the special case where $f(x)$ is homogeneous and the projective hypersurface $f(x)=0$ is non-singular has been worked out as an example in [2]. In the following we shall consider another special case where $f(x)=0$ defines an irreducible plane curve C with the origin of K^2 as a singular point. For the sake of simplicity we shall assume that $f(x)$ is irreducible in $\bar{K}[[x_1, x_2]]$, in which $\bar{K}$ is the algebraic closure of K. We shall use the same notation as before except that we put

$$E_{I_1} = E, \qquad N_{I_1} = N, \qquad n_{I_1} = n.$$

We shall start by making the following observation:

We take a K-rational point of E different from the points where E intersects other components of the total transform C^* of C; we choose local coordinates (y_1, y_2) centered at that point such that $y_1=0$ gives a local equation for E. Then we can write

$$f(x) = \alpha(y)y_1^N, \qquad dx_1 \wedge dx_2 = \beta(y)y_1^{n-1}dy_1 \wedge dy_2$$

with $\alpha(0)\beta(0)\neq 0$. Let ω' denote a special quasicharacter of $K^\times$ in which $s=-n/N$ and $\chi^N=1$; then

$$\theta_\chi = \omega'(\alpha(0, y_2))|\beta(0, y_2)|_K|dy_2|_K$$

defines a complex measure θ_χ on E (minus the above-mentioned points) independently of the choice of (y_1, y_2); the corresponding (improper) integral is absolutely convergent and

$$\rho_\chi = \int_E \theta_\chi$$

defines a complex number ρ_χ. In fact the first part is entirely general and the second part depends only on the fact that n/N is smaller than similar quotients associated with other components of C^* which intersect E. We observe that $\rho_1>0$; we put $\rho_\chi=0$ if $\chi^N\neq 1$. In the following two theorems we shall tacitly assume that the support of Φ does not contain other singular points of C:

Theorem 2. *The complex power* $Z(\omega, \Phi)$ *has a meromorphic continuation to the whole s-plane and is holomorphic in* $\mathrm{Re}(s)>-n/N$ *for every* χ; *it has poles of the form*

$$\begin{cases} m_K\rho_\chi\Phi(0)/(Ns+n) & K=\boldsymbol{R}, \boldsymbol{C} \\ m_K\rho_\chi\Phi(0)/(1-q^{-(Ns+n)}) & K=p\text{-field} \end{cases}$$

on $\mathrm{Re}(s)=-n/N$.

Theorem 3. *If K is* $\boldsymbol{R}$ *or* $\boldsymbol{C}$, *we put* $\lambda=n/N$; *then we have*

$$F_\Phi^*(i^*) \approx c\cdot\Phi(0)|\gamma i^*|_K^{-\lambda}+\cdots$$

as $|i^*|_K\to\infty$, *in which*

$$\frac{1}{2}Nc = \begin{cases} \Gamma(\lambda)/(2\pi)^\lambda\cdot\left(\rho_1\cos\left(\frac{1}{2}\pi\lambda\right)+i\rho_{\mathrm{sgn}}\sin\left(\frac{1}{2}\pi\lambda\right)\mathrm{sgn}(\gamma i^*)\right) & K=\boldsymbol{R} \\ (\Gamma(\lambda)/(2\pi)^\lambda)^2\cdot\rho_1\sin(\pi\lambda) & K=\boldsymbol{C}. \end{cases}$$

If K is a p-field, we define $\lambda \bmod 2\pi i/\log q$ *by the condition that* $\alpha=q^\lambda$ *satisfies* $\alpha^N=q^n$; *then we have*

$$F_\Phi^*(i^*) \approx \sum_\lambda c_\lambda\cdot\Phi(0)|i^*|_K^{-\lambda}+\cdots$$

as $|i^*|_K\to\infty$, *in which*

$$Nc_\lambda = g_1\rho_1\alpha^{d+1}(1-q\alpha^{-1})/(1-\alpha^{-1})+\sum_{\chi\neq 1} g_{\chi^{-1}}\rho_\chi\alpha^{d+e_\chi}\cdot\chi(ac(i^*)).$$

In order to prove Th. 2 we have only to feed the information in Th. 1 into our general theory and Th. 3 follows from Th. 2 ; cf. [1]–II, Th. 2 ; see also [2], § 2.

References

[1] Igusa, J.: Complex powers and asymptotic expansions I, Functions of certain types, Crelles J. Math. **268/269** (1974), 110–130; II, Asymptotic expansions, ibid. **278/279** (1975), 307–321.
[2] Igusa, J.: On a certain Poisson formula, Nagoya Math. J. **53** (1974), 211–233.
[3] Zariski, O.: Algebraic Surfaces, Ergeb. der Math., Springer (1932) ; Chelsea (1948).
[4] Zariski, O.: Studies in equisingularity III, Saturation of local rings and equisingularity, Amer. J. Math. **90** (1968), 961–1023.

The Johns Hopkins University

(Received August 12, 1975)

Micro-Local Calculus of Simple Microfunctions

M. Kashiwara

In 1969 [2], M. Sato pointed out the importance of micro-local analysis, the analysis on cotangent bundles, and since that time, this subject has increasingly revealed its importance. The transformations called quantized contact transformations or Fourier integral operators and studied by Maslov, Egorov, Hörmander, Sato, Kawai, the author and others, make the theory of linear differential equations very transparent. In particular, the Fourier integral operators, introduced by L. Hörmander [1], are very interesting. Of particular note is his study of the properties of the functions given by

$$\int a(x, \theta) e^{\sqrt{-1}\varphi(x,\theta)} d\theta$$

with a phase function $\varphi(x, \theta)$, and his development of the theory of their principal symbols. It is this basis we rely on when developing the theory of holonomic systems of micro-differential equations. In fact, a function of this type is characterized as a solution of a simple holonomic system of micro-differential equations and its principal symbol controls the solution completely. The present note is a summary of the theory of principal symbols in the "microfunction" category.

1. Micro-Differential Operators

Let X be a complex manifold of dimension n, and let T^*X be its cotangent bundle. We take a local coordinate system $(z_1, \cdots, z_n)$ on X and $(z_1, \cdots, z_n, \zeta_1, \cdots, \zeta_n)$ on T^*X such that $\omega = \sum \zeta_j dz_j$ is the fundamental 1-form on T^*X. For $\lambda \in \mathbf{C}$, the sheaf $\mathcal{E}^{(\lambda)}$ on T^*X is defined in the following way : for any open set Ω in T^*X, an element of $\mathcal{E}^{(\lambda)}(\Omega)$ is a set $\{P_{\lambda+j}(z, \zeta)\}_{j \in \mathbf{Z}}$ such that

o) $P_{\lambda+j}(z, \zeta)$ is a holomorphic function defined on Ω homogeneous of degree $\lambda+j$ with respect to ζ.

i) For any compact set K in Ω, there exists a constant R_K such that
$$\sup_K |P_{\lambda+j}(z, \zeta)| \leq (-j)!\, R_K^{-j} \qquad \text{for } j < 0.$$

ii) For any compact set K in Ω and $\varepsilon > 0$, there exists $C_{K,\varepsilon}$ such that
$$\sup_K |P_{\lambda+j}(z, \zeta)| \leq C_{K,\varepsilon}\, \varepsilon^j / j! \qquad \text{for } j > 0.$$

$\{P_{\lambda+j}(z, \zeta)\}$ is usually denoted by $\sum P_{\lambda+j}(z, D)$. We can define the product R of $P = \sum P_{\lambda+j}(z, D) \in \mathcal{E}^{(\lambda)}$ and $Q = \sum Q_{\mu+k}(z, D) \in \mathcal{E}^{(\mu)}$ as

$$R_{\lambda+\mu+l}(z,\zeta)=\sum_{l=j+k-|\alpha|}\frac{1}{\alpha!}(D_\zeta^\alpha P_{\lambda+j}(z,\zeta))(D_z^\alpha Q_{\mu+k}(z,\zeta))$$

which is a section of $\mathcal{E}^{(\lambda+\mu)}$. Here α runs in $\boldsymbol{Z}_+^n$. D_z^α means $\partial^{\alpha_1+\cdots+\alpha_n}/\partial z_1^{\alpha_1}\cdots\partial z_n^{\alpha_n}$. Therefore, $\mathcal{E}=\mathcal{E}^{(0)}$ is a sheaf of $\boldsymbol{C}$-algebras. For $P=\sum P_{\lambda+j}(z,D)$, the largest $\lambda+j$ such that $P_{\lambda+j}(z,\zeta)\neq 0$, is called the order of P and denoted by $\mathrm{ord}(P)$. $\mathcal{E}^f$ is the subring of $\mathcal{E}^{(0)}$ of micro-differential operators of finite order. We denote by $\mathcal{E}(\lambda)$ the subsheaf of $\mathcal{E}^{(\lambda)}$ consisting of $P=\sum P_{\lambda+j}$ such that $P_{\lambda+j}=0$ for $j>0$.

Let $\mathcal{J}$ be a left Ideal of $\mathcal{E}^f$ locally generated by a finite number of micro-differential operators. The Ideal $\bar{\mathcal{J}}$ of the Ring $\mathcal{O}_{T^*X}$ of holomorphic functions on T^*X generated by principal symbols of micro-differential operators is called the *symbol Ideal* of $\mathcal{J}$. The set of common zeros of the functions in $\bar{\mathcal{J}}$ is called the *characteristic variety* of the system of micro-differential equations $\mathcal{J}u=0$.

It is known (see [3]) that the characteristic variety V is always involutory, that is, for any two functions f, g vanishing on V their Poisson bracket $\{f,g\}$ vanishes on V. Therefore, the codimension of the characteristic variety does not exceed the dimension n of X. We say that the system of micro-differential equations

$$\mathcal{J}u=0$$

is *holonomic* if the codimension of the characteristic variety coincides with n.

The holonomic system is very important because many functions appearing in Mathematics and Physics satisfy the holonomic systems and, since the dimension of solutions of holonomic systems is finite, the holonomic systems give many informations on the solutions.

A holonomic system is called *simple* if its characteristic variety Λ is non-singular and the symbol Ideal coincides with the ideal of functions vanishing on Λ. Now, let M be a real analytic manifold, and X its complexification. The conormal bundle T_M^*X of M, identified with $\sqrt{-1}\,T^*M$, is embedded in the cotangent bundle T^*X of X. Let S^*M be the cotangent sphere bundle $(T^*M-M)/\boldsymbol{R}^+$ and let γ be the projection $\sqrt{-1}\,T^*M-M\to\sqrt{-1}\,S^*M$. In [3] we defined the sheaf $\mathcal{C}_M$ of microfunctions on $\sqrt{-1}\,S^*M$ and the sheaf $\mathcal{B}_M$ of hyperfunctions on M. In this note, we modify them slightly. Let us consider the presheaf on $\sqrt{-1}\,T^*M$

$$\begin{aligned}U\rightsquigarrow\{(f,u):\ & f\in\Gamma(U\cap M,\mathcal{B}_M),\\ & u\in\Gamma(\gamma(U-M),\mathcal{C}_M)\text{ such that}\\ & \mathrm{sp}(f)=u\quad\text{on}\quad\pi^{-1}(U\cap M)\}.\end{aligned}$$

The sheaf associated to this presheaf is denoted by $\check{\mathcal{C}}_M$. If U is a cone (i.e. invariant under multiplication by $\boldsymbol{R}^+$), then $\Gamma(U,\check{\mathcal{C}}_M)$ coincides with the set of sections of this presheaf on U. $\check{\mathcal{C}}_M$ shares the property

$$\check{\mathcal{C}}_M|_M=\mathcal{B}_M$$
$$\check{\mathcal{C}}_M|_{\sqrt{-1}\,T^*M-M}=\gamma^{-1}\mathcal{C}_M.$$

The important fact is that $\mathcal{E}_X^{(\lambda)}$ operates on $\check{\mathcal{C}}_M$, that is, there exists a canonical bilinear sheaf homomorphism

$$\mathcal{E}_X^{(\lambda)}|_{\sqrt{-1}\,T^*M}\times\check{\mathcal{C}}_M\to\check{\mathcal{C}}_M$$

such that $P(Q(u))=(PQ)(u)$ for $P\in\mathcal{E}_X^{(\lambda)}$ and $Q\in\mathcal{E}_X^{(\mu)}$.

Therefore, if $\mathcal{J}u=0$ is a system of micro-differential equations, we can consider a microfunction solution, that is, a microfunction u satisfying $\mathcal{J}u=0$. It is evident that the support of a microfunction solution is contained in the characteristic variety of the system $\mathcal{J}u=0$.

In the sequel, $\check{\mathcal{C}}_M$ *is abbreviated by* $\mathcal{C}_M$.

2. Symbol of Simple Microfunctions

Let u be a microfunction, and Λ a Lagrangian manifold in $\sqrt{-1}\,T^*M$, that is, an n-dimensional submanifold in $\sqrt{-1}\,T^*M$ on which the fundamental 1-form ω vanishes.

If u satisfies a simple holonomic system of micro-differential equations with the complexification of Λ as its characteristic variety, u is called a *simple microfunction* supported on Λ.

At a generic point of Λ, the projection $\Lambda\to M$ is of constant rank, that is, there exists an open dense subset Λ' of Λ such that the dimension of the image of $T_x\Lambda\to T_xM$ is locally constant on $x\in\Lambda'$. Then, locally, Λ' coincides with the conormal bundles $\sqrt{-1}\,T_N^*M$ of a submanifold N of M. Let us choose a local coordinate system $(x_1,\cdots,x_n)$ on M such that N is given by $x_1=\cdots=x_l=0$. Then, a simple microfunction u supported on Λ is written in the form

$$P(x,D)\delta(x_1,\cdots,x_l)$$

with $P\in\mathcal{E}(\lambda)$ such that $\sigma_\lambda(P)|_\Lambda\not\equiv 0$. We will call

$$\frac{1}{(2\pi)^{l/2}}\sigma(P)\Big|_\Lambda\sqrt{\frac{|d\xi_1\cdots d\xi_l dx_{l+1}\cdots dx_n|}{|dx_1\cdots dx_n|}}$$

the *principal symbol* of u and denote it by $\sigma_\Lambda(u)$. This is regarded as a section of $V_\Lambda^{\otimes 1/2}\otimes V_M^{\otimes(-1/2)}$, in which V_Λ (resp. V_M) is the sheaf of densities on Λ (resp. on M) with real analytic functions as coefficients. Since they are oriented, we can consider $V_\Lambda^{\otimes 1/2}$ (or $V_M^{\otimes 1/2}$). $\sigma_\Lambda(u)$ is invariant under coordinate transformations, because $\sqrt{|d\xi_1\cdots d\xi_l dx_{l+1}\cdots dx_n|/|dx_1\cdots dx_n|}$ is $\sqrt{|d\xi_1\cdots d\xi_l|/|dx_1\cdots dx_l|}$, and $|d\xi_1\cdots d\xi_l|$ is transformed in the same manner as $|dx_1\cdots dx_l|^{-1}$ under coordinate transformations. Therefore, $\sqrt{|d\xi_1\cdots d\xi_l dx_{l+1}\cdots dx_n|/|dx_1\cdots dx_n|}$ is transformed in the same way as $|dx_1\cdots dx_l|^{-1}$. On the other hand $\delta(x_1,\cdots,x_l)$ is changed in the same way as $|dx_1\cdots dx_l|^{-1}$. Therefore, we can define canonically the principal symbol $\sigma_\Lambda(u)$ of u on Λ' as a section of $V_\Lambda^{\otimes 1/2}\otimes V_M^{\otimes(-1/2)}$. The homogeneous degree of $\sigma_\Lambda(u)$ with respect to ξ is called the *order* of u and denoted by $\mathrm{ord}_\Lambda(u)$. $\mathrm{ord}_\Lambda(u)$ is therefore equal to $\lambda+l/2$. $\sigma_\Lambda(u)$ cannot be prolonged to a real analytic section of $V_\Lambda^{\otimes 1/2}\otimes V_M^{\otimes(-1/2)}$ over all Λ, in general. However, there exists a line bundle L_Λ, introduced by Hörmander, isomorphic to $V_\Lambda^{\otimes 1/2}\otimes V_M^{\otimes(-1/2)}$ on Λ', and $\sigma_\Lambda(u)$ can be prolonged on L_Λ over all Λ.

We will describe L_Λ. In the following, we write $V^{1/2}$ for $V^{\otimes 1/2}$.

For any $p\in\sqrt{-1}\,T^*M$, $V=T_p(\sqrt{-1}\,T^*M)$ has a structure of imaginary symplectic vector space, that is, there exists a purely imaginary valued non-degenerate skew-symmetric bilinear form E on V. If ω is the fundamental 1-form on $\sqrt{-1}\,T^*M$ (which is purely imaginary valued), E is given by

$$E(\nu_1, \nu_2) = \langle d\omega, \nu_1 \wedge \nu_2 \rangle .$$

For a linear subspace W of V, the orthogonal complement $W^\perp$ is, by definition, $\{\nu \in V; E(\nu, \omega)=0 \text{ for any } \omega \in W\}$. W is called isotropic (resp. involutory, Lagrangian) if $W^\perp \supset W$ (resp. $W^\perp \subset W$, $W^\perp = W$). Because $\dim W^\perp + \dim W = \dim V = 2n$, $\dim W = n$ for a Lagrangian plane W. For a triplet of Lagrangians $\lambda_1, \lambda_2, \lambda_3$ of V, we define the index $\tau(\lambda_1, \lambda_2, \lambda_3)$ as follows: $Q(x_1, x_2, x_3) = E(x_1, x_2) + E(x_2, x_3) + E(x_3, x_1)$ $(x_\nu \in \lambda_\nu)$; this is considered as a purely imaginary valued quadratic form on the $3n$-dimensional vector space $\lambda_1 \oplus \lambda_2 \oplus \lambda_3$. $\tau(\lambda_1, \lambda_2, \lambda_3) = \sqrt{-1}\ \mathrm{sgn}\ (Q/\sqrt{-1})$. Here, sgn represents the signature of the quadratic form, that is, the difference of the number of positive eigen-values and that of the negative eigen-values. τ has the properties expressed in the following propositions.

Proposition. o) $\tau(\lambda_1, \lambda_2, \lambda_3) \in \sqrt{-1}\,\boldsymbol{Z}$.

i) *τ is alternating with respect to the permutation of the λ_j's, that is,* $\tau(\lambda_1, \lambda_2, \lambda_3) = -\tau(\lambda_2, \lambda_1, \lambda_3) = -\tau(\lambda_1, \lambda_3, \lambda_2)$.

ii) *τ satisfies the chain condition:* $\tau(\lambda_2, \lambda_3, \lambda_4) - \tau(\lambda_1, \lambda_3, \lambda_4) + \tau(\lambda_1, \lambda_2, \lambda_4) - \tau(\lambda_1, \lambda_2, \lambda_3) = 0$.

iii) *If λ_j moves such that* $\dim(\lambda_1 \cap \lambda_2)$, $\dim(\lambda_2 \cap \lambda_3)$ *and* $\dim(\lambda_3 \cap \lambda_1)$ *are invariant, then* $\tau(\lambda_1, \lambda_2, \lambda_3)$ *does not change.*

By these properties, for N Lagrangians $\lambda_1, \cdots, \lambda_N$,

$$\tau(\lambda_1, \lambda_2, \mu) + \tau(\lambda_2, \lambda_3, \mu) + \cdots + \tau(\lambda_{N-1}, \lambda_N, \mu) + \tau(\lambda_N, \lambda_1, \mu)$$

does not depend on the choice of a Lagrangian μ. We will denote it by $\tau(\lambda_1, \cdots, \lambda_N)$. This has the similar properties, which are derived easily from the properties of τ.

i) $\tau(\lambda_1, \cdots, \lambda_N) = \tau(\lambda_2, \cdots, \lambda_N, \lambda_1) = -\tau(\lambda_N, \cdots, \lambda_1)$.

ii) *If λ_j moves such that* $\dim(\lambda_1 \cap \lambda_2), \cdots, \dim(\lambda_N \cap \lambda_1)$ *are invariant, then* $\tau(\lambda_1, \cdots, \lambda_N)$ *does not change.*

The line bundle L_Λ on a Lagrangian manifold Λ in $\sqrt{-1}\ T^*M$ can be defined by this index τ. For $p \in \Lambda$, we denote by $\lambda_\Lambda(p)$ (resp. $\lambda_M(p)$) the tangent space $T_p\Lambda$ (resp. the tangent space of the fibre $\pi^{-1}\pi(p)$ through p), which is a Lagrangian plane in $T_p(\sqrt{-1}\ T^*M)$. Let S_p be the space of all Lagrangian planes in $T_p(\sqrt{-1}\ T^*M)$ transversal to $\lambda_\Lambda(p)$ and $\lambda_M(p)$. Set $S = \bigcup_p S_p$. S is a manifold with the canonical projection $t: S \to \Lambda$. For any open set U in Λ the set of real analytic sections of L_Λ is, by definition, the set of the real analytic maps f from $t^{-1}(U)$ to $V_\Lambda^{-1/2} \otimes V_M^{-1/2}$ satisfying

$$\begin{aligned} f(\mu_1) &= \exp\frac{\pi}{4}(\tau(\lambda_\Lambda(p), \mu_1, \lambda_M(p)) - \tau(\lambda_\Lambda(p), \mu_2, \lambda_M(p)))f(\mu_2) \\ &= \exp\frac{\pi}{4}\tau(\lambda_\Lambda(p), \mu_1, \lambda_M(p), \mu_2)f(\mu_2) \end{aligned}$$

for $\mu_1, \mu_2 \in S_p$. On Λ', L_Λ is canonically isomorphic to $V_\Lambda^{1/2} \otimes V_M^{-1/2}$ by the isomorphism $V_\Lambda^{1/2} \otimes V_M^{-1/2} \ni g \mapsto f \in L_\Lambda$ given by

$$f(\mu) = \exp\Big(\frac{\pi}{4}\tau(\lambda_\Lambda(p), \mu, \lambda_M(p))\Big) g,$$

because $\dim(\lambda_\Lambda(p) \cap \mu) = \dim(\mu \cap \lambda_M(p)) = 0$ and $\dim(\lambda_\Lambda(p) \cap \lambda_M(p))$ is locally constant, hence $\tau(\lambda_\Lambda(p), \mu, \lambda_M(p))$ is a locally constant function.

We can prove the following theorem.

Theorem. *A principal symbol $\sigma_\Lambda(u)$ of a simple microfunction can be continued to a real analytic section on L_Λ.*

The principal symbol is important because a solution of a simple holonomic system is completely controlled by its principal symbol in the following sense.

Theorem. *Let $\mathcal{J}u = 0$ be a simple holonomic system with the complexification of Λ as its characteristic variety.*

i) *Locally, there exists only one microfunction solution up to constant multiples.*

ii) *If two microfunction solutions u_1 and u_2 of $\mathcal{J}u = 0$ have the same principal symbol, then $u_1 = u_2$.*

The following theorem is concerned with the integral transformation with a simple microfunction as kernel.

Theorem. *Let M and N be two real analytic manifolds, Λ_N (resp. Λ) be a Lagrangian manifold in $\sqrt{-1}\, T^*N$ (resp. $\sqrt{-1}\, T^*(M \times N)$). Let p_1 (resp. p_2^a) be the projection $\sqrt{-1}\, T^*(M \times N) \to \sqrt{-1}\, T^*M$ (resp. $\sqrt{-1}\, T^*N$) defined by $(x, y: \sqrt{-1}(\xi, \eta)) \to (x, \sqrt{-1}\xi)$ (resp. $(y, -\sqrt{-1}\eta)$). Suppose that*

i) *$p_2^a|_\Lambda : \Lambda \to \sqrt{-1}\, T^*N$ is transversal to Λ_N (see the remark following this theorem).*

ii) *$\Lambda \cap p_2^{a-1}(\Lambda_N)$ is isomorphic to a submanifold Λ_M in $\sqrt{-1}\, T^*M$ by the projection p_1.*

Then, for any simple microfunction $v(y)$ (resp. $K(x, y)$) supported on Λ_N (resp. Λ), $u(x) = \int K(x, y) v(y) dy$ is a simple microfunction on M supported on Λ_M and satisfies

$$\operatorname{ord}_{\Lambda_M}(u) = \operatorname{ord}_\Lambda(K) + \operatorname{ord}_{\Lambda_N}(v) - \dim N/2.$$

$$\sigma_{\Lambda_M}(u) = (2\pi)^{\frac{\dim N}{2}} \sigma_\Lambda(K) \sigma_{\Lambda_N}(v) |dy| / \sqrt{|d\eta dy|},$$

that is,

$$(1) \qquad \sigma_{\Lambda_M}(u)(\mu_M) = (2\pi)^{\frac{\dim N}{2}} \exp\frac{\pi}{4}\tau(\lambda_{\Lambda_M} \times \lambda_{\Lambda_N}^a, \mu_M \times \mu_N^a, \lambda_M \times \lambda_N, \mu, \lambda_\Lambda)$$

$$\sigma_\Lambda(K)(\mu) \sigma_{\Lambda_M}(v)(\mu_N)\, |dy| / \sqrt{|d\eta dy|}.$$

for any Lagrangian planes μ (resp. μ_M and μ_N) transversal to $\lambda_{M \times N}$ and λ_Λ (resp. λ_M and λ_{Λ_M}, λ_N and λ_{Λ_N}).

Remark. Let X and Y be two manifolds and Z be a submanifold of Y. Then a map $f: X \to Y$ is said to be transversal to Z if, for any $x \in f^{-1}(Z)$, $T_x X \to T_{f(x)} Y / T_{f(x)} Z$ is surjective. In this case, $f^{-1}Z$ is a submanifold of X and $T_x f^{-1}Z$ is the kernel of $T_x X \to T_{f(x)} Y / T_{f(x)} Z$. Therefore, $V_X / V_{f^{-1}Z} = V_Y / V_Z$.

Remark. The meaning of the formula (1) is as follows: $\sigma_{\Lambda_N}(v)(\mu_N)|dy|$ belongs to $V_{\Lambda_N}^{1/2} \otimes V_N^{-1/2} \otimes V_N = V_{\Lambda_N}^{1/2} \otimes V_N^{1/2}$. Since $V_{\Lambda_N} \otimes V_{\sqrt{-1}\, T^*N}^{-1} = V_{p_2{}^{a-1}\Lambda_N \cap \Lambda} \otimes V_\Lambda^{-1} = V_{\Lambda_M} \otimes V_\Lambda^{-1}$, $\sigma_{\Lambda_N}(v)(\mu_N)|dy|/\sqrt{|dydy|}$ belongs to $V_{\Lambda_N}^{1/2} \otimes V_\Lambda^{-1/2} \otimes V_N^{1/2}$.
Therefore, the right hand side of (1) belongs to

$$(V_\Lambda^{1/2} \otimes V_{M\times N}^{-1/2}) \otimes (V_{\Lambda_M}^{1/2} \otimes V_\Lambda^{-1/2} \otimes V_N^{1/2}) = V_{\Lambda_M}^{1/2} \otimes V_M^{-1/2}.$$

Remark. $\sqrt{-1}\, T^*(M \times N)$ is identified with $\sqrt{-1}\, T^*M \times \sqrt{-1}\, T^*N$ by p_1 and p_2^a. For Lagrangian planes λ and μ in $T(\sqrt{-1}\, T^*M)$ and $T(\sqrt{-1}\, T^*N)$ respectiviely, the product $\lambda \times \mu^a$ signifies the corresponding Lagrangian plane in $\sqrt{-1}\, T^*(M \times N)$ under this identification.

Remark. A real analytic function $f(x)$ on M can be regarded as a microfunction with the support in the zero section $\sqrt{-1}\, T_M^*M$. If $f(x)$ does not vanish anywhere, $f(x)$ is a simple microfunction supported on $\sqrt{-1}\, T_M^*M$.

References

[1] Hörmander, L.: Fourier integral operators I, Acta Math. **127** (1971), 79–183.
[2] Sato, M.: Hyperfunctions and partial differential equations, Proc. Intern. Conf. on Functional Analysis and Related Topics, Univ. of Tokyo Press, 1969, 91–94.
[3] Sato, M., Kawai, T. and Kashiwara, M.: Microfunctions and pseudo-differential equations, Proc. of a Conference of Katata, Lecture Notes in Mathematics 287, Springer 1973.

Department of Mathematics
Nagoya University

(Received March 1, 1976)

A Note on Steenrod Reduced Powers of Algebraic Cocycles

S. Kawai

In the present note we prove the following theorem on the Steenrod reduced power of an algebraic cocycle.

Theorem. *If z is an algebraic cocycle on a non-singular projective variety, then the Steenrod square $Sq^{2k}(z)$ and the Steenrod reduced power $P^k(z)$ are also algebraic, and $Sq^{2k+1}(z)$ vanishes.*

This is an analogue of Atiyah-Hirzebruch [1].

1. Blowing up of an Algebraic Cocycle

Let Y be a non-singular projective variety and N a complex vector bundle over Y. We denote by $g : \boldsymbol{P}(N) \rightarrow Y$ the fibre bundle associated to N with projective space as fibre and by L_N the canonically defined line bundle on $\boldsymbol{P}(N)$ such that its restriction to a fibre of $\boldsymbol{P}(N)$ is associated to a hyperplane of the projective space. The following theorems are proved by Grothendieck [2].

Theorem 1. *The homomorphism $g^* : H^*(Y, \boldsymbol{Z}) \rightarrow H^*(\boldsymbol{P}(N), \boldsymbol{Z})$ of the cohomology rings is injective, $\boldsymbol{Z}$ being the ring of integers. If we denote by h_N the first Chern class of L_N and identify $H^*(Y, \boldsymbol{Z})$ with its image by g^*, then $H^*(\boldsymbol{P}(N), \boldsymbol{Z})$ is a free module over $H^*(Y, \boldsymbol{Z})$ with base $(1, h_N, h_N^2, \cdots, h_N^{r-1})$, where r is the rank of N. Moreover letting $c_i(N)$ be the Chern class of N, the following identity holds in $H^*(\boldsymbol{P}(N), \boldsymbol{Z})$:*

$$h_N^r = -c_1(N)h_N^{r-1} - c_2(N)h_N^{r-2} - \cdots - c_r(N).$$

Theorem 2 *The above theorem holds in the category of non-singular projective algebraic varieties, when cohomology rings are replaced by Chow rings.*

Let η be a k-dimensional unitary bundle on a differentiable manifold V and E (resp. A) the associated bundle with $(2k-1)$-sphere (resp. $2k$-disk) as fibre. We consider A to be a manifold with boundary E and V to be contained in A as centers of fibres. Let $q : A \rightarrow V$ be the projection of the fibre bundle and ψ be the injection of

V into A. The following fundamental theorem is obtained by Thom [5].

Theorem 3 *There is a class U of $H^{2k}(A, E; Z)$ such that the homomorphism*

$$\varphi : H^{i-2k}(V, Z) \xrightarrow{q^*} H^{i-2k}(A, Z) \xrightarrow{U\cdot} H^i(A, E; Z)$$

is an isomorphism and $\psi^*(U)$ is the Chern class $c_k(\eta)$. Moreover φ coincides with the Gysin homomorphism $G\psi_*$ associated to $\psi : V \to A$.

As a corollary to this theorem we have the following proposition.

Proposition 4. *Let X and Y be a compact complex manifold and its complex submanifold of complex codimension r. If we denote by $N(Y, X)$ the normal bundle of Y in X, and if we let i be the injection map of Y into X, then we have*

$$i^*Gi_*(u) = uc_r(N(Y, X)), \qquad \text{for} \quad u \in H^*(Y, Z),$$

where Gi_* is the Gysin homomorphism induced by the injection i.

Proof. We consider the tubular neighborhood N of Y in X and apply Theorem 3. Denoting by $\dot{N}$ the boundary of N and letting

$$\mu : (N, \dot{N}) \to (X, X - \text{int } N),$$
$$r : (X, \phi) \to (X, X - \text{int } N)$$

be the injection maps, then we have by excision the isomorphism

$$\mu^* : H^*(X, X - \text{int } N) \simeq H^*(N, \dot{N}),$$

and we infer readily that

$$(*) \qquad Gi_*(u) = r^* \circ \mu^{*-1} \circ \varphi(u) = r^* \circ \mu^{*-1}(q^*(u) \cdot U),$$

where φ is the Thom isomorphism associated to the sphere bundle $\dot{N}$ over Y and $q : N \to Y$ is the projection as fibre bundle. Therefore from the commutative diagram

$$\begin{array}{ccccc} H^*(X, X-\text{int } N) & \xrightarrow{\mu^*} & H^*(N, \dot{N}) & & \\ r^* \downarrow & & \downarrow & \psi^* & \\ i^* : \quad H^*(X) & \to & H^*(N) & \to & H^*(Y), \end{array}$$

we see

$$i^*Gi_*(u) = i^*r^*\mu^{*-1}(q^*(u) \cdot U) = u \cdot \psi^*(U) = uc_r(N(Y, X)),$$

since $q \circ \psi$ is the identity.

Now let $f : \tilde{X} \to X$ be the blowing up of a non-singular projective variety X with non-singular center Y of codimension r and let g be the restriction of f to $\tilde{Y} = f^{-1}(Y)$.

$$\begin{array}{ccc} \tilde{X} & \xrightarrow{f} & X \\ \cup j & & \cup \\ \tilde{Y} & \xrightarrow{g} & Y \end{array}$$

Then $g : \tilde{Y} \to Y$ is the fibre bundle $\boldsymbol{P}(N(Y, X))$ with the normal bundle $N(Y, X)$ and the restriction $j^*([\tilde{Y}])$ of the line bundle $[\tilde{Y}]$ is $L^{-1}_{N(Y,X)}$, where j is the injection of $\tilde{Y}$ in $\tilde{X}$. We prove the following well-known fact.

Proposition 5. *The homomorphism*

$$\Phi(x+\sum_{\alpha\geqq 1} u_{s-2\alpha}) = f^*(x) + \sum_{\alpha\geqq 1} Gj_*(g^*(u_{s-2\alpha}))\tilde{y}^{\alpha-1}$$

gives an isomorphism

$$\Phi : H^s(X, Z) \oplus H^{s-2}(Y, Z) \oplus H^{s-4}(Y, Z) \oplus \cdots \simeq H^s(\tilde{X}, Z),$$

where $\tilde{y}$ is the cohomology class dual to $\tilde{Y}$ and Gj_* is the Gysin homomorphism induced by the injection j.

Proof. In the commutative diagram of the cohomology groups

$$\begin{array}{ccccccccccc} \cdots \to & H^{s-1}(Y) & \to & H^s(X, Y) & \to & H^s(X) & \to & H^s(Y) & \to & H^{s+1}(X, Y) & \to \cdots \\ & \downarrow & & \wr\downarrow f^* & & \downarrow g^* & & \downarrow & & \wr\downarrow & \\ \cdots \to & H^{s-1}(\tilde{Y}) & \to & H^s(\tilde{X}, \tilde{Y}) & \to & H^s(\tilde{X}) & \xrightarrow{j^*} & H^s(\tilde{Y}) & \to & H^{s+1}(\tilde{X}, \tilde{Y}) & \to \cdots \end{array}$$

with exact rows, we replace the first row by

$$\cdots \to H^s(X, Y) \to H^s(X) \oplus \sum H^{s-2\alpha}(Y) \to$$
$$H^s(Y) \oplus \sum H^{s-2\alpha}(Y) \to H^{s+1}(X, Y) \to \cdots$$

where the homomorphisms on added parts are identities and trivial ones. By Theorem 1 we see that map

$$\Psi(u_s + \sum u_{s-2\alpha}) = g^*(u_s) + \sum g^*(u_{s-2\alpha})(-h_{N(\tilde{Y},\tilde{X})})^\alpha$$

gives the isomorphism

$$\Psi : H^s(Y) \oplus \sum H^{s-2\alpha}(Y) \simeq H^s(\tilde{Y}).$$

The following diagram with exact rows

$$\begin{array}{ccccccccc} \cdots \to H^s(X, Y) & \to & H^s(X) \oplus \sum H^{s-2\alpha}(Y) & \to & H^s(Y) \oplus \sum H^{s-2\alpha}(Y) & \to & H^{s+1}(X, Y) \to \cdots \\ \wr\downarrow & & \downarrow \Phi & & \downarrow \Psi & & \wr\downarrow \\ \cdots \to H^s(\tilde{X}, \tilde{Y}) & \to & H^s(\tilde{X}) & \xrightarrow{j^*} & H^s(\tilde{Y}) & \to & H^{s+1}(\tilde{X}, \tilde{Y}) \to \cdots \end{array}$$

is commutative, since $c_1(N(\tilde{Y}, \tilde{X})) = j^*(\tilde{y}) = -h_{N(\tilde{Y},\tilde{X})}$ and hence by Proposition 4 we have

$$j^*(Gj_*(g^*(u_{s-2\alpha}))\tilde{y}^{\alpha-1}) = g^*(u_{s-2\alpha})j^*(\tilde{y})^\alpha = g^*(u_{s-2\alpha})(-h_{N(\tilde{Y},\tilde{X})})^\alpha.$$

Therefore we infer readily by the "five lemma" that Φ is an isomorphism.

Now we consider an algebraic cocycle z dual to an irreducible subvariety Z of X of codimension s which contains Y properly. Let $\tilde{Z}$ be the strict transform of Z by the blowing up f and $\tilde{z}$ the cohomology class dual to $\tilde{Z}$. Then we have

Proposition 6. *If we put*

$$\Phi^{-1}(\tilde{z}) = x + \sum u_{s-2\alpha},$$

then $x=z$ and *$u_{s-2\alpha}$ are algebraic cocycles.*

Proof. From the identity

$$\tilde{z} = f^*(x) + \sum_\alpha Gj_*(g^*(u_{s-2\alpha}))\tilde{y}^{\alpha-1},$$

we obtain

$$j^*(\tilde{z}) = j^*f^*(x) + \sum_\alpha g(u_{s-2\alpha})(-h_{N(\tilde{Y},\tilde{X})})^\alpha.$$

By assumption $\tilde{Z}$ and $\tilde{Y}$ intersect properly and hence a linear combination of the irreducible components of the intersection $\tilde{Z} \cap \tilde{Y}$ represents the cohomology class $j^*(\tilde{Z})$. From Theorem 2 we infer readily that $u_{s-2\alpha}$ are algebraic. On the other hand we have

$$z = Gf_*(\tilde{z}) = Gf_*(f^*(x) + \sum Gj_*(g^*(u_{s-2\alpha}))\tilde{y}^{\alpha-1}) = x.$$

Remark. If Z is non-singular, then we have in the cohomology ring $H^*(\tilde{X}, \boldsymbol{Z}_p)$, Z_p being the prime field of characteristic p,

$$\tilde{z} = f^*(z) + \sum (-1)^\alpha Gj_*(g^*(c_{s-\alpha}(N(Z, X))))\tilde{y}^{\alpha-1}.$$

Proof. In this case the restriction of f to $\tilde{Z}$ is the blowing up of Z with the center Y and the intersection $\tilde{Z} \cap \tilde{Y}$ is identified with $\boldsymbol{P}(N(Y, Z))$. On the other hand we have the following exact sequence of vector bundles on Y.

$$0 \to N(Y, Z) \to N(Y, X) \to N(Z, X) \to 0.$$

Seeing that $j^*(\tilde{z})$ is the dual of $\tilde{Z} \cap \tilde{Y}$, it is sufficient to prove the following fact. Let

$$0 \to E \to F \to G \to 0$$

be an exact sequence of complex vector bundles on a complex manifold Y with E of rank r and G of rank s, and let $j: \boldsymbol{P}(E) \to \boldsymbol{P}(F)$ be the canonical injection. Putting $X = \boldsymbol{P}(F)$ and $Z = \boldsymbol{P}(E)$, we denote by z the dual of Z in X. Then we have in the cohomology ring $H^*(X, \boldsymbol{Z}_p)$

$$z = h_F^s + c_1(G)h_F^{s-1} + \cdots + c_{s-1}(G)h_F + c_s(G).$$

A proof of this is as follows. We put

$$z = c_0 h_F^s + c_1 h_F^{s-1} + \cdots + c_{s-1}h_F + c_s, \qquad c_i \in H^{2i}(Y, \boldsymbol{Z}_p),$$

and determine the coefficient c_i by Poincaré duality. Let $\varphi: X \to Y$ and $\psi: Z \to Y$ be the canonical projections of the fibre bundles.

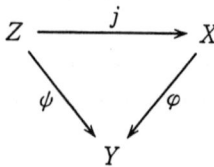

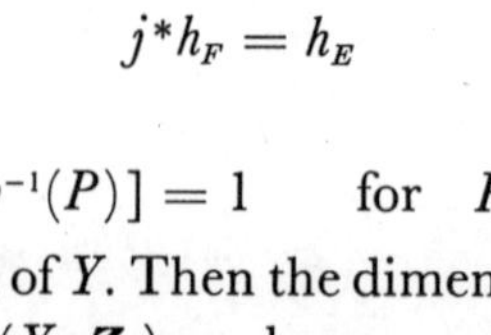

It is clear that

$$j^* h_F = h_E$$

and

$$h_Y^{r-1}[\psi^{-1}(P)] = 1 \qquad \text{for} \quad P \in Y.$$

Let n be the complex dimension of Y. Then the dimension of X is equal to $n+r+s-1$ and for any element d_n of $H^{2n}(Y, \boldsymbol{Z}_p)$ we have

$$d_n h_F^{r-1} Z[X] = j^*(d_n h_F^{r-1})[Z] = d_n h_E^{r-1}[Z] = d_n[Y].$$

Similarly we have

$$d_n h_F^{r-1} \sum_{i=0}^{s} c_i h_F^{s-i}[X] = d_n c_0[Y],$$

since $d_n c_i = 0$ for $i > 0$. Therefore we have

$$d_n[Y] = c_0 d_n[Y],$$

which implies

$$c_0 = 1.$$

Now we prove by induction on i that $c_i = c_i(G)$. If we put

$$\begin{aligned} F(x) &= 1 + c_1(F)x + c_2(F)x^2 + \cdots, \\ E(x) &= 1 + c_1(E)x + c_2(E)x^2 + \cdots, \\ G(x) &= 1 + c_1 x + c_2 x^2 + \cdots, \end{aligned}$$

then by Theorem 1 and an easy computation we have for any $d_{n-i} \in H^{2(n-i)}(Y, \boldsymbol{Z}_p)$

$$\begin{aligned} & d_{n-i} h_F^{r+i-1}(h_F^s + c_1 h_F^{s-1} + \cdots + c_{s-1} h_F + c_s)[X] \\ & = \left(d_{n-i} \times \text{the coefficient of } \frac{G(x)}{F(x)} \text{ of degree } i\right)[Y], \end{aligned}$$

and

$$d_{n-i} h_F^{r+i-1} Z[X] = \left(d_{n-i} \times \text{the coefficient of } \frac{1}{E(x)} \text{ of degree } i\right)[Y].$$

We have by the assumption of induction

$$G(x) \equiv 1 + c_1(G)x + c_2(G)x^2 + \cdots, \quad \operatorname{mod}(x^{i-1}),$$

and by a property of Chern class

$$F(x) = E(x)(1 + c_1(G)x + c_2(G)x^2 + \cdots).$$

Hence, considering the coefficient of the power series

$$\frac{1}{E(x)}\left(\frac{G(x)}{1 + c_1(G)x + c_2(G)x^2 + \cdots} - 1\right),$$

we obtain

$$d_{n-i}(c_i - c_i(G))[Y] = 0.$$

Therefore by Poincaré duality we have

$$c_i = c_i(G).$$

2. Proof of Theorem

In Thom [5] the following theorem is obtained.

Theorem 7 *Let X and Z be a manifold and its submanifold and let $j: Z \to X$ be the injection map. Then we have*

$$Sq^k z = Gj_* w_k(N(Z, X)),$$

where z is the dual of Z and $w_k(N(Z, X))$ is the Stiefel-Whitney class of the normal bundle of Z in X.

From this theorem we see imediately the following fact.

Corollary. *If Z is a non-singular subvariety of a non-singular projective variety X and z is the dual of Z, then $Sq^{2k+1}z=0$ and $Sq^{2k}z$ is an algebraic cocycle.*

In a similar manner we can prove the following proposition.

Proposition 8. *Let X and Z be a complex manifold and its complex submanifold and let $q_k(N(Z, X))$ be the Wu class* (cf. Milnor [4]) *of the normal bundle of Z in X. Then we have*

$$P^k z = Gj_*(q_k(N(Z, X))).$$

Proof. Letting N be a tubular neighborhood of Z in X and $\dot{N}$ its boundary, we consider N (resp. $\dot{N}$) to be a fibre bundle over X associated to $N(Z, X)$ with disk (resp. sphere) as fibre. Let s be the codimension of Z in X and let

$$\varphi : H^{*-2s}(Z) \to H^{*-2s}(N) \to H^*(N, \dot{N})$$

be the Thom isomorphism, then we have by definition

$$q_k(N(Z, X)) = \varphi^{-1}P^k\varphi(1).$$

Hence from (*) in the proof of Proposition 4 we see

$$P^k z = P^k Gj_*(1) = r^*\mu^{*-1}P^k\varphi(1) = r^*\mu^{*-1}\varphi(q_k(N(Z, X))) = Gj_*(q_k(N(Z, X))).$$

Corollary. *If Z is a non-singular subvariety of a non-singular projective variety X and z is the dual of Z, then $P^k z$ is also algebraic.*

Proof. By Wu (cf. Milnor [4]) the Wu class of a complex vector bundle is a polynomial of its Pontrjagin class.

Now we prove our theorem by induction on the dimension of ambient variety X. If Z is an irreducible subvariety with singularities, by Hironaka [3] there exists a finite succession of monoidal transformations $\{f_i : X_{i+1} \to X_i\}_{i=0}^{m}$ with centers Y_i, where $X_0 = X$, such that if Z_{i+1} is the strict transform of Z_i by f_{i+1} with $Z_0 = Z$, then Z_i contains Y_i properly and Z_m is a non-singular subvariety of X_m. By the above, our theorem holds for the cohomology class z_m dual to Z_m. We are to apply Propositions 5 and 6. Put $f = f_i$ and $\tilde{Y} = f^{-1}(Y_i)$ and let $j : \tilde{Y} \to X_{i+1}$ be the injection map. If z_ν is the cohomology class dual to Z_ν, then by Propositions 5 and 6 we have

$$z_{i+1} = f^*(z_i) + \sum_\alpha Gj_*(u_{s-2\alpha})\tilde{y}^{\alpha-1},$$

where $\tilde{y}$ is the dual of $\tilde{Y}$ and $u_{s-2\alpha}$ is an algebraic cocycle. By the property of P^k we have

$$P^k \tilde{z}_{i+1} = f^*(P^k z_i) + \sum_{\alpha} \sum_{\nu} P^{k-\nu}(Gj_*(u_{s-2\alpha}))P^{\nu}(\tilde{y}^{\alpha-1}).$$

Hence we have

$$P^k z_i = Gf_*(P^k \tilde{z}_{i+1} - \sum_{\alpha,\nu} P^{k-\nu}(Gj_*(u_{s-2\alpha}))P^{\nu}(\tilde{y}^{\alpha-1})),$$

since $Gf_* \circ f^*$ is the identity. Clearly the image of an algebraic cocycle by Gysin homomorphism is also algebraic. Therefore by the induction assumption it is sufficient to prove the following proposition.

Proposition 9. *We have for* $w \in H^*(\tilde{Y})$

$$P^k Gj_*(w) = Gj_*(P^k w) + Gj_*(P^{k-1}w)\tilde{y}^{p-1},$$

and

$$Sq^k Gj_*(w) = Gj_*(Sq^k w) + Gj_*(Sq^{k-2}w)\tilde{y}.$$

Proof. From (*) in the proof of Proposition 4, we have

$$\begin{aligned} P^k Gj_* w &= r^*\mu^{*-1}(\sum_{\nu} P^{k-\nu}q^*(w)\cdot P^{\nu}U) = r^*\mu^{*-1}(P^k q^*(w)\cdot U + P^{k-1}q^*(w)\cdot U^p) \\ &= r^*\mu^{*-1}(q^*(P^k(w))\cdot U + q^*(P^{k-1}(w))\cdot U\cdot U^{p-1}) \\ &= Gj_*(P^k w) + Gj_*(P^{k-1}w)\tilde{y}^{p-1}. \end{aligned}$$

Considering torsion it is clear that $Sq^1 U = 0$. Hence similarly we have

$$Sq^k Gj_* w = Gj_*(Sq^k w) + Gj_*(Sq^{k-2}w)y.$$

Thus we have proved our theorem completely.

References

[1] Atiyah, M. and Hirzebruch, F. : Analytic cycles on complex manifolds, Topology, **1** (1962), 25–45.

[2] Grothendieck, A. : Sur quelques propriétés fondamentales en théorie des intersections, Séminaire Chevalley (1958), Anneaux de Chow et applications.

[3] Hironaka, H. : Resolution of singularities of an algebraic variety over a field of characteristic zero, Ann. of Math., **79** (1964), 109–326.

[4] Milnor, J. W. (and J. D. Stasheff) : Characteristic Classes, Ann. of Math. Studies (1974).

[5] Thom, R. : Espaces fibrés en sphères et carrés de Steenrod, Ann. Sci. École Norm. Sup., **69** (1952), 109–182.

[6] Thom, R. : Quelques propriétés globales des variétés différentiables, Comment. Math. Helv., **28** (1954), 17–86.

Departmet of Mathematics
Rikkyo University

(Received February 20, 1976)

Polynomial Growth C^{∞}-de Rham Cohomology and Normalized Series of Prestratified Spaces

N. Sasakura

0. Introduction

1. In this note we will mainly be concerned with polynomial growth properties of C^{∞}-differential forms related to real analytic varieties. We announced basic results on analytic de Rham cohomology in $[4]_{1\text{-}4}$. This note should be read as a continuation of these four notes. Details of the present note are too long to be included here, and will appear elsewhere.

The purpose of this note is as follows :

(I) To introduce the notion of *p.g.*[1] *simple prestratification* for analytic varieties and for prestratified spaces of certain types (§ 1).

(II) To discuss relations between the notion of p.g. simple prestratification and that of *normalized series* of prestratified spaces introduced in $[4]_4$ (§ 2).

2. To explain the notion in (I), let M be a C^{∞}-manifold, and let $\mathcal{A}=\{A_{\lambda}\}_{\lambda}$ be an open covering of M. Recall that $\mathcal{A}$ is called *simple* if, for any $A_{\lambda_1}, \cdots, A_{\lambda_t} \in \mathcal{A}$ such that $A_{\lambda_1 \cdots t}(=\bigcap_s A_{\lambda_s}) \neq \phi$, $A_{\lambda_1 \cdots t}$ is contractible (cf. A. Weil [8]). The existence of simple coverings for C^{∞}-manifolds was used as a basic tool for the proof of the C^{∞}-de Rham theorem (cf. [8]).

Now the notion of *p. g. simple prestratification* is a combination of the notions of prestratification (of a topological space), p. g. property of C^{∞}-differential forms (related to an analytic variety)[2] and simple covering (of a C^{∞}-manifold).

Roughly, the role of p. g. prestratification in our study of p. g. properties of C^{∞}-differential forms related to analytic varieties is similar to that in [8] of simple coverings in the C^{∞}-de Rham theorem.

Actually the author's starting point in C^{∞}-aspects of the discussions on analytic de Rham cohomology is to find a suitable substitute for the notion of simple covering in the study of p. g. properties of C^{∞}-differential forms related to analytic varieties.

The discussion on attaching p. g. simple prestratifications to an analytic variety, so that the prestratifications closely reflect properties of the variety, is

1) "p. g." ="polynomial growth"

2) In this note "analytic variety" and "analytic function" mean always "real analytic" ones. We abbreviate sometimes "analytic variety" as "variety".

most important in our study of p. g. C^∞-de Rham cohomology theory.

Remark. The algebraic pattern of our proof of the p. g. C^∞-de Rham theorem (Theorem 2.2) is, in principle, parallel to that of the proofs of the C^∞-de Rham theorem for C^∞-manifolds (cf. [1], [8]) : let V be an analytic variety, and let $\mathcal{S}$ be a p. g. simple prestratification of V. Then the study of p. g. properties of C^∞-differential forms attached to V is *localized* to the study of p. g. properties of C^∞-differential forms attached suitably to sequences of strata (cf. $[4]_3$, [5]).

However, the localization step[1] mentioned above is *not* entirely parallel to the localization step[2] in the standard proof of the C^∞-de Rham theorem (cf. [1], [8]). The difference seems to be non-neglisible, and will be discussed elsewhere.

3. Concerning the contents in (II), let V be an analytic variety in a bounded domain in a euclidean space $\boldsymbol{R}^n$, and let $(\mathfrak{N}, \mathfrak{F})$ be a normalized series attached to V. (See $[4]_4$.)

The basic result of (II) is that, under a suitable distance condition $(\mathfrak{N}, \mathfrak{F})$[3], p. g. simple prestratifications and other data, which are useful in the study of p. g. properties of V, are obtained from $(\mathfrak{N}, \mathfrak{F})$ in a simple fashion. The above fact would also justify the introduction of the normalized series in analytic de Rham cohomology theory.

1. P. g. Simple Prestratificatian[4]

We introduce, in § 1, the notion of *p. g. simple prestratification* for topological spaces of certain types. We also discuss certain properties of such a prestratification. Details of this section, except the definition of simple and p. g. simple prestratifications, will be found in [5].

We will assume in § 1 that every stratum is a C^∞-manifold, and that every prestratification is a finite set.

1. 1. Simple Prestratification

Let M be a C^∞-manifold, and let V be a subset of M. Moreover, let $\mathcal{S}$ be a prestratification of V. We denote by $\mathcal{S}_C$, $\mathcal{S}_O$ and $\mathcal{S}_{CO}$, respectively, the C-, O- and CO-sets of $\mathcal{S}$ (cf. $[4]_3$, § 1, [5]).

(i) For a collection $\mathcal{N}=\{N(S_\lambda)\,;\, S_\lambda\in\mathcal{S}\}$ of neighborhoods in M of strata of $\mathcal{S}$ and

---

1) cf. $[4]_3$.

2) Let M be a C^∞-manifold, and let $\mathcal{A}$ be a suitable open covering of M. Then the C^∞-de Rham theory for M is a formal consequence of C^∞-de Rham theory for the intersections of elements of $\mathcal{A}$ in a well-known manner. We may say this fact as "C^∞-de Rham cohomology for M is localized to that of the intersections of elements of $\mathcal{A}$."

3) See § 2. The condition given in § 2 is satisfied by normalized series, which appear in our geometric applications.

4) We use freely the notions in $[4]_{3,4}$.

for an element $X \in \mathcal{S}_C$, $\mathcal{S}_O$ or $\mathcal{S}_{CO}$, we use the symbol $N(X)$ for the following manifold.

(a)$_1$ If $X \in \mathcal{S}_C$, then $N(X) = \bigcup_\lambda N(S_\lambda)$, where $S_\lambda \in X$.

(a)$_2$ If $X = S_{\lambda_1 \cdots \iota} \in \mathcal{S}_O$, then $N(X) = \bigcap_s N(S_{\lambda_s})$.

(a)$_3$ If $X = (\mathcal{T}, m, S_{\lambda_1 \cdots \iota}) \in \mathcal{S}_{CO} (\subset \mathcal{S}_C \times \boldsymbol{Z}^+ \times \mathcal{S}_O)$, then $N(X) = N(\mathcal{T}_m) \cap N(S_{\lambda_1 \cdots \iota})$. (See [4]$_3$.)

(ii) Let $\mathcal{N} = \{N(S_\lambda) ; S_\lambda \in \mathcal{S}\}$ be a collection of neighborhoods[1] in M of strata of $\mathcal{S}$. Using the symbols introduced in (i), we first introduce the following definitions.

Definition 1. 1. We say that $\mathcal{N}$ is a *C^∞-thickening* of $\mathcal{S}$ if the following are valid.

(1. 1)$_1$ For any $X \in \mathcal{S}_C$, $\mathcal{S}_O$ or $\mathcal{S}_{CO}$, $N(X)$ is paracompact.

(1. 1)$_2$ If $N(S_\lambda) \cap N(S_\mu) \neq \phi$, then $S_\lambda > S_\mu$ or $S_\mu > S_\lambda$.

(1. 1)$_3$ If $N(S_\lambda) \cap S_\mu \neq \phi$, then $S_\mu > S_\lambda$.

(1. 1)$_4$ For any $\mathcal{T} \in \mathcal{S}_C$ and $X \in \mathcal{T}_O$, the natural homomorphism $i^* : H^*(N(X); \boldsymbol{R}) \to H^*(N(X) \cap |\mathcal{T}|; \boldsymbol{R})$[2] induced from the inclusion $i : N(X) \cap |\mathcal{T}| \hookrightarrow N(X)$ is bijective.

Definition 1. 2. We say that $\mathcal{N}$ is a *simple C^∞-thickening* of $\mathcal{S}$, if $\mathcal{N}$ is a C^∞-thickening of $\mathcal{S}$ and if, for any $S_{\lambda_1 \cdots \iota} \in \mathcal{S}_O$, the following are valid.

(1. 2)$_1$ $N(S_{\lambda_1 \cdots \iota})$ is contractible.

(1. 2)$_2$ There exists a retraction[3] $\tau_{1 \cdots \iota}$ of $N(S_{\lambda_1 \cdots \iota})$ to $N(S_{\lambda_1 \cdots \iota}) \cap S_{\lambda_\iota}$ such that, for any $S_\nu > S_{\lambda_\iota}$, the following hold :

(1. 2)$'_{2.1}$ $\tau_{1 \cdots \iota}((0, 1] \times (N(S_{\lambda_1 \cdots \iota}) \cap S_\nu)) \subset S_\nu$.

(1. 2)$'_{2.2}$ $\tau_{1 \cdots \iota}(0 \times (N(S_{\lambda_1 \cdots \iota}) \cap S_\nu)) = N(S_{\lambda_1 \cdots \iota}) \cap S_{\lambda_\iota}$.

Remark. Let $\mathcal{N}$ be as in the beginning of (ii). Assume that $\mathcal{N}$ satisfies (1. 1)$_{1\text{-}3}$ and (1. 2)$_{1,2}$. Then $\mathcal{N}$ satisfies (1. 1)$_4$ and is a C^∞-thickening of $\mathcal{S}$.

(iii) Now we introduce the following

Definition 1. 3. We say that $\mathcal{S}$ is *simple* if there exists a simple C^∞-thickening $\mathcal{N}$ of $\mathcal{S}$.

1. 2. C^∞-de Rham C. C. I.

Let M be a C^∞-manifold, and let V be a subset of M. Moreover, let $\mathcal{N} = \{N(S_\lambda) ; S_\lambda \in \mathcal{S}\}$ be a C^∞-thickening of $\mathcal{S}$.

Let $X \in \mathcal{S}_C$, $\mathcal{S}_O$ or $\mathcal{S}_{CO}$. Using the symbol $N(X)$ in the same sense as in § 1. 1, we denote by $\Omega(N(X))$ the cochain complex of C^∞-differential forms defined on $N(X)$.

1) In this note we use the terminology "neighborhood" for "open neighborhood".
2) $|\mathcal{T}|$ denotes the support of $\mathcal{T}$.
3) We use the terminology "retraction" for "strong deformation retract".

We write the collection $\{\Omega(N(X))\ ;\ X\in\mathcal{S}_C, \mathcal{S}_O \text{ or } \mathcal{S}_{CO}\}$ as $\Omega(\mathcal{S}, \mathcal{N})$ and call it *the C^∞-de Rham collection attached to* $(\mathcal{S}, \mathcal{N})$.

Next let $\mathcal{S}_D$ denote the D-set of $\mathcal{S}$. (See [5], [6].) Moreover, let $Y\in\mathcal{S}_D$. We then attach a homomorphism $K(Y)$ of cochain complex to Y by means of intersection and union relations of $N(X)$'s, where $X\in\mathcal{S}_C, \cdots$ (For the details see [5], [6].) The homomorphism $K(Y)$ is determined uniquely by $(\mathcal{S}, \mathcal{N})$ and Y. We write the collection $\{K(Y)\ ;\ Y\in\mathcal{S}_D\}$ as $\mathcal{K}(\Omega(\mathcal{S}, \mathcal{N}))$. Then, by the standard Mayer–Vietoris sequences applied to $N(X)$, $X\in\mathcal{S}_C, \cdots$, the pair $(\Omega(\mathcal{S}, \mathcal{N}), \mathcal{K}(\Omega(\mathcal{S}, \mathcal{N})))$ is easily checked to be a C. C. I. attached to[1] $\mathcal{S}$. (Namely the above pair satisfies exact sequences of Mayer–Vietoris types required in (1. 3) in $[4]_3$.) We call the above pair *the C^∞-de Rham* C. C. I. *attached to* $(\mathcal{S}, \mathcal{N})$.

Letting $(\Omega(\mathcal{S}, \mathcal{N}), \mathcal{K}(\Omega(\mathcal{S}, \mathcal{N})))$ be as above, we have easily the following

Proposition 1. 1. *For any $\mathcal{T}\in\mathcal{S}_C$ we have a natural isomorphism*:

(1. 3) $\quad H^*(|\mathcal{T}|\ ;\ \boldsymbol{R})\cong H^*(\Omega(N(\mathcal{T})))$.

Remark. Let $\tilde{\Omega}_M$ denote the sheaf over M of C^∞-differential forms. Then, for any $X\in\mathcal{S}_C, \cdots, \Omega(N(X))=\Gamma(N(X), \tilde{\Omega}_M)$, *and the single sheaf $\tilde{\Omega}_M$ suffices for all the arguments in* § 1. 2. We note that the contents of § 1. 2, Proposition 1. 1 as well as the fact that $(\Omega(\mathcal{S}, \mathcal{N}), \mathcal{K}(\Omega(\mathcal{S}, \mathcal{N})))$ is a C. C. I. attached to $\mathcal{S}$, are derived from portions of standard arguments in the proof of the C^∞-de Rham theorem, and do not contain anything essentially new. In the arguments below, we will discuss a type of de Rham theory in which p. g. properties of the differential forms in question appear; there arise phenomena which do not seem to appear in the standard proofs of the C^∞-de Rham theorem.

1. 3. P. g. Adequate Prestratified Space and p. g. Simple Prestratification

Let $\boldsymbol{R}^n$ be a euclidean space of dimension n. We mean by a "triplet in $\boldsymbol{R}^n$" a collection $Q=(U, V, \mathcal{S}_0)$ of a bounded domain U in $\boldsymbol{R}^n$, an analytic variety V in U and a prestratification $\mathcal{S}_0$ of (U, V). (See $[4]_4$.) We assume, in § 1. 3, that every stratum is an analytic manifold.

1. 3. 1. Let $Q=(U, V, \mathcal{S}_0)$ and $Q'=(U', V', \mathcal{S}'_0)$ be triplets in $\boldsymbol{R}^n$ such that Q' is a *d-envelop* of Q (cf. $[4]_4$).

We first introduce the following

Definition 1. 4. A *family $\mathcal{F}$ of d-comparison functions for* (Q, Q') is a collection of the following.

(a)$_1$ $\quad \{f(S'_\lambda)\ ;\ S'_\lambda\in\mathcal{S}'_0$ such that $\dim S'_\lambda\leqq n-1\}$.

(a)$_2$ $\quad \{g(S'_\mu)\ ;\ S'_\mu\in\mathcal{S}'_0$ such that fron $S'_\mu\neq\phi\}$.

1) For the notion of C. C. I., see $[4]_3$ and [5]. We remark that "cohomology theory for C. C. I. attached to $\mathcal{S}$" is, roughly, a synonym for "cohomology theory for $\mathcal{S}$ to which the localization steps (explained in § 0) are applied".

(a)$_3$ $\{h(S'_\lambda, \mathcal{T}'_\upsilon)\ ;\ (S'_\lambda, \mathcal{T}'_\upsilon) \in \mathcal{S}'_0 \times \mathcal{S}'_{0,C}$ such that $\dim \mathcal{T}'_\upsilon \leqq n-1$ and $S'_\lambda \in \mathcal{T}'_\upsilon\}$.

In the above $f(S'_\lambda)$, $g(S'_\mu)$ and $h(S'_\lambda, \mathcal{T}'_\upsilon)$ are analytic functions in U' and must satisfy the following conditions:

(a)$'_1$ $f(S'_\lambda)(P) \sim d(P, S'_\lambda)$ in $N_\sigma(S_\lambda$, fron $S'_\lambda)$[1].

(a)$'_2$ $g(S'_\mu)(P) \sim d(P$, fron $S'_\mu)$ in $N_\sigma(S_\mu$, fron $S'_\mu)$.

(a)$'_3$ $h(S'_\lambda, \mathcal{T}'_\upsilon)(P) \sim d(P, |\mathcal{T}'_\upsilon|)$ in $N_\sigma(S_\lambda$, fron $S'_\lambda)$.

Here σ is a suitable element in $\boldsymbol{R}^+ \times \boldsymbol{R}^+$. Moreover, we denote by $S_\lambda, \cdots$ the intersections $S'_\lambda \cap U, \cdots$.

Using the above definition we introduce the following

Definition 1. 5. A *p. g. adequate prestratified space* $\mathfrak{P}$ in $\boldsymbol{R}^n$ is a collection $(Q, Q', \mathcal{F})$, where Q, Q' are triplets in $\boldsymbol{R}^n$ such that Q' is a *d*-envelop of Q and $\mathcal{F}$ is a family of comparison functions for (Q, Q'). The single condition on $\mathfrak{P}$ is as follows:

(1. 4) (Q, Q') satisfies *d-separation condition.* (See [4]$_4$.)

1. 3. 2. Let $\mathfrak{P}=(Q, Q', \mathcal{F})$ be a p. g. adequate prestratified space in $\boldsymbol{R}^n$. We write Q, Q' explicitly as $Q=(U, V, \mathcal{S}_0)$ and $Q'=(U', V', \mathcal{S}'_0)$.

(i) Let $\mathcal{N}=\{N(S_\lambda)\ ;\ S_\lambda \in \mathcal{S}_0\}$ be a C^∞-thickening of $\mathcal{S}_0$. We then introduce the following definitions.

Definition 1. 6$_1$. We say that $\mathcal{N}$ is *p. g. adequate* if, for each $S_\lambda \in \mathcal{S}_0$, there exist elements $\sigma, \sigma' \in \boldsymbol{R}^+ \times \boldsymbol{R}^+$ such that

(1. 5)$_1$ $N_\sigma(S_\lambda$, fron $S'_\lambda) \cap U \subset N(S_\lambda) \subset N_{\sigma'}(S_\lambda$, fron $S'_\lambda) \cap U$.

In the above $S'_\lambda \in \mathcal{S}'_0$ such that $S'_\lambda \cap U = S_\lambda$.

Definition 1. 6$_2$. We say that $\mathcal{N}$ is *p. g. simple* if the following are valid.

(1. 6)$_1$ $\mathcal{N}$ is simple and p. g. adequate.

(1. 6)$_2$ For any $S_{\lambda_1 \cdots t} \in \mathcal{S}_0$, we have:

(1. 6)$'_2$ $H^q(\Omega_{\text{p.g.}}(N(S_{\lambda_1 \cdots t})\ ;\ \text{fron } S'_{\lambda_t})) \cong 0 \quad (q \geqq 1)$.

In the above $\Omega_{\text{p.g.}}(N(S_{\lambda_1 \cdots t})\ ;\ \text{fron } S'_{\lambda_t})$ denotes the cochain complex consisting of all the C^∞-differential forms φ's in $N(S_{\lambda_1 \cdots t})$ such that φ is *of polynomial growth*[2] *with respect to* fron S'_{λ_t}. (Of course S'_{λ_t} in (1. 6)$'_2$ denotes the element of $\mathcal{S}'_0$ such that $S'_{\lambda_t} \cap U = S_{\lambda_t}$.)

(ii) Next let $\mathcal{N}=\{\mathcal{N}_j\}_{j=1}^\infty$ be a direct system (with respect to inclusions) of C^∞-thickenings of $\mathcal{S}_0$.

Definition 1. 7$_1$. We say that $\mathcal{N}$ is *p. g. adequate* if we have:

(1. 7)$_1$ For any $j \in \boldsymbol{Z}^+$, $\mathcal{N}_j$ is *p. g. adequate.*

(1. 7)$_2$ Let $\sigma \in \boldsymbol{R}^+ \times \boldsymbol{R}^+$. Then there exists an element $j \in \boldsymbol{Z}^+$ such that

(1. 7)$'_2$ $N_j(S_\lambda) \subset N(S_\lambda$, fron $S'_\lambda) \cap U$ for any $S_\lambda \in \mathcal{S}_0$, where we denote by S'_λ the element in $\mathcal{S}'_0$ such that $S'_\lambda \cap U = S_\lambda$.

---

1) Let f_1, f_2 be real-valued functions defined in a topological space Z. We write $f_1 \sim f_2$ if there exist elements (c_1, c_2) and $(c_1', c_2') \in \boldsymbol{R}^+ \times \boldsymbol{R}^+$ such that $c_1 \cdot f_1(P)^{c_2} \leqq f_2(P) \leqq c_1' f_1(P)^{c_2'}$ for any $P \in Z$.

2) cf. [4]$_1$.

Definition 1. 7_2. We say that $\mathcal{N}$ is *p. g. simple* if the following are valid:

$(1.8)_1$ $\mathcal{N}$ is p. g. adequate and, for any $j \in \boldsymbol{Z}^+$, $\mathcal{N}_j$ is p. g. simple.

$(1.8)_2$ Let $S_{\lambda_1\cdots t} \in \mathcal{S}_{0,O}$. Then, for any $j, j' \in \boldsymbol{Z}^+$ such that $j<j'$, $N_{j'}(S_{\lambda_1\cdots t})$ is a strong deformation retract of $N_j(S_{\lambda_1\cdots t})$.

Remark. Let $\mathcal{N}=\{\mathcal{N}_j\}_{j=1}^{\infty}$ be a p. g. adequate (p. g. simple) direct system of C^∞-thickening of $\mathcal{S}_0$. To emphasize the role of $\mathfrak{P}$, we sometimes call $\mathcal{N}$ a p. g. adequate (p. g. simple) direct system of C^∞-thickenings of $\mathfrak{P}$.

(iii) Now we introduce the following

Definition 1. 8. We say that $\mathfrak{P}$ is *p. g. simple* if there exists a p. g. simple direct system of C^∞-thickenings of $\mathfrak{P}$.

When there is no fear of confusion we use the terminology '$\mathcal{S}_0$ is p. g. simple' as a synonym for '$\mathfrak{P}$ is p. g. simple'.

1. 4. P. g. C^∞-de Rham C. C. I.

Let $\boldsymbol{R}^n$ be a euclidean space, and let $\mathfrak{P}=(Q, Q', \mathcal{F})$ be a p. g. adequate prestratified space in $\boldsymbol{R}^n$. Moreover, let $\mathcal{N}=\{\mathcal{N}_j\}_{j=1}^{\infty}$ be a p. g. adequate direct system of C^∞-thickenings of $\mathfrak{P}$. We write Q, Q' explicitly as $Q=(U, V, \mathcal{S}_0)$ and $Q'=(U', V', \mathcal{S}_0')$. Moreover, we write the functions in $\mathcal{F}$ as $f(S_\lambda')$, $g(S_\mu')$, $h(S_\eta', T_\upsilon')$, $\cdots$, where $S_\lambda', \cdots \in \mathcal{S}_0'$ and $\mathcal{T}_\upsilon' \in \mathcal{S}_{0,C}'$. (cf. § 1. 3.)

For any $j \in \boldsymbol{Z}^+$ we denote by $(\Omega(\mathcal{S}_0, \mathcal{N}_j), \mathcal{K}(\Omega(\mathcal{S}_0, \mathcal{N}_j)))$ the C^∞-de Rham C. C. I. attached to $(\mathcal{S}_0, \mathcal{N}_j)$. For simplicity we write $(\Omega(\mathcal{S}_0, \mathcal{N}_j), \mathcal{K}(\Omega(\mathcal{S}_0, \mathcal{N}_j)))$ as $(\Omega_j, \mathcal{K}_j)$.

1. 4. 1. C^∞-de Rham C. C. I. $(\hat{\Omega}(\mathfrak{P}, \mathcal{N}), \hat{\mathcal{K}}(\mathfrak{P}, \mathcal{N}))$ attached to $(\mathfrak{P}, \mathcal{N})$. Let $X \in \mathcal{S}_{0,C}$, $\mathcal{S}_{0,O}$ or $\mathcal{S}_{0,CO}$. For any j, j' such that $j<j'$, let $\rho_{jj'}: \Omega(N_j(X)) \to \Omega(N_{j'}(X))$ denote the natural homomorphism induced from the inclusion $i_{jj'}: N_j'(X) \hookrightarrow N_j(X)$. We denote by $\hat{\Omega}(X)$ the direct limit: $\lim_{j\to\infty} \Omega_j(X)$. Moreover, we write the collection $\{\hat{\Omega}(X)\,;\, X \in \mathcal{S}_{0,C}, \mathcal{S}_{0,O}, \mathcal{S}_{0,CO}\}$ as $\hat{\Omega}(\mathfrak{P}, \mathcal{N})$. Now apply an obvious limit process to the system $\{\mathcal{K}_j\}_{j=1}^{\infty}$ of collections of homomorphism[1]. We then have a collection $\hat{\mathcal{K}}(\mathfrak{P}, \mathcal{N})=\{\hat{K}(Y)\,;\, Y \in \mathcal{S}_{0,D}\}$ of homomorphisms of cochain complexes related to $\hat{\Omega}(\mathfrak{P}, \mathcal{N})$. It is easily checked that the pair $(\hat{\Omega}(\mathfrak{P}, \mathcal{N}), \hat{\mathcal{K}}(\mathfrak{P}, \mathcal{N}))$ is a C. C. I. attached to $\mathcal{S}_0$. We call this C. C. I. *the C^∞-de Rham C. C. I. attached to* $(\mathfrak{P}, \mathcal{N})$.

1. 4. 2. P. g. C^∞-de Rham C. C. I. $(\hat{\Omega}_{\text{p.g.}}(\mathfrak{P}, \mathcal{N}), \hat{\mathcal{K}}_{\text{p.g.}}(\mathfrak{P}, \mathcal{N}))$ attached to $(\mathfrak{P}, \mathcal{N})$. $(\text{i})_1$ We first recall that, for any $\mathcal{T} \in \mathcal{S}_{0.C}$, fron $\mathcal{T}=|\bar{\mathcal{T}}|-|\mathcal{T}|$ is closed. (See $[4]_3$.) Next let $X=(\mathcal{T}, m, S_{\lambda_1\cdots t}) \in \mathcal{S}_{0,CO}$. Then $\mathcal{T}_m(S_{\lambda_1\cdots t}) \in \mathcal{S}_{0,C}$ (cf. $[4]_3$). We then define:

$(1.9)_1'$ fron $X=$ fron $\mathcal{T}_m(S_{\lambda_1\cdots t})$.

Thirdly let $X=S_{\lambda_1\cdots t} \in \mathcal{S}_{0,O}$. We then define:

$(1.9)_2'$ fron $X=$ fron S_{λ_t}.

1) For the precise definition of $\hat{\mathcal{K}}$, see [5], [6].

(i)$_2$ Now let $X \in \mathcal{S}_{0,C}$, $\mathcal{S}_{0,O}$ or $\mathcal{S}_{0,CO}$. Moreover, let $j \in \boldsymbol{Z}^+$. We define a cochain complex $\Omega_{\mathrm{p.g.}j}(X)$ by

(1.9) $\Omega_{\mathrm{p.g.}j}(X) = \Omega_{\mathrm{p.g.}}(N_j(X), \text{fron } X')$.[1]

We denote by $\hat{\Omega}_{\mathrm{p.g.}}(X)$ the direct limit : $\lim_{j\to\infty} \hat{\Omega}_{\mathrm{p.g.}j}(X)$. We write the collection $\{\hat{\Omega}_{\mathrm{p.g.}}(X)\ ;\ X \in \mathcal{S}_{0,C}, \mathcal{S}_{0,O} \text{ or } \mathcal{S}_{0,CO}\}$ as $\hat{\Omega}_{\mathrm{p.g.}}(\mathfrak{P}, \mathcal{N})$. We call this collection the p. g. C^∞-de Rham collection attached to $(\mathfrak{P}, \mathcal{N})$. Then $\hat{\Omega}_{\mathrm{p.g.}}(\mathfrak{P}, \mathcal{N})$ is a *subcollection of* $\hat{\Omega}(\mathfrak{P}, \mathcal{N})$ (cf. [5]). Moreover, it is easy to see that $\hat{\mathcal{K}}(\mathfrak{P}, \mathcal{N})$ *preserves* $\hat{\Omega}_{\mathrm{p.g.}}(\mathfrak{P}, \mathcal{N})$. We write $\hat{\mathcal{K}}(\mathfrak{P}, \mathcal{N})$ also as $\hat{\mathcal{K}}_{\mathrm{p.g.}}(\mathfrak{P}, \mathcal{N})$.

Now the basic fact in § 1.4 is as follows :

Theorem 1. 1. *The pair* $(\hat{\Omega}_{\mathrm{p.g.}}(\mathfrak{P}, \mathcal{N}), \hat{\mathcal{K}}_{\mathrm{p.g.}}(\mathfrak{P}, \mathcal{N}))$ *is a* C. C. I. *attached to* S_0.

Of course the key point in Theorem 1. 1 is that $\hat{\mathcal{K}}_{\mathrm{p.g.}}(\mathfrak{P}, \mathcal{N})$ satisfies exact sequences of Mayer-Vietoris types(cf. [4]$_3$). We call the C. C. I. $(\hat{\Omega}_{\mathrm{p.g.}}(\mathfrak{P}, \mathcal{N}), \hat{\mathcal{K}}_{\mathrm{p.g.}}(\mathfrak{P}, \mathcal{N}))$ *the p. g. C^∞-de Rham* C. C. I. *attached to* $(\mathfrak{P}, \mathcal{N})$.

By Theorem 1. 1, polynomial growth properties of C^∞-differential forms related to $(\mathfrak{P}, \mathcal{N})$ can be localized : in view of Lemma 1. 1, [4]$_3$, we have the following

Corollary to Theorem 1. 1. *Assume that, for each* $X \in \mathcal{S}_{0,O}$, *we have.*

(1. 10)$_1$ $H^*(\hat{\Omega}_{\mathrm{p.g.}}(X)) \cong \lim_{j\to\infty} H^*(N_j(X)\ ;\ \boldsymbol{R})$.

Then we have, for any $\mathcal{T} \in \mathcal{S}_{0,C}$, *the natural isomorphism* :

(1. 10)$_2$ $H^*(|\mathcal{T}|\ ;\ \boldsymbol{R}) \cong H^*(\hat{\Omega}_{\mathrm{p.g.}}(\mathcal{T}))$.

From the above we know the following

Theorem 1. 2. (*P. g. C^∞-de Rham theorem*) *Assume that $\mathcal{N}$ is p. g. simple. Then, for any* $\mathcal{T} \in \mathcal{S}_{0,C}$, *we have* :

(1. 11) $H^*(|\mathcal{T}|\ ;\ \boldsymbol{R}) \cong H^*(\hat{\Omega}_{\mathrm{p.g.}}(\mathcal{T}))$.

Remark. Let $\mathcal{T} \in \mathcal{S}_{0,C}$. Then our discussions on p. g. properties of C^∞-differential forms related to $\mathcal{T}$ are done in terms of $\Omega_{\mathrm{p.g.}j}(\mathcal{T})$'s $(j=1, \cdots)$ and $\hat{\Omega}_{\mathrm{p.g.}}(\mathcal{T})$. Now let $X \in \mathcal{S}_{0,O}$. We attached $\Omega_{\mathrm{p.g.}j}(X)$ $(j=1, \cdots)$ and $\hat{\Omega}_{\mathrm{p.g.}}(X)$ to X. We remark that the natures of the p. g. properties of $\Omega_{\mathrm{p.g.}j}(\mathcal{T})$ $(\hat{\Omega}_{\mathrm{p.g.}}(\mathcal{T}))$ and $\Omega_{\mathrm{p.g.}j}(X)$ $(\hat{\Omega}_{\mathrm{p.g.}}(X))$ are in general different. The cochain complexes $\Omega_{\mathrm{p.g.}j}(X)$'s are not obtained from $\Omega_{\mathrm{p.g.}j}(\mathcal{T})$'s by formal procedure.

In this point the pattern of the localization step (cf. § 0) in the p. g. C^∞-de Rham theorem is different from that in the arguments in § 1. 3.

Using terminology in the sheaf theory, we indicate the difference mentioned above in the following fashion. First note that we can attach a sheaf, denoted by $\tilde{\Omega}_{\mathrm{p.g.}}(\mathcal{T})$, to $\mathcal{T}$ in a standard manner, so that $\hat{\Omega}_{\mathrm{p.g.}}(\mathcal{T})$ is a 'global object'[2] of $\tilde{\Omega}_{\mathrm{p.g.}}(\mathcal{T})$.

1) If $X=\mathcal{T}(=\cup S_\lambda) \in \mathcal{S}_{0,C}$, $X'=\mathcal{T}'(=\cup S_\lambda')$, where $S_\lambda' \in \mathcal{S}_0'$ such that $S_\lambda' \cap U = S_\lambda$, $\cdots$.

2) E. g., elements of $H^0(N(\mathcal{A})\ ;\ \tilde{\Omega}_{\mathrm{p.g.}}(\mathcal{T}))$, where $\mathcal{A}$ is a suitable covering of $|\bar{\mathcal{T}}|$ and $N(\mathcal{A})$, is a nerve of $\mathcal{A}$.

Now let $X \in \mathcal{T}_{0,O}$. For simplicity we assume that X is of the form : $X=S_\lambda$, where $S_\lambda \in \mathcal{T}$[1]. Then we can attach a sheaf, denoted by $\tilde{\Omega}_{\text{p.g.}}(X)$, to X so that $\hat{\Omega}_{\text{p.g.}}(X)$ is a 'global object' of $\hat{\Omega}_{\text{p.g.}}(X)$ in a similar sense to the above. We note that $\tilde{\Omega}_{\text{p.g.}}(X)$ is *generally different from the restriction of* $\tilde{\Omega}_{\text{p.g.}}(\mathcal{T})$ *to* X.

The above explanations indicate the difference between the localization step in the C^∞-de Rham theorem (without p. g. condition) and that in the p. g. C^∞-de Rham theorem (cf. Remark in § 1. 2).

Our definition of p. g. C^∞-de Rham C. C. I. has the following advantages.

(a) We obtain the p. g. properties of C^∞-differential forms *simultaneously* for all the elements in $\mathcal{S}_{0,O}$.

(b) The p. g. condition imposed on each $\hat{\Omega}_{\text{p.g.}}(X)$; $X \in \mathcal{S}_{0,O}$ is meaningful for the study of p. g. properties related to both the 'global' data $\mathcal{T}$'s in $\mathcal{S}_{0,C}$ and the 'local' datum X itself.

2. P. g. Simple Prestratification and Normalized Series

Let $\boldsymbol{R}^n(x)$ be a euclidean space with a system $(x)=(x_1, \cdots, x_n)$ of coordinates. Moreover, let $(\mathfrak{N}, \mathfrak{F})$ be a *normalized series* of prestratified spaces in $\boldsymbol{R}^n(x)$, where $\mathfrak{N}$ is an admissible series in $\boldsymbol{R}^n(x)$ and $\mathfrak{F}$ is a representation datum of $\mathfrak{N}$ (For the definition of normalized series see $[4]_4$.)

We fix the data $\boldsymbol{R}^n(x)$ and $(\mathfrak{N}, \mathfrak{F})$ in § 2.

The purpose of § 2 is to discuss the relations between the notions of p. g. simple prestratifications and normalized series. First we define, for $(\mathfrak{N}, \mathfrak{F})$, a pair $(\mathfrak{N}^*, \mathfrak{F}^*)$ of series of prestratified spaces and of collections of analytic functions in a simple manner. The pair $(\mathfrak{N}^*, \mathfrak{F}^*)$ is determined uniquely by $(\mathfrak{N}, \mathfrak{F})$ and has certain interesting properties which the normalized series $(\mathfrak{N}, \mathfrak{F})$ does not share. The basic fact on $(\mathfrak{N}^*, \mathfrak{F}^*)$ is Theorem 2. 1, which states p. g. simple properties of $(\mathfrak{N}^*, \mathfrak{F}^*)$. Theorem 2. 1 is most important in the discussion of p. g. properties of C^∞-differential forms related to analytic varieties.

Now we fix notation related to $(\mathfrak{N}, \mathfrak{F})$ in the following manner : We write $\mathfrak{N}$ as (y, Q, Q'), where y is a system of coordinates[2] of $\boldsymbol{R}^n(x)$ and $Q=\{Q^j\}_{j=1}^n$, $Q'=\{Q'^j\}_{j=1}^n$ are series of triplets in $\boldsymbol{R}^n(x)$. We write the triplets Q^j, Q'^j as $(U^j, V^j, \mathcal{S}_0^j)$ and $(U'^j, V'^j, \mathcal{S}_0'^j)$, $j=1, \cdots, n$. Moreover, we denote by $\mathcal{S}^j(\mathcal{S}'^j)$ the prestratification of $V^j(V'^j)$ induced from $\mathcal{S}_0^j(\mathcal{S}_0'^j)$, $j=1, \cdots, n$. Furthermore, we write $\mathfrak{F}$ as $\{\mathfrak{F}^j\}_{j=1}^n$, where $\mathfrak{F}^j=\{\mathfrak{f}(S_\lambda'^j) ; S_\lambda'^j \in \mathcal{S}'^j\}$ is a representation datum of $\mathcal{S}'^j$. Let $j \in [1, \cdots, n]$. We write, for each $S_\lambda'^j \in \mathcal{S}'^j$, the representation datum $\mathfrak{f}(S_\lambda'^j) \in \mathfrak{F}^j$ as $(\mathfrak{f}(S_\lambda'^j), \mathfrak{f}'(S_\lambda'^j))$, where $\mathfrak{f}(S_\lambda'^j)$ and $\mathfrak{f}'(S_\lambda'^j)$ are respectively the $\boldsymbol{M}$- and $\boldsymbol{I}$-components of $\mathfrak{f}(S_\lambda'^j)$. (See [5].)

1) For a 'general' element $X=S_{\lambda_1\cdots_t} \in \mathcal{T}_{0,O}(t \geqq 2)$ we can attach a sheaf, denoted by $\tilde{\Omega}_{\text{p.g.}}(X)$, to S_{λ_t} in a natural manner. Then similar explanations to the case, where X is of the form $X=S_\lambda$, $S_\lambda \in \mathcal{T}$, are possible for the general case where X is of the form : $X=S_{\lambda_1\cdots_t}(t \geqq 2)$.

2) We write $(y_1, \cdots, y_j)$ as (y^j). Moreover, we denote by $\boldsymbol{R}^j(y^j)$ the linear subspace, defined by $y_{j+1}=\cdots=y_n=0$, of $\boldsymbol{R}^n(y^n)$.

2. 1. Minimal Vertical Refinement of a Normalized Series

1. We first introduce the following definitions.

Definition 2. 1_1. A *vertical refinement* of the series $\{\mathcal{S}_0^j\}_{j=1}^n(\{\mathcal{S}_0'^j\}_{j=1}^n)$ is a series $\{\mathcal{S}_0^{*j}\}_{j=1}^n(\{\mathcal{S}_0'^{*j}\}_{j=1}^n)$ of finite collections $\mathcal{S}_0^{*j}(\mathcal{S}_0'^{*j}), j=1, \cdots, n$, of analytic manifolds in $U^j(U'^j)$ such that the following are valid :

$(2.1)_1$ $U^j(U'^j)$ is the disjoint union of all the elements of $\mathcal{S}_0^{*j}(\mathcal{S}_0'^{*j}), j=1, \cdots, n$.

$(2.1)_2$ $\mathcal{S}_0^{*j}(\mathcal{S}_0'^{*j})$ is a refinement of $\mathcal{S}_0^j(\mathcal{S}_0'^j), j=1, \cdots, n$.

$(2.1)_3$ For each $S_\gamma^{*j-1} \in \mathcal{S}_0^{*j-1}(S_\gamma'^{*j-1} \in \mathcal{S}_0'^{*j-1})$ the inverse image $\pi_{j-1j}^{-1}(S_\gamma^{*j-1}) \cap U^j(\pi_{j-1j}^{-1}(S_\gamma'^{*j-1}) \cap U^j)$ is the union of elements in $\mathcal{S}_0^{*j}(\mathcal{S}_0'^{*j}), j=2, \cdots, n$.

Definition 2. 1_2. A vertical refinement $\{S_0^{*j}\}_{j=1}^n(\{S_0'^{*j}\}_{j=1}^n)$ of the series $\{\mathcal{S}_0^j\}_{j=1}^n$ $(\{\mathcal{S}_0'^j\}_{j=1}^n)$ is called *minimal* if the following is valid :

$(2.1)_4$ For any vertical refinement $\{\bar{\mathcal{S}}_0^{*j}\}_{j=1}^n(\{\bar{\mathcal{S}}_0'^{*j}\}_{j=1}^n)$ of the series $\{\mathcal{S}_0^j\}_{j=1}^n$ $(\{\mathcal{S}_0'^j\}_{j=1}^n)$, $\bar{\mathcal{S}}_0^{*j}(\bar{\mathcal{S}}_0'^{*j})$ is a refinement of $\mathcal{S}_0^{*j}(\mathcal{S}_0'^{*j}), j=1, \cdots, n$.

Then the following proposition is easily checked.

Proposition 2. 1. (1) *There exists one and only one minimal vertical refinement* $\{\mathcal{S}_0^{*j}\}_{j=1}^n(\{\mathcal{S}_0'^{*j}\}_{j=1}^n)$ *of* $\{\mathcal{S}_0^j\}_{j=1}^n(\{\mathcal{S}_0'^j\}_{j=1}^n)$.

(2) *For any* $S_\lambda'^{*j} \in \mathcal{S}_0'^{*j}$, $S_\lambda'^{*j} \cap U^j \in \mathcal{S}_0^{*j}, j=1, \cdots, n$.

(3) *Denote by* Rs^{*j} *the map* : $\mathcal{S}_0'^{*j} \ni S_\lambda'^{*j} \longrightarrow \mathcal{S}_0^{*j} \ni S_\lambda'^{*j} \cap U^j, j=1, \cdots, n$. *Then* Rs^{*j} *is bijective.*

In the sequel of § 2, we use the symbol $\{\mathcal{S}_0^{*j}\}_{j=1}^n(\{\mathcal{S}_0'^{*j}\}_{j=1}^n)$ for the minimal vertical refinement of $\{\mathcal{S}_0^j\}_{j=1}^n(\{\mathcal{S}_0'^j\}_{j=1}^n)$. Let $j \in [1, \cdots, n]$. We denote by $\mathcal{S}^{*j}(\mathcal{S}'^{*j})$ the collection $\{S_\alpha^{*j} \in \mathcal{S}_0^{*j} ; S_\alpha^{*j} \subset V^j\}$ $(\{S_\alpha'^{*j} \in \mathcal{S}_0'^{*j} ; S_\alpha'^{*j} \subset V'^j\})$.

Remark. Let $j \in [2, \cdots, n]$, and let $S_\gamma^{*j-1} \in \mathcal{S}_0^{*j-1}$. We use the symbols $\mathcal{S}^{*j}(S_\gamma^{*j-1})$ and $\tilde{\mathcal{S}}^{*j}(S_\gamma^{*j-1})$for the collections $\{S_\alpha^{*j} \in \mathcal{S}^{*j}; \pi_{j-1j}(S_\alpha^{*j})=S_\gamma^{*j-1}\}$ and $\{\tilde{S}_\alpha^{*j} \notin \mathcal{S}^{*j}; \pi_{j-1j}(\tilde{S}_\alpha^{*j})=S_\gamma^{*j-1}\}$ (cf. § 4. 2, [5]). By similar argements as those done for normalized series (cf. § 4. 2, [5]), we can define orders[1] : ↑ for $\mathcal{S}^{*j}(S_\gamma^{*j-1})$ and $\tilde{S}^{*j}(\mathcal{S}_\gamma^{*j-1})$. Moreover, from the assertion of biholomorphicity on normalized series (cf. $(6)_2$, $[4]_4$) we know the following :

$(2.2)_1$ For any $S_\alpha^{*j} \in \mathcal{S}^{*j}(S_\gamma^{*j-1})$, $\pi_{j-1j} : S_\alpha^{*j} \to S_\gamma^{*j-1}$ is biholomorphic (cf. $(6)_2$, $[4]_4$).

$(2.2)_2$ For any $S_\beta^{*j} \in \tilde{\mathcal{S}}^{*j}(S_\gamma^{*j-1})$, S_β^{*j} is diffeomorphic to $S_\gamma^{*j-1} \times (0, 1)$.

(We remark that, for $\mathcal{S}_0'^{*j}$s, entirely similar facts as above are valid.)

Note that, for $S_\lambda^j \in \mathcal{S}_0^j - \mathcal{S}^j$, $\pi_{j-1j}(S_\lambda^j)$ is, in general, not an element of $\mathcal{S}_0^{j-1}$. We do not formulate the corresponding facts to $(2.2)_2$ for normalized series. The property in $(2.2)_2$ is one reason why the investigation of the series $\{\mathcal{S}_0^{*j}\}_{j=1}^n$ is easier than that of the series $\{\mathcal{S}_0^j\}_{j=1}^n$.

1) The orders : ↑ are defined by means of the natural order on the real line $\boldsymbol{R}(y_j)$.

Let $j \in [1, \cdots, n]$. We write the collections $(U^j, V^j, \mathscr{S}_0^{*j})$ and $(U'^j, V'^j, \mathscr{S}_0'^{*j})$ as Q^{*j} and Q'^{*j}. Moreover, we write the series $\{Q^{*j}\}_{j=1}^n$ and $\{Q'^{*j}\}_{j=1}^n$ as Q^* and Q'^*. We then introduce :

Definition 2. 1$_3$. We call the collection (y, Q^*, Q'^*) *the minimal vertical refinement* of the admissible series $\mathfrak{R}$.

We use, in the sequel of § 2, the symbol $\mathfrak{R}^*$ for (y, Q^*, Q'^*).

The following properties of $\mathfrak{R}^*$ are checked without difficulty.

Proposition 2. 2. *For any* $j \in [1, \cdots, n]$, $\mathscr{S}_0^{*j}$ *and* $\mathscr{S}_0'^{*j}$ *satisfy the frontier condition.*

Proposition 2. 3. (*Strong going down property of* $\mathfrak{R}^*$) *Let* $(S_\alpha^{*j}, S_\gamma^{*j-1}, S_\delta^{*j-1}) \in \mathscr{S}^{*j} \times \mathscr{S}_0^{*j-1} \times \mathscr{S}_0^{*j-1}$ *such that*

(2. 3)$_1$ $\pi_{j-1j}(S_\alpha^{*j}) = S_\gamma^{*j-1}$ *and* $S_\gamma^{*j-1} > S_\delta^{*j-1}$, $j = 2, \cdots, n$.

Then there exists a unique stratum $S_\beta^{*j} \in \mathscr{S}_0^{*j}$ *so that*

(2. 3)$_2$ $\pi_{j-1j}(S_\beta^{*j}) = S_\delta^{*j-1}$ *and* $S_\alpha^{*j} > S_\beta^{*j}$.

Remark. An entirely similar fact to Proposition 2. 2 is valid for $\mathscr{S}'^{*j}$'s.

Proposition 2. 3 is stronger than the going down theorem for normalized series (§ 4, [5]) in the point that we claim the uniqueness of S_β^{*j}. Proposition 2. 3 is a consequence of the existence of certain neighborhoods associated with[1] $\mathfrak{R}^*$. (Compare Proposition 2. 6.)

2. We will attach, to each $S_\alpha'^{*j} \in \mathscr{S}_0'^{*j}$, a set of functions and a system of coordinates inductively on $j=1, \cdots, n$:

(i) Let $j=1$. Then note that $\mathscr{S}_0'^{*1} = \mathscr{S}_0'^1$. We then put :

(2. 4)$_1$ $(\mathfrak{f}^*(S_\alpha'^{*1}), y(S_\alpha'^{*1})) = (\mathfrak{f}(S_\alpha'^{*1}), \phi)$ or (ϕ, y_1) according to whether dim $S_\alpha'^{*j} = 0$ or 1.

In the right side of (2. 4)$_1$ we regard $S_\alpha'^{*1} \in \mathscr{S}_0'^{*1}$ as an element of $\mathscr{S}_0'^1$.

(ii) Let $j \in [2, \cdots, n]$. We assume that we attached, to each $S_\alpha'^{*\tilde{j}} \in \mathscr{S}_0'^{*\tilde{j}}$, a set $\mathfrak{f}^*(S_\alpha'^{*\tilde{j}})$ of functions and a system $y(S_\alpha'^{*\tilde{j}})$ of coordinates, $\tilde{j} \in [1, \cdots, j-1]$. Let $S_\alpha'^{*j} \in \mathscr{S}_0'^{*j}$.

(ii)$_1$ If dim $S_\alpha'^{*j} =$ dim $S_\gamma'^{*j-1} + 1$, where $S_\gamma'^{*j-1} = \pi_{j-1j}(S_\alpha'^{*j})$, then we put :

(2. 4)$_{2.1}$ $\mathfrak{f}^*(S_\alpha'^{*j}) = \pi_{j-1j}^* \mathfrak{f}^*(S_\gamma'^{*j-1})$, $y(S_\alpha'^{*j}) = y(S_\gamma'^{*j-1}) \cup \{y_j\}$.

(ii)$_2$ Assume that dim $S_\alpha'^{*j} =$ dim $\pi_{j-1j}(S_\alpha'^{*j})$. Then $S_\alpha'^{*j} \in \mathscr{S}'^{*j}$. We denote by $S_\lambda'^j$ the stratum of $\mathscr{S}'^j$ such that $S_\lambda'^j \supset S_\alpha'^{*j}$. Moreover, let j_α denote the element of $[1, \cdots, j-1]$ such that the following is valid.

(2. 4)$'_{2.2}$ $\pi_{\tilde{j}j}(S_\alpha'^{*j}) \subset V'^{\tilde{j}}$, $\tilde{j} = j_\alpha + 1, \cdots, j-1$ and $\pi_{j_\alpha j}(S_\alpha'^{*j}) \not\subset V'^{j_\alpha}$.

We then put :

(2. 4)$_{2.2}$ $\mathfrak{f}^*(S_\alpha'^{*j}) = \mathfrak{f}'(S_\lambda'^j) \cup \pi_{j_\alpha j}^* \mathfrak{f}^*(S_\alpha'^{*j_\alpha})$ and $y(S_\alpha'^{*j}) = y(S_\alpha'^{*j_\alpha})$.

In the above we denote $\pi_{j_\alpha j}(S_\alpha'^{*j})$ by $S_\alpha'^{*j_\alpha}$.

Let $S_\alpha'^{*j} \in \mathscr{S}_0'^{*j}, j=1, \cdots, n$. We call $\mathfrak{f}^*(S_\alpha'^{*j})$ and $y(S_\alpha'^{*j})$ respectively *the representation datum* of $S_\alpha'^{*j}$ and *the system of coordinates* of $S_\alpha'^{*j}$. Note that $y(S_\alpha'^{*j})$ provides a system

1) By Proposition 2. 1 we use freely terminologies for prestratified spaces for $\mathfrak{R}^*$.

of local parameters at any $P'^{*j}_\alpha \in S'^{*j}_\alpha$.

We write the collection $\{\mathfrak{f}^*(S'^{*j}_\alpha)\ ;\ S'^{*j}_\alpha \in \mathcal{S}'^{*j}_0\}$ as $\mathfrak{F}^{*j}$ and call it *the representation datum* of $\mathcal{S}'^{*j}_0$. We call, moreover, the series $\mathfrak{F}^* = \{\mathfrak{F}^{*j}\}^n_{j=1}$ *the representation datum* of the series $\{\mathcal{S}'^{*j}_0\}^n_{j=1}$. We then introduce the following

Definition 2. 1_4. We call the pair $(\mathfrak{R}^*, \mathfrak{F}^*)$ *the minimal vertical refinement* of $(\mathfrak{R}, \mathfrak{F})$.

2. 2. Basic Properties of the Minimal Vertical Refinement $(\mathfrak{R}^*, \mathfrak{F}^*)$

In § 2. 2. *we assume that* Q'^{*j} *is a d-envelop of* Q^{*j} *and that* (Q^{*j}, Q'^{*j}) *satisfies d-separation condition*[1], $j=1, \cdots, n$. (See $[4]_4$.)

2. 2. 1. Neighborhoods and certain sets attached to strata in $\mathfrak{R}^*$.

1. Let $S^{*j}_\alpha \in \mathcal{S}^{*j}$, $j=2, \cdots, n$. We denote dim S^{*j}_α by n^j_α. We assume that $n^j_\alpha \leqq j-2$. Moreover, let σ', σ'' and c be elements of $\boldsymbol{R}^+ \times \boldsymbol{R}^+$.

Let $P^{*j}_\alpha \in S^{*j}_\alpha$. We denote by $L(P^{*j}_\alpha)$ the linear subspace, defined by $(\mathrm{a})_0$ soon below, of $\boldsymbol{R}^j(y^j)$.

$(\mathrm{a})_0$ $\ y_{\bar{j}} = y_{\bar{j}}(P^{*j}_\alpha)$, where $y_{\bar{j}}$ exhaust all the coordinates in $\{y_j\} \cup y(S^{*j}_\alpha)$.

We denote by $T'_{\sigma'}(P^{*j}_\alpha)$ the subset of $L(P^{*j}_\alpha)$ such that $Q^{*j}_\alpha \in \boldsymbol{R}^j(y^j)$ is in $T'_{\sigma'}(P^{*j}_\alpha)$ if and only if

$(\mathrm{a})_1$ $\ \sum_{\bar{j}} |y_{\bar{j}}(Q^{*j}_\alpha) - y_j(P^{*j}_\alpha)| < \sigma' d(P^{*j}_\alpha, \text{fron}\ S'^{*j}_\alpha)$, where $y_{\bar{j}} \in (y^j) - \{\{y_j\} \cup y(S'^{*j}_\alpha)\}$.

We denote by $T'_{\sigma'}(S^{*j}_\alpha)$ the collection : $\bigcup_{P_\alpha{*j}} T'_{\sigma'}(P^{*j}_\alpha)$, where $P^{*j}_\alpha \in S^{*j}_\alpha$. Next let $Q^{*j}_\alpha \in T'_{\sigma'}(S'^{*j}_\alpha)$. We denote by $T''_{\sigma''}(Q^{*j}_\alpha)$ and $W''_c(Q^{*j}_\alpha)$ the open segments in $\boldsymbol{R}^j(y^j)$ such that $R^{*j}_\alpha \in T''_{\sigma''}(Q^{*j}_\alpha)$ or $W''_c(Q^{*j}_\alpha)$ according to whether the following are valid :

$(\mathrm{a})_{2.1}$ $\ y^{j-1}(R^{*j}_\alpha) = y^{j-1}(Q^{*j}_\alpha)$, $\quad |y_j(R^{*j}_\alpha) - y_j(Q^{*j}_\alpha)| < \sigma'' \cdot d(P^{*j}_\alpha,\ \text{fron}\ S'^{*j}_\alpha)$.

$(\mathrm{a})_{2.2}$ $\ y^{j-1}(R^{*j}_\alpha) = y^{j-1}(Q^{*j}_\alpha)$, $\quad |y_j(R^{*j}_\alpha) - y_j(Q^{*j}_\alpha)| < c \cdot d(P^{*j}_\alpha, Q^{*j}_\alpha)$.

In the aboves $P^{*j}_\alpha \in S^{*j}_\alpha$ such that $T'_{\sigma'}(P^{*j}_\alpha) \ni Q^{*j}_\alpha$.

We denote by $T_{\sigma'\sigma''}(S^{*j}_\alpha)$ and $W_{\sigma'c}(S^{*j}_\alpha)$ the collections : $\bigcup Q^{*j}_\alpha T''_{\sigma''}(Q^{*j}_\alpha)$ and $\bigcup Q^{*j}_\alpha W''_c(Q^{*j}_\alpha)$, where $Q^{*j}_\alpha \in T'_{\sigma'}(S^{*j}_\alpha)$.

Letting S^{*j}_α as in the beginning of § 2. 2. 1 we have the following propositions :

Proposition 2. 4. *We have the equivalence :*

$(2.\ 5)_1$ $\ \{T_{\sigma'\sigma''}(S^{*j}_\alpha)\}_{\sigma'\sigma''} \sim \{N_\delta(S^{*j}_\alpha, \text{fron}\ S'^{*j}_\alpha)\}_\delta$.

For the meaning of the symbol $\sim$, *see* $[4]_4$.

Proposition 2. 5. *There exist elements* σ', σ'' *and* $c \in \boldsymbol{R}^+ \times \boldsymbol{R}^+$ *such that*

$(2.\ 5)_2$ $\ T_{\sigma'\sigma''}(S^{*j}_\alpha) \cap V^j \subset W_{\sigma'c}(S^{*j}_\alpha)$.

The neighborhoods T's and W's as above are basic tools in the investigation of p. g. properties of $(\mathfrak{R}^*, \mathfrak{F}^*)$ (cf. $[4]_4$, [5]) : by Proposition 2. 4 we can replace neighborhoods $N_\delta(S^{*j}_\alpha$, fron $S'^{*j}_\alpha)$'s by $T_{\sigma'\sigma''}(S^{*j}_\alpha)$. Note that, in the inductive argu-

1) Let $P \in \boldsymbol{R}^n(x)$, and let $\boldsymbol{V}$ be a germ of variety at P. Then there exists a normalized series $(\mathfrak{R}, \mathfrak{F})$ attached to V (cf. $[4]_4$) so that the conditions mentioned here are valid for the minimal vertical refinement of $(\mathfrak{R}, \mathfrak{F})$.

ments of $(\mathfrak{R}^*, \mathfrak{F}^*)$ on $j=1, \cdots, n$, the neighborhoods T's are more suitable than N's.

Let S_α^{*j} be as in Proposition 2. 5, and let $S_\beta^{*j} > S_\alpha^{*j}$. Then Proposition 2. 5 gives *quantitative informations* on the behavior of S_β^{*j} around S_α^{*j}.

Proposition 2. 5 is proved on the basis of *the higher discriminant condition* (cf. $(9)_3$, $[4]_4$) and the inequality of Lojasiewicz. The proof of Proposition 2. 5 is very elementary. However, we point out that almost all p. g. properties of $(\mathfrak{R}^*, \mathfrak{F}^*)$ (cf. $[4]_4$, [5]) are proven with the essential aids of Proposition 2. 5.

2. Let $S_\alpha^{*j} \in \mathcal{S}_0^{*j}$, and let $Z(S_\alpha^{*j})$ be a neighborhood of S_α^{*j} in U^j. We then introduce the following

Definition 2. 2. We say that $Z(S_\alpha^{*j})$ has *strong connectivity property* for $\mathcal{S}_0^{*j}$ if the following are valid.

$(2.6)_1$ S_α^{*j} is a deformation retract of $Z(S_\alpha^{*j})$.

$(2.6)_2$ For any $S_\beta^{*j} \in \mathcal{S}_0^{*j}$ such that $S_\beta^{*j} > S_\alpha^{*j}$,

$(2.6)_{2.1}$ $Z(S_\alpha^{*j}) \cap S_\beta^{*j}$ is connected,

and

$(2.6)_{2.2}$ S_α^{*j} is a deformation retract of $\{Z(S_\alpha^{*j}) \cap S_\beta^{*j}\} \cap S_\alpha^{*j}$.

We have the following

Proposition 2. 6. (*Strong connectivity condition for* $(\mathfrak{R}^*, \mathfrak{F}^*)$)

Let $S_\alpha^{*j} \in \mathcal{S}_0^{*j}, j=1, \cdots, n$. Then there exists a direct system $\{Z_t(S_\alpha^{*j})\}_{t=1}^\infty$ of neighborhood of S_α^{*j} such that

$(2.7)_1$ $\{Z_t(S_\alpha^{*j})\}_{t=1}^\infty \sim \{N_\delta(S_\alpha^{*j}, \text{fron } S_\alpha^{*j})\}_\delta$.

$(2.7)_2$ *For any $t \in Z^+$, $Z_t(S_\alpha^{*j})$ has strong connectivity condition for $\mathcal{S}_0^{*j}$.*

Remark. In Definition 2. 2 the connectivity condition $(2.6)_{2.1}$ is basic. This condition is a peculiar advantage of the minimal vertical refinement $(\mathfrak{R}^*, \mathfrak{F}^*)$. If we want to formulate for normalized series corresponding facts to Proposition 2. 6, we should replace $(2.6)_{2.1}$ by a weaker condition (cf. § 4. 2, [5]). Proposition 2. 6 is used in the proof of Theorem 2. 1 in the next subsection.

2. 2. 2. P. g. simple C^∞-thickenings of the minimal vertical refinement $(\mathfrak{R}^*, \mathfrak{F}^*)$ Let $j \in [1, \cdots, n]$. Moreover, let $S_\tau'^{*j} \in \mathcal{S}_0'^{*j}$ and $(S_\mu'^{*j}, \mathcal{T}_\upsilon'^{*j}) \in (\mathcal{S}_0'^{*j}, \mathcal{S}_{0,C}'^{*j})$ such that the conditions in $(a)_{2,3}$ in Definition 1. 4 are valid. We then define $g(S_\tau'^{*j})$ and $h(S_\mu'^{*j}, \mathcal{T}_\upsilon'^{*j})$ in the following fashion :

$(a)_1$ $g(S_\tau^{*j}) = \Pi|f(S_\alpha'^{*j})|$, where $S_\alpha'^{*j} \subset \text{fron } S_\tau'^{*j}$.

$(a)_2$ $h(S_\mu'^{*j}, \mathcal{T}_\upsilon'^{*j}) = \Pi|f(S_\beta'^{*j})|$, where $S_\beta'^{*j} \in \mathcal{T}_\upsilon'^{*j}$ such that $S_\beta'^{*j} > S_\mu'^{*j}$.

We write the collection $\{f(S_\lambda'^{*j}), g(S_\tau'^{*j}), h(S_\mu'^{*j}, \mathcal{T}_\upsilon'^{*j})\}$, where $S_\lambda'^{*j} \in \mathcal{S}_0'^{*j}$ and $S_\tau'^{*j}$'s $(S_\mu'^{*j}, \mathcal{T}_\upsilon'^{*j})$'s are as just above, as $\tilde{\mathfrak{F}}^{*j}$. We then have the following

Lemma 2. 1. *For any $j \in [1, \cdots, n]$, $(Q^{*j}, Q'^{*j}, \tilde{\mathfrak{F}}^{*j})$ is a p. g. adequate prestratified space in $\boldsymbol{R}^j(y^j)$.*

2. 2. 3. P. g. simple property of the minimal vertical refinement $(\mathfrak{R}^*, \mathfrak{F}^*)$. We introduce the following definitions :

Definition 2. 3$_1$. A *C^∞-thickenings of* $(\mathfrak{R}^*, \mathfrak{F}^*)$ is a series $\{\mathcal{N}^j\}_{j=1}^n$ of *C^∞-thickenings* $\mathcal{N}^j$ of $\mathcal{S}_0^{*j}$, $j=1, \cdots, n$, such that, for any $S_\alpha^{*j} \in \mathcal{S}_0^{*j}$ $(j=2, \cdots, n)$,

$$(2.8)_1 \quad \pi_{j-1j}(N^j(S_\alpha^{*j})) = N^{j-1}(\pi_{j-1j}(S_\alpha^{*j})).$$

Definition 2. 3$_2$. A *p. g. simple direct system of C^∞-thickenings of* $(\mathfrak{R}^*, \mathfrak{F}^*)$ is a series $\{\mathcal{N}^j\}_{j=1}^n$ of p. g. simple direct systems $\mathcal{N}^j = \{\mathcal{N}_t^j\}_{t=1}^\infty$ of C^∞-thickenings of $(Q^{*j}, Q'^{*j}, \tilde{\mathfrak{F}}^{j*})$, $j=1, \cdots, n$, such that the following is valid.

$(2.8)_2$ For any $t \in \boldsymbol{Z}^+$, the series $\{\mathcal{N}_t^j\}_{j=1}^n$ is a C^∞-thickening of $(\mathfrak{R}^*, \mathfrak{F}^*)$.

Now the basic fact in § 2 is as follows :

Theorem 2. 1. *There exists a p. g. simple direct system of C^∞-thickenings of* $(\mathfrak{R}^*, \mathfrak{F}^*)$.

References

[1] Godement, R. : Topologie algébrique et théorie des faisceaux, Actualités Sci. Ind., Hermann, Paris, 1958.
[2] Lojasiewicz, S. : Sur la probleme de division, Studia Math., **18** (1959), 87–136.
[3] ——: Triangulation of semi-analytic sets, Annali della Scuola Norm. Sup. Pisa, **18** (1964), 449–473.
[4] Sasakura, N. : Complex analytic de Rham cohomology I, II, III and IV, Proc. Japan Acad., **49**(1973), 718–722, **50** (1974), 292–295, **51** (1975), 7–11, **51** (1975), 535–539 (cited as $[4]_1, \cdots, [4]_4$).
[5] ——: De Rham Cohomology and stratifications, to appear.
[6] ——: Differential forms and stratifications, Seminar note 240. R. I. M. S., Kyoto (1974), 152–234.
[7] Thom, R. : Ensembles et morphismes stratifies, Bull. Amer. Math. Soc., **75** (1969), 240–284.
[8] Weil, A. : Sur la theoreme de Rham, Comm. Math. Helv., **26** (1952), 119–145.
[9] Sasakura, N. : Divisible and asymptotic behaviors of coherent sheaves, Seminar note 192. R. I. M. S., Kyoto (1973), 37–327.

Department of Mathematics
Tokyo Metropolitan University

(Received January 14, 1976)

Index

Note. Italicized page numbers refer to entries under 'References' only.